Changing River
Channels

Changing River Channels

Edited by
ANGELA GURNELL *and* **GEOFFREY PETTS**
University of Birmingham, UK

JOHN WILEY & SONS
Chichester · New York · Brisbane · Toronto · Singapore

Other Wiley Editorial Offices

John Wiley & Sons, Inc., 605 Third Avenue,
New York, NY 10158-0012, USA

Jacaranda Wiley Ltd, 33 Park Road Milton,
Queensland 4064, Australia

John Wiley & Sons (Canada) Ltd, 22 Worcester Road,
Rexdale, Ontario M9W 1L1, Canada

John Wiley & Sons (SEA) Pte Ltd, 37 Jalan Pemimpin #05-04,
Block B, Union Industrial Building, Singapore 2057

Library of Congress Cataloging-in-Publication Data

Changing river channels / edited by Angela Gurnell & Geoffrey Petts.
p.   cm.
Includes bibliographical references and index.
ISBN 0-471-95727-5
1. River channels.   I. Gurnell, A.M. (Angela M.)   II. Petts, G.E.
(Geoffrey, E.)
GB561.C46   1995
551.48'3—dc20   95–3062
                                                              CIP

British Library Cataloguing in Publication Data

A catalogue record for this book is available from the British Library

ISBN 0-471-95727-5

Typeset in 10/12pt Times by Acorn Bookwork, Salisbury, Wiltshire
Printed and bound in Great Britain by Bookcraft (Bath) Ltd
This book is printed on acid-free paper responsibly manufactured from sustainable
forestation, for which at least two trees are planted for each one used for paper production.

Contents

List of Contributors vii

Preface ix

Foreword   D.E. Walling xi

Selected Publications by K.J. Gregory xv

 1   Changing River Channels: The Geographical Tradition 1
     G.E. Petts

PART I   TEMPORAL AND SPATIAL DIMENSIONS 25

 2   Changes of River Channels in Europe During the Holocene 27
     L. Starkel

 3   Holocene Channel and Floodplain Change: A UK Perspective 43
     A.G. Brown

 4   Channel Networks: Progress in the Study of Spatial and Temporal
     Variations of Drainage Density 65
     V. Gardiner

 5   Processes of Channel Planform Change on Meandering Channels
     in the UK 87
     J.M. Hooke

 6   Channel Cross-sectional Change 117
     C.C. Park

PART II   PROCESSES OF CHANGE 147

 7   Suspended Sediment Yields in a Changing Environment 149
     D.E. Walling

 8   Bedload Transport and Changing Grain Size Distributions 177
     B. Gomez

 9   Catchment Sediment Budgets and Change 201
     S.W. Trimble

10   River Channel Change: The Role of Large Woody Debris 217
     E.A. Keller and A. MacDonald

11 Vegetation Along River Corridors: Hydrogeomorphological Interations 237
   A.M. Gurnell

PART III INFORMATION FOR THE MANAGEMENT OF CHANGE 261

12 Information Flow for Channel Management 263
   M.J. Clark

13 Investigating Change in Fluvial Systems Using Remotely Sensed Data 277
   E.J. Milton, D.J. Gilvear and I.D. Hooper

14 Information from Topographic Survey 303
   S.R. Downward

15 Information from Channel Geometry–Discharge Relations 325
   G. Wharton

16 River Channel Classification for Channel Management Purposes 347
   P.W. Downs

PART IV MANAGEMENT FOR CHANGE 367

17 River Channel Restoration: Theory and Practice 369
   A. Brookes

18 Towards a Sustainable Water Environment 389
   J.L. Gardiner

19 Fluvial Geomorphology and Environmental Design 413
   M.D. Newson

Index 433

Contributors

Andrew Brookes
National Rivers Authority—Thames Region, Kings Meadow House, Kings Meadow Road, Reading RG1 8DQ, UK

A.G. Brown
Department of Geography, University of Exeter, Amory Building, Rennes Drive, Exeter EX4 4RJ, UK

M.J. Clark
Department of Geography, University of Southampton, Southampton SO9 5NH, UK

P.W. Downs
Department of Geography, University of Nottingham, University Park, Nottingham NG7 2RD, UK

S.R. Downward
School of Geography, Kingston University, Kingston Upon Thames KT1 2EE, UK

I.D. Hooper
School of Applied Sciences, University of Wolverhampton, Wulfruna Street, Wolverhampton WV1 1SB, UK

J.L. Gardiner
National Rivers Authority—Thames Region, Kings Meadow House, Kings Meadow Road, Reading RG1 8DQ, UK

V. Gardiner
Department of Environmental and Geographical Studies, Roehampton Institute, Southlands College, Wimbledon Parkside, London SW15 5NN, UK

D.J. Gilvear
Department of Environmental Science, University of Stirling, Stirling FK9 4LA, UK

B. Gomez
Department of Geography and Geology, Indiana State University, Terre Haute, Indiana 47809, USA

A.M. Gurnell
School of Geography, University of Birmingham, Edgbaston, Birmingham B15 2TT, UK

J.M. Hooke
Department of Geography, University of Portsmouth, Buckingham Building, Lion Terrace, Portsmouth PO1 3HE, UK

E.A. Keller
Environmental Studies and Geological Sciences, University of California, Santa Barbara, California 93106, USA

A. MacDonald
PTI Environmental Studies and Geological Sciences, University of California, Santa Barbara, California 93106, USA

E.J. Milton
Department of Geography, University of Southampton, Southampton SO9 5NH, UK

M.D. Newson
Department of Geography, University of Newcastle upon Tyne, Newcastle upon Tyne NE1 7RU, UK

C.C. Park
Department of Geography, University of Lancaster, Lancaster LA1 4YB, UK

G.E. Petts
School of Geography, University of Birmingham, Edgbaston, Birmingham B15 2TT, UK

L. Starkel
Department of Geomorphology and Hydrology, Institute of Geography, Polish Academy of Sciences, 31-018 Cracow, sw Jana 22, Poland

S.W. Trimble
Department of Geography, University of California, Los Angeles, 405 Hilgard Avenue, Los Angeles, California 90024-1524, USA

D.E. Walling
Department of Geography, University of Exeter, Amory Building, Rennes Drive, Exeter EX4 4RJ, UK

G. Wharton
Department of Geography, Queen Mary and Westfield College, University of London, Mile End Road, London E1 4NS, UK

Preface

The idea for these books arose at an IAHS meeting in Oslo, co-convened by Des Walling and Jim Bogen, when two of us (IDLF and AMG) were sharing a small bottle of very expensive Norwegian beer and discussing the remarkable contributions made by Ken Gregory and Des Walling to catchment research over more than 30 years. Indeed, 1994 marked the 21st anniversary of the first publication of their book, *Drainage Basin: Form and Process* (Gregory, K. J. and Walling, D.E. (1973), Edward Arnold, London). Their contributions have not only been to the understanding of catchment processes but also to the inspiration of future generations of researchers, who are now occupying a number of academic positions in UK and overseas universities, colleges and research organisations.

The very strong presence of their former postgraduate students at conferences throughout the world testifies to the important impact that both Ken and Des have had not only on the global academic community but also on the research training from which all four of us, and many others, have benefited in the Geography Departments at the Universities of Exeter and Southampton. Their contribution, however, goes well beyond formal training of ourselves and other postgraduate students. At Exeter and Southampton, postgraduates benefited from the presence of many international visitors, who were passing through or spending time working with Des and Ken. It is a testament to their academic stimulus and friendship that many of these visitors have been keen to contribute to these two volumes.

It seems appropriate that four former PhD students whom Ken Gregory and Des Walling successfully supervised to completion should have co-edited these books. Though we have all developed our own areas of research since the 1970s, it is undoubtedly the case that our approach to catchment research has been influenced to a great extent by three years of research training under their guidance.

The long-kept secret of the publication of these volumes has benefited from the availability of electronic mail (which neither of them have yet learnt to break into!), to an anonymous postman delivering parcels to Bruce at home in Exeter rather than the Geography Department, and to Edna Pellett's undercover operation at Goldsmiths' College. As a result of everybody's efforts to keep this project a secret, we are confident that, while both Ken and Des were enthusiastically aware of the books for the *other*, neither knew anything of their *own* book until they received copies at a joint reception on publication of the two volumes. This secrecy, up to the presentation, has also been due to the tremendous support given by John Wiley & Sons, who agreed not to advertise these volumes before the official launch.

We are indebted to many individuals at the Universities of Birmingham, Coventry, Exeter and Southampton, particularly the cartographers who worked so hard to produce the final versions from less-than-perfect original diagrams.

Ian Foster   Angela Gurnell
Geoffrey Petts   Bruce Webb

Foreword

DES WALLING
University of Exeter

In 1992, Professor Ken Gregory left the Department of Geography at the University of Southampton, where he had been based for the previous 16 years, to become the Warden of Goldsmiths College, University of London. Despite this move to 'higher things' and the many pressing problems facing his College, Ken has, as one might expect, still found time to maintain a very active involvement in national and international activities in the field of geomorphology and palaeohydrology. In September, 1994, for example, he hosted the UK meeting of the INQUA Commission on Global Continental Palaeohydrology, with the associated GLOCOPH Symposium at Southampton and one-day meeting at the Geological Society of London. He continues to serve as the UK National Correspondent to the International Geographical Union and is a Vice-President of the recently formed INQUA Commission on Global Continental Palaeohydrology. The move to Goldsmiths College has, however, inevitably meant a lessening of his direct involvement with field research and the end of a long succession of postgraduate research students, stretching from 1966 through to the present, who have benefited from his inspiration, guidance and supervision. In most careers, this transition comes at retirement and it is sometimes marked by the publication of a *festschrift* containing papers by former students and colleagues, which pays tribute to the contribution made. In Ken's case this contribution has been both outstanding and highly distinguished and his many former research students, colleagues and co-workers were anxious that such an opportunity to recognise his achievements should not be lost. They have collaborated in producing this volume in order to mark his outstanding and essentially unique contribution to the development, both nationally and internationally, of what he himself has termed hydrogeomorphology. Much of his work has focused on *river channels*, and *change*, both short-term and long-term, is a theme that has figured large in his thinking and writing over the past 30 years. A volume concerned with *Changing River Channels* was thus seen as a fitting tribute to his distinguished career as a university teacher and researcher.

Many may not be aware that Ken's early work was not directly concerned with rivers. As a research student at University College, London, working under Professor Eric Brown, he first turned his attention to the deglaciation of eastern Eskdale in Yorkshire. This found him mapping and interpreting glacial drainage systems, but he also indulged a more general interest in the morphometric analysis of landforms and this saw the beginning of his longstanding interest in river net-

Changing River Channels. Edited by Angela Gurnell and Geoffrey Petts.
© 1995 John Wiley & Sons Ltd.

works and particularly drainage density. From this beginning, he went on to make important and fundamental contributions to the analysis of drainage networks, ranging from the study of dry valley networks and dambo systems, through the recognition of drainage density as a dynamic entity, to the concepts of drainage network volume and power.

The early 1960s saw Ken arrive at Exeter where he was to spend some 14 years as successively, Assistant Lecturer, Lecturer and Reader, before moving to a Chair at Southampton in 1976. This was the time when interest in monitoring the rate of contemporary geomorphological processes was awakening and he was quick to recognise the potential for coupling process investigations with studies of longer-term landscape evolution and the value of exploring the links between form and process. He was soon engrossed in the vigil network, crest stage gauges and vee-notch weirs, and he established his first small instrumented catchment at Swinesloose Gully in south-east Devon in 1966. This was soon to be followed by other catchments in that area and by the urbanising Rosebarn catchment. His own house was located on the watershed of that basin and he could almost lay claim to having had a weir at the bottom of his garden! His interest in catchment studies flourished at Exeter and was maintained and expanded further after his move to Southampton. There he established the Highland Water catchment and promoted work on urban hydrology in the Lordshill catchment. Again one can point to an impressive and valuable series of contributions on many aspects of drainage basin behaviour which has continued to the present. What is probably his most recent published paper reports work on bank erosion undertaken in the Highland Water catchment that he originally established in the mid-1970s, and there are undoubtedly more to come!

Whilst his catchment studies emphasised the more hydrological aspects of hydro-geomorphology, Ken consistently balanced this with a wide-ranging interest in the geomorphological context. River channels always held a particular attraction and he has maintained a longstanding interest in river channel change. This has provided the background for important and valuable contributions to our understanding of river channel dynamics and adjustments, including the role of vegetation and woody debris and the impacts of land-use change, reservoir development, and river channelisation. Through this interest in river channel change, Ken also became increasingly aware of the potential for linking an understanding of contemporary form–process relationships and adjustment dynamics to the morphological and sedimentological evidence of past river behaviour, and this led to a growing interest in palaeohydrological reconstruction. He has played a major and pivotal role, both nationally and internationally, in promoting palaeohydrology as a sub-discipline and in contributing to the development of its methodology. Along with Leszek Starkel and John Thornes, he was responsible for the establishment and successful completion of the Fluvial component of the IGCP Project 158 entitled 'Palaeohydrology of the Temperate Zone in the last 15 000 years' which extended from 1978 to 1987. This success has led to the recent establishment of the INQUA Commission on Global Continental Palaeohydrology in which he continues to play a key role as Vice-President.

Ken's distinguished and wide-ranging contribution to the field of hydro-geomorphology is clearly evidenced by his impressive and indeed formidable list of research papers. These have been paralleled by numerous books and edited volumes, including *Drainage Basin Form and Process*, *River Channel Changes*, *Background to Palaeohydrology* and *The Nature of Physical Geography*, to name but a few, which have provided essential reading for most undergraduate courses in physical geography over the past two decades. His high scientific standing, both nationally and internationally, has been marked by the award of several medals, including the coveted Founder's Gold Medal of the Royal Geographical Society which he received in 1993. Ken's contribution must, however, be seen as extending well beyond his papers, books and medals, impressive as these are. There is also a more human dimension represented by the large number of former research students who have profited greatly from his enthusiasm, inspiration, guidance, encouragement and advice, and who have established successful careers in the field of hydro-geomorphology, in universities and higher education establishments, in research institutions and in commercial organisations. Many of us owe this success to Ken's ability to identify key new research areas and themes, his encouragement to pursue these, and his generosity in leaving us to profit from their potential, whilst he moved on to explore new avenues.

Producing a volume of this nature, has inevitably involved a great deal of behind-the-scenes negotiation and effort and I feel sure that all the contributors would wish to echo my sincere thanks to Angela Gurnell and Geoff Petts, who took an idea and brought it to fruition in record time. As probably the only contributor to have had the privilege to benefit from Ken's tutelage and friendship as both an undergraduate and postgraduate and as a colleague, and as his first research student, it is a great personal pleasure to take this opportunity, on behalf of all the contributors, to record our sincere gratitude to Ken for his inspiration and guidance over the past three decades and to dedicate this volume to him in recognition of his outstanding and distinguished contribution to developing and promoting the study of hydro-geomorphology in the UK and on the international scene. In all his endeavours Ken has been strongly supported by his wife Chris and by his family. Before the days of word processors and computerised bibliographies, for example, Chris typed and proof-read many of his manuscripts, and his children were often co-opted to collate and transcribe material for his extensive card index. This support has doubtless continued in a multiplicity of ways and in recognising their important contribution to Ken's achievements we also extend our thanks and best wishes to Chris, Caroline, Sarah and John.

Selected Publications by Ken Gregory

BOOKS

Southwest England, Nelson, 1969 (with A. H. Shorter and W. L. D. Ravenhill).
Exeter Essays in Geography, University of Exeter, 1971, 258pp. (edited with W.L.D. Ravenhill).
Drainage Basin Form and Process, Edward Arnold, 1973, 456pp. (with D.E. Walling).
Fluvial Processes in Instrumented Watersheds, Institute of British Geographers Special Publication No. 6, 1974, 196pp. (edited with D.E. Walling).
River Channel Changes, Wiley, 1977, 450pp. (edited).
Geomorphological Processes, Dawson, 1979, Butterworth, 1980, 312pp. (with E. Derbyshire and J.R. Hails).
Man and Environmental Processes, Dawson, 1979, Butterworth, 1980, 276pp. (edited with D.E. Walling).
Atlas of Drought in Britain 1975–76, Institute of British Geographers Special Publication, 1980, 88 pp. (edited with J.C. Doornkamp and A.S. Burn).
Background to Palaeohydrology, Wiley, 1983, 506pp. (edited).
The Nature of Physical Geography, Edward Arnold, 1985, 276pp.
The Encyclopaedic Dictionary of Physical Geography, Blackwell, 1985, 528pp. (edited with A. S. Goudie, B.W. Atkinson, I.G. Simmons, D.R. Stoddart and D. Sugden).
Energetics of Physical Environment, Wiley, 1987, 172pp. (edited).
Human Activity and Environmental Processes, Wiley, 1987, 466pp. (edited with D.E. Walling).
Palaeohydrology in Practice, Wiley, 1987, 370pp. (edited with J. Lewin and J.B. Thornes).
Horizons in Physical Geography, Macmillan, 1988 (edited with M.J. Clark and A.M. Gurnell).
The Earth's Natural Forces, Oxford University Press, New York, 1990, 256 pp. (edited).
The Guinness Guide to the Restless Earth, Guinness Publishing, 1991, 256pp. (edited).
Temperate Palaeohydrology of the last 15 000 years, Wiley, 1991, 548pp. (edited with L. Starkel and J.B. Thornes).
Global Continental Palaeohydrology, Wiley, 1995, 352pp. (edited with L. Starkel and V.R. Baker).

PAPERS

The deglaciation of eastern Eskdale, Yorkshire. *Proceedings of the Yorkshire Geological Society*, 1962, **33**, 363–380.
Proglacial lake Eskdale after sixty years. *Transactions Institute of British Geographers*, 1965, **36**, 19–162.
A glacial drainage system near Fishguard, Pembrokeshire. *Proceedings of the Geologists Association*, 1965, **76**, 275–282 (with D.Q. Bowen).
The description of relief in field studies of soils. *Journal of Soil Science*, 1965, **16**, 16–30 (with L.F. Curtis and J.C. Doornkamp).
Aspect and landforms in north-east Yorkshire. *Builetyn Peryglacjalny*, 1966, **15**, 115–120.
Data processing and the study of landform. *Zeitschrift fur Geomorphologie*, 1966, **10**, 237–263 (with E.H. Brown).
Dry valleys and the composition of the drainage net. *Journal of Hydrology*, 1966, **4**, 327–340.

Fluvioglacial deposits between Newport and Cardigan. *Occasional Publication No. 3, British Geomorphological Research Group*, 1966, 25–28 (with D.Q. Bowen).

Pleistocene geomorphology I. *20th International Geographical Congress Proceedings*, London, 1967, 230–236 (with R.S. Waters).

The composition of the drainage net. Morphometric analysis of maps. *Occasional Publication No. 4, British Geomorphological Research Group*, 1968, 9–12.

The variation of drainage density within a catchment. *Bulletin of the International Association of Scientific Hydrology*, 1968, **13**, 61–68 (with D.E. Walling).

Instrumented catchments in south-east Devon. *Transactions of the Devon Association*, 1968, **100**, 247–262 (with D.E. Walling).

Geomorphology. In: *Exeter and its Region*, ed. F. Barlow, Exeter, 1969, pp. 27–42.

The measurement of the effects of building construction on drainage basin dynamics. *Journal of Hydrology*, 1970, **11**, 129–144 (with D.E. Walling).

Rainfall–streamflow relationships for an instrumented catchment. *Area*, 1970, No. 3, pp. 65–66.

Drainage density changes in South-West England. In: *Exeter Essays in Geography*, ed. W.L.D. Ravenhill and K.J. Gregory, 1971, pp. 33–53.

Field measurements in the drainage basin. *Geography*, 1971, **54**, 277–292 (with D.E. Walling).

Fluvial processes in small instrumented watersheds in the British Isles. *Area*, 1973, **5**, 97–103 (with D.E. Walling).

The geomorphologist's approach to instrumented watersheds. In: *Fluvial Processes in Instrumental Watersheds*, Institute of British Geographers Special Publication No. 6, 1974, pp. 1–6 (with D.E. Walling).

Streamflow and building activity. In: *Fluvial Processes in Instrumented Watersheds*, Institute of British Geographers Special Publication No. 6, 1974, pp. 107–122.

The adjustment of channel capacity downstream from a reservoir. *Water Resources Research*, 1974, **10**, 870–873 (with C.C. Park).

Lateral variations in pebble shape in North West Yorkshire. *Sedimentary Geology*, 1974, **12**, 237–248 (with R.A. Cullingford).

Drainage density and climate. *Zeitschrift fur Geomorphologie*, 1975, **19**, 287–298 (with V. Gardiner).

Stream channel morphology in North West Yorkshire. *Revue de Geomorphologie Dynamique*, 1976, **25**, 63–72 (with C.C. Park).

Drainage networks and climate. In: *Climate and Landforms*, ed. E. Derbyshire, Wiley, 1976, pp. 299–315.

Changing drainage basins. *Geographical Journal*, 1976, **132**, 237–247.

Bankfull determination and lichenometry. *Search*, 19767, 99–100.

Pragmatic topology of river channels. *International Geography '76*, Moscow, 1976, **1**, 158–160.

The development of a Devon gully and man. *Geography*, 1976, **60**, 77–82 (with C.C. Park).

Channel capacity and lichen limits. *Earth Surface Processes*, 1976, **1**, 273–285.

Drainage basin adjustments and man. *Geographica Polonica*, 1977, **34**, 155–174.

Progress in portraying the physical landscape. *Progress in Physical Geography*, 1977, 1–22 (with V. Gardiner).

Stream network volume: an index of channel morphometry. *Bulletin of the Geological Society of America*, 1977, **88**, 1975–1980.

Fluvial geomorphology. *Progress in Physical Geography*, 1977, **1**, 345–351.

Till ridges in Wensleydale, Yorkshire. *Proceedings of the Geologists Association*, 1978, **89**, 67–79 (with R.A. Cullingford).

A physical geography equation. *National Geographer*, 1978, **12**, 137–141.

Fluvial processes in British basins. In: *Geomorphology: Present Problems and Future Prospects*, ed. C. Embleton, D. Brunsden and D.K.C. Jones, Oxford University Press, 1978, 40–72.

Fluvial geomorphology. *Progress in Physical Geography* 1978, **2**, 346–352.

Palaeohydrological changes in the Temperate Zone in the last 15 000 years. *United Kingdom Contribution to the International Geological Correlation Programme*, The Royal Society, 1978, pp. 119–123.

Hydrogeomorphology: how applied should we become? *Progress in Physical Geography*, 1979, **3**, 84–100.

Drainage network volumes and precipitation in Britain. *Transactions, Institute of British Geographers*, 1979, NS **4**, 1–11 (with J.C. Ovenden).

Drainage network power. *Water Resources Research*, 1979, **15**, 775–777.

Fluvial geomorphology. *Progress in Physical Geography*, 1979, 3, 274–282.

The permanence of stream networks in Britain. *Earth Surface Processes*, 1980, **5**, 47–60 (with J.C. Ovenden).

Fluvial geomorphology. *Progress in Physical Geography*, 1980, **4**, 421–430.

Updating geomorphology: if it moves, measure it! *Teaching Geography*, 1980, **5**, 170–174.

The Severn Basin: sample basin review. *Bulletin de l'Association Francaise pour l'Etude du Quatanaire*, 1980, **17**, 61–64 (with J.B. Thornes).

Physical geography from the newspaper. *Geography*, 1981, **66**, 42–52 (with R.F. Williams).

Present day characteristics. In: *Palaeohydrology of River Basins*, ed. L. Starkel and J.B. Thornes, British Geomorphological Research Group Technical Bulletin 28, 1981, pp. 8–24.

Fluvial geomorphology. *Progress in Physical Geography*, 1981, **5**, 409–419.

Drainage basin control. *Geographical Magazine*, 1981, **54**, 104–107 (with I. Douglas).

Fluvial geomorphology: less uncertainty and more practical application? *Progress in Physical Geography*, 1982, **6**, 427–438.

River power. In: *Papers in Earth Studies*, ed. B.H. Adlam, C.R. Fenn and L. Morris, Geobooks, 1982, 1–20.

Land use change, flood frequency and channel adjustments. In: *Gravel-bed Rivers: Fluvial Processes, Engineering and Management*, ed. R.D. Hey, J.C. Bathurst and C.R. Thorne, Wiley, 1982, pp. 757–782 (with J.R. Madew).

Drainage density in rainfall–runoff modelling. In: *Rainfall–Runoff Relationship*, ed. V.P. Singh, Water Resources Publications Colorado, 1982, 449–476 (with V. Gardiner).

Physical geography techniques: a self-paced course. *Journal of Geography in Higher Education*, 1982, **6**, 123–131 (with M.J. Clark).

Human activity and palaeohydrology: a review. In: *Palaeohydrology of the Temperate Zone*, ed. S. Kozarski, Quaternary Studies in Poland, 1983, **4**, pp 73–80.

An assessment of river channelization in England and Wales. *The Science of the Total Environment*, 1982, **27**, 97–111 (with A. Brookes and F.H. Dawson).

Hydrogeomorphology downstream from bridges. *Applied Geography*, 1983, 3, 145–159 (with A. Brookes).

River channel forms, processes and metamorphosis. In: *Studies in Quaternary Geomorphology*, ed. D.J. Briggs and R.S. Waters, Geobooks, 1983, pp. 19–30.

Fluvial geomorphology. *Progress in Physical Geography*, 1983, 7, 385–396.

Why rivers change their course. *Geographical Magazine*, 1984, **56**, 120–128.

The influence of vegetation on stream channel processes. In: *Field experiments in Geomorphology*, ed. D.E. Walling and T.P. Burt, Geobooks, 1984, 515–535 (with A.M. Gurnell).

Fluvial geomorphology. *Progress in Physical Geography*, 1984, **8**, 421–430.

The impact of river channelization. *Geographical Journal*, 1984, **151**, 53–74 (with D.L. Hockin, A. Brookes and M.P. Brooker).

Detrended correspondence analysis of heathland vegetation: the identification of runoff contributing areas. *Earth Surface Processes and Landforms*, 1985, **10**, 343–351 (with A.M. Gurnell, S. Hollis and C.T. Hill).

The permanence of debris dams related to river channel processes. *Hydrological Sciences Bulletin*, 1985, **30**, 371–381 (with A.M. Gurnell and C.T. Hill).

Progress in Palaeohydrology. Special issue of *Earth Surface Processes and Landforms*, 1985, **10**, 203–304 (with V. Gardiner and L. Starkel).

Fluvial geomorphology—process explicit and implicit? *Progress in Physical Geography*, 1985, **9**, 414–424.

Temperate landscapes. In: *Handbook of Engineering Geomorphology*, ed. P.G. Fookes and P.R. Vaughan, Blackie, 1986, 87–108.

Human impact on the fluvial environment. In: *IGCP 158, Palaeohydrological Changes in the Temperate Zone in the Last 15 000 Years*, ed. J.J. Gaillard, Lund, Sweden, 1986, 9.

Power to compare. *Geographical Magazine*, 1986, **58**, 468–473 (with C.T. Hill).

Water table level and contributing area: the generation of runoff in a heathland catchment.

In: *Conjunctive Water Use*, Proceedings of the Budapest Symposium, July 1986. International Association of Hydrological Sciences Publication 156, 1986, pp. 87–95 (with A.M. Gurnell).

Vegetation characteristics and the generation of runoff: Analysis of an experiment in the New Forest. *Hydrological Processes*, 1987, **1**, 125–141 (with A.M. Gurnell).

Hydrogeomorphology of Alpine proglacial areas. In: *Glacio-Fluvial Sediment Transfer*, ed. A.M. Gurnell and M.J. Clark, Wiley, 1987, pp. 87–107.

The use of Landsat multispectral scanner data for the analysis and management of flooding in the river Severn, England. *Environmental Management*, 1987, **11**, 695–701 (with A.G. Brown and E.J. Milton).

Environmental effects of river channel changes. *Regulated Rivers: Research and Management*, 1987, **1**, 358–363.

Vegetation influences upon river channel form and process. In: *Biogeomorphology*, ed. H.A. Viles, Blackwell, 1988, pp. 11–42 (with A.M. Gurnell).

Curriculum development and geomorphology. *Journal of Geography in Higher Education*, 1988, **12**, 21–30.

Energetics of environment. *Geography Review*, 1988, **2**(1), 31–35.

Channelization, river engineering and geomorphology. In: *Geomorphology and Environmental Planning*, ed. J.M. Hooke, Wiley, 1988, pp. 145–767 (with A. Brookes).

Impact of the '87 storm. *Geography Review*, 1989, **2**(4), 13–16.

Changes in urban stream channels in Zimbabwe. *Regulated Rivers: Research and Management*, 1989, **4**, 27–42 (with J.R. Whitlow).

River discharge estimated from channel dimensions. *Journal of Hydrology*, 1989, **106**, 365–376 (with G. Wharton, N.W. Arnell and A.M. Gurnell).

Have global hazards increased? *Geography Review*, 1990, **4**(2), 35–38 (with H. Rowlands).

How integrated is drainage basin management? *Environmental Management*, 1991, **15**, 299–309 (with P.W. Downs and A. Brookes).

Palaeohydrological results from the Severn basin and future research requirements. *Quaternary Studies in Poland*, 1991, 8.

Changing physical environment and changing physical geography. *Geography*, 1992, **77**, 323–333.

Vegetation and river channel process interactions. In: *River Conservation and Management*, ed. P.J. Boon, P. Calow and G.E. Petts, Wiley, 1992, pp. 255–270.

Coarse woody debris in stream channels in relation to river channel management in woodland areas. *Regulated Rivers: Research and Management*, 1992, **7**, 117–136 (with R.J. Davis).

Identification of river channel change due to urbanisation. *Applied Geography*, 1992, **12**, 299–318 (with R.J. Davis and P.W. Downs).

The sensitivity of river channels in the landscape system. In *Landscape Sensitivity*, ed. D.S.G. Thomas and R.J. Allison, Wiley, 1993, pp. 15–30 (with P.W. Downs).

Spatial distribution of coarse woody debris dams in the Lymington Basin, Hampshire, UK. *Geomorphology*, 1993, **6**, 207–224 (with R.J. Davis and S. Tooth).

Debris jams in river channels: to clear or not to clear? *Geography Review*, March 1993, **6**(4), 2–6 (with R.J. Davis).

The perception of riverscape aesthetics: An example from two Hampshire rivers. *Journal of Environmental Management*, 1993, **39**, 171–185 (with R.J. Davis).

Stability of the pool–riffle sequence in changing river channels. *Regulated Rivers: Research and Management*, 1994, **9**, 35–43 (with A.M. Gurnell, C.T. Hill and S. Tooth).

A new distinct mechanism of river bank erosion in a forest catchment. *Hournal of Hydrology*, 1994, **157**, 1–11 (with R.J. Davis).

Classification of river corridors: issues to be addressed in developing an operational methodology. *Aquatic Conservation*, 1994, **4**, 219–231 (with A.M. Gurnell and P. Angold).

The role of dead wood in aquatic habitats in forests. In: *Forests and Water*, ed. A.J. Lowe, Proceedings of the 1994 Spring Discussion Meeting of the Institute of Chartered Foresters, in press (with A.M. Gurnell and G.E. Petts).

The role of coarse woody debris in forest aquatic habitats: implications for management. *Aquatic Conservation*, in press (with A.M. Gurnell and G.E. Petts).

1   Changing River Channels: The Geographical Tradition

GEOFFREY E. PETTS

School of Geography, University of Birmingham, UK

INTRODUCTION

If it be admitted that the little stream has worn out the gutter in which it runs, it is hard to deny that the larger stream has not done similar work on a larger scale. The whole affair is indeed a mere question of time.

(Huxley, 1880, p. 139)

The study of changing river channels is the domain of fluvial geomorphology which, in the UK, has been the province of geographers. Traditionally, the geographical approach involves a method of enquiry founded in field observations, and the recognition of relationships between forms and processes, and between types of channel change and local environmental variables. This approach is clearly seen in the works of geographers, geologists and engineers during the 18th and 19th centuries. Essays on physical geography at that time typically included sections on 'agents of change' and, in his *Physiography*, T.H. Huxley (1880) consolidated field observations of fluvial processes and process–form relationships into a coherent approach for examining rivers within river basins. He also advanced the need to consider long time-scales, following the diffusion of the notion of evolution after the publication in 1859 of Darwin's *Origin of Species*. Others incorporated aspects of both uniformitarianism and catastrophism into approaches to studying landscape change:

Our notice of physical geography would be incomplete without some account of the phenomena of change which are in constant operation, and by the agency of which considerable alterations—some of them slow and gradual, so as only to exhibit the result when tested by the experience of a long term of years; others sudden and violent in their mode of action—are affected in the aspect of the earth's surface.

(Hughes, 1878, p. 33)

In such essays, discussions of changing river channels typically are confined to descriptions of catastrophic floods (or '*debacles*') and deltas. It was more than 100 years after the publication of the *Origin of Species* that modern geographical fluvial geomorphology established a scientific basis for the study of 'change', integrating measurements of short-term process mechanics and channel dynamics over historical

Changing River Channels. Edited by Angela Gurnell and Geoffrey Petts.
© 1995 John Wiley & Sons Ltd.

time-scales with observations on long-term channel evolution. This 'century of foundation' (Gregory, 1985) was dominated by the works of W.M. Davis (1850–1934) and G.K. Gilbert (1843–1918). The Davisian approach advanced a framework to classify any landscape according to its evolutionary stage (as youthful, mature or old age) and offered a trilogy for the understanding of landscape in terms of structure, process and time (Gregory, 1985, p. 29). The approach not only complemented uniformitarianism but also provided the basis for the historical interpretation of landforms. However, modern fluvial geomorphology was to be equally influenced by the work of Gilbert whose studies in the western USA described physical erosive processes and derived a system of laws governing landform changes (Chorley *et al.*, 1964).

The publication of *Fluvial Processes in Geomorphology* by Leopold *et al.* in 1964 heralded the establishment of modern fluvial geomorphology with an emphasis on processes and mechanisms of morphological change, achieved through the acquisition of new techniques for measurement and analysis. However, in USA, this development led to a progressive reinforcement of discrepancies, first, between those concerned with process dynamics and those with interests in long-term landscape change (Ritter, 1988) and secondly between problem-oriented studies and method-orientated ones (Baker, 1988).

In the UK, geographers were confronted with the same dichotomies of purpose. Paradoxically, however, diversification within geography, within physical geography and within fluvial geomorphology led to integration at all levels. The strength of this integration within fluvial geomorphology was due in large part to the timely publication in 1973 of *Drainage Basin Form and Process* in which K.J. Gregory and D.E. Walling (p. 3) demonstrated that in the context of drainage basins and their fluvial processes: 'apparently diverse approaches are united by focusing attention upon present systems, their content, mechanics and spatial variation'.

Integration within geomorphology in general and studies of changing river channels in particular, also reflects the important role played by the British Geomorphological Research Group (BGRG) in providing the necessary fora for the free exchange of ideas and minimising the tensions that will always evolve between advancing specialisms in a rapidly changing discipline. This role is well illustrated by *River Channel Changes*, edited by Gregory in 1977, the subject of which emerged as the theme for a BGRG symposium. The book includes papers on process dynamics, channel geometry, channel pattern, drainage networks and hillslope processes, theoretical deductions and modelling. *River Channel Changes* established the foundation for developments in research on changing river channels over the next two decades.

This volume on the theme of 'Changing River Channels', focuses on the geographical approach that over a period of 30 years has been—and continues to be—advanced and fostered by Ken Gregory. This chapter, in introducing the subsequent contributions, summarises the development of the 'geographical approach', briefly reviews recent advances, and assesses the prospects for the future. Twenty years ago, *Drainage Basin Form and Process* (Gregory and Walling, 1973) promoted the development of studies of landform–process relationships 'because these provide results for understanding the past, for estimating the future, and for application to other fields' (p.9). An additional reason for such studies of changing river channels has evolved over the past 20 years, namely, the application of the knowledge generated to problems of river management. Thus, this volume develops in four sections: discussion of (1) the temporal and spatial dimensions for studying changing river channels and (2) the processes of change, lead into consideration of (3) information for, and (4) approaches to, managing changing river channels.

All the chapters in the book have been written by former students and co-workers of Ken Gregory, and they reflect the strength of the continuing geographical tradition. There are many other students and colleagues who could have contributed to this volume; their omission is regrettable but the practicalities of publishing inevitably meant that the number of contributions had to be limited. We hope they consider the volume is a fitting tribute to their mentor and friend.

THE GEOGRAPHICAL APPROACH

Fluvial geomorphology is a field science; classification and description are at the heart of this science. The modern 'geographical', or spatial–analytical (Baker, 1988), approach to studies of changing river channels applies classification and description to establishing functional relationships between landforms, processes and other environmental variables over a range of scales encompassing the region, the river basin and the reach over periods of time from less than a year to 10 000 years. The approach evolved as conceptual and analytical frameworks were developed to structure the interpretation of field observations between 1950 and 1970. Advances in measurement and a focus on process studies became the pervading features of many developments during this period. These advances were made through both functional studies which sought repeatable and predictable relationships between form and process, and realistic approaches which attempted to elucidate the mechanisms and underlying structures responsible for the process–form relationships. The geographical tradition has functional studies as its core but the major advances have resulted from the integration of both approaches.

The conceptual framework

Advancement of fluvial geomorphology during the two decades prior to the publication of *Drainage Basin Form and Process* was largely in response to major developments founded in the Columbia school of geomorphology in the USA, led by A.N. Strahler. First, Strahler (1952) advocated the need for a system of geomorphology grounded in basic principles of mechanics and fluid dynamics. Thus, geomorphological processes would be (p.923):

> treated as manifestations of various types of shear stresses, both gravitational and molecular, acting upon any type of earth material to produce the varieties of strain, or failure, which we recognise as the manifold processes of weathering, erosion, transportation and deposition.

This realistic view was to evolve rapidly through links with engineering science to underpin process-based studies of changing river channels.

Secondly, modern studies of long-term channel changes developed from the geological tradition of fluvial geomorphology and were founded in the work of Leopold and Miller (1954) whose studies of the alluvial chronology of valleys in Wyoming highlighted the interaction of climate, vegetation and runoff in determining channel form and process. They also illustrated different fluvial responses of mountain and plains rivers. 'Palaeohydrology' was given considerable impetus by a series of papers by S.A. Schumm during the mid 1960s, notably papers on river channel adjustment to altered hydrological regime (1968a) and hydrological controls of terrestrial sedimentation (1968b) over a Quaternary time-scale. Later, Schumm's contributions were consolidated and he elucidated the approach in the context of the fluvial system (Schumm, 1977).

Thirdly, the seminal work of Schumm and Lichty (1965) on 'time, space and causality in geomorphology' initiated a new attitude towards changing river channels by proposing a method to reconcile apparently alternative views – process-based studies and investigations of long-term environmental change. Their central theme focuses on the idea that as the dimensions of time and space change, cause–effect relationships may be obscured or even reversed. They illustrated this idea with reference to rivers (Table 1.1) such that the status of a geomorphic variable is seen to change according to the time-scale being considered.

This rationalization of approaches to long-term evolution and short-term change was reinforced and developed by three other lines of research. First, Wolman and Miller (1960) had demonstrated the role of the magnitude and frequency of geomorphic processes. Having explained that above the level of competence, sediment transport can be expressed over the time-scale under consideration, they showed that despite the catastrophic changes that may be associated with extreme, rare floods, in the long term the largest proportion of sediment transported by rivers is carried by flows that occur on average once or twice each year. Secondly, Schumm (1968b) highlighted the importance of stability thresholds for channel change and demonstrated the relative roles of intrinsic and extrinsic erosion thresholds in determining the effects of floods (Schumm, 1973). Progressive change resulting from low-magni-

Table 1.1   The changing status of river variables (based on Schumm and Lichty, 1965)

River variables	Geologic (10^6 years)	Modern (10^3 years)	Present (10^{-1} years)
Time	Independent		
Geology	Independent	Independent	Independent
Climate	Independent	Independent	Independent
Relief	Dependent	Independent	Independent
Valley dimensions	Dependent	Independent	Independent
Vegetation	Dependent	Independent	Independent
Palaeohydrology	Dependent	Independent	Independent
Mean discharge (water and sediment)		Independent	Independent
Channel morphology		Dependent	Independent
Instream hydraulics			Dependent

tude, frequent processes, was shown to move channel form to a condition of increasing instability with time; the combination of extrinsic and intrinsic thresholds leading to complex sequences of channel changes following disturbance. This was subsequently illustrated in the dramatic example of Douglas Creek, Colorado (Womack and Schumm, 1977), in which channel response to overgrazing in the catchment was characterised by a complex sequence of erosional and deposition events reflecting a catchment-scale response to initial arroyo cutting. The third development in understanding the geomorphological effectiveness of events of different magnitude and frequency came after the publication of *Drainage Basin Form and Process* but had an important impact on subsequent research. This development addressed the persistence of channel changes and demonstrated the need to consider the rate of recovery of channel morphology following a disturbance (Wolman and Gerson, 1978). Thus, the long-term persistence of channel changes caused by low-frequency, high-magnitude events in semi-arid catchments was shown mainly to reflect the weakness of recovery processes between 'effective' flood events in contrast to rivers in temperate basins.

The fourth major influence on the development of fluvial geomorphology initiated during the 1950s and 1960s was the introduction of the general systems approach in physical geography by R.J. Chorley (1962) who was significantly influenced by earlier works by the Columbia school. The approach was later elaborated by Chorley and Kennedy (1971). In advocating an open-system approach, Chorley identified a number of important values and, with regard to subsequent developments of research on changing river channels, three may be emphasised: the approach depends upon the universal tendency towards adjustment of form and process; it directs attention towards the essentially multivariate character of geomorphic phenomena; it fosters a dynamic approach to geomorphology to complement the historical one. As noted by Gregory (1985, p. 146) the impact of the systems approach on physical geography can be attributed to the ways in which it 'could rationalize physical geography endeavour, and could catalyse the introduction of new concepts especially concerned with temporal change.'

Conflict or integration?

The fundamental characteristic of the geographical approach is the focus on unravelling the range of factors that influence fluvial landforms over different temporal and spatial scales. Despite the advances in systems approaches, measurement techniques, and quantitative methods to the analysis of spatial patterns since the 1960s, the fundamental geographical approach to fluvial geomorphology has been accused of being unreputable, a view discussed by Baker (1988), Ritter (1988), and Smith (1993) and worthy of only summary here. Research involving complex mathematical procedures or detailed laboratory experimentation has tended to be regarded as having greater prestige than field-based research. In part this may reflect the uncertain and technically unsophisticated nature of field research, subject to unpredictable weather or other field conditions and reliant on robust and often rather insensitive equipment. The important contributions to fluvial geomorphology of advances in process studies, effected principally and increasingly by mathematical and stochastic models with an emphasis on method, must not be underestimated. However, there are a number of levels of increasing complexity for examining changing river channels and all have their place in a well-founded college of scientific endeavour.

To quote Smith (1993, pp. 258–259): 'Most great geomorphic discoveries were serendipitous, usually landform anomalies were noted, the question "why" was asked, and then either a slow evolution of an idea or a flash of inspiration explained causes.' As the first step in all studies of the natural environment, the explanation of ideas and concepts in descriptive terms is a valuable precursor to experimental and theoretical investigations. The demonstration of empirical regularities or identification of anomalies often initiates the search for the natural generating mechanisms (Richards, 1990). Furthermore, this level of analysis is valuable for specific problems where the number of assumptions required to produce useful models means that the highest levels of investigation are inappropriate.

Baker (1988) identified two groups of fluvial geomorphologists: geomorphogenists who apply the spatial–analytical approach and deductive reasoning, and their opposites, the geomorphotechnicians, whose research is based in detailed measurement of process, the inductive method, and predictive numerical modelling. In isolation, both extremes offer limited contributions to scientific knowledge, being prone in the former to qualitative generalisation and in the latter to bypassing understanding to achieve elegant predictions (Baker, 1988). The integration of these approaches to achieve a coherent understanding of fluvial forms and processes, as advocated by Gregory and Walling (1973), remains at the core of the geographical approach.

The roots of the geographical approach to studies of changing river channels

The geographical approach with its emphasis on field-based, empirical studies devoted to the understanding of the fluvial system, has its roots embedded in the observational science practiced by engineers and geologists during the 18th and 19th centuries. First, engineers established basic process–form relationships and used their 'theories' in river management. Then, during the 19th century, geologists used similar relationships as the key to the past.

The engineering tradition

The roots of our modern understanding of fluvial geomorphology lie especially in Italy during the 17th century. These evolved (1) following the work of Galileo in the mid 16th century which led to the formulation of the laws governing water flow and sediment transport, and (2) in response to the practical needs for flood control, land drainage, and navigation (Wheeler, 1893). A commission, appointed by Pope Innocent XII in 1693 to investigate the increasing flood hazard on rivers passing through Bologna, Ferrara and Romagna, included an engineer named Guglielmini. His report 'Natura de Fiumi', published in 1697, laid down fundamental proposals on the effects of transporting and eroding power of water in rivers. Gregory (1976) named Guglielmini the first 'fluvial geomorphologist'.

In applying his 'laws', Guglielmini advised great caution especially when shortening rivers by cutting off bends, and that before making such cuts a perfect knowledge of the soil through which the river passed should be obtained. In Britain, the engineering tradition was advanced by Sir John Rennie who in 1837 brought for-

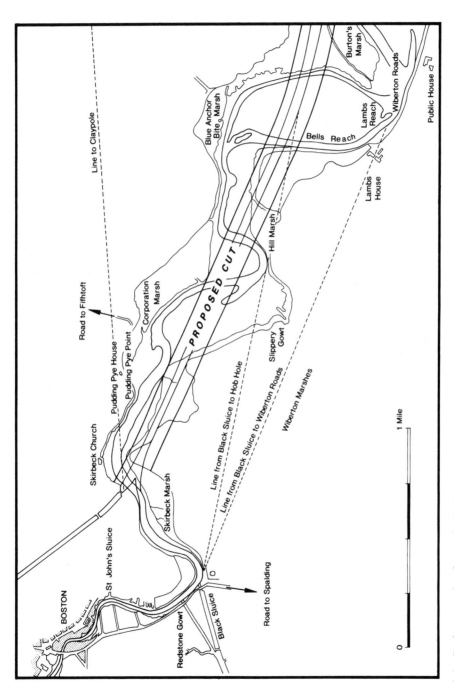

Figure 1.1 Early example of a 'trained' channel: a section of the lower River Witham, Lincolnshire, UK, in 1822 and the proposed channelisation scheme (after Rennie, 1822)

ward a scheme for training by fascine-work the Ouse, Nene, Welland and Witham rivers in the fens of eastern England (Wheeler, 1893). Earlier, Rennie (1822) provided one of the first reports of changing river channels in Britain: the reduction of average channel width along two reaches of the lower River Witham from 28 to 23 m and from 102 to 64 m in only 22 years. In response to this 'evil' Rennie proposed the canalisation of the river (Figure 1.1). However, it was another British engineer, Jessop, whose observations on rivers had earlier led to a 'theory' of fluvial geomorphology which he applied to the River Trent in 1782 (p.1):

> ... Where rivers run through a Country where the soil is pure Clay, Loam or any Thing of light and homogeneous Quality, they are always very Deep, and in general Narrow; on the Contrary, where they run through a Soil that has in its Composition a considerable Mixture of Sand, Gravel or other hard Matter, they always become wide or shallow, and more or less so in Proportion to the Quantity of hard Matter; for as they are continually changing their Situation, where not bounded by the Rocks, or defended by Works of Art, the lighter Soil is carried away, or laid on to the gaining Sides of the River, while the heavier Parts are deposited in the Bottom, and the Bottom becomes relatively hard; thus the effect of the Widening becomes a Cause of its Continuance, for as any certain Quantity of Water must have a certain Capacity in the River equal to the Discharge of it, when it is debarred from Depth by the hardness of the Bottom, it will have it in width, unless the sides are made harder than the Bottom ... the hard Matter . . . forms a Sort of natural Dams over which the Fall of Water is often very considerable, while the intermediate Parts of the River ... run deep and languid...

Such observations on fluvial landforms and processes became incorporated into practical schemes for improving the courses of the rivers for navigation purposes. Two other British engineers deserve note in this context: W.A. Brooks' treatise on *Improvement of the Navigation of Rivers* published in 1841 and Captain E.K. Calver's *Conservation and Improvement of Tidal Rivers* published in 1853. During this period in the USA, surveys of the lower Mississippi and Ohio rivers were used to apply the same principles to protect the delta from inundation and to improve navigation with special regard to the bars at the mouths of the Mississippi (Ellett, 1853). These works provide a clear statement on the state-of-the-science in the mid 19th century.

The geological tradition

The first half of the 19th century also witnessed the establishment of the geological tradition within fluvial geomorphology with the publication of Charles Lyell's *Principles of Geology* (in 1830) which presented a series of

> preliminary essays to explain the facts and arguments which lead me to believe that the forces operating upon and beneath the earth's surface may be the same, both in kind and degree, as those which at remote epochs have worked out geological changes (1850, p. iv)

As President of the Geological Society of London, Lyell was influential in securing the acceptance of the uniformitarian ideas of Hutton and Playfair. The fundamental tradition of fluvial geomorphology as an observational science is again clearly demonstrated, for example with reference to the Mississippi's downstream variation of channel form (Lyell, 1850, p. 210) and meander evolution following cutoff (p. 212):

> No river affords a more striking illustration of the law ... than an augmentation of volume does not occasion a proportional increase in surface ... The Mississippi is half a mile wide at its junction with the Missouri, the latter being also of equal width; yet the united waters have only, from their confluence to the mouth of the Ohio, a medial width of about half a mile.

> As soon as the river has excavated the new passage, bars of sand and mud are formed at the two points of junction with the old bend, which is soon entirely separated from the main river by a continuous mud-bank covered with wood.

He continues to observe the significance of these processes for river ecology in the face of human impact:

> The old bend then becomes a semicircular lake of clear water, inhabited by large gar-fish, alligators and wild fowl, which the steam-boats have nearly driven away from the main river.

Two further examples illustrate the awareness of the link between river channels, hillslopes and riparian zones. First, with regard to the Po he notes (p. 208):

> *Mountain torrents ... have become more turbid since the clearing away of forests, which once clothed the southern flanks of the Alps. It is calculated that the mean rate of advance of the delta of the Po on the Adriatic between the years 1200 and 1600 was 25 yards or metres a year, whereas the mean annual gain from 1600 to 1804 was 70 metres.*

Secondly, with reference to the rafts or masses of floating trees on the Mississippi which collected during the flood season to form natural bridges reaching entirely across the stream (p. 213):

> Several acres [of floodplain] at a time, thickly covered with wood, are precipitated into the stream.

> One of the largest of these [rafts] was called the raft of the Atchafalaya ... the drift trees collected in about thirty-eight years previous to 1816 formed a continuous raft, no less than ten miles in length, 220 yards wide, and eight feet deep.

Many of the questions asked by geologists and engineers during the 19th century remain important foci for research today, as illustrated by the subsequent chapters in this volume. Most advances have been made in conceptualising problems and in understanding the factors that constrain our development of coherent theory, rather than in the formulation of general models to explain the dynamics of changing river channels over appropriate time-scales. Nevertheless, important progress has been made over the past two decades.

ADVANCES OVER THE PAST 20 YEARS

Subsequent to the publication of *River Channel Changes* (Gregory, 1977) three sub-disciplines evolved: namely, process mechanics, channel dynamics and palaeohydrol-ogy. These relate to three clearly defined, but interrelated, spatial and temporal scales for the analysis of change (Figure 1.2). Their subsequent development is described by similar histories which emphasise collective advancement rather than isolationism. In large part they have all been motivated by the need to advance knowledge to contribute in the elaboration of future environmental changes and especially for a more complete understanding of the human impact so that pre-ventative, mitigative and restorative measures can be developed. The rate of research development reflects the advancement in techniques and methods, and the response of scientists to opportunities presented by natural and artificial events.

Process mechanics

One of the major factors influencing scientific advancement in all disciplines is the development of new technology. Developments in sensory and sampling technology and the advancement of data loggers (see Parr, 1994) has led to increasingly inten-sive real-time monitoring programmes so that short-term variations, flood responses, and sediment and solute sourcing can be investigated and modelled (Carling, 1992; Walling and Webb, 1992; Webb and Walling, 1992; and see Chapter 7). The devel-opment of small electromagnetic current meters supported by high-resolution depth sounders has allowed detailed studies of flow and sediment dynamics (Carling, 1992). One novel development has been the Photo-Electronic Erosion Pin (PEEP) system to provide au automatic, quasi-continuous, erosion monitoring technique to quantify the erosional and depositional impact of individual flow events (Lawler, 1992).

Following these technical advances, studies of small-scale phenomena have devel-oped rapidly around three themes at the interface of fluid mechanics, hydraulic engi-neering and fluvial geomorphology, namely: sediment transport, turbulent flow and bedform development (e.g. Hoey, 1992; Robert, 1993). Recent advances have inves-tigated the dynamics of the three primary components of flow resistance: individual grains (Hassan and Church, 1992), clusters of particles or small bedforms (Hassan and Reid, 1990; Clifford *et al.*, 1992; Reid *et al.*, 1992), and large-scale bedforms such as pools and riffles (Carling, 1991). Such studies have made important con-tributions to understand the conditions for initiating bedload transport, the domi-nant process linking channel change and hydraulic conditions (Chapter 8).

A second group of studies focus on the relationships between channel bed mor-phology and sedimentary composition, and particle mobility. Particular attention has been given to flow patterns, sediment transport and channel morphology at river channel confluences (Best, 1986; Ashmore *et al.*, 1992). Changes of bed morphology by bedform migration in sand-bed streams (Gomez *et al.*, 1990) and bar migration in braided gravel-bed rivers (Ashmore, 1991; Hoey and Sutherland, 1991) have been shown to explain pulses of bed-material transport. The advancement of gravel sheets or lobes has been shown to be an important mechanism of bed change across the range of flow events (Laronne and Duncan, 1992).

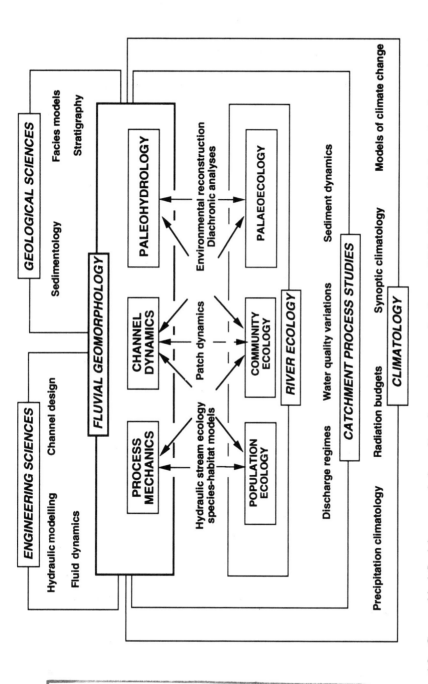

Figure 1.2 Geographical fluvial geomorphology: its primary structure showing established links with the engineering and geological sciences, its fundamental relationship with catchment process studies (elaborated in Foster *et al.*, 1995) which are closely linked to studies of climatology and climate change, and new links with river ecology

A third focus for research on process dynamics has been the mechanisms of bank erosion (Chapter 5). Two mechanisms are generally considered to account for the retreat of channel banks: the removal of individual particles or fine layers from the bank surface and mass failures. Davis and Gregory (1994) describe a third mechanism, the development of subsurface cavities after washout of gravels, succeeded by slow subsidence of a segment of the former bank. Techniques for measuring riverbank erosion and lateral channel change have been reviewed by Lawler (1993). Because of the significance of bank erosion for river-managers, particular efforts have been directed to the development of models of riverbank instability that give due regard to bank geometry changes due to toe scour and lateral erosion (Osman and Thorne, 1988; Thorne and Osman, 1988) and to the development of tension cracks (Darby and Thorne, 1994).

Channel dynamics

Considerable progress has been made in understanding the adjustment of channel cross-sections (Chapter 5) and the dynamics of channel pattern (Chapter 6). Developments have also been made in modelling the evolution of meso-scale depositional forms such as channel confluence bars (Best, 1986), and slackwater bars (Nanson and Page, 1983); bar-bend theory, for example, has successfully predicted the formative conditions of alternate bars, point bars and meander bends (Rhoads and Welford, 1991). The dynamics of braided rivers has attracted particular interest (Best and Bristow, 1993) and studies of now rare anastomosing channels (Harwood and Brown, 1993; Knighton and Nanson, 1993) and other stable divided channels may be especially important given their possible widespread occurrence prior to early human impact (Brown *et al.*, 1995). Surveys of channel adjustments show complex patterns of channel change in response to natural environmental variations, such as alternating flood- and drought-dominated regimes (Erskine and Warner, 1988) and inherited valley-floor conditions (Warner, 1992), and to human impacts (Hooke and Redmond, 1992; Gurnell *et al.*, 1995).

The complexity of channel response has been explained by different relationships between channel form and discharge due to changing bed materials, bank sediments and channel vegetation (Ebisemiju, 1991). Field documentation of progressive changes in channel pattern and associated bar forms along rivers with strong energy gradients (Ferguson and Ashworth, 1991) has demonstrated the sensitivity of channel morphology to changes of channel slope in gravel-bed streams. However, time-lags in morphological responses as channels and drainage networks adjust to changes of discharge and/or sediment loads add to the complexity of channel forms that may be observed. The integration of studies of contemporary channel dynamics and of channel changes over historic time-scales is developing our understanding of the mechanisms of channel change that operate at the catchment scale (Petts, 1979, 1987; Petts *et al.*, 1989).

Palaeohydrology

In Europe, modern palaeohydrology, having a focus on the Holocene, again evolved from *River Channel Changes* (Gregory, 1977). A 10-year (1977–1986) research programme on fluvial palaeohydrology for the International Geological Correlation Programme (IGCP Project 158), was initiated by Leszek Starkel—a contributor to Gregory's volume—with the aim of reconstructing climatic variations over the last 15 000 years. The British contribution was guided by Gregory whose interests were founded in studies of underfit valleys and drainage network dynamics (Chapter 4). The programme stimulated a very significant increase in research on the palaeohydrology of the temperate zone and this is described in three major publications (Gregory, 1983; Gregory et al., 1987; Starkel et al., 1991). Recently, Macklin and Lewin (1993), from a synthesis of available information on river alluviation during the Holocene in Britain, have identified eight episodes of regional and country-wide alluviation and demonstrated the synchrony of these episodes in Europe and the USA, particularly over the past 5000 years, suggesting the dominance of climatic rather than anthropogenic influences. However, in detail, across Europe river channels have experienced a complex evolution during the Holocene related to geological history (Chapter 2).

The way in which palaeohydrological analysis complements that obtained by field and theoretical research in contemporary fluvial systems to suggest new approaches to hydraulic behaviour is discussed by Thornes and Gregory (1991). The main unifying problem is the need to determine the sensitivity of channels, and the different components of channel morphology at different scales of analysis, to adjustment in response to changes of discharge and/or sediment (Chapter 3). A primary link with the channel dynamics subdiscipline is through the focus on stability–instability relations and the nature of the transitions between them. This is particularly seen in contemporary studies of human impacts on rivers during the industrial era of the last 200 years (Knox, 1987; Petts et al., 1989; Macklin et al., 1992).

Opportunities seized

A major constraint to the advancement of knowledge on changing river channels is the time-scale and complexity of channel adjustment in response to climate changes. Opportunities for gaining new insights into a range of fluvial geomorphological processes include natural storm events and human impacts that provide field laboratories for geomorphologists by causing major system disturbance. Disturbances may be classified as three types (Underwood, 1994); *pulse disturbances* are acute, short-term episodes (e.g. storm events); *press disturbances* are sustained impacts (e.g. below dams); and *catastrophes* are major, often planned, destructions of physical features (e.g. canalisation). Although there are some problems with this classification, it provides a useful framework for considering opportunities to investigate the nature, time-course and possible responses of channel forms. Gregory made particular use of such investigations to elucidate a range of channel change mechanisms. These included studies of channel changes below dams (Gregory and Park, 1974), urban areas (Gregory and Park, 1976a and b; Gregory and Whitlow, 1989; Gregory et al., 1992), and bridges (Gregory and Brookes, 1983); the effect of channelisation (Gregory et al., 1984); and the impact of storm events (Gregory, 1992).

Extreme floods have always attracted the attention of fluvial geomorphologists and research has been consolidated into the discipline of 'flood geomorphology' (see Baker et al., 1988). Case studies of rare large floods provide opportunities to investigate the factors contributing to the variable impact of flood-induced channel and floodplain changes (e.g. the 1990 flood event on the River Tay, Scotland: Gilvear and Harrison, 1991). Such studies have immediate applications in river and flood-plain management, as well as providing fundamental information on flood deposits and opportunities to investigate recovery mechanisms. At a smaller scale, artificial spates produced by controlled reservoir releases have provided opportunities to investigate channel processes especially with regard to channel bed stability and the effectiveness of flushing flows (Petts *et al.*, 1985; Gilvear, 1987; Sear 1992).

Needs to model and predict channel changes have encouraged historical studies over periods of 10 to 200 years (Petts *et al.*, 1987). Some types of human impacts

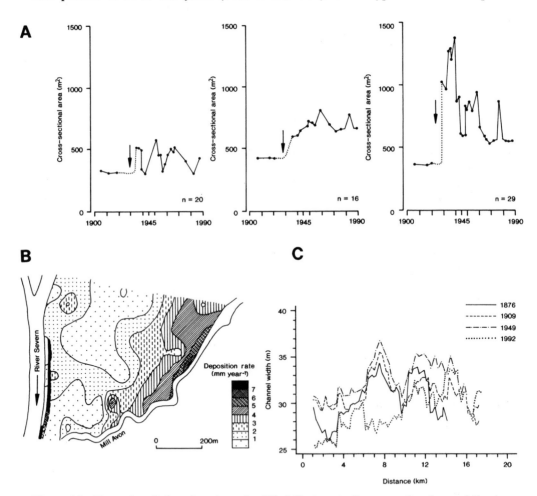

Figure 1.3 Examples of changing channels: (A) Adjustment of cross-sectional area following weir construction (indicated by arrows) at sites approximately 500 m below three locks on the lower River Murray, Australia (after Thoms and Walker, 1993). (B) Estimates of floodplain deposition rates during the period 1986–1989 based on the spatial pattern of Chernobyl-derived ^{134}Cs inventories in a reach of the River Severn, UK (after Walling *et al.*, 1992). (C) Trends in channel width in one reach of the River Dee identified using a GIS-based data-handling methodology (after Gurnell *et al.*, 1995)

have provided particular opportunities to examine historic average rates of processes (Chapter 9). Syntheses of documentary and cartographic information with field evidence of channel forms and sedimentological characteristics have elucidated the factors influencing the complex sequence of morphological responses over relaxation times of more than 100 years (e.g. to flow regulation: Petts, 1979; to channelization: Brookes, 1987; and to tin mining: Knighton, 1991; Figure 1.3A). Other studies have benefited from artificial 'events', the most notable of which has been the use of caesium-137 as a medium-term tracer for studying sediment routing and investigating rates of fluvial sedimentation on floodplains (e.g. Walling and Bradley, 1989). Caesium-137 is a man-made radionuclide that first occurred in global fallout as a result of nuclear weapons testing in the atmosphere that began in 1952 and reached a maximum rate of deposition in 1963, the rate of the Nuclear Test Ban Treaty. An additional short-term input occurred in 1986 as a result of the Chernobyl accident, and for studies this has provided added possibilities for quantifying recent rates of deposition (e.g. Figure 1.3B).

CHALLENGES FOR THE FUTURE

The subsequent chapters in this volume clearly demonstrate the dominance of 'problem-oriented' themes (cf. method-orientated ones) which share a common ambition to improve our capability for predictive modelling. However, universal theories of channel development and change are yet to emerge. First, the contrasting spatial and temporal scales within which individual researches are conducted will continue to constrain the formulation of a single model. Secondly, the development of iterative and spatially-distributed, form–process–form feedback models may achieve explanations at appropriate scales (Richards, 1990) but the resolution of 'scale-linkage' problems (Rhoads, 1990) remains the greatest challenge for studies of changing river channels into the next century. Thirdly, despite numerous attempts to establish two- or three-dimensional models of channel morphology in relation to sediment and water discharge, the assumptions required to overcome mathematical indeterminacy and the lack of an adequate bedload equation remain major constraints (Thornes and Gregory, 1991). Nevertheless, as demonstrated by the three papers in the final section of this book, the geographical approach to studying changing river channels within the context of their drainage basins has considerable application to the solution of practical problems extending from local channel restoration to catchment management and planning. Studies of process mechanics are important for developing mechanistic models of changing river channels, but progress needs to be made with reach-scale studies if this link is to be fully understood.

Return to large rivers

Ironically, the emphasis in recent research on streams and small rivers is in contrast to early studies of changing river channels which focused on large rivers. The motive of these early studies was to improve navigation, and the techniques of hydrographic survey had been established for the study of coastal and estuarine shipping routes. Over the past 30 years, in search for a more experimental approach, fluvial geomorphologists followed a reductionist trend, although major sampling campaigns since the late 1970s have provided quantitative information for some large rivers such as the Amazon (e.g. Meade *et al.*, 1985).

Today, the application of remote-sensing techniques is leading to important developments in fluvial geomorphology (Chapter 13), not the least important of which is a return to studies of large rivers. Thus, Gilvear *et al.* (1995) used image analysis techniques on multispectral imagery and aerial photography to map changes in channel morphology in a shallow gravel-bed river with a history of placer mining, and Mertes (1994) analysed Landsat data for sediment concentrations in surface waters to investigate floodplain sedimentation along the central Amazon. Such approaches may be complimented by the development of ground-penetrating radar, allowing examination of subsurface stratigraphy, sedimentary structures and facies to depths of 30 m in favourable conditions without excavation (Jo and Smith, 1991).

The development of remote-sensing technology has necessarily been paralleled by developments in data analysis and interpretation (Chapter 14). Thus, Gurnell *et al.* (1994) successfully applied a geographical information system (GIS)-based data-handling methodology to analyse channel planform change on the River Dee (UK) meanders between 1876 and 1992 (Figure 1.3C). Five map sources and one set of air photographs provided the data for analysis; all scale adjustments were handled within the GIS. This approach not only allowed the estimation of a variety of indices of change but also supported the estimation of the potential errors associated with digitising the channel boundary locations and registering the historical data sources to a common base.

Linking geomorphology and ecology

New directions for research on changing river channels is being stimulated by closer integration between geomorphologists and ecologists (Figure 1.2). First, fluvial geomorphologists, in the UK largely in response to the work of Gregory (Gregory and Gurnell, 1984; Gregory, 1992), have developed an interest in the role of vegetation in channel dynamics (Chapters 3 and 11) not least because of the recognition of woody debris in determining channel forms and processes (Chapter 10). However, the roles of fluvial geomorphology in the dynamics of communities, populations and species (Statzner and Higler, 1986; Statzner *et al.*, 1988; and see Carling, 1992) and in zoogeographic studies of aquatic and floodplain biota (e.g. Bravard *et al.*, 1986; Petts *et al.*, 1989, 1992; Greenwood *et al.*, 1991) are also attracting increased attention.

The Ecology of Running Waters by H.B.N. Hynes (published in 1970) and later 'The stream and its valley' (Hynes, 1975) provided the catalysts for the recognition of fluvial geomorphology as the basis of freshwater ecology. However, interaction and collaboration between fluvial geomorphologists and freshwater ecologists were slow to evolve. Whilst geomorphologists and hydrologists were establishing the river and its drainage basin as their fundamental unit of study, ecologists focused more on lakes and other standing-water bodies. This is explained by Ryder and Pesendorfer (1989) by two facts: rivers represent only 0.004% of the world's freshwaters and in the northern hemisphere, where most research has taken place, there is an inordinately high ratio of lakes to large rivers. Moreover, the clearly circumscribed form of ponds and lakes make them more amenable for ecological research than long, morphologically complex and highly dynamic rivers (the very characteristics that were attractive to geomorphologists and of concern to engineers!). The two directions of current research in fluvial geomorphology by ecologists evolved, first, following the 'hydraulic stream ecology' approach (Statzner *et al.*, 1988), which is based on the simple observation that hydraulic conditions (i.e. the interaction of flow and channel form) determine where organisms live, and secondly in recognition of the role of channel dynamics over recent and historical time-scales in sustaining biodiversity of the river corridor (Pautou and Décamps, 1985; Salo *et al.*, 1986; Petts, 1990; La Chavanne, 1995).

Applications in river management

In his preface to *River Channel Changes*, Gregory concluded (p. 7) that changing river channels merit investigation for a number of reasons amongst which the human significance is paramount, but it was more than a decade before river managers came to recognise this strategic role for fluvial geomorphologists. For 200 years, the approach to the natural environment and river channels in particular focused on exploitation and control; in Britain, as across Europe and throughout much of the USA, rivers were canalised and regulated by dams for navigation, flood-control and water-resource developments (Petts *et al.*, 1989; Cosgrove and Petts, 1990). Today, environmental restoration and sustainability of ecosystems are two themes at the centre of the modern 'person-within-environment' paradigm. Uniquely, fluvial geomorphology offers approaches to planning for sustainability within catchments (Chapters 18 and 19)—its success depends on our ability to communicate our science to the user community (Chapter 12).

The continued development of problem-oriented approaches, especially those focusing on large rivers and links between geomorphology and ecology will lead to the increasing recognition of the value of geographical enquiry for river management. This is reflected especially by advances in river restoration and floodplain management (Chapter 17). Understanding why and how river channels change is vital if restoration strategies are to be developed further (Chapters 16 and 17). Experimental perturbations of fluvial systems under controlled conditions represent the most effective approach to gaining the necessary understanding of the complex and interactive mechanisms of channel change, over the range of spatial and temporal scales, for the development of coherent predictive models. Ironically, much of our developing understanding of why changing river channels have a particular behaviour has emanated from studies of systems perturbed by human impacts. Knowledge about the ecological importance of the different fluvial forms that determine the habitat structure and patchiness of channels and floodplains must underpin policies for environmentally-sensitive management of river corridors. Brookes (1992) suggests that there is considerable scope for applying fluvial geomorphological principles to restore channelised streams where modified channels are adjacent to land undergoing a change of use, or where flood-alleviation or agricultural schemes were previously over-designed.

In *Drainage Basin Form and Process* Gregory and Walling (1973) advanced a geographical approach which facilitates consideration of interrelationships and encourages integrated studies of the fluvial system. They envisaged a new golden era of geomorphology founded in developments of methods for measuring drainage basin processes and representing landforms, and for relating the two. The achievements over the past two decades in drainage basin studies fostered by Des Walling are reviewed in a companion volume (Foster *et al.*, 1995). The contributions to this volume reflect the advancements in knowledge of changing river channels inspired by Ken Gregory. Together, the volumes testify to the success of their 'do-it-yourself' kit (Gregory and Walling, 1973, p. 400) in fostering the geographical tradition which will surely provide a foundation for improved management of rivers and drainage basins for the 21st century.

REFERENCES

Ashmore, P.E. (1991) How do gravel-bed rivers braid? *Canadian Journal of Earth Sciences* **28**, 326–341.

Ashmore, P.E., Ferguson, R.I., Prestegaard, K., Ashworth, P. and Paola, C. (1992) Secondary flow in anabranch confluences of a braided, gravel-bed stream. *Earth Surface Processes and Landforms* **17**, 299–311.

Baker, V.R. (1988) Geological fluvial geomorphology. *Geological Society of America Bulletin* **100**, 1157–1167.

Baker, V.R., Kochel, R.C. and Patton, P.C. (eds) (1988) *Flood Geomorphology*. Wiley Interscience, New York, 503pp.

Best, J.L. (1986) The morphology of channel confluences. *Progress in Physical Geography* **10**, 157–174.

Bravard, J.P., Amoros, C. and Pautou, G. (1986) Impacts of civil engineering works on the succession of communities in a fluvial system: a methodological and predictive approach applied to a section of the Upper Rhone River. *Oikos* **47**, 92–111.

Brookes, A. (1987) *Channelized Rivers*. Wiley, Chichester.

Brookes, A. (1990) Restoration and enhancement of engineered river channels: some European experiences. *Regulated Rivers: Research and Management* **5**, 45–56.

Brookes, A. (1992) Recovery and restoration of some engineered British river channels. In: Boon, P.J., Calow, P. and Petts, G.E. (eds), *River Conservation and Management*, Wiley, Chichester, 337–352.

Brooks, W.A. (1841) *Treatise on the improvement of the Navigation of Rivers*. John Weale, London, 154pp.

Brown, A.G., Keough, M.K. and Rice, R.J. (1995) Floodplain evolution in the East Midlands, UK: the Lateglacial and Flandrian alluvial record from the Soar and Nene valleys. *Philosophical Transactions, Royal Society, London, A* (in press).

Calver, E.K. (1853) *The Conservation and Improvement of Tidal Rivers*. John Weale, London, 101pp.

Carling, P.A. (1990) Particle over-passing on depth-limited gravel bars. *Sedimentology* **37**, 345–355.

Carling, P.A. (1991) An appraisal of the velocity reversal hypothesis for stable pool–riffle sequences in the River Severn, England. *Earth Surface Processes and Landforms* **16**, 19–31.

Carling, P.A. (1992) In-stream hydraulics and sediment transport. In: Calow, P. and Petts, G.E. (eds), *The Rivers Handbook*, Blackwell Scientific, Oxford, Vol. 1, 101–125.

Chorley, R.J. (1962) *Geomorphology and General Systems Theory*. United States Geological Survey, Professional Paper, 500B.

Chorley, R.J. and Kennedy, B.A. (1971) *Physical Geography: A Systems Approach*. Prentice-Hall, London.

Chorley, R.J., Dunn, A.J. and Beckinsale, R.P. (1964) *The History of the Study of Landforms: Volume 1, Geomorphology before Davis*. Methuen, London, 678pp.

Clifford, N.J., Robert, A. and Richards, K.S. (1992) Estimation of flow resistance in gravel-bedded rivers: a physical examination of the multiplier of roughness length. *Earth Surface Processes and Landforms* **17**, 111–126.

cosgrove, D. and Petts, G. (eds) (1990) *Water, Engineering and Landscape*. Belhaven, London, 214pp.

Darby, s.E. and Thorne, C.R. (1994) Prediction of tension crack location and riverbank erosion hazards along destabilized channels. *Earth Surface Processes and Landforms* **19**, 233–245.

Davis, R.J. and Gregory, K.J. (1994) A new distinct mechanism of river bank erosion in a forested catchment. *Journal of Hydrology* **157**, 1–11.

Ebisemiju, F.s. (1991) Some comments on the use of spatial interpolation techniques in studies of man-induced river channel changes. *applied Geography* **11**, 21–34.

Ellett, C. (1853) *The Mississippi and Ohio Rivers*. Lippincott, Grambo and Co., Philadelphia, 367pp.

Erskine, W.D. and Warner, R.F. (1988) Geomorphic effects of alternating flood- and drought-dominated regimes on the N.S.W. coastal rivers. In: Warner, R.F. (ed.), *Fluvial Geomorphology in Australia*, Academic Press, Sydney, 223–244.

Ferguson, R. and Ashworth, P. (1991) Slope-induced changes in channel character along a gravel-bed stream: the Allt Dughaig, Scotland. *Earth Surface Processes and Landforms* **16**, 65–82.

Foster, I.D.L., Gurnell, A.M. and Webb, B. (eds) (1995) *Sediment and Water Quality in River Catchments*. Wiley, Chichester (in press).

Gilvear, D.J. (1987) Suspended solids transport within regulated rivers experiencing periodic reservoir releases. In: Craig, J.F. and Kemper, J.B. (eds), *Regulated Streams: Advances in Ecology*, Plenum, New York, 245–256.

Gilvear, D.J. and Harrison, D.J. (1991) Channel change and the significance of floodplain stratigraphy: the 1990 flood event, lower Tay, Scotland. *Earth Surface Processes and Landforms* **16**, 753–762.

Gilvear, D.J., Waters, T.M. and Milner, A.M. (1995) Image analysis of aerial photography to quantify changes in channel morphology and instream habitat following placer mining in interior Alaska. *Freshwater Biology* (in press).

Gomez, B., Hubbell, D.W. and Stevens, H.H. (1990) At-a-point bedload sampling in the presence of dunes. *Water Resources Research* **26**, 2717–2731.

Greenwood, M.T., Bickerton, M.A., Castella, E., Large, A.R.G. and Petts, G.E. (1991) The use of Coleoptera (Arthropoda:Insecta) for patch characterisation of the floodplain of the River Trent, UK. *Regulated Rivers: Research and Management* **6**, 4.

Gregory, K.J. (1976) Changing drainage basins. *Geographical Journal* **132**, 237–247.

Gregory, K.J. (ed.) (1977) *River Channel Changes*. Wiley, Chichester, 448pp.

Gregory, K.J. (ed.) (1983) *Background to Palaeohydrology*. Wiley, Chichester, 486pp.

Gregory, K.J. (1985) *The Nature of Physical Geography*. Edward Arnold, London, 262pp.

Gregory, K.J. (1992) Vegetation and river channel process interactions. In: Boon, P.J., Calow, P. and Petts, G.E. (eds), *River Conservation and Management*, Wiley, Chichester, 255–270.

Gregory, K.J. and Brookes, A. (1983) Hydrogeomorphology downstream of bridges. *Applied Geography* **3**, 145–159.

Gregory, K.J. and Gurnell, A.M. (1984) The influence of vegetation on stream channel processes. In: Walling, D.E. and Burt, T.P. (eds), *Field Experiments in Geomorphology*. Geobooks, Norwich, 515–535.

Gregory, K.J. and Park, c.C. (1974) Adjustment of river channel capacity downstream from a reservoir. *Water Resources Research* **10**, 870–873.

Gregory, K.J. and Park, C.C. (1976a) Stream channel morphology in northwest Yorkshire. *Revue de Geomorphologie Dynamique* **25**, 63–72.

Gregory, K.J. and Park, C.C. (1976b) The development of a Devon gully and man. *Geography* **61**, 77–82.

Gregory, K.J. and Walling, D.E. (1973) *Drainage Basin Form and Process*. Edward Arnold, London, 458pp.

Gregory, K.J. and Whitlow, J.R. (1989) Changes in urban stream channels in Zimbabwe. *Regulated Rivers: Research and Management* **4**, 27–42.

Gregory, K.J., Hockin, D.L., Brookes, A. and Brooker, M.P. (1984) The impact of river channelization. *Geographical Journal* **151**, 53–74.

Gregory, K.J., Gurnell, A.M. and Hill, C.T. (1985) The permanence of debris dams related to river channel processes. *Hydrological Sciences Bulletin* **30**, 371–381.

Gregory, K.J., Lewin, J. and Thornes, J.B. (eds) (1987) *Palaeohydrology in Practice: A River Basin Analysis*. Wiley, Chichester, 370pp.

Gregory, K.J., Davis, R.J. and Downs, P.W. (1992) Identification of river channel change due to urbanisation. *Applied Geography* **12**, 299–318.

Gurnell, A.M., Downward, S.R. and Jones, R. (1995) Channel planform change on the River Dee meanders, 1876–1992. *Regulated Rivers: Research and Management* (in press).

Harvey, A.M. (1991) The influence of sediment supply on the channel morphology of upland streams: Howgill Fells, northwest England. *Earth Surface Processes and Landforms* **16**, 675–684.

Harwood, K. and Brown, A.G. (1993) Fluvial processes in a forested anastomosing river: flood partitioning and changing flow patterns. *Earth Surface Processes and Landforms* **18**, 741–748.

Hassan, M.A. and Church, M. (1992) The movement of individual grains in the streambed. In: Billi, P., Hey, R.D., Thorne, C.R. and Tacconi, P. (eds), *Dynamics of G Ravel-bed Rivers*. Wiley, Chichester, 159–176.

Hassan, M.A. and Reid, I. (1990) The influence of microform bed roughness elements on flow and sediment transport in gravel bed rivers. *Earth Surface Processes and Landforms* **15**, 739–750.

Hey, R.D. (1994) Environmentally-sensitive river engineering. In: Calow, P. and Petts, G.E. (eds), *The Rivers Handbook*, Blackwell Scientific, Oxford, vol. 2, 337–362.

Hoey, T. (1992) Temporal variations in bedload transport rates and sediment storage in gravel-bed rivers. *Progress in Physical Geography* **16**(3), 319–338.

Hoey, T.B. and Sutherland, A.J. (1991) Channel morphology and bedload pulses in braided rivers: a laboratory study. *Earth Surface Processes and Landforms* **15**, 717–737.

Hooke, J.M. and Redmond, C.E. (1992) Causes and nature of river planform change. In: Billi, P., Hey, R.D., Thorne, C.R. and Tacconi, P. (eds), *Dynamics of Gravel-bed Rivers*, Wiley, Chichester, 558–571.

Hughes, W. (1878) *The Treasury of Geography*. Longman, Green and Co., London, 892pp.

Huxley, T.H. (1880) *Physiography: An Introduction to the Study of Nature*, 3rd edition. Macmillan and Co., London, 384pp.

Hynes, H.B.N. (1970) *The Ecology of Running Waters*. University of Toronto Press, toronto.

Hynes, H.B.N. (1975) The stream and its valley. *Verhandlungen de Internationalen Vereinung fur theoretische und angewandte Limnologie* **19**, 1–15.

Jessop, W. (1782) *Report of William Jessop, Engineer, on a Survey of the River Trent, in the Months of August and September 1782, relative to a Scheme for improving its Navigation*. Burbage & Son, Nottingham.

Jo, H.M. and Smith, D.G. (1991) Ground penetrating radar of northern lacustrine delta.s *Canadian Journal of Earth Sciences* **28**, 1939–1947.

Knighton, A.d. (1991) Channel bed adjustment along mine-affected rivers of north-east Tasmania. *Geomorphology* **4**, 205–219.

Knighton, A.D. and Nanson, G.C. (1993) Anastomosis and the continuum of channel patterns. *Earth Surface Processes and Landforms* **18**, 613–626.

Knox, J. (1987) Historical valley floor sedimentation in the Upper Mississippi valley. *Annals Association American Geographer* **77**, 224–244.

La Chavanne, J-L. (1995) *Biodiversity of River Margins*. Unesco, Paris (in press).

Laronne, J.B. and Duncan, M.J. (1992) Bedload transport paths and gravel bar formations. In: Billi, P., Heym, R.D., Thorne, C.r. and Tacconi, P. (eds), *Dynamics of Gravel-bed Rivers*, Wiley, Chichester, 177–204.

Lawler, D.M. (1992) Process dominance in bank erosion systems. In: Carling, P.A. and Petts, G.E. (eds), *Lowland Floodplain Rivers: Geomorphological Perspectives*. BGRG Symposia Series, Wiley, Chichester, 117–144.

Lawler, D.M. (1993) The measurement of river bank erosion and lateral channel change: a review. *Earth Surface Processes and Landforms* 18, 777–821.

Leopold, L.B. and Miller, J.P. (1954) *A Postglacial Chronology for some Alluvial valleys in Wyoming*. US Geological Survey Water Supply Paper 1261, 1–90.

Leopold, L.b., Wolman, M.G. and Miller, J.P. (1964) *Fluvial Processes in Geomorphology*. Freeman, San Francisco, 522pp.

Lyell, C. (1850) *Principles of Geology*. John Murray, London, 8th edition, 811pp. First edition published 1830.

Macklin, M.G. and Lewin, J. (1993) Holocene river alluviation in Britain. *Zeitschrift fur Geomorphologie N.F. Suppl.-Bd.* 88, 109–122.

Macklin, M.G., Rumsby, B.T. and Newson, M.D. (1992) Historical floods and vertical accretion of fine-grained alluvium in the Lower Tyne Valley, Northeast England. In: Billi, P., Hey, R.D., Thorne, C.R. and Tacconi, P, (eds), *Dynamics of Gravel-bed Rivers*, Wiley, Chichester, 574–589.

Meade, R.H., Dunne, T., Richey, J.G., de M. Santos, U. and Salati, E. (1985) Storage and remobilization of suspended sediment in the lower Amazon River of Brazil. *Science* 228, 488–490.

Mertes, L.A.K. (1994) Rates of floodplain sedimentation on the central Amazon River. *Geology* 22, 171–174.

Nanson, G.C. and Page, K. (1983) Lateral accretion of fine-grained concave benches on meandering rivers. In: Collinsson, J.D. and Lewin, J. (eds), *Modern and Ancient Fluvial Systems*, Blackwell, Oxford, 133–143.

Osman, A.M. and Thorne, C.R. (1988) Riverbank stability analysis I: Theory. *Journal of Hydraulic Engineering* 114, 134–150.

Parr, W. (1994) Water-quality monitoring. In: Calow, P. and Petts, G.E. (eds), *The Rivers Handbook*, Blackwell Scientific, Oxford. Vol. 2, 124–143.

Pautou, C. and Décamps, H. (1985) Ecological interactions between the alluvial forests and hydrology of the Upper Rhone. *Archives fur Hydrobiologie* 104, 13–37.

Petit, F. (1990) Evaluation of grain shear stress required to initiate movement of particles in natural rivers. *Earth Surface Processes and Landforms* 15, 135–148.

Petts, G.E. (1979) Complex response of river channel morphology subsequent to reservoir construction. *Progress in Physical Geography* 3, 329–362.

Petts, G.E. (1984) *Impounded Rivers*. Wiley, Chichester, 326pp.

Petts, G.E. (1987) Time-scales for ecological change in regulated rivers. In: Craig, J.F. and Kemper, J.B. (eds), *Regulated Rivers: Advances in Ecology*. Plenum, New York, 257–266.

Petts, G.E. (1990) Forested river corridors: a lost resource. In: Cosgrove, D. and Petts, G. (eds), *Water, Engineering and Landscape*. Belhaven, London, 12–34.

Petts, G.E. and Greenwood, M.T. (1985) Channel changes and invertebrate faunas below Nant-y-Moch dam, River Rheidol, Wales, UK. *Hydrobiologia* 122, 65–80.

Petts, G.E. and Pratts, J.D. (1983) Channel changes following reservoir construction on a lowland English river. *Catena* 13, 305–320.

Petts, G.E., Foulger, T.R., Gilvear, D.J., Pratts, J.D. and Thoms, M.C. (1985) Wave movement and water quality variations during a controlled release from Kielder Reservoir, North Tyne River, UK. *Journal of Hydrology* 80, 371–389.

Petts, G.E., Moller, H. and Roux, A.L. (eds) (1989) *Historical Change of Large Alluvial Rivers: Western Europe*. Wiley, Chichester, 356pp.

Petts, G.E., Large, A.R.G., Greenwood, M.T. and Bickerton, M.A. (1992) Floodplain assessment for restoration and conservation: linking hydrogeomorphology and ecology. In: Carling, P. and Petts, G.E. (eds), *Lowland Floodplain Rivers: Geomorphological Perspectives*, Wiley, Chichester, 217–234.

Reid, I., Frostick, L.E. and Brayshaw, A.C. (1992) Microform roughness elements and the selective entrainment of particles in gravel-bed rivers. In: Billi, P., Hey, R.D., Thorne, C.R. and Tacconi, P. (eds), *Dynamics of Gravel-bed Rivers*. Wiley, Chichester, 253–276.

Rennie, J. (1822) *Report Concerning the Improvement of Boston Haven*, 29th June. London.

Rhoads, B.L. (1990) Hydrologic characteristics of a small desert mountain stream: implications for short-term magnitude and frequency of bedload transport. *Journal of Arid Environments* **18**, 151–163.

Rhoads, B.L. and Welford, M.R. (1991) Initiation of river meandering. *Progress in Physical Geography* **15**, 127–156.

Richards, K. (1990) Real geomorphology. *Earth Surface Processes and Landforms* **15**, 195–197.

Ritchie, J.C. and McHenry, J.R. (1990) Application of radioactive fallout caesium-137 for measuring soil erosion and sediment accumulation rates and patterns: a review. *Journal of Environmental Quality* **19**, 215–233.

Ritter, D.F. (1988) Landscape analysis and the search for geomorphic unity. *Bulletin Geological Society of America* **100**, 160–171.

Robert, A. (1993) Bed configuration and microscale processes in alluvial channels. *Progress in Physical Geography* **17**(2), 123–136.

Ryder, R.A. and Pesendorfer, J. (1989) Large rivers are more than flowing lakes: a comparative review. In: Dodge, D.P. (ed.), *Proceedings of the International Large River Symposium (LARS)*, Canadian Special Publication of fisheries and Aquatic Sciences **106**, pp. 65–85.

Salo, J., Kalliola, R., Hakkinen, I., Makinen, Y., Niemala, P., Puhakka, M. and Coley, P.D. (1986) River dynamics and the diversity of Amazon lowland forest. *Nature* **322**, 254–258.

Schumm, S.A. (1968a) River Adjustment to Altered Hydrological Regimen—Murrumbidgee River and Palaeochannels, Australia. *United States Geological Survey, Professional Paper* 598.

Schumm, S.A. (1968b) Speculations concerning palaeohydrologic controls of terrestrial sedimentation. *Bulletin Geological Society of America* **79**, 1573–1588.

Schumm, S.A. (1973) Geomorphic thresholds and complex response of drainage systems. In: Morisawa, M. (ed.), *Fluvial Geomorphology*, Publications in Geomorphology, SUNY, Binghampton, 299–309.

Schumm, S.A. (1977) *The Fluvial System*. Wiley Inter-Science, New York.

Schumm, S.A. and Lichty, R.W. (1965) Time, space and causality in geomorphology. *American Journal of Science* **263**, 110–119.

Sear, D.A. (1992) Impact of hydroelectric power releases on sediment transport processes in pool–riffle sequences. In: Billi, P., Hey, R.D., Thorne, C.R. and Tacconi, P. (eds), *Dynamics of Gravel-bed Rivers*, Wiley, Chichester, 629–650.

Smith, D.G. (1993) Fluvial geomorphology: where do we go from here? *Geomorphology* **7**, 251–262.

Starkel, L., Gregory, K.J. and Thornes, J.B. (eds) (1991) *Temperate Palaeohydrology*. Wiley, Chichester, 548pp.

Statzner, B. and Higler, B. (1986). Stream hydraulics as a major determinant of benthic invertebrate zonation patterns. *Freshwater Biology* **16**, 127–139.

Statzner, B., Gore, J.A. and Resh, V.H. (1988) Hydraulic steam ecology: observed patterns and potential applications. *Journal North American Benthological Society* **7**, 307–360.

Strahler, A.N. (1952) Dynamic basis of geomorphology. *Bulletin Geological Society of America* **63**, 923–937.

Thoms, M.C. and Walker, K.F. (1993) Channel changes associated with two adjacent weirs on a regulated lowland alluvial river. *Regulated Rivers: Research and Management* **8**, 271–284.

Thorne, C.R. and Osman, A.M. (1988) Riverbank stability analysis II: Applications. *Journal of Hydraulic Engineering* **114**, 151–172,

Thornes, J.B. and Gregory, K.J. (1991) Unfinished business: a continuing agenda. In: Starkel, L., Gregory, K.J. and Thornes, J.b. (eds), *Temperate Palaeohydrology*. Wiley, Chichester, 521–536.

Underwood, A.J. (1994) Spatial and temporal problems with monitoring. In: Calow, P. and Petts, G.E. (eds), *The Rivers Handbook*. Blackwell Scientific, Oxford, Vol. 2, 101–123.

Walling, D.E. and Bradley, S.B. (1989) Rates and patterns of contemporary floodplain sedimentation: a case study of the River Culm, Devon, UK. *Geojournal* **19**, 53–62.

Walling, D.E. and Webb, B.W. (1992) Water quality I. Physical characteristics. In: Calow, P. and Petts, G.E. (eds), *The Rivers Handbook*, Volume 1, Blackwell Scientific, Oxford, 48–71.

Walling, D.E., Quine, T.A. and He, Q. (1992) Investigating contemporary rates of floodplain sedimentation. In: Carling, P.A. and Petts, G.E. (eds), *Lowland Floodplain Rivers: Geomorphological Perspectives*. Wiley, Chichester, 165–184.

Warner, R.F. (1992) Floodplain evolution in a New South Wales coastal valley. Australia: spatial process variations. *Geomorphology* **4**, 447–458.

Webb, B.W. and Walling, D.E. (1992) Water quality II. Chemical characteristics. In: Calow, P. and Petts, G.E. (eds), *The Rivers Handbook*, Volume 1, Blackwell Scientific, Oxford, 73–100.

Wheeler, W.H. (1893) *Tidal Rivers*. Longmans, Green and Co., London, 467pp.

Wolman, M.G. and Gerson, R. (1978) Relative scales of time and effectiveness of climate in watershed geomorphology. *Earth Surface Processes* **3**, 189–208.

Wolman, M.G. and Miller, J.P. (1960) Magnitude and frequency of forces in geomorphic processes. *Journal of Geology* **68**, 54–74.

Womack, W.r. and Schumm, S.A. (1977) Terraces of Douglas Creek, north-western Colorado: an example of episodic erosion. *Geology* **5**, 72–76.

Part I

TEMPORAL AND SPATIAL DIMENSIONS

2 Changes of River Channels in Europe During the Holocene

LESZEK STARKEL

Institute of Geography, Polish Academy of Sciences, Cracow, Poland

INTRODUCTION

The main types of river channel reflect the features of the hydrological regime and the sediment load, particularly whether bedload or suspended load is dominant (Schumm, 1977, 1981). A major control on sediment load is the type and density of vegetation cover, which influences sediment supply as well as the stability of river banks. The nature of the interaction between the hydrological regime, sediment load and vegetation also varies according to whether the river channels form part of the perennial, intermittent or ephemeral components of the drainage network (Gregory, 1977). The different types of river channels are associated with typical alluvial bodies and floodplains (Allen, 1977).

Examination of palaeochannels dating back to the Late Vistulian and Holocene indicate the dramatic changes in vegetation, river discharge and sediment load during the transitional phase, the stability of the full Holocene which was characterised by dense forests, and the rapid reaction of the fluvial system to human interference during the last few centuries (Klimek and Starkel, 1974; Gregory, 1977; Kozarski, 1983; Starkel, 1983; Petts *et al.*, 1989). In the areas left by deglaciation, new channel patterns were created and the rapid environmental change caused fossilisation of the former sandur and ice marginal streamways, which became drained by underfit streams (Dury, 1964).

The map of Europe shows a great variety of landscapes (Figure 2.1), each of which has undergone different palaeogeographic changes during the last 10–13 ka. This great diversity is accentuated in the channels of the largest rivers, which pass through several zones, as well as in the smaller ones which may possess a sequence of different types of reach (Starkel, 1990, 1991a). This paper tries to show the role of various factors in the complicated evolution of river channels during the Holocene, against the background of the existing diversity.

THE DIVERSITY OF THE EXISTING RIVER CHANNELS IN EUROPE

European rivers are controlled by various climatic factors extending from the zone of the subarctic tundra to the Mediterranean, and from the oceanic to the con-

Changing River Channels. Edited by Angela Gurnell and Geoffrey Petts.
© 1995 John Wiley & Sons Ltd.

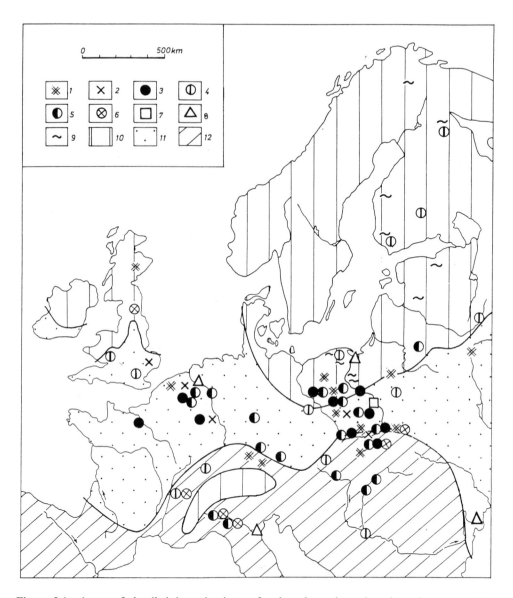

Figure 2.1 Areas of detailed investigations of palaeochannels and main palaeogeographic zones in Europe. 1. pre-Bolling braided channels and underfit streams (drained earlier by meltwaters); 2. Younger Dryas braided channels; 3. Late Vistulian palaeomeanders (Bölling, Alleröd, Younger Dryas); 4. Holocene palaeomeanders (single); 5. several generations of Holocene palaeomeanders; 6. braided pattern from 17th to 19th century; 7. anastomosing palaeochannels; 8. deltas with anastomosing and other palaeochannels; 9. new channels composed of various reaches (after deglaciation); 10. zone of deglaciation with superimposed newly established drainage pattern; 11. former periglacial zone; 12. mountains and the subsiding basins of the Alpine orogeny

tinental hydrologic regimes (Sundborg and Jansson, 1991). The existing channels of large and small streams, especially in the West and Mid European countries, are mainly channelised and embanked (Petts *et al.*, 1989). In addition, many of the natural channels' patterns have been stabilised by incision.

Longitudinal channel profile

In the classic three-component division of the channel profile, the upper portion is the zone of water and sediment production, the middle one of transfer, and the lower one of deposition (Schumm, 1977). With increasing drainage area, the channel cross-section and sediment load increase and lateral migration through erosion and accretion characterise channel dynamics (Froehlich *et al.*, 1977; Gregory and Maizels, 1991). Distinctive types of channels are characteristic of these separate river courses (Figure 2.2). For the forest zone of Europe (excluding its eastern part) the most typical channel systems are of the mountain–lowland catchment type. In their upper courses straight and sinuous channels prevail, which are incised or confined. In the mountain or upland foreland aggradation is observed. Short braided reaches

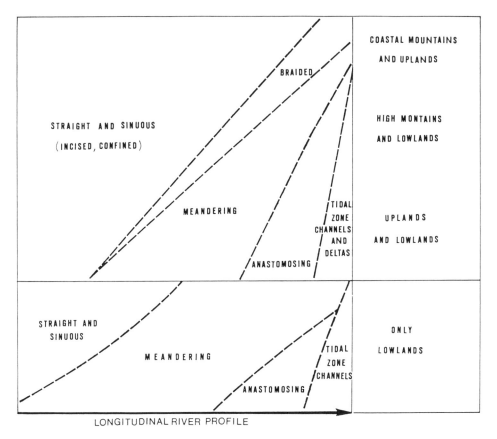

Figure 2.2 Typical channel forms in the longitudinal profiles of European rivers (depending on type of relief)

are replaced downstream by meandering channels, and next by anastomosing ones, finally passing into deltas or channels of the tidal zone (Figure 2.2). However, in reality, the picture is much more complicated. For example, on the way from the mountains to the sea, rivers may pass various upland and basin areas, which disturb the simple sequence of reaches and their history in the past (Starkel, 1979, 1990; Teisseyre, 1991).

In the more uniform lowland river basins, the upper course is not so distinct. In the headwater areas, the rivers may start from lakes and swamps. Meandering and anastomosing patterns prevail through most of the river course. However, even in these rivers, depending on the changes in the channel slope and the resistance of the bottom deposits, there may be erosional and braided reaches (Falkowski, 1975; Rotnicki, 1991).

In general, the repetition of the segment types is very frequent in the longitudinal profile. Furthermore, the longest rivers, passing through various climatic zones (e.g. Volga) and with a complicated palaeogeographic history, may reflect this diversity in their hydrologic regime and sediment delivery characteristics.

The role of inherited landscapes

At the European scale, a substantial component in the diversity of river network and channel types results from geological history; in particular, whether they have been subject to older (tectonics, lithology) or to younger (ice sheets and the periglacial zone during the Pleistocene) influences. Thus, across Europe we may distinguish three distinct regions or belts (cf. Figure 2.1): (a) northern Europe, which was glaciated during the last cold stage and so has a fresh postglacial relief and a young river network, (b) the former periglacial zone, mainly covered by older glaciations, and (c) southern Europe, which is influenced by Tertiary orogenic movements which produced mountains and plateaux, separated by subsiding depressions.

Northern Europe is characterised by river channels formed after deglaciation (Mudel and Raukas, 1991) which, with time, have incorporated a network of reaches of varying genetic origin and channel gradient.

The central belt has a channel pattern developed on gentle gradients. In the wide valley floors, meandering and anastomosing rivers are prevalent. In the northern and central belts, the character of channels is modified from west to east by the presence of the ice-jam floods in the east (Sundborg and Jansson, 1991).

The southern zone represents a mosaic of incised, braided and meandering channels. In the Mediterranean part, with its distinct dry season (whose effects have been enhanced by deforestation), the frequent braided channels are associated with ephemeral flood flows (Thornes, 1976).

THE FACTORS INFLUENCING RIVER CHANNEL TRANSFORMATION IN THE PAST

Evolution of the river channels may be influenced by various external factors, such as climatic, tectonic, eustatic and anthropogenic ones, which influence the river gradient, runoff and sediment load. There may also be autogenic changes: the process of ageing and progressive adaptation of features inherited from tectonic uplift (in the distant past) or deglaciation (in the more recent past). Falkowski (1975) calls such immature river channels 'young ones', in contrast with free meanders, developed on flat mature floodplains with thick alluvial sequences.

The evolution of river channels in the past may be investigated by using various methods. In bedrock reaches, which mainly occur in the upper courses of rivers, the reconstruction of changes is based on a very careful study of slack-water deposits and other signs of channel transformation during extreme floods (Baker, 1983; Baker et al., 1988). In the middle and lower courses, the examination of past changes is simpler, since it is possible to study the palaeochannels formed by cutoffs or by avulsions that are preserved on floodplains or are buried. For specific time intervals, characteristics of the palaeochannels help to reconstruct the palaeodischarge and sediment load (Schumm, 1968; Maizels, 1983; Rotnicki, 1981; Gregory and Maizels, 1991).

Climatic changes

In the extraglacial parts of Europe, in the wide valley floors filled with alluvium, it has been suggested that during the close of the Pleistocene, a general decline occurred in flood frequency and in sediment delivery (Schumm, 1965, 1977). Depending on the leading factor, either the channel-forming discharge declined first (1) or the sediment load decreased (2)—(cf. Starkel 1983):

$$Q_W^- > Q_{S^-} = w^- \ d^- \ L^- \ s^- \ P^+ \tag{1}$$

$$Q_W^- < Q_{S^-} = w^- \ d^+ \ L^- \ s^- \ P^+ \tag{2}$$

Where Q_w is water discharge, Q_s is sediment load, w is channel width, d is channel depth, L is meander wavelength, s is slope, P is sinuosity, and $+$ or $-$ represent, respectively, an increase or decrease.

The effects of those changes appear in the transformation of river channels from a braided to a meandering form (Schumm, 1965), which in many valleys is expressed in a two-step change with large lateglacial palaeomeanders in the transitional phase (Upper Vistula basin—Falkowski, 1975; Starkel et al., 1982; Szumanski, 1983; Warta and Prosna valleys—Kozarski and Rotnicki, 1977). The existence of several systems of channels from various phases of the Holocene helps to reconstruct several periods with a higher flood frequency (Starkel, 1983; Kalicki, 1991) and to connect changes during recent centuries with the human impact (Klimek and Starkel, 1974; Szumanski, 1977; Petts et al., 1989).

The tectonic factor

Separate tendencies in channel evolution are present in uplifting and subsiding areas. The former are characterised by channels that continue downcutting into the bedrock or into the periglacial deposits accumulated during the last cold stage (Starkel, 1990). Rivers on the emerged shields and platforms show incision throughout their length, as for example in the rivers of south-west Finland (Mansikkaniemi, 1991).

The courses of rivers which traverse resistant blocks (e.g. the Danube), reflect the existence of local erosional bases (Starkel, 1990).

In subsiding basins, alluvial fans dominate the forelands with very frequent chan-

32

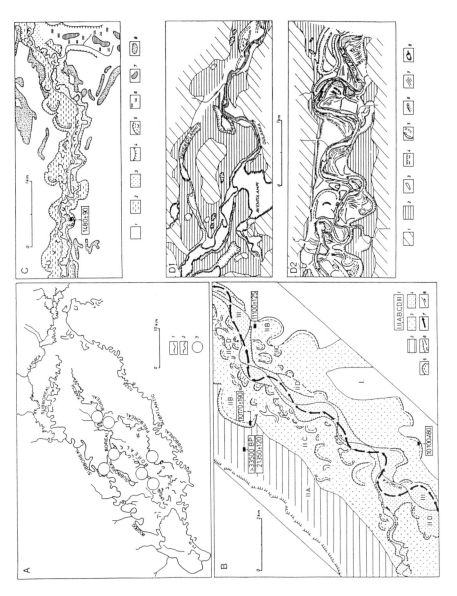

Figure 2.3

nel bifurcations and the avulsion of braided as well as meandering channels. Such features have been described in the Hungarian Basin (Vaskovsky, 1977; Borsy and Felegyhazi, 1983, Figure 2.3A), the Swiss Plateau (Wohlfarth and Amman, 1991), the Poo Plain (Cremaschi and Marchesini, 1978; Braga and Gervasoni, 1989) as well as in small intermontane basins (Baumgart-Kotarba, 1991–1992).

The lithological factor

Channels which develop over various lithological units may reflect such influences, even over short distances, through changes in their characteristics. River channels cut in resistant rocks are narrow, with a high gradient; upstream and downstream there are reaches with a low gradient and a braided or meandering pattern of aggrading rivers (cf. Rhone valley—Roux *et al.*, 1989). Therefore, the development of such narrowing and widening may be independent of climatic changes (cf. Danube valley near Regensburg—Buch, 1988). This is also valid in the case of epigenetic reaches superimposed on a surface left after deglaciation (cf. Oulanka valley in Finland—Koutaniemi, 1979; Koutaniemi and Ronkainen, 1983). In some cases a relatively resistant layer of ferruginous clay or coarse boulders is enough to disturb the process of free meandering and to change the character of the river channel (Falkowski, 1975; Maizels, 1983). In the marginal parts of mountain and upland areas, the flat valley floors are too narrow, and there confined meanders may be observed (Lewin and Brindle, 1977), a situation frequently found in the Carpathian foothills (Dauksza *et al.*, 1982). The palaeochannels preserved in such reaches may not be used for palaeohydrological retrodiction.

Deglaciation

During the retreat of the Scandinavian ice sheet on the vast European lowlands, a new drainage pattern started to form. New areas were progressively incorporated into these drainage networks. Depressions of various origin were utilised: subglacial

---

Figure 2.3 Examples of palaeochannels in various regions.
A. Changes of the Tisza river channel in the north-east part of the Pannonian Plains during the Holocene (after Borsy and Felegyhazi, 1983, simplified). 1. Tisza and other existing river channels; 2. Tisza palaeochannels; 3. sites providing pollen spectra documenting the age of palaeochannels.
B. Fragment of the Wisloka river at the foreland of the Carpathians (after Starkel *et al.*, 1982, simplified). 1. symbols for terrace and floodplain levels (e.g. I, Pleniglacial); 2. pre-Late-glacial level (IIA); 3. Holocene floodplains (IIB–D); 4. actual floodplain level of braiding from 18th to 19th century); 5. palaeochannels; 6. undercuttings and terrace edges; 7. course of Wisloka in 1980.
C. Part of the geomorphological map of the Mienia river valley, a tributary of the lower Vistula (after Andrzejewski, 1994, simplified). 1. terraces of Vistula river; 2. higher terrace of Mienia river; 3. floodplain of Mienia river; 4. terrace edges; 5. palaeochannels; 6. swampy plains; 7. kame terrace (of former lake); 8. dunes.
D. Fragments of the Oulanka river with lakes (D1) and with confined meanders (D2) (after Koutaniemi, 1979, simplified and reproduced with permission of Professor L. Koutaniemi). 1. bedrock, morainic and glacifluvial plateaux; 2. valley-train delta formation; 3. kettle hole; 4. undercuts and terrace edges; 5. palaeochannels; 6. bars; 7. river terraces with scrolls and bar ridges; 8. lakes and rivers

channels, dead ice depressions, sandur plains and ice marginal streamways. These were connected by overflow segments. Such complex river valleys have been described in northern Poland (Koutaniemi and Rachocki, 1981; Florek and Florek, 1986, Andrzejewski, 1994, Figure 2.3C), where they are characterised by transfluent lakes and a lack of distinct channels in wide swampy depressions. The southern zone of glacial deposition is very different from the northern zone of glacial erosion where step-like systems of river channels are composed of gorge sections cut into bedrock alternating with flat reaches where the rivers flow across plains or are incised into glacial, glacifluvial, lacustrine or marine deposits (Koutaniemi, 1979, 1991; Mudel and Raukas, 1991). In the Oulanka valley, several millennia were needed to form the mature reach in which, since 3–2 ka BP, the free meandering channel has developed (Figure 2.3D).

In the zones of former meltwater outflow at the glacial margins or in the transfluent areas, small underfit streams developed (Falkowski, 1975; Kvasov, 1987). Some of these wide swampy streamways do not show any clear remains of palaeochannels; and it is probable that in these depressions, before deforestation and drainage of bogs, there were no distinct stream channels (Żurek, 1984).

Marine transgression

The continuous rise of the sea level up to 6–5 ka BP caused the shortening of river courses and the shifting of the aggradation segment upstream. Impact on the upper reaches of rivers was very limited; in the case of the smaller rivers of the Baltic coast it did not exceed a few kilometres (Florek and Florek, 1986). In the case of the tidal mouths of rivers, its influence was also limited (Burrin and Jones, 1991), but on large rivers, with a high sediment load, extensive deltas were formed. Their evolution was associated with alternating fluctuations of the Flandrian transgression, river floods and sediment load. Detailed investigations have shown multiple changes of the channel pattern, both in anastomosing and in small distributary channels in the Rhine delta (Pons et al., 1963), the Poo delta (Veggiani, 1974), the Danube delta (Ghenea and Mihailescu, 1991) and others. In the case of the Rhine, during phases of an increased flood frequency (and probable lowering of the sea level), channels were cut in the organic and clay deposits of the perimarine facies, and were dated in the Rhine area by van der Woude (1981) at 6.4–6.1 ka, 5.3–4.6 and 4.1–3.4 ka BP.

The anthropogenic factor

Forest clearance and overgrazing caused an acceleration of runoff and sediment delivery to river channels. The oldest early-Neolithic changes in runoff are dated at 7–5 ka BP (Starkel, 1987). In Europe, the first distinct changes of river channels with a tendency towards braiding, were described in the Mediterranean area during the Roman period (Vita-Finzi, 1969) and at the transition to the Medieval period (Budel, 1977). Cutting of trees on the Vistula floodplain (Kalicki and Starkel, 1987) accelerated the lateral migration of the river channel and induced aggradation. Extensive agriculture and higher density cart-roads during the 16th to 19th centuries caused more frequent flooding and higher sediment loads. Especially in the mountain and upland forelands, a transformation of the river courses from meandering to straight and braided ones followed in response to the changed discharge and sediment regime (Szumanski, 1977; Braga and Gervasoni, 1989). Channel embankments as well as the modification of channels by narrowing and straightening were introduced in an attempt to control these tendencies, but resulted in simultaneous downcutting of river channels (Gregory, 1977; Klimek, 1987; Petts *et al.*, 1989).

EVOLUTION OF RIVER CHANNELS FROM THE LATEGLACIAL TO THE 19TH CENTURY

The Late Vistulian (13–10 ka BP) was a phase of considerable changes of river channels in Central Europe. There were two threshold changes: one occurring at about 13–12 ka BP, and the other at the transition to the Holocene at about 10 ka BP. The former is indicated by a change from a braided pattern to large meanders with dimensions characteristic of channel-forming discharges that are three to eight times greater than at present. It is associated with the retreat of the permafrost, the development of a dense cover of vegetation with open woodlands, and a change in the runoff regime with a decline in snowmelt and ice-jam floods. This change is well documented in the Warta valley (Kozarski, 1983) as well as in the Vistula catchment (Szumanski, 1983; Starkel, 1990), Scheldt valley (Kiden, 1991) and others. In the rivers which drain higher mountains with higher bedload transport, this change either occurred later, with a delay of 2–3 ka (Soła and Dunajec—tributaries of the upper Vistula, cf. Starkel, 1990), or the large meanders which developed during the Younger Dryas cooling were transformed again into a braided pattern, such as has been documented in the Vistula valley (Kalicki, 1991), tributaries of the Rhine (Heine, 1982) and in eastern England (Rose and Boardman, 1983). In the middle course of the Vistula the late Vistulian anastomosing pattern has stabilised (Sarnacka, 1987).

At the beginning of the Holocene there followed the next rapid change, which has been described for the upper and middle Vistula basin (Falkowski, 1975; Szumanski, 1983; Gębica and Starkel, 1987; Starkel, 1991b, 1994; Starkel *et al.*, 1991), Warta valley (Kozarski, 1983, 1991), Mark valley (Bohncke and Vandenberghe, 1991) and many others (Figure 2.3B). The channels which were abandoned between 9.5 and 8.0 ka BP had a width and a depth indicating channel-forming discharges from two to three times lower than the present ones (Figure 2.4). The simultaneous absence of wide point-bar zones (so typical of the lateglacial channels) and presence of clay overbank deposits, indicates the stabilisation of channels. In a particular case in the lower San valley one tortuous meander is preserved, which was abandoned during the Boreal time, superimposed, and incised into a large lateglacial palaeomeander (Szumanski, 1986).

When the tendency to a concentration of flow in a single-thread channel occurred in most valleys (Burrin and Jones, 1991), in some smaller valleys and in abandoned ice marginal streamways overgrown with peatbogs, it is likely that permanent concentrated flow stopped and that during heavy rain or snowmelt, water may have flowed across the entire floodplain. This has been described by Huybrechts (1985) in northern Belgium, where in a small valley he was unable to find any channel, and it may be supported by Zurek (1984), who studied the wide marginal Biebrza valley in

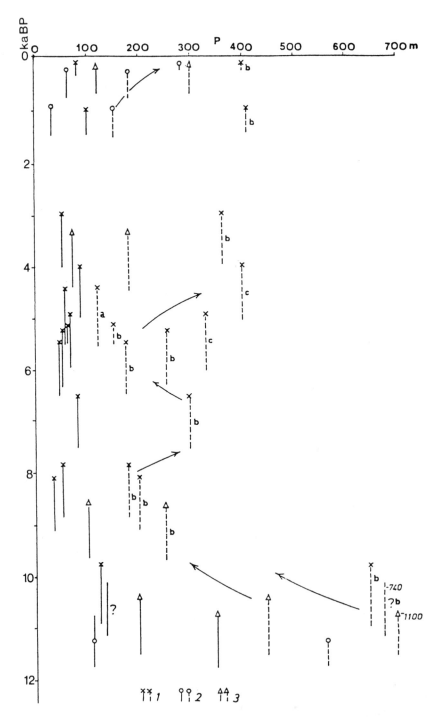

Figure 2.4 Palaeochannel parameters in the catchment of the upper Vistula (after various sources—cf. Starkel, 1990). Unbroken line—channel width, dashed line—meander radius. 1. Dated palaeochannel fill in Vistula valley (a—in Cracow gate, b—in western part of Sandomierz Basin, c—in central part of the Sandomierz basin); 2. palaeochannel in Wisloka valley; 3. palaeochannel in San valley. Arrows indicate the change of tendency (mainly after avulsion)

north-east Poland. However, the pattern of overbank flow was probably also influenced by beaver dams (Coles, 1992).

In larger valley floors, abandoned or buried under loams, several generations of palaeochannels may be identified, indicating several phases of increased fluvial activity and avulsions during the Holocene. These phases are especially distinct in the forelands of the Alps and the Carpathians, and are concentrated in the following intervals: 8.8–8.0, 6.5–6.0, 5.4–5.0, 4.7–4.3, 3.3–3.0, 2.4–1.8, 1.2–0.9 and 0.4–0.2 ka BP (cf. Fink, 1977; Starkel, 1990, 1991a; Kalicki, 1991; Wohlfarth and Amman, 1991). These phases coincide with the main periods of buried black oaks (Becker, 1982), glacier advances (Patzelt, 1977), lake level fluctuations (Magny, 1993), landslide reactivation (Starkel, 1985) and other events.

Abandoned channels have also been found in lowland catchments (Paulissen, 1973; Schrimer, 1983). Especially notable are phases at 8.5–8.0 ka BP and c. 5 ka BP, which have been identified in the Polish Lowland (Turkowska, 1988; Kozarski, 1991), in the Neman basin (Gaigalas and Dvareckas, 1987), Western Dvina and further to the east (Chebotarieva et al., 1965). Many palaeochannels are dated at the Atlantic–Subboreal transition, probably reflecting major disturbances in the fluvial regime. In Finland several abandoned channels have been formed dating from the late Atlantic and indicating discharges 30% lower than the present ones (Aario and Castren, 1969; Koutaniemi and Ronkainen, 1983). The comparison of palaeochannel parameters from various phases shows (Figure 2.4) that higher channel-forming discharges prevailed during the early Atlantic and early Subboreal as well as commencing during the Medieval period. A detailed examination of channel parameters from the last millennium (Starkel, 1983) and from the close of the Atlantic phase (Gębica and Starkel, 1987) supports the hypothesis of a new channel system originating during the phase with frequent floods, which was preceded by the straightening and incision of meandering channels.

Forest clearance on floodplains and an increase in soil erosion in southern Poland during the late Roman and Medieval times caused a tendency towards aggradation and, in some reaches, the formation of braided channels (Starkel et al., 1982; Kalicki and Starkel, 1987). However, similar tendencies have been identified in the Byelorussian river valleys, where there was no developed agriculture, which indicates that the main influencing factor was climatically-controlled higher flood frequency (Kalicki, 1993).

DISCUSSION

Changes in channel patterns across Europe were not simultaneous and depended on the combination of various factors. The sequences of changes identified in cuts and fills and palaeohydrological reconstructions may only be made in valley reaches with a mature river and floodplain system, favourable to the preservation of indicators of change. The upper mountain reaches with a continuous tendency to downcutting do not preserve evidence of any clear changes relating to climatic fluctuations during the Holocene until the period of direct human intervention.

In the reconstruction of palaeodischarges based on channel parameters, a major unknown factor is whether one, two or several channels or branches existed simultaneously. In all reconstructions developed from meandering channels, it is assumed that only a single thread channel exists (Rotnicki, 1991). Yet, braided channel patterns in Siberia and other regions show the functioning of parallel meandering and braided channels in one valley segment, and in wide valley floors two active meandering channels have been found with an anastomosing tendency. The braided and meandering segments may also shift upstream and downstream in time. There are a number of other open questions inherent in the retrodiction procedure, which are discussed by Gregory (1983) in his introductory paper to *Background to Palaeohydrology*.

Another problem is identification of the time interval during which the specific palaeochannels were formed. It is well known that a new channel may be created during one extreme event (Baker *et al.*, 1988) and that meanders may shift very rapidly (Knighton, 1977). However, more frequently a mature palaeochannel pattern with a sinuosity reaching 200–300% develops over at least decades and more probably over centuries. Therefore, it seems to be more correct to relate such abandoned palaeomeander systems (e.g. the Vistula river downstream of Cracow (Gebica and Starkel, 1987)) to longer and more stable periods.

The time of channel abandonment is usually identified by dating the basal layer of the organic fill. Such dating is more difficult if this base is not organic. Furthermore, a part of the channel deposits up to the first lag horizon may have been deposited during the floods just after the abandonment by the river, and in such cases the channel cross-section will appear relatively small. The example of the infilled early Holocene tortuous meander from the San valley is very instructive in this context (Szumansnki, 1986). In the tangle of palaeomeanders on the Hungarian and on the Poo Plains, the repeated incorporation of abandoned channels into the active drainage pattern was probably not a particularly rare event. Taking into consideration the ideal meandering pattern of various systems, the author wishes to suggest that the large palaeochannels of the Hungarian rivers, which have been ascribed to the Preboreal phase by Gabris (1985) with reconstructed channel-forming discharges 9.5 times greater than at present, probably should be related to the close of the Younger Dryas. Similarly, channels ascribed to the Boreal, having discharges three times lower than the present discharges, were probably formed by rivers during the Preboreal–early Boreal phase. Following similar arguments the reconstructed great palaeochannel in the Prosna valley which has been associated with a channel-forming discharge four times greater than present (the fill has been dated at 9330 ± 140 years BP—cf. Rotnicki, 1991) may represent the transition from the Younger Dryas to Preboreal. Future detailed investigations of different palaeochannel types and changes in their parameters in various parts of Europe should help to elaborate the links between river channel evolution and other climatic indicators such as glacier advances, lake-level fluctuations, vegetation changes, etc.

REFERENCES

Aario, R. and Castren, V. (1969) The northern discharge channel of ancient Paijanne and the palaeohydrology of the Atlantic period. *Bulletin of the Geological Society of Finland* **41**, 3–20.
Allen, J.R.L. (1977) Changeable rivers: some aspects of their mechanics and sedimentation. In *River Channel Changes*, ed. K.J. Gregory, Wiley, Chichester, 15–45.

Andrzejewski, L. (1994). Ewolucja systemu fluwialnego doliny dolnej Wisły w późnym vistulianie i holocenie na podstawie wybranych dolin jej dopływów. *Rozprawy UMK*, Toruń, 1–112.

Baker, V.R. (1983) Paleoflood hydrological analysis from slack-water deposits. *Proceedings of the Holocene Symposium, Quaternary Studies in Poland* **4**, 19–26.

Baker, V.R., Kochel, R.C. and Patton, P.C. (eds) (1988) *Flood Geomorphology*. Wiley, Chichester, 503pp.

Baumgart-Kotarba, M. (1991–1992) Rozwoj geomorfologiczny Kotliny Orawskiej pod wplywem ruchów tektonicznych. *Studia Geomorphologica Carpatho-Balcanica* **25–26**, 3–28.

Becker, B. (1982) Dendrochronologie und Palaeokologie subfossiler Baumstämme aus Flussablagerungen, ein Beitrag zur nacheiszeit-lichen Auenentwicklung im südlichen Mitteleuropa. *Mitteilungen der Komission fur Quartärforschung*, Österreichische Akademie der Wissenschaften, 5, Wien, 120pp.

Bohncke, S. and Vanderberghe, J. (1991) Palaeohydrological development in the Southern Netherlands during the last 15 000 years. In *Temperate Palaeohydrology*, ed. L. Starkel, K.J. Gregory and J.B. Thornes, Wiley, Chichester, 253–281.

Borsy, Z. and Felegyhazi, E. (1983) Evolution of the network of water courses in the North-ern–Eastern part of the Great Hungarian Plain from the end of the Pleistocene to our days. *Quaternary Studies in Poland* **4**, 115–124.

Braga, G. and Gervasoni, S. (1989) Evolution of the Poo river: an example of the application of historic maps. In: *Historical Changes of Large Alluvial Rivers, Western Europe*, ed. G.E. Petts, H. Moller and A.L. Roux, Wiley, Chichester, 113–126.

Buch, M.W. (1988) Spätpleistozäne und holozäne fluviale Geomorphodynamic im Donautal zwischen Regensburg und Straubing. *Regensburger Geographische Schriften* **21**, 1–197.

Büdel, J. (1977) *Klima–Geomorphologie*. Gebruder, Borntrager Berlin–Stuttgart, 304pp.

Burrin, P.J. and Jones, D.K.C. (1991) Environmental processes and fluvial response in a small temperate zone catchment: a case study of the Sussex Ouse Valley, Southeast England. In: *Temperate Palaeohydrology*, ed. L. Starkel, K.J. Gregory and J.B. Thornes, Wiley, Chichester, 217–252.

Chebotarieva, N.S., Malgina, E.A., Devirts, A.L. and Dobkina, E.I. (1965) On the age of river terraces in the north-western part of the Russian plain. In: *Upper Pleistocene and Holocene Palaeogeography and Chronology in the Light of Radiocarbon Dating*, Publ. House Nauka, Moscow, 51–66 (in Russian).

Coles, B. (1992) Further thoughts on the impact of beaver on temperate landscapes. In: *Alluvial Archeology in Britain*, ed. S. Needham and M.G. Macklin, Oxbow Monograph 27, Oxford, 93–99.

Cremaschi, M. and Marchesini, A. (1978) L'Evoluzione di un tratto di puanura Padana (prov. Reggio e Parma) in rapporto agli insediamednti ed alla struttura geologica tra il XV sec. a.C. ed il sec. XI d.C. *Archeologia Medievale* **5**, 542–570.

Dauksza, L., Gil, E. and Soja, R. (1982) The Holocene and present evolution of the mountainous reach of the Ropa river valley. In: *Geographical Studies IG i PZ PAN*, Special Issue 1, Warsaw, 21–39.

Dury, G.H. (1964) Principles of Underfit Streams, *US Geological Survey, Professional Paper 452A, 67pp*.

Falkowski, E. (1975) Variability of channel processes of lowland rivers in Poland and changes of the valley floors during the Holocene. *Biuletyn Geologiczny Universytetn*, Warszawa **19**, 45–78.

Fink, J. (1977) Jüngste Schotterakkumulationen im österreichischen Donauabschnitt. *Erdwiss. Forschung* **13**, 190–211.

Florek, E. and Florek, W. (1986). Age and development of the Slupia river floodplain terrace, Pomerania, Poland. *Quaternary Studies in Poland* **7**, 5–24.

Froehlich, W., Kaszowski, L. and Starkel, L. (1977) Studies of present-day and past river activity in the Polish Carpathians. In: *River Channel Changes*, ed. K.J. Gregory, Wiley, Chichester, 410–428.

Gabris, G. (1985) An outline of the paleohydrology of the Great Hungarian Plain during the Holocene. In: *Environmental and Dynamic Geomorphology*, ed. M. Pecsi, Academia Budapest, 61–75.

Gaigalas, A. and Dvareckas, V. (1987) Geomorphological structure and development of river valley during the Last Glaciation and Holocene in the south peribaltic area. In: *Palaeohydrology of Temperate Zone, Rivers and Lakes*, eds. A. Raukas and L. Saarse Valgus, Estonian Academy of Sciences, Tallin, 99–110.

Gebica, P. and Starkel, L. (1987) The evolution of the Vistula river valley at the northern margin of the Niepołomice Forest during last 15 000 years. In: *Evolution of the Vistula River Valley during the Last 15 000 Years, vol. II*, Geographical Studies, Special Issue 4, 71–86.

Ghenea, C. and Mihailescu, N. (1991) Palaeogeography of the Lower Danube valley and the Danube delta during the last 15 000 years. In: *Temperate Palaeohydrology*, ed. L. Starkel, K.J. Gregory and J.B. Thornes, Wiley, Chichester, 343–364.

Gregory, K.J. (1977) *River Channel Changes*. Wiley, Chichester, 450pp.

Gregory, K.J. (1983) Introduction. In: *Background to Palaeohydrology*, ed. K.J. Gregory, Wiley, 3–23.

Gregory, K.J. and Maizels, J.K.. (1991) Morphology and sediments: typological characteristics of fluvial forms and deposits. In: *Temperate Palaeohydrology*, ed. L. Starkel, K.J. Gregory and J.B. Thornes, Wiley, Chichester, 31–59.

Heine, K. (1982) Das Mündugsgebiet der Ahr im Spät-Würm und Holozän. *Erdkunde* **31**(1), 1–11.

Huybrechts, W. (1985) Morfologische evolutie van de rivervlakte van der Mark (Geraardsbergen) tijdens de leatste 20 000 jaar. PhD thesis, Vrije Universiteit Brussels, 250pp.

Kalicki, T. (1991) The evolution of the Vistula river valley between Cracow and Niepolomice in Late Vistulian and Holocene times. In: *Evolution of the Vistula River Valley during the Last 15 000 years*, part IV, Geographical Studies, Special Issue 6, 11–37.

Kalicki, T. (1993) Studia nad poźnoglacjalna i holoceńska ewolucja wybranych dolin rzecznych na Bialorusi (Studies on Lateglacial and Holocene evolution of some river valleys in Byełorussia) *Folia Geographica*, vol. 24–25, *Series Geographica Physica*, Kraków, 83–84.

Kalicki, T. and Starkel, L. (1987) The evolution of the Vistula river valley of Cracow during the last 15 000 years. In: *Evolution of the Vistula River Valley during the Last 15 000 Years*, vol. II, Geographical Studies, Special Issue 4, 51–70.

Kiden, P. (1991) The Lateglacial and Holocene evolution of the Middle and Lower River Scheldt, Belgium. In: *Temperate Palaeohydrology*, ed. L. Starkel, K.J. Gregory and J.B. Thornes, Wiley, Chichester, 283–299.

Klimek, K. (1987) Man's impact on fluvial processes in the Polish Western Carpathians. *Geografiska Annaler* **69**A(1), 221–226.

Klimek, K. and Starkel, L. (1974) History and actual tendency of floodplain development at the border of the Polish Carpathians. *Nachr. Akad. Gottingen, Rep. of Com. of IGU Present-day Geomorphic Processes*, 185–195.

Knighton, A.d. (1977) Short-term changes in hydraulic geometry. In: *River Channel Changes*, ed. K.J. Gregory, Wiley, Chichester, 101–119.

Koutaniemi, L. (1979) Outline of the development of relief in the Oulanka river valley, Northeastern Finland. *Acta Universitatis Ouluensis A* **82**, Geol. 3, 29–38.

Koutaniemi, L. (1991) Glacio-isostatically adjusted paleohydrology, the rivers Ivalojoki and Oulankajoki, Northern Finland. In: *Temperate Palaeohydrology*, ed. L. Starkel, K.J. Gregory and J.B. Thornes, Wiley, Chichester, 65–78.

Koutaniemi, L. and Rachocki, A. (1981) Palaeohydrology and landscape development in the middle course of the Radunia basin, North Poland. *Fennia* **159**(2), 335–342.

Koutaniemi, L. and Ronkainen, R. (1983) Palaeocurrents from 5000 and 1600–1500 BP in the main rivers of the Oulanka basin, North Eastern Finland. *Quaternary Studies in Poland* **4**, 145–158.

Kozarski, S. (1983) River channel changes in the middle reach of the Warta valley, Great Poland Lowland. *Quaternary Studies in Poland* **4**, 159–169.

Kozarski, S. (1991) Warta—a case study of a lowland river. In: *Temperate Palaeohydrology*, ed. L. Starkel, K.J. Gregory and J.B. Thornes, Wiley, Chichester, 189–215.

Kozarski, S. and Rotnicki, K. (1977) Valley floors and changes of river channel pattern in the North Polish Plain during the Late-Wurm and Holocene. *Questiones Geographicae 4*, 51–93.

Kvasov, D. (1987) The late Quaternary history of the Volga river. In: *Paleohydrology of the Temperate Zone, Vol. I, Rivers and Lakes*, Institute of Geology, Estonian Academy of Sciences, Tallin, 43–55.

Lewin, J. and Brindle, J.B. (1977) Confined meanders. In: *River Channel Changes*, ed. K.J. Gregory, Wiley, Chichester, 221–233.

Magny, M. (1993) Holocene fluctuations of lake levels in the French Jura and Sub-Alpine ranges and their implications for past general circulation patterns. *The Holocene* 3(4), 306–313.

Maizels, J.K. (1983) Channel changes, paleohydrology and deglaciation: evidence from some lateglacial sandur deposits, northeast Scotland. *Proceedings of the Holocene Symposium, Quaternary Studies in Poland* 4, 171–187.

Mansikkaniemi, H. (1991) Regional case studies in Southern Finland with reference to glacial rebound and Baltic regression. In: *Temperate Palaeohydrology*, ed. L. Starkel, K.J. Gregory and J.B. Thornes, Wiley, Chichester, 79–104.

Müdel, A. and Raukas, A. (1991) The evolution of the river systems in the East Baltic. In: *Temperate Palaeohydrology*, ed. L. Starkel, K.J. Gregory and J.B. Thornes, Wiley, Chichester, 365–292.

Patzelt, G. (1977) Der zeitliche Ablauf und das Ausmass postglaziale Klimaschwankungen in den Alpen. In: *Dendrochronologie und postglaziale Klimaschwankungen in Europa, Erdwissenschaftliche Forschung* 13, Wiesbaden, 249–259.

Paulissen, e. (1973) De morfologie en de kwartairstraatigrafie van de Maasvallei in Belgisch Limburg. *Verhandelungen Konigliche Vlaamse Academie Klasse Wetenschappen*, 35, 127, Brussel, 266pp.

Petts, G.E., Moller, H. and Roux, A.L. (eds) (1989) *Historical Changes of Large Alluvial Rivers: Western Europe.* Wiley, Chichester, 355pp.

Pons, L.J., Jelgersma, S., Wiggers, A.J. and de Jong, J.D. (1963) Evolution of the Netherlands coastal area during the Holocene. *Verhandelungen Konigliche Nederland Geologishe Mijn. Genoots*, Geol. Serie 21, 2, str. 197–208.

Rose, J. and Boardman, J. (1983) River activity in relation to short-term climatic deterioration. *Quaternary Studies in Poland* 4, 189–198.

Rotnicki, K. (1983) Modelling past discharges of meandering rivers. In: *Background to Palaeohydrology*, ed. K.J. Gregory, Wiley, Chichester, 321–354.

Rotnicki, K. (1991) Retrodiction of palaeodischarges of meandering and sinuous alluvial rivers and its palaeohydroclimatic implications. In: *Temperate Palaeohydrology*, ed. L. Starkel, K.J. Gregory and J.B. Thornes, Wiley, Chichester, 431–471.

Roux, A.L., Bravard, J.P., Amoros, C. and Pautou, G. (1989) Ecological changes of the French upper Rhone river since 1750. In: *Historical Changes of Large Alluvial Rivers: Western Europe*, ed. G.E. Petts, H. Moller and A.L. Roux, Wiley, Chichester, 323–350.

Sarnacka, Z. (1987) The evolution of the Vistula river valley between the outlets of Radomka and Swider rivers during Last Glacial and Holocene. In: *Evolution of the Vistula River Valley during the Last 15 000 years*, vol. II, Geographical Studies, Special Issue 4, 131–150.

Schirmer, W. (1983) Criteria for the differentiation of late Quaternary river terraces. *Quaternary Studies in Poland* 4, 199–205.

Schumm, S.A. (1985) Quaternary palaeohydrology. In: *The Quaternary of the United States*, ed. H.E. Wright and D.G. Frey, Princeton University Press, 783–794.

Schumm, S.A. (1968) *River Adjustment to Altered Hydrological Regime—Murrumbidgee River and Palaeochannels, Australia.* United States Geological Survey, Professional paper 598, 62pp.

Schumm, S.A. (1977) *The Fluvial System.* Wiley Inter-Science, New York, 338pp.

Schumm, S.A. (1981) Evolutional and response of the fluvial system, sedimentological implications. *SEPM Special Publication* No. 31, 19–29.

Starkel, L. (1979) Typology of river valleys in the temperate zone during the last 15 000 years. *Acta Univ. Ouluensis* A **82**, Geol. 3, 9–18.

Starkel, L. (1983) The reflection of hydrological changes in the fluvial environment of the temperate zone during the last 15 000 years. In: *Background to Palaeohydrology*, ed. K.J. Gregory, Wiley, Chichester, 213–234.

Starkel, L. (1985) The reflection of the Holocene climatic variations in the slope and fluvial deposits and forms in the European mountains. *Ecologia Mediterranea* **11**, 91–97.

Starkel, L. (1987) Man as a cause of sedimentologic changes in the Holocene. *Striae* **26**, 5–12.

Starkel, L. (1990) *Evolution of the Vistula river valley during the last 15 000 years*, vol. III. *Geographical Studies, Special Issue 5*, 220pp.

Starkel, L. (1991a) Long-distance correlation of fluvial events in the temperate zone. In: *Temperate Palaeohydrology*, ed. L. Starkel, K.J. Gregory and J.B. Thornes, Wiley, 473–493.

Starkel, L. (1991b) Environmental changes at the Younger Dryas–Preboreal transition and during the early Holocene: some distinctive aspects in central Europe. *The Holocene* **1**(1), 234–242.

Starkel, L. (1994) Frequency of floods during the Holocene in the Upper Vistula Basin. *Studia Geomorphologica Carpatho-Balcanica* 27–28, 3–16.

Starkel, L., Gebica, P., Niedzialkowska, E. and Podgórska-Tkacz, A. (1991) Evolution of both the Vistula floodplain and lateglacial–early Holocene palaeochannel systems in the Grobla Forest (Sandomierz Basin). In: *Evolution of the Vistula River Valley during the Last 15 000 Years*, part IV, *Geographical Studies, Special Issue 6*, 87–99.

Starkel, L., Klimek, K., Mamakowa, K. and Starkel, L. (1982) The Wistoka valley in the Carpathian Foreland during the Late Glacial and Holocene. *Geographical Studies, Special Issue 1*, Warsaw, 41–56.

Sundborg, A. and Jansson, M. (1991) Hydrology of rivers and river regimes. In: *Temperate Palaeohydrology*, ed. L. Starkel, K.J. Gregory and J.B. Thornes, Wiley, Chichester, 13–29.

Szumański, A. (1977) Changes in the course of the Lower San channel in XIX and XX centuries and their influence in the morphogenesis of its floodplain. In: *Studia Geomorphologica Carpatho-Balcanica*, vol. XI, Krakow, 139–154.

Szumański, A. (1983) Palaeochannels of large meanders in the river valleys of the Polish Lowland. In: *Quaternary Studies in Poland* **4**, 207–216.

Szumański, A. (1986) Postglacjalna ewolucja i mechanizm transformacji dna doliny dolnego Sanu, *Kwartalnik AGH. Geologia* **12**, 1.

Teisseyre, A.K. (1991) Klasyfikacja rzek w świetle analizy systemu fluwialnego i geometrii hyudraulicznej (Sum. River classification in the light of analysis of the fluvial system and hydraulic geometry), Prace Geologiczne-Mineralogiczne 22, Wrocław, 209pp.

Thornes, J.B. (1976) *Semi-arid Erosional System: Case Studies from Spain*. London School of Economics, Geography Dept, Paper 7, 96pp.

Turkowska, K. (1988) Evolution des valleas fluviatiles sur le Plateau de Lodz au cours du Quaternaire tardif. In: *Acta Geographica Lodziensia* **57**, 157pp.

Van der Woude, J.D. (1981) *Holocene Palaeoenvironmental Evolution of a Perimarine Fluviatile Area*. Vrije Universitet, Amsterdam, 112pp.

Vaskovsky, J. (1977) *Kvarter Slovenska* (*Quaternary of Slovakia*). Geologicky Ustav. D. Stura, Bratislava.

Veggiani, A. (1974) Le variazioni idrografiche del basso corso del fiume P. negli ultimi 3000 anni, Estratto de *PADUSA—Rivista del Centro Polesano di Studi Storici Archeologici et Etnografici—Rovigo*, 1–2, pp. 1–22.

Vita-Finzi, C. (1969) *The Mediterranean Valley*. Cambridge University Press.

Wohlfarth, B. and Amman, B. (1991) The history of the Aare river and the forealpine lakes in Western Switzerland. In: *Temperate Palaeohydrology*, ed. L. Starkel, K.J. Gregory and J.B. Thornes, Wiley, Chichester, 301–318.

Zurek, S. (1984) Relief, geological structure and hydrography of the Biebrza ice-marginal valley. *Polish Ecological Studies* **10**(3–4), 239–251.

3 Holocene Channel and Floodplain Change: A UK Perspective

A.G. BROWN

Department of Geography, University of Exeter, UK

The aim of this chapter is to review key concepts and recent research on the Holocene record of channel and floodplain change using sedimentary data. Emphasis is placed upon the identification of change, both intrinsic and extrinsic, and the coupling of channel and floodplain. The spatial scale covered is basin/regional, rather than global, which is covered elsewhere (Gregory and Starkel, 1995). The focus is predominantly on the temperate zone and the UK in particular.

A BRIEF HISTORY OF IDEAS

The middle and lower reaches of rivers are characterised by tracts of relatively flat land adjacent to the channel which is periodically flooded and forms the vast majority of the river's banks, i.e. a floodplain. This is not, however, universal; there are rivers without floodplains (non-alluvial rivers in bedrock gorges) and floodplains without rivers (so-called underfit or dry valleys). It may seem self-evident that when the two are present the river produced the floodplain and therefore a change in river behaviour will result in some change to the floodplain. While this is true it begs the twin questions of mechanism and rate—or how and when? What are the key channel processes responsible for the formation and destruction of floodplains and at what rates and over what time-scales do they operate? Ideas about both questions have a history. Prior to the late 19th century it was, not that unreasonably, believed that floods were solely responsible for the formation of floodplains. In 1906 Fenneman pointed out that floodplains could be constructed without floods through the lateral deposition of channel sediments. In a conceptually advanced if empirically unproven paper Melton (1936) recognised a spectrum of floodplain–channel systems varying from those dominated by overbank sediments to those dominated by lateral sediments, a continuum also seen in the more recent classification of river-floodplains by Nanson and Croke (1992) (see also Chapter 11). The classic paper by Wolman and Leopold (1957) showed that floodplains could be 80% lateral deposits and postulated the prime cause to be the reduced overbank frequency associated with continued overbank deposition. This presupposes no bed aggradation and is typical of relatively high stream power systems. Work on the Dane, a relatively small river in the West Midlands, has shown how a combination of meandering, avulsion, aggradation and incision has created a set of terraces during the late Holocene (Hooke and Harvey, 1983). Leopold and Wolman's ideas dominated thinking in the 1960s and early 1970s despite many data which suggested that overbank sedimentation could be significant and that rivers had not reworked substantial parts of their floodplain in thousands of years. Even the monitoring of channel changes suggested that most lowland rivers in the UK were characterised by stability rather than high migration rates (Hooke, 1977), although rates are higher in piedmont zones due to higher stream power (Ferguson, 1981). Stene (1980) was one of the first to 'exhume' vertical accretion revealing significant vertical accretion on Holocene terraces in Alberta, and Brackenridge (1984) similarly incorporated both vertical and lateral accretion processes in his model of the evolution of the Duck River, Tennessee. Since the early 1980s far more attention has been paid to the variety of floodplain deposits, both channel bed and bank (Taylor and Woodyer, 1978), overbank and channel infill (Lewis and Lewin, 1983; Brown et al., 1994) and sub-aerial/aeolian deposits. Alongside this there has been increased interest in the reconstruction of past hydrologies (palaeohydrology, Gregory, 1983; Gregory et al., 1987; Starkel et al., 1991) with one of the obvious sources of data for changes in external controls being changes in alluvial sedimentation. One of the desired results of modelling river–floodplain systems is an ability to predict the contribution of different sediment types to the floodplain fill under steady-state conditions and predict variations in past external variables under transient conditions (Howard, 1992)

FORMS OF CHANGE

Changing channel forms and floodplain change

In the 1960s it was realised that each channel type produced its own relatively distinct sedimentology (Allen, 1965). This work has been extended in the case of relatively simple meandering rivers, wandering and braided systems (Bridge and Leeder, 1979; Carson, 1984; Billie et al., 1992). Melton (1936) and later Schumm (1960, 1977) related channel form and particularly width/depth (W/D) ratio to the nature of bed and bank material. Schumm's (1960) relationship suggested that W/D decreased with increasing silt/clay percentage in bed and banks. This has been incorporated into general accounts of channel and floodplain metamorphosis (Schumm, 1977). Where palaeochannels are well preserved an independent test is possible. Brown and Keough (1992a) showed that a change in palaeochannel W/D accompanied the change in bank material from gravel to cohesive silt and clay. For example one early Holocene palaeochannel perimeter was over 60% sand and gravel but the equivalent modern channel perimeter is under 30%. The percentage silt–clay of palaeochannel perimeters plotted against W/D ratio does support a positive relationship; however, an alternative explanation is vertical accretion of silt–clay onto the old gravel banks, i.e. if channels remain on gravel beds but accrete silty clay banks the percentage perimeter which is composed of silt–clay will be directly related to the W/D ratio. Brown and Keough (1992a) provide an approximation for trapezoidal channels. This supports the stable-bed aggrading-banks model (discussed later in this chapter) and many observations of rivers with residual bed material particularly common in regions where rivers run over Pleistocene gravels. Changes in the shape and capacity of channels may be the result of changes in water and sediment loads supplied to the reach and/or they may be compensated for (in continuity terms) by changes in the number of channels.

Changing channel numbers and floodplain change

Because most modern rivers have one channel it has commonly been assumed that this is the 'normal' equilibrium state, except in steep valleys where braiding is common. However, the single meandering/sinuous channel is often an artifact of late Holocene channel change and channelisation (Petts, 1989). The low-slope equivalent of braiding which in planform terms is anastomosing/anabranching does, however, remain common in most environments; boreal (Smith and Smith, 1980), semi-arid (Rust, 1981) and humid tropics (Coleman, 1969). This is partly because in these environments there remain large rivers which have not been as modified by human activity as those in the temperate zone. The number of channels of an alluvial river should be seen as a major channel adjustment; it is more than another degree of freedom (Hey, 1978) as each channel retains its (nine) degrees of freedom and adjustment can take place through a combination of changes to channel form. This is a further aspect of the indeterminate behaviour of alluvial channels (Richards, 1982). Channels may bifurcate through the processes of avulsion, neck-cutoff and blind anabranching/headwall erosion (Harwood and Brown, 1993; Knighton and Nanson, 1993). This has profound effects on floodplain velocity patterns and sedimentation. First, near channels higher velocities and channel edge vortices will be more common and because deposition varies with distance from the channel (traction load deposition decreasing in relation to suspended/wash load) more of the floodplain will be covered by levee deposits. Secondly, scour is likely to be more important with more opportunities for water to leave the channel at high velocity and more areas of convergent flow on the floodplain. In some cases this will result in further dissection of the floodplain (Harwood and Brown, 1993). It is a combination of bank and overbank deposition, avulsion and meandering by individual channels that best explains the 'parcel-type' stratigraphy that is so common in many temperate low-slope floodplain valleys (Needham, 1985, 1989; Brown et al., 1994). Changes in the number of active and dead channels (at low to intermediate discharges) also have important repercussions for palaeohydrological reconstruction where, if multiple channels are not recognised, large errors would result. A change in channel numbers can be caused by both factors intrinsic to the channel/floodplain and extrinsic factors. Intrinsic factors include vegetation, especially large organic debris (Keller and Swanson, 1979; Gregory et al., 1985; Harwood and Brown, 1993; Chapter 10), trees (Sigafoos, 1964; Davis and Gregory, 1994) and beaver activity, which helps to explain why anastomosing rivers are more common in afforested floodplains in the temperate zone. A second intrinsic factor is quasi-braiding with island/islet creation from a dissected point bar or through the stabilisation of a mid-channel bar, a common occurrence in aggrading reaches. Critical extrinsic controls, or the combinations of extrinsic controls responsible for changes in channel numbers are poorly understood but are likely to include a change in discharge, especially if it is too rapid for channel adjustment, and changes in load causing either bed aggradation and/or a rate of levée deposition leading to instability. A channel with bed and levées above the height of palaeochannels on the floodplain surface will always be liable to avulsion and the formation of an anastomosing pattern. The relative rate of hydrological and channel change will be discussed later in this chapter as will the importance of inherited forms.

INTRINSIC CONTROLS

Downstream variations and flood regime

Due to variations in slope and discharge, channel characteristics change downstream in a systematic manner (Richards, 1982). However, although floodplains also change downstream this is less regular and not directly equatable with downstream hydraulic geometry. The result is that floodplain sediments may be variable reach to reach and therefore a full survey of the longitudinal variation of floodplains is required in order to assess their relationship to extrinsic factors (e.g. Burrin and Scaife, 1984). Rivers in previously glaciated terrain are particularly variable with river captures, confining Pleistocene terraces and gorges, common in lower reaches. Other complicating factors are changes in geology, often from hard rocks in the upper reaches to soft rocks in lower reaches, and neotectonic influences, particularly common in upper and middle reaches. The only generalisation that can be made is that, as with channel sediments, Holocene floodplain fills tend to fine downstream; terraces also become less common due to a tendency to net downstream aggradation.

The Perry, a medium-sized catchment in Midland England, is a good example of the complexity of downstream change in floodplain sediments caused by bedrock and the Pleistocene history of the region (Brown, 1990). In a comparison of four macro-reaches considerable diachrony was found with the basal age of the top two metres varying from 1000 years to 9000 years BP. The trend, a decrease in age downstream, is explained by the basin-and-gorge bedrock profile inherited from the Devensian glaciation of the area (Figure 3.1). This is reflected in a positive relationship between the mean floodplain accumulation rate and the *SL* index (slope-stream length product, Figure 3.2). This work showed that different macro-reaches had adjusted in different ways to long-term catchment changes and this had changed the sediment conveyance characteristics of the basin during the Holocene irrespective of changing inputs.

Larger floodplain systems may show these effects to even greater degrees due to differences in their networks and hence catchment area/discharge relationships. Recent work on the middle reaches of the Trent in Central England has shown how it has migrated across its floodplain with braided/anastomosing channels reworking the Devensian gravels and in many places producing a 10 000 + year hiatus in the sedimentary sequence (Salisbury, 1992). This is in contrast to the lower reaches of all other large British rivers such as the Thames (see Needham, 1985) or the Severn (Brown, 1983a, 1987a) and is much more characteristic of the larger European rivers such as the Maine and Wesser (Becker and Schirmer, 1977) or Vistula (Kalicki and Krapiec, 1991). It is hypothesised that the reason for this is the unique downstream discharge trend of the Trent which receives large floods from upland catchments

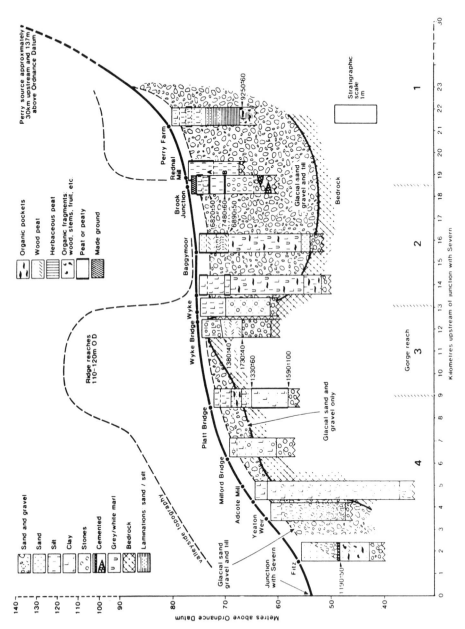

Figure 3.1 The stratigraphic long profile and bedrock topography of the Perry valley. From Brown (1990). Reproduced by permission of John Wiley and Sons

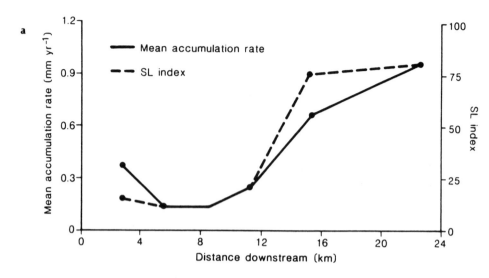

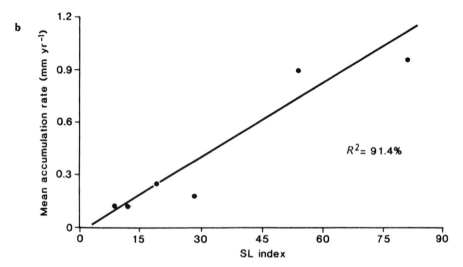

Figure 3.2 Mean floodplain accumulation rate and *SL* index with distance downstream in the Perry valley. From Brown (1990). Reproduced by permission of John Wiley and Sons

which join the main channel well below its headwaters. This is evident in the analysis of historic floods and was probably even more significant earlier in the Holocene, especially in the Little Ice Age and Medieval Climatic Optimum.

Each reach of a valley has differences in both intrinsic and extrinsic factors, varying from the importance of colluvial inputs to the slope of the frequency/discharge curve, and some of these factors are discussed below. The complex response of basins associated with coeval incision and aggradation will also lead to variations reach to reach (Schumm, 1977). This variation valley to valley, reach to reach, is one of the confusing factors in attempts to correlate fluvial response to climate change over large areas (Brown, 1991).

Vegetation and floodplain formation

Since floodplains are ecosystems with both physical and biotic components they are subject to changes associated with intrinsic controls. Vegetation is one of the most important components of natural floodplain systems acting through channel processes and through overbank dynamics (see also Chapter 11). Both bankside, levée, and backswamp vegetation is believed to have significant effects on the velocity of flows, the deposition of sediment and floodplain topography through such processes as tree-throw and the occurrence of floodplain deflation. Animals have rarely been considered an important part of the floodplain system, although they have been recognised as a significant component of temperate slope systems (Imeson *et al.*, 1980). This neglect has occurred because geomorphologists have failed to recognise that floodplains (and channels) are ecosystems, indeed they are arguably the most clearly defined ecosystems that occur in all climatic regions. They naturally have extremely high biodiversity and a complex trophic structure including animals of all orders. Several animals are known to have significant effects on floodplains, especially large herbivores and rodents (particularly beavers).

Before deforestation, vegetation dynamics must have had a much greater effect on fluvial processes than is the case today. Studies in some of the few remaining forested wilderness areas show the importance of trees through the creation of debris dams (Keller and Swanson, 1979; Gregory *et al.*, 1985, Chapter 10), and changing the patterns of channel erosion and deposition (Nanson, 1981). The development of floodplain and bank vegetation would have increased bank resistance and increased channel stability producing a tendency towards bank aggradation and vegetation associated channel change. There is palaeoecological evidence of Holocene changes in floodplain vegetation and soil conditions. A site on the River Stour in western Central England (Figure 3.3) has accumulated peat almost continuously from *c.* 9800 years BP until recent times. The relatively constant peat accumulation rate was controlled by the rise in the floodplain watertable. Pollen and macrofossil evidence shows that the local floodplain underwent a vegetation succession from birch (*Betula* sp.) to willows (*Salix* sp.) to alder (*Alnus glutinosa*) during the Pre-boreal and Boreal (Figure 3.4; Brown, 1988). These trees have different ecological requirements, with birch being a pioneer adapted to growth on nutrient-poor sand and gravels, willows being phreatophytes with twigs easily taking root into mud over sand and gravels, and lastly alder which is a hygrophyte adapted to fine sediment, waterlogged conditions and a relatively high C/N ratio. Alder then forms a dominant component of most floodplain woodlands in Britain until removed by human activity in the middle to late Holocene. Small rivers would not have had the power or size to migrate freely in such dense floodplain woodland and a more common pattern would have been multiple channels which shifted by avulsion. Channels under about 20–25 m in width would also have been prone to damming by debris and by beaver activity, creating ponds and by-pass channels (Harwood and Brown, 1993).

The role of vegetation in overbank deposition has only received incidental attention and it has been assumed that its role is primarily through increasing floodplain

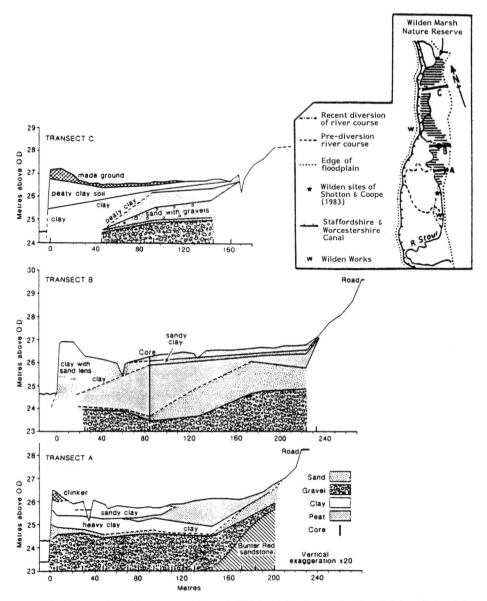

Figure 3.3 A stratigraphic cross-section of Wilden Marsh on the floodplain of the River Stour. Please note that the stratigraphy of cross section C does not extend the full length of the topographic survey line in the inset. From Brown (1988). Reproduced by permission of the Trustees of *New Phytologist*

resistance to erosion and changing floodplain roughness, inducing turbulence producing scour around trees and bushes and deposition in the downstream zone of velocity reduction. Gupta and Fox (1974) have argued that, under natural conditions, there is a relationship between vegetation height on floodplains and the recurrence interval of large floods; certainly tree-throw whether induced by flood or wind is an

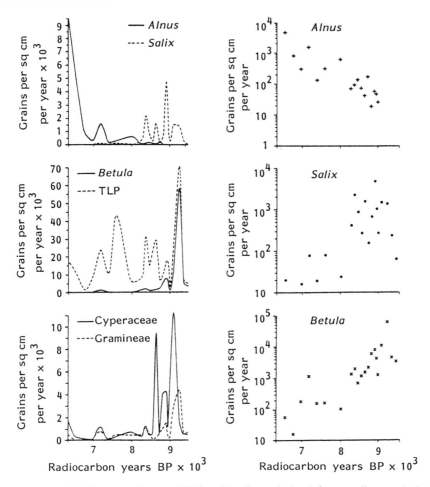

Figure 3.4 Floodplain succession at Wilden Marsh as derived from pollen analysis (TLP-total land pollen sum). From Brown (1988). Reproduced by permission of the Trustees of *New Phytologist*

important process on wooded floodplains. However, vegetation, including grasses and herbs, has other effects on the flood water column during floods, including increasing the surface area for suspended sediment deposition and possibly by flood scavenging and inducing flocculation (Brookes and Brown, 1995).

EXTRINSIC CONTROLS

Allogenic change can take many forms, it may directly reflect change in the regional or even the global environment, or it may be indirect through adjustments external to the river and floodplain but part of internal adjustment within the drainage basin as a whole. An example of the latter is the impact of landslides on channels and floodplains.

Landslides, colluviation and floodplain change

Landslides are most common in the upland reaches of floodplains or in gorges, and they may be triggered by heavy rainfall and/or snow reflecting climate change, or by alterations to the infiltration capacity of slopes caused by land use changes. Earthquakes are another important cause of landsliding in tectonically active areas such as the Carpathians (Starkel, 1960). An exceptionally well-known example from a less dynamic environment was the landslide which in 1773 dammed the River Severn and caused channel avulsion at Coalbrookdale in the Ironbridge Gorge, West Midlands, England (Waters, 1949). The slip which was 290 m wide caused the river level to rise nearly 4 m upstream and fall 2 m downstream (at Bridgnorth). It is difficult to assess the long-term effects of such events on channel bed profiles and sediment storage but they are unlikely to be negligible.

In the lower reaches of floodplains landslides and colluviation generally only affect the floodplain margins where the topographic edge of the floodplain is blurred. As the floodplain continues to accumulate vertically so overbank and colluvial sediments interdigitate providing extremely valuable sites for palaeoenvironmental reconstruction (Brown and Keough, 1992). The differentiation of colluvial inputs from alluvial sediments can under favourable geological conditions be achieved using mineral magnetics (Allen, 1992; Brown, 1992). In practice, the floodplain fill of the middle reaches (the lower reaches are generally too wide) of valleys can contain as much as 20% colluvial sediment (Lattman, 1960). The indistinct or completely obscured terrace edges of many British floodplains are the result of colluvial deposition which limits flooding thus increasing flood depth per unit area of floodplain. It is not clear whether increased deposition from suspended sediment from the greater water column depth can compensate as this depends upon the processes of overbank deposition.

The Lateglacial metamorphosis of floodplains and its inheritance

Several studies have revealed braided–wandering–meandering transitions in the Lateglacial in response to changing discharge and sediment supply (Rose et al., 1980; Brown et al., 1994; Brown, 1995). Several sites on the River Nene, Central England, have revealed channel change in the Lateglacial. At Raunds a channel was cut into gravels of Lateglacial stage Ic (presumed Bølling) and subsequently filled with organic sediments during the Lateglacial interstadial (Allerød equivalent). The relatively common occurrence of cross-bedded sands and gravels of equivalent age suggests channel migration, probably by meandering channels, during the Lateglacial interstadial. Evidence from a site only 4 km upstream indicates even greater fluvial activity in the Lateglacial. At Ditchford the entire sub-alluvial gravel was deposited between c. 11 200 and 10 200 years BP (Figure 3.5). Other sites in the Nene valley also suggest greater reworking of the Pleistocene gravels in the upper part of the middle valley than in the lower valley. This is probably due to the reduction of slope caused by Lateglacial floodplain erosion by meltwater discharges, with relatively little upstream gravel supply, resulting in downstream aggradation. However, this model does not apply universally and not even to all of Central England. Evidence from Mountsorrel on the Soar valley, and only 60 km to the north-

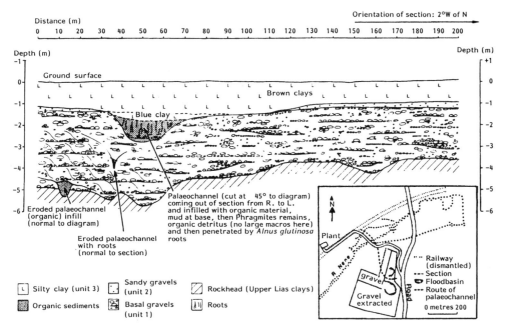

Figure 3.5 The stratigraphy of the Ditchford site on the River Nene. From Brown *et al.*, (1994). Reproduced by permission of The Royal Society

west of Ditchford, indicates that initial deposition of the sub-alluvial gravels commenced just before 28 000 years BP and deposition could have continued until as late as 10 200 years BP. The two basins are in most ways very similar with the exception that the Upper and Middle Nene is orientated generally W–E while the smaller Soar valley is orientated S–N. Although this may have had effects on the snowmelt and permafrost regime in this area, the reasons for the differences in fluvial history are not known.

A major feature of the very Lateglacial and early Holocene is an increase in channel abandonments and a later phase of abandonments occurs after *c.* 4000 years BP (Holyoak and Seddon, 1984; Brown and Keough, 1992b). Other features of Holocene lowland floodplain stratigraphy include existence of extensive areas of palaeo-landsurface, evidence of subaerial processes, such as tree-throw topography and levée formation. The parcels (*sensu* Needham, 1989) of layered stratigraphy are separated by palaeochannels, which occur at approximately the same elevation. It is virtually impossible to prove that more than one channel may have been operated simultaneously in the past, but there are several lines of evidence which suggest this may have been the case:

1. There is a lack of channel migration indicated by the stratigraphy and this suggests that avulsion was the predominant mechanism of channel change.
2. Channel morphology is generally different to contemporary channels with a higher width/depth ratio (Brown, 1987b) and often a much lower channel capacity than is explainable on climatic grounds alone.

3. The frequent occurrence of funery monuments of middle Holocene age built deliberately on 'islands' and the frequent preservation of bridges, causeways and mills at both sides of these islands.
4. The not uncommon occurrence today of reaches with two or three sinuous channels of approximately equal channel dimensions.

Early Holocene floodplains were in high disequilibrium with the discharges supplied to them, the floodplain surfaces were composed of sand and gravel and were irregular with abundant palaeochannels. The channels were floored by sediments too coarse for the flow to easily rework and there was an excess of channel capacity. The fine sediment supplied to these channels produced the first phase of alluviation, which is rather spatially restricted in most valleys (Macklin and Lewin, 1994).

Climatic change

Variations in the rate of alluviation and cycles of deposition/incisions have long been used as evidence of changes in the frequency/magnitude of rainfall especially in semi-arid regions (Vita-Finzi, 1969, 1976). Although complicated by sediment supply factors and vegetation growth in the floodplain the episodic/cyclical nature of variations in semi-arid rainfall does seem to be an overriding factor in semi-arid fills in areas not significantly altered by human activity (Haynes, 1968). However, in temperate regions these complicating factors increase in importance and the fluctuations in climate during the Holocene have been of a lesser magnitude and less critical in terms of plant growth. The most convincing studies of climatically driven alluviation in Britain come from upper and middle reaches of rivers with high stream powers. This is because these channels tend to produce Holocene terraces which are not subsequently buried or destroyed by erosion, and flood laminations which are not obscured by pedogenesis. Periods of increased flood magnitudes have been identified by Macklin et al. (1992) which were responsible for channel and floodplain metamorphosis in the British Uplands. Periods of alluviation in Britain caused by wetter climate have been correlated with Wendland and Bryson's (1974) climatic discontinuities and the fluvial record from North America (Macklin and Lewin, 1994). Given the varying importance of vegetation cover, soil development (altering erodibility) and human activity the reality is likely to be far more complex and teleconnections may not necessarily reflect causal links. However, where stream power is high and sediment abundant a climatic signal is likely to be visible in the alluvial record as illustrated by the studies of the redistribution of heavy metals along with the fine overbank sediments of the river Tyne (Macklin et al., 1992). Changes in the sediment regime of Central European rivers have generally been ascribed to Holocene climatic change, with fluctuations in glacial meltwater runoff and snowmelt being more important than in maritime provinces of Europe (Starkel, 1991).

Palaeoclimatic interpretation of floodplain deposits has for some time been more advanced from the less disturbed floodplains and catchments of the Mid-West and Central USA (Haynes, 1968; Knox, 1984) and the interpretation of flood slack-water deposits in non-alluvial reaches has led to the development of a subdiscipline within fluvial geomorphology (Costa, 1986; Baker, 1991).

Land use change

Studies of lowland alluviation have tended to focus more on catc
particularly deforestation and arable agriculture. Robinson and Lam
the late Holocene alluviation of the Thames floodplain as predominantly a response
to changes in land use. Brown and Barber (1985) attempted to refine the relationship by using a small basin where the sediment and land use data (derived from
pollen analysis) would be spatially coincident. The study showed a large increase in
sediment deposition after the deforestation of the catchment slopes and the adjacent
terraces. The lag observed is interpreted by Brown and Barber (1985) as being due
to slope storage but it may also be due to an element of transport dependency upon
climatic events. It should be remembered that numerous contemporary and lake-
based studies illustrate the sometimes dramatic effects of changes in land use upon
sediment output (Blackie et al., 1980; Dearing et al., 1990).

The distinction between climate and land use is a false dichotomy with propo-
nents stressing different elements of the erosion–transport equation which must, in
the case of silt and clay, include both terms for availability (e.g. erosivity and erodibility) and transport energy (entrainment and conveyance). In the case of fine cohesive sediments it is not possible to distinguish between the two sets of processes as
aggregates need to be detached to be entrained and effective grain size is dependent
upon the constant formation and destruction of aggregates which is governed by
shear forces and electrochemical bonding. An alternative approach to the separation
of climatic from catchment components in the alluvial record is the analysis of contrasting basins within homogenous hydroclimatic areas such as mountain ranges.

MODELLING CHANNEL AND FLOODPLAIN CHANGE

Inferential stratigraphic models

One of the most common methods of explaining the formation of landforms is
through the inference of processes or sequence of processes from sediments utilising
sediment transport theory and contemporary process–form relationships. This is
essentially how fining-upward units and epsilon-cross-stratification came to be associated with meandering channels, and low-angle cross-stratification, tabular bedding
and shallow channels associated with braided rivers.

The combination of the Lateglacial channel data and early Holocene stratigraphy
suggests a metamorphosis from many shallow gravel-bedded channels to fewer
deeper and relatively narrower channels with flow continuity and any change in discharge being accommodated by a combination of increased channel capacities and a
decrease in channel numbers. This is the basis of the stable-bed aggrading-banks
(SBAB) conceptual model (Brown et al., 1994). In this model the channel bed is an
inherited Lateglacial gravel unit with the addition of cohesive banks which aggrade
along with the aggradation of the floodplain (Figure 3.6). The approximate con-
stancy of the return period of bankfull is maintained by the increasing discharge
caused by the abandonment through siltation of secondary channels. A similar stratigraphy is seen in other lowland floodplains in the UK such as the Ouse and Cuck-

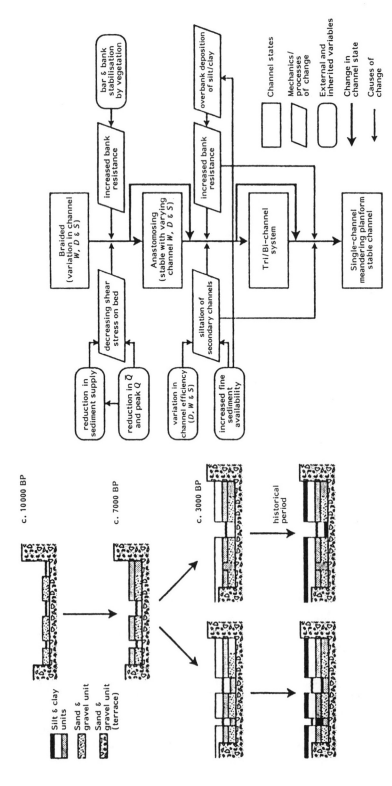

Figure 3.6 The stable-bed aggrading-banks model of lowland floodplain formation. From Brown *et al.* (1994). Reproduced by permission of The Royal Society

mere (Burrin and Scaife, 1984), the Thames (Robinson and Lambrick, 1984; Needham, 1985) and the Gipping (Rose *et al.*, 1980). The stratigraphic data suggest that in these low-energy systems the fluvial change from the Lateglacial to the Holocene may have been a transition from braiding, possibly through a meandering phase, to lower gradient anastomosing systems which subsequently developed into meandering single-channel systems.

Floodplain sedimentation/budget models

Two approaches have been taken to modelling floodplain sediment budgets; one is the measurement of contemporary or recent rates of deposition by monitoring or known-rate additions methods (see below) and the second is the calculation of longer-term rates from dated alluvial sequences. The results of both have been related to land use and climatic data (see also Chapter 9).

Dating is critical and the value of long-term rates is dependent on the accuracy and precision of the dating method. Alluvial deposits can be dated by most radiometric techniques, palaeomagnetics and thermoluminescence/optimal stimulation luminescence but each method has its own ranges of applicability and bias through sampling. In order to model the basin response quantitatively, volumes must be calculated and allowance made for losses to the system. This approach is typified by the work of Trimble (Trimble, 1983) based upon the earlier surveys by Happ (1940) in Coon Creek. The study highlights the lags in erosion–deposition, the role of slope bases and floodplains as stores, the success of soil conservation measures and the resulting disequilibrium between the erosion rate and suspended sediment output from a basin. This approach has been used to model the storage of mine tailings and contaminants (Marron, 1992). In these studies and Prehistoric/Medieval equivalents (Brown and Barber, 1985; Bork, 1989) no attempt is made to separate-out the climatic and land use causal elements as both were assumed to be operating; however, land use was, and is, under human control and in lowland systems effectively controls sediment availability through vegetation cover and agriculture (Boardman, 1986). An intermediate technique bridging the dated-sequence–monitored-rates gap is the use of known-rate additions methods. In theory, if the rate of addition of anything to the surface of an accretion floodplain is known and it is either indestructible or has a known rate of loss/decay then, by measuring its volume in a known volume of floodplain sediment, the rate of accretion of that sediment can be calculated. Such additions which have a known rate of addition and decay include the shortlived isotopes, especially ^{137}Cs. This isotope has now been used extensively on floodplains and provides the best measurements of short- to medium-term rates. Work by Walling *et al.* (1992) and on the floodplains of the River Severn at Tewkesbury has shown how spatially variable the flood deposition is across a relatively uniform segment of floodplain. Further modelling work is required to relate these measurements to longer-term rates and overbank dynamics.

Due to the physical problems of data collection and greater interest in channel as opposed to extra-channel sedimentation, until recently few measurements have been made of floodplain deposition after floods. Exceptions to this are the analysis of catastrophic events such as the Lynmouth flood (Dobbie and Wolfe, 1953) and studies of large rivers (Carleson and Runnels, 1952; Kesel *et al.*, 1974) and occasional observations on relatively small floodplains (Gomez and Sims, 1981; Brown, 1983b). These all illustrate the importance of floodplain hydraulics in the variable deposition of sediments both in terms of calibre and quantity. More recently Walling and Bradley (1989) have measured overbank deposition and have been able to relate it to both suspended sediment conveyance loss and ^{137}Cs profiles. The modelling of overbank deposition is in its early stages but is now being considered as an essential component of channel–floodplain models.

Channel and floodplain models

The comprehensive modelling of channels and floodplains is in its early stages but the interaction of in-channel suspended load and floodplain deposition has been modelled using a steady-state diffusion analogy by James (1985) and Pizzuto (1987). These models output curves of representative thickness of various grain sizes above 63 μ with distance from a channel (Figure 3.7). Large floodplains with complex topographies have yet to be modelled due to the need to include large variations in diffusivity. Marriott (1992) used textural analysis of a flood of the River Severn, England, in order to test James's (1985) model and found it satisfactory in relative terms. These models predict textural variation due to sand transport which is velocity dependent and which in simple situations can be predicted. However, the quantity of silt and clay deposited on the floodplain is far less predictable as sedimentation is due to several processes including initial suspended sediment concentrations, flocculation, vegetation-associated deposition, flood-water infiltration and depressional storage, none of which are analogous to diffusion. Interest in the interaction between floodplain dynamics and sedimentation/erosional patterns has stimulated the development of finite element models for flood flows (Gee *et al.*, 1990). Bates *et al.* (1992) have developed a two-dimensional finite element model for depth-averaged overbank flows (Figure 3.8). These models are capable of incorpor-

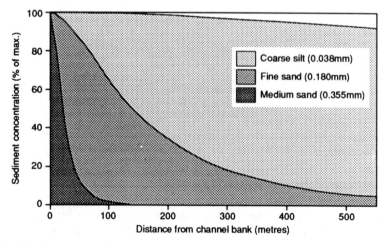

Figure 3.7 The deposition of sand with distance away from the channel predicted by the James model. Redrawn from Marriott (1992). Reproduced by permission of John Wiley and Sons

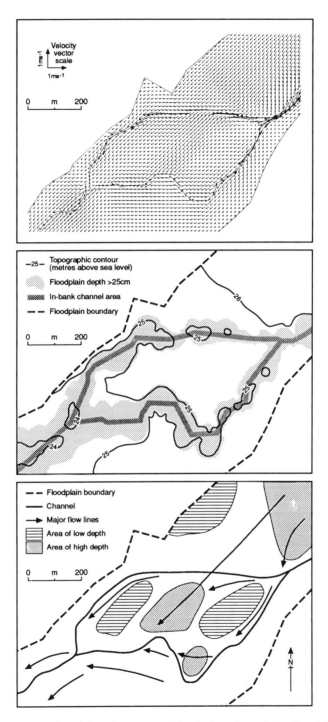

Figure 3.8 The output of a finite element model of the River Culm floodplain. Redrawn from Bates *et al.* (1992). Top, simulation of velocity vectors at peak flow of the 1 in 1 year recurrence interval; middle, floodplain topography and the simulated depths for the same event and, bottom, a schematic diagram of the flow field information. Reproduced by permission of John Wiley and Sons

ating complex topographies and multiple channels. When coupled with suitable flood deposition models testable simulations of floodplain formation and change may be possible.

UNCOUPLED FLOODPLAINS AND CHANNELS

Given the scale of rivers with floodplains and the long relaxation times involved in fluvial processes it is unlikely that most floodplains are in equilibrium with the rivers that flow through them, and disequilibrium may be the rule rather than the exception (Nanson, 1986). This is not to say that river processes and floodplain processes are unrelated, they are clearly not, but that they adjust on different time-scales to a different frequency distribution of effective events. It seems likely that many systems undergo pseudo-cycles in which periods of channel incision are associated with floodplain stability or effective aggradation alternates with periods when the channel is aggrading faster than the rate of alluviation. Both states are unstable in the medium to long term and result in cycles of channel/floodplain aggradation which would be the equivalent of a long-term steady state if it were not for the fact that the independent variables are in a constant state of flux. These cycles therefore have a complex relationships to cyclical changes in the driving variables such as the frequency and duration of storm events (e.g. Hurst phenomena). Due to the changing geometry of floodplain and channel the same event, in discharge terms, can have quite different effects on both the floodplain and channel. This partial uncoupling of the river and floodplain has been greatly increased by human actions including deliberately increasing flood deposition (warping and flood meadows), reducing flood deposition through the construction of embankments (in some cases exceeding the probable maximum flood) and dredging, the prevention of avulsion and migration by embanking and revetting, and infilling of secondary channels. If not prevented completely floods are greatly modified (Lewin et al., 1978), changing the relationship between extrinsic variables and floodplain sedimentation. Managed floodplains are much less sensitive to climatic or land use change than they would naturally be, although more work is required on the possible effects of changes in the flood frequency magnitude curve due to global warming (Arnell et al., 1994).

REFERENCES

Allen, J.R.L. (1965) A review of the origin and characteristics of recent alluvial sediments. *Sedimentology* **5**, 89–191.
Allen, M.J. (1992) Products of erosion and the prehistoric land-use of the Wessex chalk. In: *Past and Present Soil Erosion*, ed. M. Bell and J. Boardman, Oxbow Monograph 22, 37–52.
Arnell, N.W., Jenkins, A. and George, D.G. (1994) *The Implications of Climate Change for the National Rivers Authority*. Institute of Hydrology Report for the National Rivers Authority, London, HMSO.
Baker, V.R. (1991) A bright future for old flows. In: *Temperate Palaeohydrology*, ed. L. Starkel, K.J. Gregory and J.B. Thornes, Wiley, Chichester, 497–520.
Bates, P.D., Anderson, M.G., Baird, L., Walling, D.E. and Simm, D. (1992) Modelling flood-

plain flows using a two-dimensional finite element model. *Earth Surface Processes and Landforms* **17**, 575–588.

Becker, B. and Schirmer, W. (1977) Palaeoecological study on the Holocene valley development of the river Main, Southern Germany. *Boreas* **6**, 303–321.

Billi, P., Hey, R.D., Thorne, C.R. and Tacconi, P. (eds) (1992) *Dynamics of Gravel-bed Rivers*. Wiley, Chichester.

Blackie, J.R., Ford, E.D., Horne, J.E.M., Kinsman, D.J.J., Last, F.T. and Moorhouse, P. (1980) *Environmental Effects of Deforestation: An Annotated Bibliography*. Freshwater Biological Association, Occasional Publication 10.

Boardman, J. (1986) The context of soil erosion. In: *Soil Erosion*, ed. C.P. Burnham and J.I. Pitman, SEESOIL vol. 3, King's College, 2–13.

Bork, H-R. (1989) Soil erosion during the last millennium in Central Europe and its significance within the geomorphodynamics of the Holocene. In: *landforms and Landform Evolution in West Germany* (ed.), F. Ahnert, Catena Supplement 15, Cremlingen-Destedt, 107–120.

Brackenridge, R. (1984) Alluvial stratigraphy and radiocarbon dating along the Duck river Tennessee: implications regarding floodplain origin. *Bulletin of the Geological Society of America* **95**, 9–25.

Bridge, J.S. and Leeder, M.R. (1979) A simulation model of alluvial stratigraphy. *Sedimentology* **26**, 617–644.

Brooks, A. and Brown, A.G. (1995) Vegetation and the deposition of fine sediment from overbankflows (in prep.).

Brown, A.G. (1983a) Floodplain deposits and accelerated sedimentation in the lower Severn basin. In: *Background to Palaeohydrology* (ed.) K.J. Gregory, Wiley, Chichester, 375–398.

Brown, A.G. (1983b) An analysis of the overbank deposits of a flood at Blandford Forum Dorset, England. *Revue de Geomorphologie Dynamique* **32**, 95–99.

Brown, A.G. (1987a) Long-term sediment storage in the Severn and Wye catchments. In: *Palaeohydrology in Practice*, ed. K.J. Gregory, J. Lewin and J.B. Thornes, Wiley, Chichester, 307–332.

Brown, A.G. (1987b) Holocene floodplain sedimentation and channel response of the lower river Severn, U.K. *Zeitschrift fur Geomorphologie N.F.* **31**, 293–310.

Brown, A.G. (1988) The palaeoecology of *Alnus* (alder) and the postglacial history of floodplain vegetation: pollen percentage and influx data from the West Midlands, U.K. *New Phytologist* **110**, 425–436.

Brown, A.B. (1990) Holocene floodplain diachrony and inherited downstream variations in fluvial processes: a study of the river Perry, Shropshire, England. *Journal of Quaternary Science* **5**, 39–51.

Brown, A.G. (1991) Hydrogeomorphology and palaeoecology of the Severn basin during the last 15,000 years: orders of change in a maritime catchment. In: *Fluvial Processes in the Temperate Zone during the Last 15,000 Years*, ed. K.J. Gregory, L. Starkel and J.B. Thornes, Wiley, Chichester, 147–169.

Brown, A.G. (1992a) Slope erosion and cultivation at the floodplain edge. In: *Past and Present Soil Erosion*, ed. M. Bell and J. boardman, Oxbow Monograph **22**, 77–88.

Brown, A.G. (1992) Palaeochannels and palaeolandsurfaces: the geoarchaeological potential of some Midland (U.K.) floodplains. In: *Archaeology under Alluvium*, ed. S. Needham and M. Macklin, Oxbow Monograph **27**, 185–196.

Brown, A.G. (1995) Lateglacial–Holocene sedimentation in lowland temperate environments: Floodplain metamorphosis and multiple channel systems. *Palaeoclimate Research/Palaoklimforschung*, Vol. 14, Special Issue 9, 1–15 (in press).

Brown, A.G. and Barber, K.E. (1985) Late Holocene palaeoecology and sedimentary history of a small lowland catchment in Central England. *Quaternary Research* **24**, 87–102.

Brown, A.G. and Keough, M. (1992a) Holocene floodplain metamorphosis in the East Midlands, United Kingdom. *Geomorphology* **4**, 433–446.

Brown, A.G. and Keough, M. (1992b) Palaeo-channels, palaeo-landsurfaces and the 3-D reconstruction of floodplain environmental change. In: *Lowland Floodplain Rivers: Geomorphological Perspectives* ed. P.A. Carling and G.E. Petts, Wiley, Chichester, 185–202.

Brown, A.G., Keough, M.K. and Rice, R.J. (1994) Floodplain evolution in the East Midlands, United Kingdom: The Lateglacial and Flandrian alluvial record from the Soar and Nene valleys. *Philosophical Transactions of the Royal Society, London*, Series A, **348**, 261–293.

Burrin, P. and Scaife, R.G. (1984) Aspects of Holocene valley sedimentation and floodplain development in southern England. *Proceedings of the Geologists Association* **95**, 81–96.

Carleson, W.A. and Runnels, R.T. (1952) A study of silt deposited by the July 1951 flood, central Kansas river valley. *Transactions of the Kansas Academy of Science* **55**, 209–213.

Carson, M.A. (1984) Observations on the meandering–braiding transition, Canterbury Plains, New Zealand. *New Zealand Geographer* **40**, 89–99.

Coleman, J.M. (1969) Brahmaputra river: channel processes and sedimentation. *Sedimentary Geology* **3**, 129–239.

Costa, J.E. (1986) A history of palaeoflood hydrology in the United States, 1800–1970. *Transactions of the American Geophysical Union* **67**, 425–430.

Davis, R.J. and Gregory, K.J. (1994) A new distinct mechanism of river bank erosion in a forested catchment. *Journal of Hydrology* **157**, 1–11.

Dearing, J.A., Alstrom, K., Bergman, A., Regnell, J. and Sandgren, P. (1990) Recent and longterm records of soil erosion from Southern Sweden. In: *Soil Erosion on Agricultural Land*, ed. J. Boardman, I.D.L. Foster and J.A. Dearing, Wiley, Chichester, 173–192.

Dobbie, C.H. and Wolfe, P.O. (1953) The Lynmouth flood of August 1952. Part II. *Proceedings of the Institute of Civil Engineers* **2**, 522–588.

Fenneman, N.M. (1906) Floodplains produced without floods. *Bulletin of the American Geographical Association* **38**, 89–91.

Ferguson, R.I. (1981) Channel form and channel changes. In: *British Rivers*, ed. J. Lewin, Allen and Unwin, London, 90–125.

Gee, D.M., Anderson, M.G. and Baird, L. (1990) Large-scale floodplain modelling. *Earth Surface Processes and Landforms* **15**, 513–523.

Gomez, B. and Sims, P.C. (1981) Overbank deposits of the Narrator Brook, Dartmoor, England. *Geological Magazine* **118**, 77–82.

Gregory, K.J. (ed) (1983) *Background to Palaeohydrology*. Wiley, Chichester.

Gregory, K.J. and Starkel, L. (eds) (1995) *Global Continental Palaeohydrology*. Wiley, Chichester (in press).

Gregory, K.J., Gurnell, A.M. and Hill, C.T. (1985) The performance of debris dams related to river channel processes. *Hydrological Sciences Journal* **30**, 371–381.

Gregory, K.J., Lewin, J. and Thornes, J.B. (eds) (1987) *Palaeohydrology in Practice*. Wiley, Chichester.

Gupta, A. and Fox, H. (1974) Effects of high-magnitude floods on channel form: a case study in Maryland piedmont. *Water Resources Research* **10**, 499–509.

Happ, S. (1940) Some principles of accelerated stream and valley sedimentation. *United States Department of Agriculture Technical Bulletin* 695.

Harwood, K. and Brown, A.G. (1993) Changing in-channel and overbank flood velocity distribution and the morphology of forested multiple channel (anastomosing) systems. *Earth Surface Processes and Landforms* **18**, 741–748.

Haynes, V. (1968) Geochronology of late Quaternary alluvium. In: *Means of Correlation of Quaternary Successions*, ed. R.B. Morrison and H.E. Wright, Jr, University of Utah Press, Salt Lake City, 591–631.

Hey, R.D. (1978) Determinate hydraulic geometry of river channels. *Journal of the Hydraulics Division, American Society of Civil Engineers* **104**, 869–885.

Holyoak, D.T. and Seddon, M.B. (1984) Devensian and Flandrian fossiliferous deposits in the Nene valley, Central England. *Mercian Geologist* **9**, 127–150.

Hooke, J. (1977) The distribution and nature of changes in river channel patterns: the example of Devon. In: *River Channel Changes*, ed. K.J. Gregory, Wiley, Chichester, 265–280.

Hooke, J.M. and Harvey, A.M. (1983) Meander changes in relation to bend morphology and secondary flows. In: *Modern and Ancient Fluvial Systems*, ed. J.D. Collinson and J. Lewin, International Association of Sedimentologists Special Publication 6, 121–132.

Howard, A.D. (1992) Modelling floodplain–channel interactions. In: *Lowland Floodplain*

Rivers: A Geomorphological Perspective, ed. P. Carling and G.E. Petts, Wiley, Chichester, 265)–280.

Imeson, A.C., Kwaad, F.J.P.M. and Mucher, H.J. (1980) Hillslope processes and deposits in forested areas of Luxembourg. In: *Timescales in Geomorphology*, ed. R.A. Cullingford, D.A. Davidson and J. Lewin, Wiley, Chichester, 31–42.

James, C.S. (1985) Sediment transfer to overbank sections. *Journal of Hydraulic Research* **23**, 435–452.

Kalicki, T. and Krapiec, M. (1991) Subboreal 'black oaks' identified from the Vistula alluvia at Grabie near Cracow (south Poland). *Geologia* **17**, 155–169.

Keller, E.A. and Swanson, F. (1979) Effects of large organic debris on channel form and fluvial processes. *Earth Surface Processes* **4**, 361–380.

Kesel, R.H., Dunne, K.C., McDonald, R.C., Allison, K.R. and Spicer, B.E. (1974) Lateral erosion and overbank deposition on the Mississippi river in Louisiana caused by the 1973 flooding. *Geology* **2**, 461–464.

Knighton, A.D. and Nanson, G.C. (1993) Anastomosis and the continuum of channel pattern. *Earth Surface Processes and Landforms* **18**, 613–626.

Knox, J.C. (1984) Responses of river systems to Holocene climates. In: *Late Quaternary Environments of the United States. Vol. 2. The Holocene*, ed. H.E. Wright, Longman, London, 26–41.

Lattman, L.H. (1960) Cross section of a floodplain in a moist region of moderate relief. *Journal of Sedimentary Petrology* **30**, 275–282.

Lewin, J., Collin, R.L. and Hughes, D. (1978) Floods on modified floodplains. In: *Man's Impact on the Hydrological Cycle in Britain*, ed. G.E. Hollis, Geo Books, Norwich, 109–119.

Lewis, G.W. and Lewin, J. (1983) Alluvial cutoffs in Wales and the Borderland. In: *Modern and Ancient Fluvial Systems*, ed. J.D. Collinson and J. Lewin, International Association of Sedimentologists, Special Publication 6, 145–154.

Macklin, M.G. and Lewin, J. (1994) Holocene River Alluviation in Britain. *Zeitschrift fur Geomorphologie Suppl.* **88**, 109–122.

Macklin, M.G., Rumsby, B.T. and Newson, M.D. (1992) Historical floods and vertical accretion of fine-grained alluvium in the Lower Tyne valley, Northeast England. In: *Dynamics of Gravel-bed Rivers*, ed. P. Billi, R.d. Hey, C.R. Thorne and P. Tacconi, Wiley, Chichester, 564–580.

Marriott, S. (1992) Textural analysis and modelling of flood deposits: river Severn, U.K. *Earth Surface Processes and Landforms* **17**, 687–698.

Marron, D.C. (1992) Floodplain storage of mine tailings in the belle Fourche river system: a sediment budget approach. *Earth Surface Processes and Landforms* **17**, 675–686.

Melton, F.A. (1936) An empirical classification of floodplain streams. *Geographical Review* **26**, 593–609.

Nanson, G.C. (1981) New evidence of scroll bar formation on the Beatton river. *Sedimentology* **28**, 889–891.

Nanson, G.C. (1986) Episodes of vertical accretion and catastrophic stripping: A model of disequilibrium flood-plain development. *Geological Society of America Bulletin* **97**, 1467–1475.

Nanson, G.C. and Croke, J.C. (1992) A genetic classification of floodplains. *Geomorphology* **4**, 459–486.

Needham, S. (1985) Neolithic and Bronze age settlement on the buried floodplains of Runneymede. *Oxford Journal of Archaeology* **4**, 125–137.

Needham, S. (1989) River valleys as wetlands: the archaeological prospects. In: *The Archaeology of Rural Wetlands in Britain*, ed. J.M. Coles and B.J. Coles, English Heritage/Wetlands Archaeological Research Project, London/Exeter, 29–34.

Petts, G.E. (1989) Historical analysis of fluvial hydrosystems. In: *Historical Change of Large Alluvial Rivers: Western Europe*, ed. G.E. Petts, H. Möller and A.L. Roux, Wiley, Chichester, 1–18.

Pizzuto, J. (1987) Sediment diffusion during overbank flows. *Sedimentology* **34**, 301–317.

Richards, K. (1982) *Rivers*. Methuen, London.

Robinson, M.A. and Lambrick, G.H. (1984) Holocene alluviation and hydrology in the upper Thames basin. *Nature* **308**, 809–814.

Rose, J., Turner, C., Coope, G.R. and Bryan, M.D. (1980) Channel changes in a lowland river catchment over the last 13,000 years. In: *Timescales in Geomorphology*, ed. R.A. Cullingford, D.A. Davidson and J. Lewin, Wiley, Chichester, 159–176.

Rust, B.R. (1981) Sedimentation in an arid-zone anastomosing system: Cooper Creek, central Australia. *Journal of Sedimentary Petrology* **51**, 745–755.

Salisbury, C.R. (1992) The archaeological evidence for palaeochannels in the Trent valley. In: *Alluvial Archaeology in Britain*, ed. S. Needham and M. Macklin, Oxbow Monographs 27, 155–162.

Schumm, S.A. (1960) *The Shape of Alluvial Channels in Relation to Sediment Type*. United States Geological Survey Professional paper 352B.

Schumm, S.A. (1977) *The Fluvial System*. Wiley-Interscience, New York.

Sigafoos, R.s. (1964) *Botanical Evidence of Floods and Floodplain Deposition*. United States Geological Survey Professional Paper 485A.

Smith, D.G. and Smith, N.D. (1980) Sedimentation in anastomosing river systems: examples from alluvial valleys near Banff, Alberta. *Journal of Sedimentary Petrology* **50**, 157–164.

Starkel, L. (1960) Development of the relief of the Polish Carpathians during the Holocene. *Prace Geograficzne Instytut Geografi*, Polish Academy of Sciences **22**, 1–239.

Starkel, L. (1991) Characteristics of the temperate zone and fluvial palaeohydrology. In: *Temperate Palaeohydrology*, ed. L. Starkel, K.J. Gregory and J.B. Thornes, Wiley, Chichester, 3–121.

Starkel, L., Gregory, K.J. and Thornes, J.B. (eds) (1991) *Temperate Palaeohydrology*. Wiley, Chichester.

Stene, L.P. (1980) Observations on lateral and overbank deposition–evidence from Holocene terraces southwest Alberta. *Geology* **8**, 314–317.

Taylor, G. and Woodyer, K.G. (1978) Bank deposition in suspended load streams. In: *Fluvial Sedimentology*, ed. A.D. Miall, Canadian Society of Petroleum Geologists Memoir 5, 257–275.

Trimble, S.W. (1983) A sediment budget for Coon Creek basin in the Driftless area, Wisconsin, 1853–1977. *American Journal of Science* **283**, 454–474.

Vita-Finzi, C. (1969) *The Mediterranean Valleys. Geological Changes in Historical Times*. Cambridge University Press, Cambridge.

Vita-Finzi, C. (1976) Diachronism in Old World alluvial sequences. *Nature* **263**, 218–219.

Walling, D.E. and Bradley, S.E. (1989) Rates and patterns of contemporary floodplain sedimentation: a case study of the River Culm, Devon, UK. *Geojournal* **19**, 53–62.

Walling, D.E., Quine, T.A. and He, Q. (1992) Investigating contemporary rates of floodplain sedimentation. In: *Lowland Floodplain Rivers*, ed. P.A. Carling and G.E. Petts, Wiley, Chichester, 165–184.

Waters, B. (1949) *Severn Stream*. Dent, London.

Wendland, W.M. and Bryson, R.A. (1974) Dating climatic episodes in the Holocene. *Quaternary Research* **4**, 9–24.

Wolman, G.H. and Leopold, L.B. (1957) Floodplains: some observations on their development. *United States Geological Survey Professional paper 282C*.

4 Channel Networks: Progress in the Study of Spatial and Temporal Variations of Drainage Density

VINCE GARDINER

Roehampton Institute, Southlands College, London, UK

Could these streams have cut such valleys as they now flow through? If there be any true relation between cause and effect they could not.

(De la Beche, 1829, p. 242)

INTRODUCTION

In 1971 Gregory published an essay on drainage density changes in south-west England. This took what was then, in the United Kingdom, the emerging study of the drainage network as a dynamic entity and applied it to a well-established geomorphological problem—the existence of 'dry' and underfit (Dury, 1964) valleys—to produce a model of explanations of drainage density changes. The model embraced what were at that time exciting new ideas emerging in hydrology and geomorphology, including Gregory's own earlier work on dry valleys (1966) and changing drainage networks (Gregory and Walling, 1968), the hydrological ideas concerning types of runoff, as embodied in the work of Jamieson and Amerman (1969), and theoretical notions concerning geomorphological time-scales of Schumm and Lichty (1965). He finally presented a case study of the valley of the River Otter in southeast Devon, based upon field mapping and observation, and morphometric analysis. The essay is notable for the masterly way in which hydrological and geomorphological ideas are woven together so effectively, and for the way in which the newly emerging ideas of quantitative geomorphology, including morphometric analysis, are used as effective tools. Above all, it is notable for stimulating the study of underfit valleys as phenomena resulting from the dynamics of the drainage network, rather than as geomorphological accidents.

The present chapter takes three major themes of Gregory's 1971 essay and examines the progress which has since been made. These themes are: the distribution and explanation of underfit valleys, with particular reference to south-west England; the dynamics of drainage density; and the application of morphometric analysis to the understanding of underfit valleys. Some of the themes will be illustrated by reference to recent work in the Channel Islands. In one short chapter it is clearly impossible

Changing River Channels. Edited by Angela Gurnell and Geoffrey Petts.

to provide a full statement of progress made in all of these areas, but some broad directions of progress can at least be indicated.

DRAINAGE DENSITY—THE SPATIAL DIMENSION

Progress in studying the spatial variation of drainage density can be examined at local, regional, national and global scales.

The local scale

Shortly after Gregory's essay appeared the present author used a detailed study of the area centred on Dartmoor (Gardiner, 1971) to illustrate a method of producing drainage density maps of large areas, later reported in detail (Gardiner, 1979). For the Dartmoor area, it was shown that apparent contemporary drainage density was dependent upon rock and soil type. The extent of dry valleys, as shown by contour crenulations, demonstrated the substantial development of underfit valleys throughout the area, from about 1.3 to 1.6 km km^{-2}, irrespective of rock type. Although in 1971 little comment was made on these values, Gardiner later (1983a) revisited these data, and used the fact that about the same amount of underfit valley system existed throughout the area to argue that the fossil networks are likely to have been produced when lithological differences were rendered insignificant by permafrost, which rendered the surface impermeable and much more hydrologically homogenous. This argument had been convincingly employed by Cheetham (1976, 1980) in the Kennet catchment, a tributary of the Thames. Although attractive within the understanding of the time, the assumptions inherent in this argument must now be challenged in the light of a much better understanding of runoff processes in permafrost climates, as shown below.

Gerrard has recently (1993) taken the morphometric analysis of Dartmoor's morphology further, using an analysis based upon grid squares to assess landscape sensitivity. Drainage density is related to specific relief and the number of contour crossings. Subtle patterns of landscape sensitivity are identified, with areas of low sensitivity surrounded by highly sensitive areas. Much rarer are areas of high sensitivity surrounded and therefore protected by areas of lower sensitivity. Gerrard (1989) has also carried out a more traditional basin-based morphometric analysis of the Dartmoor area. In this, the ways in which basins are grouped together help explain how the overall drainage patterns have evolved.

The regional scale

As well as reporting on drainage densities in the Dartmoor region, Gardiner (1974, 1976, 1977, 1978) and Gregory and Gardiner (1975) reported values from north-west Devon and the whole of south-west England. The paper by Gregory and Gardiner (1975) demonstrated that, within south-west England, drainage density was related to mean annual precipitation, although the relationship was a complex one, substantially affected by lithology. The other work confirmed that there were relationships between rock type and drainage density throughout the peninsula, that drainage density was itself a consistently important component of overall drainage basin morphometry, and that drainage density provided a quantitative index of landscape character.

Much excellent work has been carried out by Walling, Webb, Foster, Park and others on hydrological and fluvial processes in south-west England, but full synthesis of the development of fluvial landscapes in the peninsula has yet to be attempted. Despite the progress made, it is salutary to note that some very basic information has yet to be established. For example, valley-side asymmetry is often cited as evidence of valley formation under periglacial conditions (French, 1976) as used, for example, by Gardiner (1983b) in Cannock Chase. However, whereas Clayden (1971) reports that in the Exeter district valleys frequently have their westerly-facing slopes steeper than the opposite slopes, Cullingford (1982) mentions that asymmetric valleys exist throughout Devon with their steeper slopes facing to the north-east. Gerrard (1988) further complicates the picture by noting that on western Dartmoor east-facing slopes are steeper than west-facing ones. This he ascribes to greater accumulation of snow in the lee of west-facing slopes, leading to increased snowmelt and snow avalanche activity on east-facing slopes, with gullies being initiated by snow avalanches. Until such basic observational uncertainties and interpretations can be resolved, overall models must clearly be tenuous.

An illustration

Some of the above themes may be briefly illustrated by a study still in progress of the Channel Islands. This archipelago is situated in the Gulf of St Malo of the English Channel, and comprises the two major islands, Jersey and Guernsey, and many smaller ones, principally Alderney, Sark and Herm. The islands were not glaciated, and their landscapes can therefore be interpreted without the complications of glacial features. However, many elements of the landscapes owe their origins to changing morphogenetic conditions throughout the Quaternary, especially climatic and sea level changes (see Keen, 1978, 1981).

The density of valleys on the islands, especially Jersey and the higher parts of Guernsey, is remarkable. Apart from the coastal fringes, underlain by blown sand and/or Holocene deposits (Jones et al., 1990), the islands are dissected by a dense network of valleys (Figure 4.1). However, it is not possible to distinguish entirely satisfactorily between the present functioning drainage network and an entirely fossil component, as runoff records are skeletal. On the short-time scale, temporal variability of flow is high. For example, almost every stream flowing towards the west coast of Jersey ceases to flow during the driest summers, such as that of 1976 (Manton, 1977). On slightly longer time-scales there are indications that the islands were formerly more humid, and streams more extensive within the present valley system. For example, travellers' accounts of Jersey in the 17th century refer to its well-watered nature, and there was sufficient water to drive many water mills. Within the present century unpublished observations by the water authority of Alderney suggest a progressive desiccation. Overall, across the islands, less than 300 mm of the annual 800–900 mm of precipitation contributes to surface runoff. Present stream channels are small, valley-side slope activity is negligible, and streams are manifestly underfit (Dury, 1964).

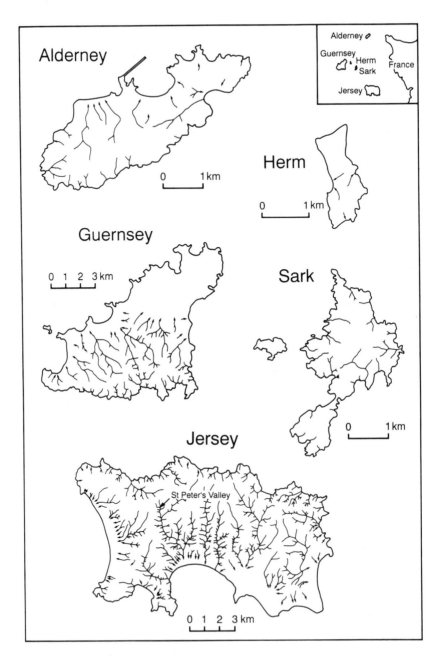

Figure 4.1 Valley networks of the Channel Islands. Arrows indicate directions of 'flow' for sections not terminating in the sea. Note that the scale differs between the larger and smaller islands

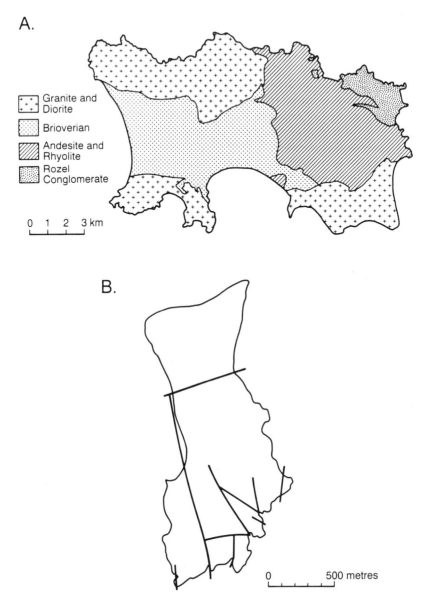

Figure 4.2 Some geological influences on the development of valley networks in the Channel Islands. A. Lithologies in Jersey. B. Major faults in Herm

There is a marked correspondence between the distribution of valley density and aspects of the geology of the islands (Figure 4.2). In Jersey (Gardiner, 1986), the highest valley densities are found on the Brioverian sediments, with the four major valley systems on this group having densities in the range 2.4 to 3.67 km km^{-2}. The granitic rocks have much lower valley densities, typically around 2.6 km km^{-2}. The volcanic group has a valley density only slightly higher. The range of variation of valley development with lithology is similar in many ways to that observed by Gregory in the Otter Valley (Gregory, 1971), and is greater than the more restricted range of variation reported from areas thought to have suffered more extreme periglaciation, for example Dartmoor (Gardiner, 1971) and the Kennet Valley (Cheetham, 1980).

Geological structure has also played an important part in controlling the detailed development of some valleys. For example, on the small island of Herm (Gardiner, 1989), the four fairly shallow valleys contain little or no discharge. Two of them are oriented along a NNW–SSE aligned fault, and may be similarly controlled by a fault yet to be detected (Figure 4.2B). On Alderney the major valley in the eastern half of the island is eroded along the line of a faulted contact between diorite and sandstone. Hydrogeological investigations (Hodgson, 1990) confirm that this is a zone of substantial groundwater, but only limited surface runoff. On Jersey there is also substantial fault guidance of several major valleys, including, for example, the upper part of St Peter's Valley.

Some aspects of valley morphology are, however, independent of geology. Figure 4.3 shows values of relative slope asymmetry for the major tributary basins trending E–W within the St Peter's Valley system. This index is:

$$\text{Relative slope asymmetry} = \frac{A - B}{(A + B)/2}$$

where A = steepest north-facing slope angle (degrees), and B = steepest south-facing slope angle (degrees).

For only two of the valleys is the index negative, and hence there is a small but persistent slope asymmetry, with north-facing slopes being steeper than the opposite ones. The underlying rocks are mainly the Jersey Shale Formation of Brioverian age. These are extensively deformed (Helm, 1984), but have no consistent structure which would explain this slope distribution with respect to aspect. Indeed, the predominant structure within the St Peter's Valley is a series of folds with N–S trending axes (Bishop and Bisson, 1989). It is therefore tempting to ascribe a periglacial

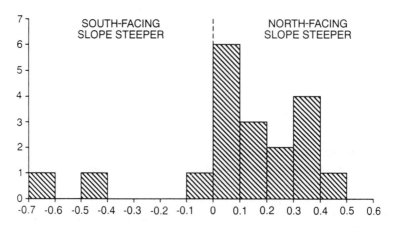

Figure 4.3 Histogram of relative slope asymmetry values for sub-basins trending E–W in the St Peter's Valley, Jersey

origin to this asymmetry, with the south-facing slopes having received a greater amount of radiation and undergone a greater degree of thawing under periglacial conditions (French, 1976), and therefore slope processes and slope lowering having been more active and effective on the south-facing slopes.

The national scale

Elsewhere in Britain, since 1971, underfit valley networks have been examined on a variety of areas and rock types, including Mercia Mudstone (Jones, 1979), Sherwood Sandstone (Gardiner, 1983b), Quaternary sands and gravels (Richards and Anderson, 1978), Chalk, sandstone and clay (Morgan, 1971; Cheetham, 1980), and a variety of igneous and sedimentary rocks in Jersey (Gardiner, 1986) and the Severn (Jones, 1982; Dawson, 1986). There has been a growing realisation that the valley has to be considered within the context of the entire fluvial system, and that the present valley is only a snapshot within a dynamically expanding and contracting entity. For example, in south-west England, Gerrard (1988) identified partially infilled gully systems on Dartmoor, similar to those reported earlier by Waters (1966) in the Haldon Hills. A fluvial fan, emanating from an underfit valley system and now buried beneath the alluvium of the Otter Valley, was reported by Gardiner (1983a). Cant (1974) carried out a painstaking investigation of the subsurface deposits of the upper Exe Valley, and Durrance (1969, 1971, 1974) used geophysical techniques to examine the subsurface morphology of the lower Exe and Teign Valleys. Dalzell and Durrance (1980) also used a subsurface geophysical investigation of the Valley of the Rocks, on the north Devon coast, to argue for a fluvial origin for this contentious (Stephens, 1966) feature. Gardiner (1983a) emphasised, with reference to national examples, that in considering morphological evidence of network expansion, sedimentological evidence of deposition should be sought wherever possible, although often buried beneath more recent deposits or transported into the marine environment. However, despite much detailed work, no national-scale synthesis has yet emerged.

The global scale

Despite the large number of individual studies carried out, only very general and tentative conclusions may be drawn concerning the relationships between drainage density and climate, on a global scale. Areas with low annual rainfall have high values of drainage density. These semi-arid areas have episodic high-intensity rainfall and sparse vegetation, and hence rapid runoff. In humid mid-latitudes drainage densities are lower and some studies have reported that there is a tendency for values not to be positively related to mean annual rainfall (Gregory and Gardiner, 1975). However, Abrahams and Ponczynski (1984) reported that, within the USA, the relationship between drainage density and precipitation intensity is an inverse one, at least in all but desert climates. They suggest that previous studies inadvertently controlled for differences in the vegetation and soil cover. Drainage density is controlled by both precipitation intensity and mean annual precipitation, via the web of controls and relationships shown in Figure 4.4. Whether the relationship found between drainage density and precipitation intensity is direct or inverse therefore depends on

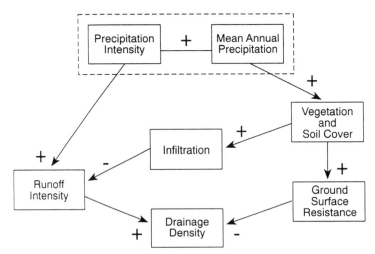

Figure 4.4 Multiple causation diagram showing the paths whereby precipitation controls drainage density (from Abrahams and Ponczynski, 1984, reproduced with permission of Elsevier Science)

which of the two control paths, to left or right of the diagram, is the dominant. They propose that the left hand path dominates in desert climates, and the right hand path elsewhere.

Particularly missing are data on drainage densities in the more extreme climates. Humid tropical zones appear to be characterised by somewhat higher values than the temperate zones, and also to have a higher range of values. Data from periglacial areas are few, but, in overviewing the limited data available, Gardiner in 1988 reversed his provisional (1983a) view, and concluded that drainage densities are generally fairly high. This is considered further below.

DRAINAGE DENSITY—THE TEMPORAL DIMENSION

Much of the early work on drainage basin morphometry regarded drainage density as an essentially static index, characteristic of the hydrological environment of the basin. However, there has been a realisation that no single value of drainage density can fully portray the drainage basin, as drainage density is an index which is itself temporally variable in response to changing hydrological conditions, on all time-scales. However, empirical evidence on the nature and variability of this assertion is not easy to obtain (Gardiner, 1982).

Short-term network dynamics

On short term time-scales the drainage network is difficult to determine instantaneously throughout a drainage basin in order to assess the extent of the network at single moments of time. Uneven spatial distributions of rainfall result in different drainage network patterns in different parts of the basin. Remote sensing or instru-

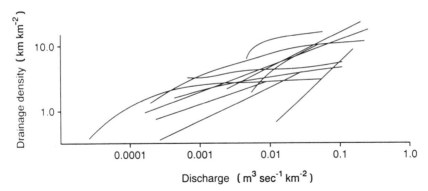

Figure 4.5 Some reported relationships between drainage density as a dynamic index and discharge. (Plotted from Dunne (1976), Gurnell (1978) and sources therein)

mental techniques are insufficiently precise, and conditions can change so rapidly that field observation is difficult and sometimes hazardous.

However, studies carried out throughout the 1970s, including some carried out or guided by Gregory, have succeeded in establishing some of the range of variations in drainage density which occurs on short time-scales (e.g. Day, 1978). Some of the ways in which discharge and hence drainage density respond to changing meteorological conditions are shown in Figure 4.5. Clearly, although increases in discharge are consistent with increases in drainage density, the precise nature of the relationships identified varies widely, according to other basin characteristics. Most authors have suggested a linear logarithmic (i.e. power function) relationship between discharge and drainage density, although a few have identified curvilinear relationships. The significance of this principle is instanced by the criticisms advanced of Dingman's (1978) paper on drainage density and streamflow by Gregory and Gardiner (1979). Dingman's paper pays little attention to the dynamic nature of the drainage network, which hampered its evaluation as a meaningful index.

One very significant control on the rate of network expansion and contraction is lithology (Gurnell, 1978; Gardiner and Gregory, 1982). Basins underlain by more-permeable rocks are characterised by low drainage densities at low flows, but high rates of network expansion with increasing discharge. Basins with less permeable lithologies are initially characterised by high drainage densities, but lower rates of network expansion with increasing discharge. Gardiner and Gregory (1982) incorporated this notion in a theoretical model developed from Carlston's (1963, 1965) model relating discharge to drainage density. The necessary implication is that a particular network can only be meaningfully envisaged as the product of a particular flow magnitude (Gardiner, 1983a).

One of the key concepts in Gregory's 1971 essay was the relationship between the drainage network and the hydrological processes operating. The complexities of this relationship are well demonstrated by a study of the hydrological evolution of drainage basins disturbed by surface mining (Ritter and Gardner, 1993). The steady-state infiltration capacity of the newly reclaimed soils is a key variable. When this is less than 3 cm hr^{-1} the dominant runoff process is infiltration-excess (Hortonian) overland flow. Phases of drainage network expansion and contraction characterise network evolution. During network extension storm hydrographs are characterised by increasing peak and total runoff and decreasing time to peak. During phases of network contraction/abstraction they are characterised by decreasing peak and total runoff, and increasing time to peak. In contrast, where infiltration capacity recovers to more than 3 cm hr^{-1}, runoff is initially dominated by infiltration-excess overland flow, but saturation overland flow becomes increasingly dominant with time. Drainage network development is limited to only skeletal network initiation and elongation, occurring whilst infiltration-excess overland flow occurs. Storm hydrographs have decreased peak runoffs and increased times to peak. Although developed for small, artificially created basins, there are clear implications for natural network development. Hillslope and channel hydrology interact to produce high drainage densities where saturation flow dominates. When infiltration capacity is modified by external changes, for example of climate, drainage density is increased when infiltration capacity is decreased, and vice-versa. Such differences have implications for longer-term evolution of fluvial landscapes. For example, Ijjasz Vasquez et al. (1992) show how simulated patterns of basin evolution differ slightly for basins dominated by Hortonian runoff as compared with those dominated by subsurface flow. Differences in runoff mechanisms also have a pronounced impact on basin hypsometric curves (Ciccacci et al., 1992).

Longer-term network evolution

The determination of drainage density variations on longer time-scales is also attended by difficulties. Morphological evidence is not always preserved, and that best preserved might be characteristic of extreme hydrological conditions rather than more typical magnitudes of discharge. For example, in south-west England, Anderson and Calver (1977), by examination of the features produced by the catastrophic Exmoor floods of 1952, concluded that the preservation of landscape features depended upon both the frequency distribution of geomorphic events and the order of the most recent ones. Evidence is also open to varying interpretations, as evidenced by the controversies concerning both single features such as the Valley of the Rocks in north Devon (Stephens, 1966; Dalzell and Durrance, 1980) and whole networks. The latter can be illustrated by the varying interpretations of the underfit valleys of the Chalk and associated rocks of south-east England as submarine canyons (Winslow, 1966) and tidal channels (Geyl, 1976), as well as more conventionally as fluvial features. Finally, evidence of fluvial evolution is difficult to date and correlate because stratigraphic relationships are often obscure or absent.

The precise nature of longer-term drainage network expansion or contraction has been examined by Ovenden and Gregory (1980). They examined cartographic, field and documentary evidence for network change in three areas of the United Kingdom. Changes within the last century were shown to be significant, and often occurred not simply as an expansion or contraction of the network per se, but by the transformation of network typology, and in particular network expansion by the replacement of ill-defined flushes by well-defined channels. Burt and Gardiner (1982) have questioned the specific interpretation of cartographic evidence, but the general principle is accepted. The concept of network typology has been further developed by Gregory (1979), and applied to networks in terms of flow resistance by Gardiner and Gregory (1982) and Ovenden and Gregory (1980).

Implicit in much early work on channel network development (e.g. Horton, 1945, and the great wealth of work this paper stimulated—see Gardiner and Park, 1978) was the assumption that surface runoff was the dominant mode of stream channel initiation and development. However, increasingly attention has been focused on sapping as an alternative mechanism. In Gomez and Mullen's (1992) experimental study, channel initiation occurred through excavation of a spring head by effluent subsurface flow. This migrates upstream, and tributary channels then develop. Uchupi and Oldale (1994) have also suggested spring sapping as the origin of steep-sided flat-floored linear valleys in Massachusetts, USA.

Dietrich and Dunne (1993) review studies relating channel heads to drainage density, pointing out that valley density is more stable than channel density because of the greater amount of geomorphological work required to excavate valleys. The identification of stream heads in early morphometric studies was initially based upon field mapping, and then upon the application of consistent criteria based upon topographic map analysis. A resurgence of interest in this topic has occurred for a number of reasons, including the growth of field and theoretical studies of the consequence of threshold conditions for channel initiation on landscape morphology (e.g. Kirkby, 1980; Montgomery and Dietrich, 1992). Models of the position of channel heads have been produced, based upon the relative magnitudes of mass-movement versus sheetwash processes. For example, Tarboton et al. (1992) identify a characteristic breakpoint in the plot of drainage area versus slope as the location at which channel-forming processes become dominant. Such relationships are used in determining the extent of drainage networks from digital elevation data, as used by, for example, Montgomery and Dietrich (1992), and Montgomery and Foufoula-Georgiou (1993).

A study illustrating some of the complexities of longer-term drainage network development is that by Van Nest and Bettis (1990), of the Buchanan drainage in central Iowa, USA. Development of this system was both episodic and time-transgressive. During the late Wisconsinan and early Holocene, headward extension was in response to base-level lowering, whereas in the late Holocene, headward extension was initiated in the headwaters, probably in response to climatic change and a consequent rise in the level of the groundwater table. The spatial dimension is therefore critical in interpreting the time-transgressive deposits, and associated phases of aggradation, in the way developed by Schumm as 'Complex Response' (Schumm, 1977). Spatial and temporal dimensions are inextricably linked in drainage network development.

Although perhaps the majority of recent work in fluvial geomorphology has emphasised the importance of processes, the role of lithology in determining network evolution over longer time-scales has not been ignored. For example, a study by Miller et al. (1990) of drainage network evolution on clastic sedimentary rocks in Indiana, USA, showed that morphometric characteristics of drainage basins reflect an intimate adjustment to the lithology underlying the basin, as well as to the extent of solutional modification of carbonates by karst processes. The ergodic hypothesis was used to suggest a schematic model of drainage network evolution in response to an increasing exposure of carbonate rocks within the catchment (Figure 4.6).

BASIN PLANVIEW   VALLEY CROSS-SECTIONS

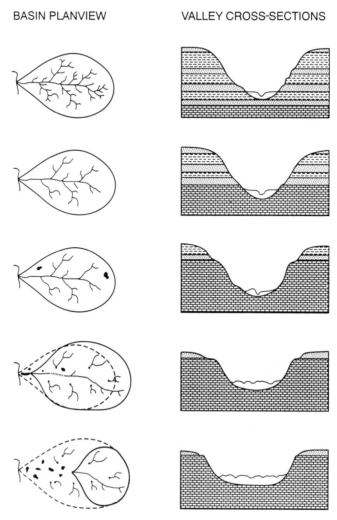

Figure 4.6 Schematic illustration of changes in landscape morphology in relation to basin evolution in an area of clastic sedimentary rocks (from Miller *et al.*, 1990, reproduced with permission from *American Journal of Science*)

The majority of studies of drainage density and its spatial and temporal variations have focused on fluvial systems dominated by surface runoff, or at least having surface runoff as a major component of flow. However, in a series of papers de Vries (1976, 1977, 1994) has developed the Groundwater Outcrop Erosional Model (GOEM) of stream network evolution. He identifies a close relationship between groundwater depth and stream network characteristics, including spacing and hence density, in the lowland area underlain by the sandy permeable Pleistocene deposits of the Netherlands. This applies for average conditions as well as to the seasonal expansion and contraction of the network of streams. The stream network is envisaged as the interface between groundwater and stream systems, and one which adapts in response to the discharge capacity required to release excess precipitation

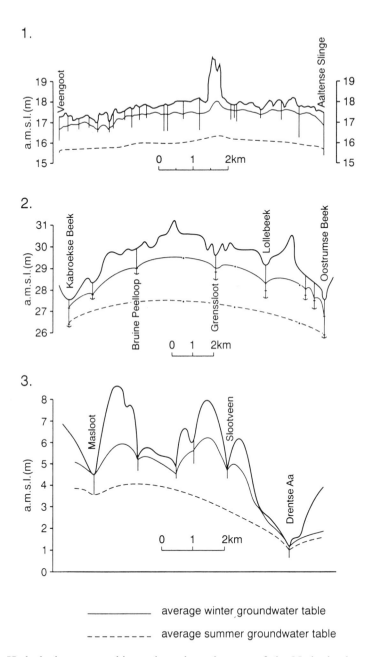

Figure 4.7 Hydrologic–topographic sections through areas of the Netherlands underlain by Pleistocene deposits, showing relationships between groundwater depth and stream density. The vertical lines indicate positions and depths of drainage channels (from de Vries, 1994, reproduced with permission of Elsevier Science)

from the groundwater–stream-water continuum. As the depth to groundwater and topographic slope reduce, stream density increases (Figure 4.7).

Drainage density has also been found to be a useful index in assessing the groundwater potential of areas. For example, Uma and Hekinde (1992) found drainage density to be one of several topographic and lithologic measurements of value in estimating the groundwater potential of parts of south-eastern Nigeria.

APPLYING DRAINAGE DENSITY: PALAEOHYDROLOGICAL INVESTIGATIONS OF UNDERFIT VALLEYS

Since Gregory's 1971 essay the field of palaeohydrology has emerged (Gregory, 1983; Gregory et al., 1987; Starkel et al., 1991). Within this, morphometric analysis has been applied to the explanation of underfit valleys, and in particular to the estimation of former discharges, as reviewed by Gardiner (1983a, 1986). The principle underpinning this approach is that drainage density is related to discharge in some systematic way. The approach is not, however, without problems.

The assumption that drainage density is related to discharge is to some extent an act of faith rather than a demonstrable truth. Whilst on the one hand relationships have been demonstrated, for example by Carlston (1963, 1965) and others summarised in Table 4.1, on the other hand many studies have failed to demonstrate such relationships. For example, Pitlick (1994) investigated relationships between peak flows and relief measures for mountainous regions in the USA, including drainage density. The mean annual flood was related to basin area and precipitation, but no significant relationship with drainage density was found. The accuracy and definition of cartographic sources, the extent to which short-term variability of drainage network extent can be taken into account, and the robustness of statistical methods and assumptions (Gardiner, 1973) are all important considerations in assessing the weight of evidence.

In the case of the Channel Islands, there are no long-term records of discharge available, and the hydrological system has been heavily modified by human activity, especially for agriculture. It is therefore difficult to assess the extent to which the valleys represent underfit or largely fossil features. However, estimates of palaeodischarges were made by Gardiner (1986) for some of the valley systems in Jersey, using Cheetham's re-evaluation of Carlston's (1963, 1965) relationships. For the easternmost valley on the Brioverian rocks this gave a palaeodischarge estimate of 1.99 m^3 s^{-1}. For the largest valley system draining the north-west igneous complex a discharge estimate of about 0.92 m^3 s^{-1} was derived. The present streams are not regularly gauged, but crude estimates of discharge were derived from data in Butler et al. (1985). These accord in general terms with estimates made from restricted duration stage data in Jones et al. (1993). Estimated palaeodischarges were thus estimated to be of the order of 1.7 to 3.1 times present discharges, which accorded broadly with the magnitude of estimates obtained for areas in southern England (e.g. Cheetham, 1980; Jones, 1982; Dawson, 1986). This work is presently being refined and expanded by Andrew Wright at Roehampton, who is working on relationships amongst a much fuller range of morphometric information for Jersey.

An assumption often made in examining underfit valleys is that under permafrost

Table 4.1 Some reported relationships between discharge and drainage density as a static index

Author	Year	Area	Discharge index	Exponent in relationship with drainage density
PEAKFLOWS				
Carlston	1963	Central and East USA	$Q_{2.33}$	2.0
Cheetham	1980	Re-evaluation of Carlston (1963)—see Gardiner (1988)	$Q_{2.33}$	1.59
Rodda	1969	UK	$Q_{2.33}$	0.81
Patton and Baker	1976	Central Texas	Q_{max}	−1.68
		Southern California	Q_{max}	−1.87
		Indiana	Q_{max}	0.56
		Appalachian Plateau	Q_{max}	−0.50
		(based upon multiple relationships)		
Murphey et al.	1977	Basin and Range Province, USA	Q_{mpi}	3.88
BASEFLOWS				
Carlston	1965	Central and East USA	Q_v	−2.0
Gregory and Walling	1968	Devon, UK	Q_{min}	2.2
Trainer	1969	Piedmont and Blue Ridge, USA	Q_v	−3.54

$Q_{2.33}$ = peak flow with recurrence interval of 2.33 years.
Q_{max} = annual maximum peak flow
Q_{mpi} = maximum peak in period of observation
Q_b = baseflow
Q_{min} = minimum flow in period of observation

conditions runoff generation occurs largely as overland flow, and the significance of rock type is suppressed because infiltration and permeability variations are negated by the permafrost. For this reason Gardiner (1983a) and Cheetham (1980), amongst others, cited uniformity of underfit valleys with respect to variations in rock type as evidence in support of formation under periglacial conditions. However, this must be questioned in the light of more recent observations on the processes of runoff generation in contemporary periglacial areas, and in the light of studies where spatial uniformity of valley development does not exist.

In periglacial areas, overland flow is particularly widespread during the spring melt period when the active zone is shallow (Woo and Steer, 1982; Wood, 1983), but saturated throughflow becomes dominant as the thaw depth increases in the summer (Woo and Steer, 1983). Summer rainfall may therefore produce only throughflow (Lewellen, 1972). French and Lewkowicz (1981) contend that this is not effective for denudation under periglacial regimes. Furthermore, it has been suggested that not all permafrost is necessarily impermeable. For example, van Everdingden (1974) reviewed studies of groundwater in Canadian permafrost regions, and Mackay (1983) showed, from both field and laboratory observations, that in summer water can move from the active layer to the frozen active layer, and even into the permafrost. Gray *et al.* (1985) suggested that infiltration is only restricted when there is an impermeable surface or near-surface layer, such as an ice lens. Younger (1989) has embraced such ideas in developing a model of Chalk dry valley formation based upon differences in permeability under periglacial conditions. He suggests that during the Devensian permafrost would prevent infiltration. However, the friction of flowing water would melt the upper layers of permafrost. This would in turn weaken the regolith and enhance stream erosion. Eventually an ice-free talik would develop, possibly remaining throughout the year, even when runoff ceases in winter.

In addition, overland flow is not an inevitable consequence of the presence of permafrost. Arctic rainfall intensities are low, and Hortonian overland flow generation is therefore rare. Saturation overland flow is due to the superpermafrost water table rising to the ground surface. Woo and Steer (1982, 1983) reported how during spring snowmelt there is a large supply of water from melting, and little evaporation loss. The thin active layer has little storage capacity, and therefore saturated overland flow is generated. Spring is the period of maximum surface flow. As summer progresses, the depth of the thaw increases, as does evaporation. The snow is depleted and rainfall is limited. The superpermafrost layer is therefore increased. The ground is not saturated, with surface flow therefore only occurring sporadically during rain. The thaw depths are spatially variable because of the insulating effects of vegetation, and the nature of the underlying materials. Seepage and re-emergence can therefore occur downslope in a complex pattern. The greater exterior link lengths of streams originating on permafrost have been explained (Bredthauer and Hock, 1979) as being due to subsurface flow exfiltrating near the base of slopes, and therefore having a short distance to travel before reaching higher order streams. Variability of runoff generation is also induced by variability in snow cover (Woo, 1976). Thus the spatial and temporal patterns of runoff generation are complex (Gardiner, 1988).

Woo and Steer (1982) have mapped the extent of surface flow in a small area during one runoff season. A highly dynamic pattern of surface runoff and hence drainage density occurs. Chacho and Bredthauer (1983) have also stressed the importance of the partial area concept. They found little contribution to runoff from the 30% of the basin not underlain by permafrost. Slaughter *et al.* (1983) have also demonstrated that permafrost-free catchments have higher summer baseflows than permafrost-dominated basins, as well as higher sediment concentrations and hence erosion rates. Zalesskiy (1976) estimated that for major flash floods in permafrost regions, only 60% of the catchment was contributing to flow. Thus the actuality of runoff in frozen regions is very far from the simple model of overland flow occurring uniformly as implicit in many early studies of fossil features in temperate regions.

CONCLUSIONS: RETROSPECT AND PROSPECT

The two decades since 1971 have seen considerable progress in establishing the extent and nature of both spatial and temporal variations of drainage density. The application of morphometric techniques to palaeohydrology has been initiated, although attendant problems have yet to be solved or circumvented. However, perhaps the most significant advance made is that geomorphological interpretations are now perhaps inevitably, but certainly inextricably, wedded to a realistic understanding of hydrological processes. The co-existence of form and process in a mutually systemic and synergenetic relationship was a key underpinning of Gregory's 1971 essay, and one which continues to provide a most fruitful basis for geomorphological understanding.

ACKNOWLEDGEMENT

In 1966 I arrived in Exeter as a naive and nervous undergraduate, vaguely contemplating careers based upon something to do with geography. Chance assigned me to Ken Gregory as tutee, and the rest, as they say, is geomorphological history! Ken Gregory's lively and challenging lectures and tutorials stimulated my interest in geomorphology as an undergraduate. His perceptive and supportive supervision both sustained and challenged me through three postgraduate years. Most importantly, his friendship has underpinned a variety of professional relationships during two further decades. Thanks, Ken.

I would also like to thank the Roehampton Institute London, Research Committee, for grants enabling fieldwork in the Channel Islands.

REFERENCES

Abrahams, A.D. and Poncynski, J.J. (1984) Drainage density in relation to precipitation intensity in the U.S.A. *Journal of Hydrology* **75**, 383–388.

Anderson, M.G. and Calver, A. (1977) On the persistence of landscape features formed by a large flood. *Transactions, Institute of British Geographers, New Series* **2**, 243–254.

Bishop, A.C. and Bisson, G. (1989) *Jersey. Description of 1:25 000 Channel Islands Sheet 2,* HMSO, London.

Bredthauer, R. and Hoch, D. (1979) *Drainage Network Analysis of a Subarctic Watershed, Caribou-Poker Creeks Research Watershed, interior Alaska.* US Army Cold Regions Research and Engineering Laboratory, Hanover, New Hampshire, Special Report 79-19.

Burt, T.P. and Gardiner, A.T. (1982) The permanence of stream networks in Britain: some further comments. *Earth Surface Processes* **7**, 327–332.

Butler, A.P., Grundy, J.D. and May, B.R. (1985) Analysis of extreme rainfalls observed in Jersey. *Meterological Magazine* **114**, 383–395.

Cant, B.C. (1974) Aspects of valley floor development in the upper Exe basin. *Reports and Transactions of the Devonshire Association for the Advancement of Science* **106**, 77–94.

Carlston, C.W. (1963) *Drainage Density and Streamflow.* United States Survey Professional Paper 422-C.

Carlston, C.W. (1965) The effect of climate on drainage density and streamflow. *International Association for Scientific Hydrology, Bulletin* **11**, 62–69.

Chacho, E.F., Jr and Bredthauer, S. (1983) Runoff from a small Subarctic watershed, Alaska. In: *Permafrost; 4th International Conference Proceedings*, 115–120.

Cheetham, G.H. (1976) Palaeohydrological investigations of river terrace gravels. In: Davidson, D.A. and Shackley, M. (eds), *Geoarchaeology: Earth Science and the Past*, Duckworth, London, 335–343.

Cheetham, G.H. (1980) Late Quaternary palaeohydrology: the Kennet valley case-study. In: Jones, D.K.C. (ed.), *The Shaping of Southern England*, IBG Spec. Pub. 11, Academic Press, 203–223.

Ciccacci, S., D'Alessandro, L., Fredi, P. and Lupia Palmieri, E. (1992) Relations between morphometric characteristics and denudational processes in some drainage basins of Italy. *Zeitschrift für Geomorphologie* **36**, 53–67.

Clayden, B. (1971) *The Exeter District*. Memoir of the Soil Survey of Great Britain.

Cullingford, R.A. (1982) The Quaternary. In: Durrance, E.M. and Laming, D.J.C. (eds), *The Geology of Devon*, 249–290, Wheatons, Exeter.

Dalzell, D. and Durrance, E.M. (1980) The evolution of the Valley of the Rocks, North Devon. *Transactions, Institute of British Geographers, New Series* **5**, 66–79.

Dawson, M.R. (1986) *Late Devensian fluvial environments of the Lower Severn Basin, U.K.* Unpublished PhD thesis, University of Leicester.

Day, D.G. (1978) Drainage density changes during rainfall. *Earth Surface Processes* **3**, 319–326.

De la Beche, H.T. (1829) Notice on the excavation of valleys. *Philosophical Magazine and Annals of Philosophy (New Series)* **6**, 241–248.

de Vries, J.J. (1976) The groundwater-outcrop erosion model; evolution of the stream network in The Netherlands. *Journal of Hydrology* **29**, 43–50.

de Vries, J.J. (1977) The stream network in The Netherlands as a groundwater discharge phenomenon. *Geologie Mijnbouw* **56**, 103–122.

de Vries, J.J. (1994) Dynamics of the interface between stream and groundwater system in lowland areas, with reference to stream net evolution. *Journal of Hydrology* **155**, 39–56.

Dietrich, W.E. and Dunne, T. (1993) The channel head. In: Beven, K. and Kirkby, M.J. (eds), *Channel Network Hydrology*, Wiley, Chichester, 175–220.

Dingman, S.L. (1978) Drainage density and streamflow: a closer look. *Water Resources Research* **14**, 1183–1187.

Dunne, T. (1976) Field studies of hillslope flow processes. In: Kirkby, M.J. (ed.), *Hillslope Hydrology*, Wiley, Chichester, 227–294.

Durrance, E.M. (1969) The buried channels of the Exe. *Geological Magazine* **106**, 174–189.

Durrance, E.M. (1971) The buried channel of the Teign estuary. *Proceedings of the Ussher Society* **2**, 299–306.

Durrance, E.M. (1974) Gradients of buried channels in Devon. *Proceedings of the Ussher Society* **3**, 111–119.

Dury, G.H. (1964) *Principles of Underfit Streams*. United States Geological Survey Professional Paper 452A, 67pp.

French, H.M. (1976) *The Periglacial Environment*. Longman, London.

French, H.M. and Lewkowicz, A.G. (1981) Periglacial slope investigations, Banks Island, Western Arctic. *Biuletyn Peryglacjalny* **18**, 25–45.

Gardiner, V. (197) A drainage density map of Dartmoor. *Reports and Transactions of the Devonshire Association for the Advancement of Science* **103**, 167–180.

Gardiner, V. (1973) Univariate distributional characteristics of some morphometric variables. *Geografiska Annaler* **54A**, 147–153.

Gardiner, V. (1974) Land form and land classification in North-west Devon. *Reports and Transactions of the Devonshire Association for the Advancement of Science* **106**, 141–153.

Gardiner, V. (1976) Land evaluation and the numerical delimitation of natural regions. *Geographia Polonica* **34**, 11–30.

Gardiner, V. (1977) Estimated drainage density and physical regions in Southwest England. *National Geographer* **12**, 115–130.

Gardiner, V. (1978) Redundancy and spatial organisation of drainage basin form indices: an empirical investigation of data from north-west Devon. *Transactions, Institute of British Geographers, New Series* **3**, 416–431.

Gardiner, V. (1979) Estimation of drainage density from topological variables. *Water Resources Research* **15**, 909–917.

Gardiner, V. (1982) Drainage basin morphometry—quantitative analysis of drainage basin form. In: Sharma, H.S. (ed.), *Perspectives in Geomorphology*, vol. II, Concept Publishing Company, New Delhi, 107–142.

Gardiner, V. (1983a) Drainage networks and palaeohydrology. In: Gregory, K.J. (ed.), *Background to Palaeohydrology*, Wiley, Chichester, 258–277.

Gardiner, V. (1983b) The relevance of geomorphometry to studies of Quaternary morphogenesis. In: Briggs, D.J. and Waters, R.S. (eds), *Studies in Quaternary Geomorphology*, Geo-Abstracts, 1–17.

Gardiner, V. (1986) Fluvial palaeohydrology: the morphometric contribution. *National Geographer* **21**, 17–31.

Gardiner, V. (1988) On the use of drainage density/discharge relationships and periglacial runoff generation models in the palaeohydrological interpretation of temperate zone dry valleys. In: Singh, S. and Tiwari, R.C. (eds), *Geomorphology and Environment*, Allahabad Geographical Society, Allahabad, 470–483.

Gardiner, V. (1989) The geomorphology of Herm. *Report and Transactions of La Société Guernesiaise* **22**, 619–627.

Gardiner, V. and Gregory, K.J. (1982) Drainage density in rainfall–runoff modeling. In: Singh, V. (ed.), *Rainfall–Runoff Relationship*, Water Resources Publications, Colorado, 449–476.

Gardiner, V. and Park, C.C. (1978) Drainage basin morphometry: review and assessment. *Progress in Physical Geography* **2**, 1–35.

Gerrard, A.J. (1988) Partially infilled gully systems on Dartmoor. *Proceedings of the Ussher Society* **7**, 86–89.

Gerrard, A.J. (1989) Drainage basin analysis of the granitic upland of Dartmoor, England. *Geographical Review of India* **51**, 1–17.

Gerrard, A.J. (1993) Landscape sensitivity and change on Dartmoor. In: Thomas, D.S.G. and Allison, R.J. (eds), *Landscape Sensitivity*, Wiley, Chichester, 49–63.

Geyl, W.F. (1976) Tidal palaeomorphs in England. *Transactions of the Institute of British Geographers, New Series* **1**, 203–224.

Gomez, B. and Mullen, V.T. (1992) An experimental study of sapped drainage network development. *Earth Surface Processes and Landforms* **17**, 465–476.

Gray, D.M., Landine, P.G. and Granger, R.J. (1985) Simulating infiltration into frozen prairie soils in streamflow models. *Canadian Journal of Earth Sciences* **22**, 464–472.

Gregory, K.J. (1966) Dry valleys and the composition of the drainage net. *Journal of Hydrology* **4**, 327–340.

Gregory, K.J. (1971) Drainage density changes in south-west England. In: Ravenhill, W.L.D. and Gregory, K.J. (eds), *Exeter Essays in Geography*, University of Exeter, Exeter, 33–54.

Gregory, K.J. (1979) Changes of drainage network composition. *Acta Universitatis Ouluensis*, Ser. A., **82**, 19–28.

Gregory, K.G. (1983) (ed.) *Background to Palaeohydrology*. Wiley, Chichester.

Gregory, K.G. and Gardiner, V. (1975) Drainage density and climate. *Zeitschrift für Geomorphologie* **19**, 287–298.

Gregory, K.J. and Gardiner, V. (1979) Comment on 'Drainage density and streamflow: a closer look' by S.L. Dingman. *Water Resources Research* **15**, 1662–1664.

Gregory, K.J. and Walling, D.E. (1968) The variation of drainage density within a catchment. *International Association for Scientific Hydrology, Bulletin* **13**, 61–68.

Gregory, K.J., Lewin, J. and Thornes, J.B. (1987) *Palaeohydrology in Practice*. Wiley, Chichester.

Gurnell, A.M. (1978) The dynamics of a drainage network. *Nordic Hydrology* **9**, 293–306.

Helm, D.G. (1984) The tectonic evolution of Jersey, Channel Islands. *Proceedings of the Geologists' Association* **94**, 1–15.

Hodgson, J. (1990) An investigation of the hydrogeology of Alderney. Unpublished MSc thesis, University of Birmingham.

Horton, R.E. (1945) Erosional development of streams and their drainage basins: hydrophysical approach to quantitative morphology. *Bulletin of the Geological Society of America* **56**, 275–370.

Ijjasz-Vasquez, E.J., Bras, R.L. and Moglen, G.E. (1992) Sensitivity of a basin evolution model to runoff production and to initial conditions. *Water Resources Research* **28**, 2733–2741.

Jamieson, D.G. and Amerman, C.R. (1969) Quick return subsurface flow. *Journal of Hydrology* **8**, 122–136.

Jones, F., Kay, D. and Wyer, M. (1993) Assessment of the bacteriological quality of bathing waters and land drainage to the Jersey coastal zone during the 1993 bathing season. Report to the Public Services Department, States of Jersey. Centre for Research into Environment and Health, University of Leeds, 20pp + appendices.

Jones, M.D. (1982) Palaeogeography and palaeohydrology of the River Severn, Shropshire, during the Late Devensian Glacial Stage and the Early Holocene. Unpublished M.Phil. thesis, University of Reading.

Jones, P.F. (1979) The origin and significance of dry valleys in south-east Derbyshire. *The Mercian Geologist* 7, 1–18.

Jones, R.L., Keen, D.H., Birnie, J. and Waton, P. (1990) *Past Landscapes of Jersey*. Société Jersiaise, 144pp.

Keen, D.H. (1978) *The Pleistocene Deposits of the Channel Islands*. Report of the Institute of Geological Sciences, 78/26.

Keen, D.H. (1981) *The Holocene Deposits of the Channel Islands*. Report of the Institute of Geological Sciences, 81/10.

Kirkby, M.J. (1980) The stream head as a significant geomorphic threshold. In: Coates, D.R. and Vitek, A.D. (eds), *Thresholds in Geomorphology*. Allen and Unwin, London, 53–73.

Lewellen, R.I. (1972) *Studies on the Fluvial Environment: Arctic Coastal Plain Province, North Alaska (2 volumes)*, published by author, Colorado, Littleton.

Mackay, J.R. (1983) Downward water movement into frozen ground, Western Arctic coast, Canada. *Canadian Journal of Earth Sciences* 20, 120–134.

Manton, P.G.K. (1977) The Great Drought of 1975–6. *Société Jersiaise Annual Bulletin* 22, 91–94.

Miller, J.R., Ritter, D.F. and Kochel, R.c. (1990). Morphometric assessment of lithologic controls on drainage basin evolution in the Crawford Upland, south-central Indiana. *American Journal of Science* 290, 569–599.

Montgomery, D.R. and Dietrich, W.E. (1992) Channel initiation and the problem of landscape scale. *Science* 255, 826–830.

Montgomery, D.R. and Foufoula-Georgiou, E. (1992) Channel Networks, some representations using a DEM. *Water Resources Research* 29, 3925–3924.

Morgan, R.P.C. (1971) A morphometric study of some valley systems on the English Chalklands. *Transactions, Institute of British Geographers* 54, 33–44.

Murphey, J.B., Wallace, D.E. and Lane, L.J. (1977) Geomorphic parameters predict hydrograph characteristics in the Southwest. *Water Resources Bulletin* 13, 25–38.

Ovenden, J.C. and Gregory, K.J. (1980) The permanence of stream networks in Britain. *Earth Surface Processes* 5, 47–60.

Patton, P.C. and Baker, V.R. (1976) Morphometry and floods in small drainage basins subject to diverse hydrogeomorphic controls. *Water Resources Research* 12, 941–952.

Pitlick, J. (1994) Relation between peak flows, precipitation and physiography for five mountainous regions in the western U.S.A. *Journal of Hydrology* 158, 219–240.

Richards, K.S. and Anderson, M.G. (1978) Slope stability and valley formation in glacial outwash deposits, North Norfolk. *Earth Surface Processes* 3, 301–318.

Ritter, J.B. and Gardner, T.W. (1993) Hydrologic evolution of drainage basins disturbed by surface mining, central Pennsylvania. *Bulletin of the Geological Society of America* 105, 101–115.

Rodda, J.C. (1969) The significance of characteristics of basin rainfall and morphology in a study of floods in the United Kingdom. In: *Unesco Symposium on Floods and their Computation*, vol. 2, International Association of Scientific Hydrology, 834–845.

Schumm, S.A. (1977) *The Fluvial System*. Wiley-Interscience, New York, 356pp.

Schumm, S.A. and Lichty, R.W. (1965) Time, space and causality in geomorphology. *American Journal of Science* 263, 110–119.

Slaughter, C.W., Hilgert, J.W. and Culps, E.H. (1983) Summer streamflow and sediment yield from discontinuous permafrost headwaters catchments. *Permafrost; 4th International Conference Proceedings*, 1172–1177.

Starkel, L., Gregory, K.J. and Thornes, J.B. (1991) *Temperate Palaeohydrology: Fluvial Processes in the Temperate Zone during the Last 15000 Years*. Wiley, Chichester.

Stephens, N. (1966) Some Pleistocene deposits in north Devon. *Biuletyn Peryglacjalnye* 15, 103–114.

Tarboton, D.G., Bras, R.L. and Rodriguez-Iturbe, I. (1992) A physical basis for drainage density. *Geomorphology* **5**, 59–76.

Trainer, F.W. (1969) *Drainage Density as an Indicator of Baseflow in Part of the Potomac River*. United States Geological Survey Professional Paper 650C, C177–C183.

Uchupi, E. and Oldale, R.N. (1994) Spring sapping origin of the enigmatic relict valleys of Cape Cod and Martha's Vineyard and Nantucket Islands, Massachusetts.*Geomorphology* **9**, 83–95.

Uma, K.O. and Kehinde, M.O. (1992) Quantitative assessment of the groundwater potential of small basins in parts of southeastern Nigeria. *Hydrological Sciences Journal* **37**, 359–74.

van Everdingen, R.O. (1974) Groundwater in Permafrost regions of Canada. In: *Permafrost Hydrology, Proceedings of a Workshop Seminar, 1974*, Ottawa, Canadian National Committee, International Hydrological Decade, 83–94.

Van Nest, J. and Bettis, E.A. III (1990) Postglacial response of a stream in Central Iowa to changes in climate and drainage basin factors. *Quaternary Research* **33**, 73–85.

Waters, R.S. (1966) The Exeter Symposium. *Biuletyn Periglacjalny* **15**, 123–149.

Winslow, J.H. (1966) Raised submarine canyons: an exploratory hypothesis. *Annals Association of American Geographers* **56**, 634–672.

Woo, M.K. (1976) Hydrology of a small Canadian high Arctic basin during the snowmelt period. *Catena* **3**, 155–168.

Woo, M.K. (1983) Hydrology of a drainage basin in the Canadian High Arctic. *Annals, Association of American Geographers* **73**, 577–596.

Woo, M.K. and Steer, P. (1982) Occurrence of surface flow on Arctic slopes, southwestern Cornwallis Island. *Canadian Journal of Earth Sciences* **19**, 2368–2377.

Woo, M.K. and Steer, P. (1983) Slope hydrology as influenced by thawing of the active layer, Resolute, NWT. *Canadian Journal of Earth Sciences* **20**, 978–986.

Younger, P.L. (1989) Devensian periglacial influences on the development of spatially variable permeability in the Chalk of southeast England. *Quarterly Journal of Engineering Geology* **22**, 343–354.

Zalesskiy, F.V. (1976) Flash flood formation in permafrost regions. *Soviet Hydrology, Selected Papers* **15**, 95–97.

5 Processes of Channel Planform Change on Meandering Channels in the UK

J.M. HOOKE

Department of Geography, University of Portsmouth, UK

Change in Platform (handwritten annotation)

INTRODUCTION

Changes in river planform are one of the most studied aspects of changing river channels, partly because of the relative ease of acquiring evidence through maps and aerial photographs and also because of the scale of change, the impact on fluvial landforms and development, and the practical problems created by bank erosion and channel mobility. Meandering channels have received the most attention, especially in Britain, not surprisingly since they are the most common type of channel pattern. Research into braided channels has increased in recent years both in process studies and in analysis of longer-term changes (Best and Bristow, 1993). Some component processes of change are common to both types of channel.

Changes in channel planform take place by erosion of the banks, by deposition within the channel and by some chute flow or avulsion, involving switching of channel position. The processes effect lateral movement and changes in form. Over the past 20 years considerable understanding has been gained of the processes of bank erosion, there have been some advances in understanding characteristics of deposition and some progress has been made in elucidating the nature of meander changes. The aim of this paper is, first, to examine the component processes, their spatial and temporal occurrences and how they bring about changes in form. Secondly, the paper examines meander patterns and evaluates the extent to which movement in individual meanders and in sets of meanders can be understood from the components, and the extent to which a more holistic approach needs to be taken.

It has emerged from work worldwide on meandering rivers that the most active rivers, those moving laterally at the highest rates, rarely exhibit a stability of form or morphological characteristics, rather they show progressive development (e.g. Hickin, 1983; Hooke, 1984). A key question still is the extent to which such behaviour is allogenic and the extent to which it is autogenic. Since channel planform is one of the most rapidly changing features of a river channel, it can be regarded as highly sensitive. Analysis of planform changes should therefore provide the opportunity to quantify sensitivity and to put figures on time-scales of adjustment to various disturbances and impulses in the fluvial system.

Changing River Channels. Edited by Angela Gurnell and Geoffrey Petts.
© 1995 John Wiley & Sons Ltd.

BANK EROSION

Various mechanisms of bank erosion have been identified (e.g. Hooke, 1979; Thorne and Tovey, 1981; Hey, 1982; Lawler, 1992). The terminology and classifications vary slightly but generally the types can be recognised by the morphological evidence of the mechanism of failure and the associated features. A wider classifi-

Table 5.1 Mechanisms of bank erosion

Process	Description	Reference
A. Bank weakening		
Pre-wetting	Wetting downwards by the precipitation front, inwards from the river or upwards by the water table. Seepage can lead to piping at sand/clay junctions	Hooke (1979), Wolman (1959)
Desiccation	Condition of high temperature and low moisture. Effects are debated but can lead to cracking and spalling	Lawler (1992) Ellis (1993)
Freeze-thaw action	Can be important in making bank more susceptible to erosion. Needle ice can produce a dusting of aggregates. Only dominant in small systems (<85 km^2). Permafrost and ice rafting effects in cold climates	Lawler (1986, 1987), Church and Miles (1982), Klimek (1989)
B. Fluvial erosion/entrainment		
Direct removal	In non-cohesive material entrainment of coarse material depends mainly on competence of flow and stress on bank; removal during peak flows. Resistance affected by vegetation, composition and state of materials. Cohesive material processes more complex. slaking of some materials	Thorne and Lewin (1979), Hooke (1979, 1980)
C. Mass Failure		
Shear, beam and tensile cantilever failures	Occur mainly on composite banks. Initial fluvial entrainment of basal coarse materials then failure of upper, fine blocks	Thorne and Tovey (1981), Hooke (1979)
Shear failures (shallow, planar, slab, rotational)	Occur in cohesive materials associated with high/increased bank angle and/or bank height, high moisture content and pore pressures. Failure after peak flow. Tension cracks may develop. Geotechnical properties of materials important	Thorne and Osman (1988)
D. Other		
Waves	May be wind or boat induced	Nanson *et al.* (1994),
Trampling	Direct effects of cattle or people	Trimble (1994)

cation of types of change with their field indicators has been used by Gregory *et al.* (1992). The various researchers have analysed the conditions under which particular mechanisms (Table 5.1) occur and some rates of erosion associated with specific mechanisms are cited in the literature. The morphological evidence combined with this information can be used as guidance in the field to assess processes and levels of activity. This is very valuable in reconnaissance, especially of practical problems associated with bank erosion, as recommended, for example, in relation to possible impacts of boats on the Thames (Hooke *et al.*, 1991). Of course, further analysis of rates should usually be carried out using historical maps and aerial photographs (Hooke and Kain, 1982; Hooke and Redmond, 1989a) to give a quantitative perspective at a site.

The results that emerge from analysis of the mechanisms, conditions and rates of bank erosion are that the highest rates of erosion in meandering streams are achieved by fluvial entrainment, mostly occurring by beam (toppling) failure. This mechanism therefore brings about the greatest amounts of planform change and is the process typical of the most laterally unstable meandering channels, such as the upper Severn and middle Dane in Britain, both of which have been recommended as fluvial SSSIs. The process is dominant on such wide channels with relatively low banks of composite stratigraphy and high sand–silt content of the fine unit (Figure 5.1a). In spite of the modelling, e.g. by Thorne and Osman (1988) and the well-known factors influencing slope stability in general, the actual occurrence of shear failures is still not fully understood. There are obviously some sites which are prone to this type of failure, notably where banks are high and material has some degree of cohesion, but sliding failures can sometimes occur on the same low, alluvial banks as toppling failures (Figure 5.1b). Field observation as well as theory indicates that this is influenced by the pore water pressures and the particular combinations of rates of rise and fall of the river level and the wetness of the banks. However, distinction should be made between the different types of sliding failure and also shear cantilever failure (Thorne and Tovey, 1981).

In terms of direct contribution to recession of river banks, frost action and desiccation cracking would appear to be negligible except where conditions are very severe, e.g. the effects of ice rafting as identified by Klimek (1989) in Poland, and drying of banks in semi-arid areas (Andrews, 1982; Ellis, 1993). The impact is greatest on banks with high clay content because these materials are most susceptible to these processes and such banks are resistant to fluvial action. The indirect contribution of frost action, desiccation and subaerial weathering is more difficult to assess without direct comparison of periods of differing conditions at particular sites and detailed data on individual events. Such data are still rare in the literature but the preparation effects of these processes may be significant (Lawler, 1992).

The spatial variability of bank erosion can be examined at various scales. At the basin scale, although overall rates of erosion generally increase with catchment size (Hooke, 1980), considerable evidence indicates that the highest rates of lateral mobility occur in the middle reaches of drainage basins, particularly in the piedmont zones (Newson, 1981; Lewin, 1987; Hooke and Redmond, 1989b). Bank erosion rates are highest where the force on the bank most greatly exceeds the resistance of the bank. In these middle reaches stream power is still quite high, indeed may be at

Figure 5.1   (a) Bank erosion by beam (toppling) failure. (b) Shear failures on alluvial bank

its maximum (Nanson and Young, 1981; Graf, 1983, Lawler, 1992), but material is alluvial and usually highly erodible because of maximum sand content in these reaches. Downstream, gradient declines and clay content tends to increase. However, Lewin (1987) did not find a high correlation between meander mobility and stream power for Welsh streams. At the scale of a reach or individual bends then slight variations in the resistance of banks affect rates of erosion. Variations in material can be found, especially in these most mobile channels because the mobility itself leaves a legacy of old channel fills in the floodplain, so there is a feedback effect. In the early classic literature on the Mississippi (Friedkin, 1945) the effect of clay materials in old ox-bows, inhibiting migration of meanders, was demonstrated. Detailed observation of several smaller, mobile channels in Britain indicates that in such environments where the materials are coarser, old channels may have the opposite effect and be more erodible. This is particularly the case where the present channel is eroding relatively new floodplain which is also less consolidated and has lower organic content. However, on active alluvial bank sections in Britain what is perhaps more remarkable is the homogeneity of material and the degree of systematic variation in rates, as is discussed below.

The effect of vegetation on bank erosion and resistance is still proving difficult to quantify, in spite of considerable research efforts (Hooke, 1977a; Amarsinghe, 1993; Masterman and Thorne, 1994). The impact of trees is complicated (Pizzuto, 1984) but several researchers (e.g. Nanson and Hickin, 1986; Murgatroyd and Ternan, 1983) observe erosion on banks under trees because of lack of grass. The evidence from both the mechanisms of erosion, where fronds of grass and soil are often left at the top of banks, and of sparse data on rates of erosion indicate that grass has a significant effect in reducing erosion and this is substantiated by evidence of banks on which fallen blocks have become vegetated and stabilised. It has been found at sites being monitored on the River Dane in north-west England that, on banks that were already highly eroding under grass, the rate increases even more when the grass is ploughed (unpublished data). Similar effects are observed on the Severn (Hooke et al., 1994) and the current increase in arable cultivation on floodplains in Britain therefore has severe implications for erosion. In general, vegetation is recommended as a method of bank protection.

Within a bend, certain patterns of locus of erosion are apparent but some temporal variability in location of maximum erosion also occurs. The classic pattern of erosion in meander bends is of maximum erosion just downstream of the apex of the bend. This is true in bends that are dominantly migrating in the downstream direction, but in bends that are growing in amplitude and amplifying across the valley then erosion must be at a maximum at the apex. That such meander behaviour is common and can be seen to be part of a longer-term sequence of meander development will be discussed in a later section. Erosion can also be spatially variable on a shorter time-scale as shown by examination of impacts of individual, frequently occurring erosion events on a series of river banks (Figure 5.2). These examples from a stream in south-west England are banks on which beam (toppling) failures are dominant and the pattern is associated with the block failures. The data also illustrate the high amounts of erosion achieved in individual events, several of which may occur in a year (Hooke, 1979).

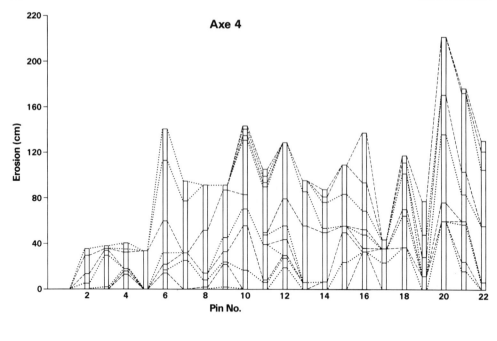

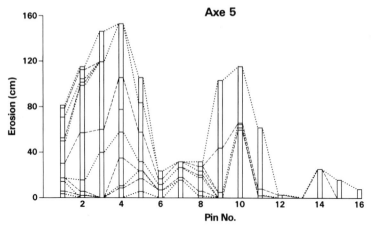

Figure 5.2 Amounts of erosion measured on erosion pins along banks at two sites on the River Axe, Devon. Dotted and dashed lines separate erosion in successive events over a two-year period

DEPOSITION

Deposition occurs where the force of flow is no longer sufficient to carry the size or amount of sediment in transport. There are various loci within a channel in which deposition commonly occurs and the combination of locations and shapes of bars has been the basis of classification of types of bars. However, the terminology is confusing and several authors have used different terms for the same features. No single classification has been widely adopted (Church and Jones, 1982). Some major

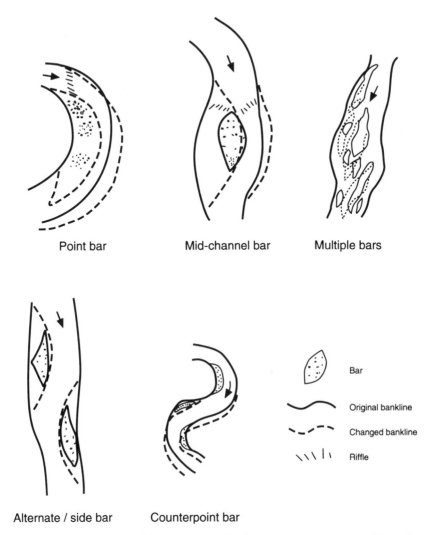

Point bar   Mid-channel bar   Multiple bars

Alternate / side bar   Counterpoint bar

Figure 5.3 Common types of bars and associated changes on active gravel-bed rivers

types of bars commonly found on active gravel-bed rivers in Britain and the way in which planform changes in association with them, are illustrated in Figure 5.3. This evidence is taken from mapping at intervals of time on various streams though, as Ferguson (1993) warns, full understanding of changes in bars should be based on three-dimensional and subsurface analysis.

Point bars are characteristic of meandering streams (Figure 5.4a) and tend to extend in the channel direction and downstream, generally keeping roughly parallel with the eroding bankline. The sediment is generally graded in both the downstream direction and horizontally with height from the waterline. The bars normally grade into the floodplain but vertical accretion may take years before the bar surface

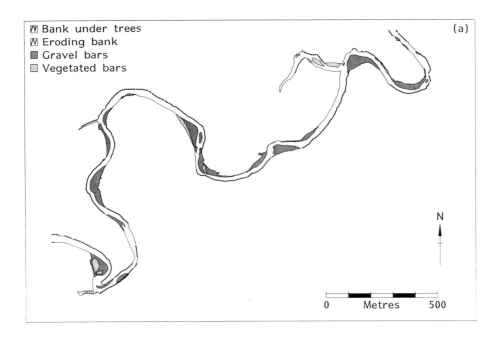

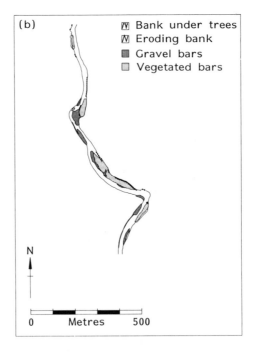

Figure 5.4 Reaches on the Upper Severn, near Caersws, mid-Wales, exhibiting (a) point bars, (b) side bars

reaches the height of mature floodplain, resulting in faceting of young floodplains. The point bars can be closely associated with riffles, the face being continuous. Thompson (1986) and Whiting and Dietrich (1993a and b), for example, consider them as one unit but in other cases the two can be separate and act independently as confirmed by bend theory (Rhoads and Welford, 1991).

Single *mid-channel* bars are a much more common component of active meandering channels than is often recognised. Mid-channel (medial) bars can be considered as the component bars of braiding but are usually multiple in that case. The bars are usually lozenge (diamond) shaped or lobate and often begin as the emergent, central parts of riffles. The sequence of evolution from framework gravel bars with flow both sides to vertical accretion and attachment to the floodplain in meandering rivers has been traced by Hooke (1986) and can be very rapid on active gravel-bed streams.

Alternate bars are a commonly recognised type occurring in straight channels and as part of the early sequence of development of meanders. They have been demonstrated in both experiments (e.g. Ackers and Charlton, 1970) and in field reaches after straightening (Lewin, 1976; Babinski, 1992). They occur in both gravel- and sand-bed channels.

Side bar is a term that has been used in various ways. As illustrated in Lewin (1978) they are not common in active streams in Britain but there is a type of bar which occurs as an attached bar in straight sections. It could be the attached stage of an alternate bar but in any field situation the evolution is not always clear and some of these bars are also rather more irregular in position. Illustrations of side bars are given in Figure 5.4b. These are in a reach which was formerly straightened. In a period of aggradation a series of side bars may coalesce to form a bench as appears to have happened on the Upper Severn near Dolwen.

Counterpoint bars (Lewin, 1983) or concave bank benches (Page and Nanson, 1982) have been identified as occurring in rapidly migrating river channels. They occur in slack-water areas where the channel is migrating especially rapidly and/or where a tight corner, e.g. up against a bedrock cliff, has developed. They may start as elongate, unattached bars but eventually form an attached bench form. Similarly, near-circular bars occasionally form in the apex of confined meanders.

Tributary confluence bars may take on the form of side or mid-channel bars. They develop in the separation zones of confluences, in single or multiple formations depending on the sediment loads and flow dynamics. Typical morphologies have been identified, for example by Best (1986).

One important aspect of bar development, highlighted by Lewin in 1978, is the need to quantify the time-scale of development of these mesoforms. This was investigated for mid-channel bars on one river by Hooke (1986) in which it was shown that a bar may go through the whole cycle from emergent gravel framework to complete incorporation in the floodplain in as little as a decade. Vegetation growth is an important component in the process of bar evolution though the extent to which deposition of fines allows growth and the extent to which growth encourages deposition remains to be disaggregated. Undoubtedly, vegetation growth helps to stabilise a bar though deposition can continue on top. Some bars do not become attached and incorporated in the floodplain and are rather longer-lived. Circumstantial historical evidence implies that this may be related to supply especially

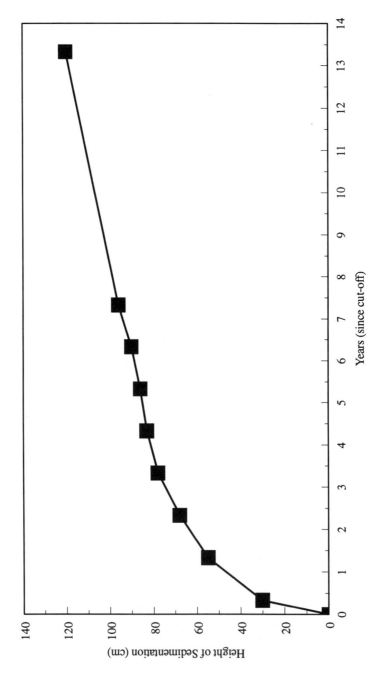

Figure 5.5 Rate of vertical accretion in a cutoff, River Bollin, north-west England

where bars have developed because of specific input of sediment, e.g. as a result of mining or tributary input.

Bar evolution at one scale and structures of deposition at a smaller scale have both been examined, but rarely have the processes and rates of accretion been examined in detail to relate the two parts together. Rates of accretion are commonly quoted averaged over long periods and derived from floodplain sites and so including overbank sedimentation. In the work on the mid-channel bars (Hooke, 1986) it was shown that rates of vertical accretion can reach 0.25–0.50 m yr^{-1}. Measurements of deposition which has taken place in several cutoffs from inception on the rivers Bollin and Dane in north-west England show how rapid infill may be where sediment supply is high and how rates vary over time. On these streams there is an abundant supply of both sand and gravel-sized material. An initial plug of coarse sediment into the upstream entrance to the cutoff channel can take place extremely rapidly; large deposits of gravel butting against fresh eroding bank faces have been observed. The evidence from one cutoff on the Dane is that the initial plug must have been deposited in a matter of hours, if not minutes. On the River Bollin after a neck cutoff in November, 1980, sand overlying gravel had been deposited to a height of 0.30 m above low water level within four months. The area at the cutoff entrance/point bar had accreted to 75% of the full height of the floodplain, within 6½ years. The annual rate of vertical accretion measured over 14 years is 8.6 cm yr^{-1}, but rates are high at first then decrease (Figure 5.5). Channel deposition on bars and in cutoffs on these gravel-bed rivers generally results in a distinct two-part alluvial unit, comprising a base of coarse gravel and an upper fine layer, three to four times as thick. Occasional gravel layers are found higher up resulting from large flood events.

CHUTE FLOW, CUTOFFS AND AVULSIONS

Cutoffs, avulsions and changes produced by chute flow have received much less attention than the more systematic processes of change. Chute flow can occur at various morphological scales from across bars, across bends to along whole reaches. The occurrence of chute flow, cutoffs and avulsions is obviously influenced by erosion and deposition and the position of scour and accretion. The occurrence is also influenced by curvature and pattern of current, height of different zones of the floodplain and the pattern of overbank or bar-top flow in floods. Various morphological types of cutoff have been identified in meandering rivers including simple neck and chute cutoffs, mobile bar, multiloop chute and multiloop neck cutoffs (Lewis and Lewin, 1983) and bend flattening (Matthes, 1948; Erskine et al., 1992).

Ferguson (1993) describes chute cutoff as an erosional process in which headwards incision occurs by flow taking a short cut across a bar. However, Ashmore (1991) argues that the increase in bar-top flow to initiate incision is often due to aggradation of the outer channel or bar rim. Large-scale avulsions have long been observed on large meandering rivers such as the Mississippi. Overbank flood flow can be diverted from the main channel for many kilometres, though more rarely does this become the main flow. Thus, chute flows at the local scale and avulsions tend to take place where the flow becomes poised over the surrounding surface and where curvature is such that at high flow the flood waters spill over to take a straighter course. However, the actual timing of chute flows, cutoffs and avulsions is more difficult to predict. The spatial and temporal occurrence of these various types of change have still received little systematic analysis, with notable exceptions (Lewis and Lewin, 1983).

FLOW PATTERNS AND BEND DEVELOPMENT

The position and amount of bank erosion which takes place depends on the forces on the bank relative to resistance. It has already been shown that it is primarily the direct fluvial forces which are of the greatest importance in effecting the greatest change in planform. The process of beam failure is dependent on entrainment of the coarse material at the base of the bank and it is therefore this threshold that exerts the main control on that type of erosion (Thorne, 1982; Nanson and Hickin, 1986). Whether a failed block remains at the base of the bank or is removed depends on the forces and sediment relations at the base of the bank (Thorne, 1982). It is evident, therefore, that the forces of flow at the base of a bank are important and these are influenced by the pattern and strength of currents in the bend or section.

Debate still surrounds the question of whether it is the primary or secondary flows which influence the pattern of erosion and meander development (Rhoads and Welford, 1991). Much work in the late 1970s and 1980s examined patterns of secondary flows in meander bends (Hey and Thorne, 1975; Bathurst, 1979; Bridge and Jarvis, 1982; Dietrich, 1987). Some differences exist between the models proposed but results of the work included evidence that the circulation differs between pools and riffles and that secondary flows have an influence on shear stresses and therefore on erosion and deposition. It was suggested (Bathurst et al., 1977) that bank erosion occurs adjacent to pools in meanders and that development of a strong bank cell with upward flow against the bank is an integral part of the mechanism of bank erosion. Most of the British workers emphasise the importance of the secondary flows (e.g. Thorne, 1982; Thompson, 1986). Much of the North American literature appears to agree that it is the primary velocity which is the main control on bank erosion, although there was discussion over whether it is the absolute near-bank velocity (Pizzuto, 1984) or the excess near-bank over mean velocity (Parker et al., 1983; Hasegawa, 1989). Howard (1992) has used near-bank velocity in a bank erosion equation as the basis for a meander development simulation model. In 1986, Lapointe and Carson stated that 'although there appears to be a consensus on the relevance of primary velocity patterns to bank erosion, the importance of secondary currents and sediment evacuation has still to be clarified'.

Work by Hooke and Harvey (1983) demonstrated that in many active meanders a sequence of development takes place in which, after a phase of migration, bends grow in amplitude; this is followed by a stage of compound development and associated with this is development of an extra riffle in the apex of the bend. Hooke and Harvey (1983) attributed this to modification of the secondary flows as the path length of the bend increases. These ideas were pursued in more detail by Thompson (1986). Further analysis of the pattern of flow and position of erosion and deposition in 100 loops on the River Dane from detailed field mapping has indicated that, although the general model of bend development and riffle formation applies, erosion is not always consistently associated with a pool and therefore with an upward-flowing secondary cell as hypothesised by Hey and Thorne (1975). Examples of individual bends (Figure 5.6) indicate that erosion can occur adjacent to riffles and this situation has persisted for several years, indicating it is not simply a lag effect. This shows that the particular pattern of secondary flow is not a prerequisite for bank erosion. Observation and measurement of flows in bends (Hooke, 1990) appear to confirm the hypothesis that it is the pattern of primary flow which is the major control on erosion. Primary velocities are much higher than secondary flows though can be slightly modified by these, and it is the strength of the primary flows adjacent to the bank which is crucial. Maximum erosion takes place where the primary current impinges directly on the bank and velocities are highest. Local velocity is influenced by the cross-sectional form and acceleration of flow, by the local gradient which is strongly influenced by the riffle at the entrance to the bend, and by the curvature of flow off the riffle or into the section. Monitoring of bends over a period of a decade has shown that the position of impingement can alter with time but that this tends to follow a consistent pattern resulting in the identified sequence of morphological changes.

In some sections it has been observed that erosion may suddenly increase in rate; this occurs in two particular circumstances. One is on straight or low curvature sections where erosion is initially negligible. Secondly, on moderate curvature bends which are migrating and then start to grow, erosion increases. This latter is now a widely recognised phenomenon (Hickin and Nanson, 1984; Hooke, 1991) and must involve a shift in locus of erosion, but the mechanics and timing remain to be clarified. Case studies indicate that the first situation usually involves alteration of the form of the riffle at the upstream end of the section such that the primary current becomes more skewed and impinges on the bank. What causes the change in riffle morphology and flow pattern is not clear but it could be associated with deposition at the end of the point–bar riffle unit. Carson (1986) indicates that this may be so from study of changes on high energy meandering streams in New Zealand. An example of this bend development and accelerated erosion is given in Figure 5.7. The section comprises a 2 m-high alluvial bank with a riffle at the upper end. In the early 1980s the bank was bare or only lightly covered in vegetation and erosion rates were of the order of a few centimetres a year. The point bar opposite was poorly developed and not very active. In 1987/8 bank erosion suddenly became much more active; large beam and shear failures took place. The erosion has resulted in the section becoming much more curved (Figure 5.7). Bushes at either end of the section have been removed and the maximum bank recession is about 12 m. Erosion is still continuing at a high rate.

Work by Whiting and Deitrich (1993a and b) to examine experimentally the mechanisms and processes of bar and pool formation in bends indicate the important influence of the riffle and bed topography and curvature at the entrance to the bend. They suggest that multiple bars in large amplitude bends are mainly due to an alternate bar-like instability, confirming the theoretical analyses of bar theory (Parker and Johannesson, 1989; Rhoads and Welford, 1991).

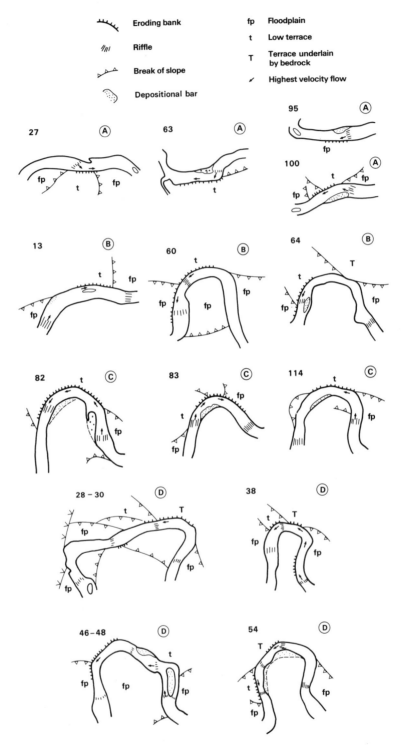

Figure 5.6 Examples of meanders showing position of bank erosion in relation to riffles and maximum primary velocity, River Dane, north-west England. A–D = types of meander

SEDIMENT SUPPLY AND BUDGETS

A key question in terms of understanding the processes of planform change is which comes first, the erosion or the deposition. Does erosion occur which then decreases shear stresses on the inside of the bend and so causes deposition, or does deposition occur which pushes the current over, thus impinging on the bank and causing erosion? The evidence from the case study above and the experimental work implies that it is slight changes in the bar form which precede erosion. Similarly, evidence from various experiments on meander development identify shoaling as an early stage which then alters the currents. However, considerable evidence from field studies indicates that formation is closely associated with bank erosion and that bank erosion is the major source of sediment in active meandering, floodplain reaches (Newson and Leeks, 1987). For erosion and removal of sediment to take place there must be excess energy and a sediment deficit and the sediment budget is illustrated by the basal end point removal situation (Thorne, 1982). Coarse sediment is only transported to the next bar downstream (Arkell et al., 1983) but in time the effects of sediment excess or deficit are transmitted.

Sediment supply obviously influences channel pattern as demonstrated by Schumm (1981) and more recently in both historical (e.g. Passmore et al., 1993) and contemporary (e.g. Harvey, 1991) transformations from meandering to braided planform. Recently, both Leopold (1992) and Friend (1993) have emphasised the importance of coarse material in influencing the morphology of river channels. Experimental studies (Hoey, 1992) and field measurements (Goff and Ashmore, 1994) have shown how pulses of sediment moving down braided channels actually cause change independently of discharge.

Evidence from the River Dane suggests that, although minor variations in riffle and bar configuration, may take place, generally erosion precedes deposition in a bend resulting in a phase of widening before deposition catches up. It is possible this is a reflection of the sequence of flow events (see below) but it would appear that the overall impetus for erosion is excess stream power. The general assumption is that the channel operates as an integrated and highly connected system and that sediment is transmitted through the system. While this may be true of suspended sediment the localised nature of coarse sediment means that it only applies over the long-term. Adjacent reaches of channel can behave in very different ways as illustrated in one example in Figure 5.10a. In another example on the Dane, there is one straight, stable reach which at both ends is juxtaposed to highly active and sinuous reaches (Hooke, 1987b). The extent to which coarse material is transported through this reach is unknown but morphological evidence (lack of bars and riffles) and theoretical calculations of sediment transport indicate it is low. It would therefore appear that in these active, gravel-bed meandering rivers erosion which takes place upstream usually supplies sediment to bars immediately downstream at least in the short-term. Storage then takes place in the bars for varying lengths of time.

EVENTS AND FLOW REGIME

The processes discussed and therefore the channel changes are mostly not continuous but take place in association with flow events above the thresholds for ero-

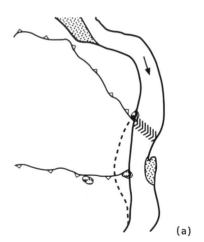

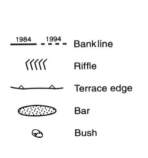

<u>1984</u> _ _1994_ _	Bankline
⟨⟨⟨⟨⟨	Riffle
⟋△⟍△⟋	Terrace edge
◌◌◌◌	Bar
◌	Bush

(a)

(b)

(c)

(d)

(e)

Figure 5.7 Rapid erosion and development of a bend on River Dane, north-west England: (a) map, (b) 1987, (c) 1988, (d) 1991, (e) 1994

sion and sediment transport. Consideration of the magnitude–frequency of events and of the nature of the flow regime is always an important component of analysis of fluvial activity and landform development. The incidence of competent and effective events varies with physiographic environment and good data on thresholds and event frequency are lacking for many environments.

The frequency of erosion varies with the erodibility of the material and the flow regime. On the active, perennial meandering rivers where bank erosion rates are high, such as the examples used here, bank erosion can take place several times a year. Thorne and Lewin (1979) indicate that, on a very mobile section of the Upper Severn, the threshold for coarse sediment entrainment and therefore for bank erosion by beam failure is about one-third bankfull level. Similarly, Hooke (1980) found that, for streams in Devon, erosion occurred at flow levels much below bankfull though amount of erosion increases generally with discharge. Monitoring of the

River Dane since 1980 provides opportunity for evaluation of the differing frequency and sequences of event from year to year. In the early 1980s the River Dane and neighbouring rivers were highly active with many bars covered with large cobbles. Since then most bars have become much finer in superficial composition and many have become stabilised by vegetation. That this is a trend and not the stage of evolution of individual bars is apparent from the large number of bars on which this is evident. Preliminary analysis (Hooke, 1990) indicates some relation to incidence of large floods but that the relationship is more complicated than this. Yearly observation of the river has shown that in years with high flows erosion is high and results in a net widening of the channel but in years of lower peak flows the events transport sand rather than cobbles and deposition becomes dominant. The bars are built up and the channels narrow. Erosion processes have a strong seasonal component due to the influence of vegetation and of soil moisture levels (Hooke, 1979; Thorne, 1990) and high summer floods are not nearly as effective as the same magnitude floods in winter. Now that many of the channel bars and some of the eroding banks on the Dane and other rivers have become stabilised by vegetation it is expected that the threshold has been raised and that it will take a larger flood to destabilise them again.

The rates of activity on these streams make them highly sensitive to flow variations. This sensitivity and the short time-scale in which significantly different responses can be detected is also illustrated from the River Severn (Hooke *et al.*, 1994). Evidence of channel planform changes using maps and aerial photographs of various dates has been compiled for the Upper Severn Geological Conservation Review site, near Caersws in mid-Wales. Comparison of movement of the channel and areas of erosion and deposition show that rates of activity were significantly higher in the period 1984–1992 than in the preceding period, 1973–1984. This coincides with change from a period of relatively low and infrequent peak flows in the 1970s to increased flow levels in the 1980s. This is due to rainfall fluctuations and the resulting changes in discharge regime overwhelming effects of river regulation and land use change (Higgs and Petts, 1988). The sensitivity of these active meandering reaches mean that they are likely to provide the earliest indications of impact of global warming and therefore long-term monitoring is recommended.

In other situations it is the large, infrequent floods which effect channel changes, notably in upland environments (e.g. Wolman and Gerson, 1978; Harvey, 1986; Newson and Macklin, 1990) and in semi-arid and arid areas (e.g. Burkham, 1972). Much of the less systematic change, particularly that which takes place by chute overflows and avulsion, occurs in such events. Such changes are much less predictable but can radically alter pattern.

MORPHOLOGICAL CHANGES

The outcome of these processes and the nature of changes in planform can be examined using the longer-term evidence of maps and aerial photographs. Compilation of the historical evidence, commonly available for about the last 150 years in Britain, allows the sequence of movement and changes in form to be elucidated and, if changes are large and maps sufficiently accurate, for rates of movement to be

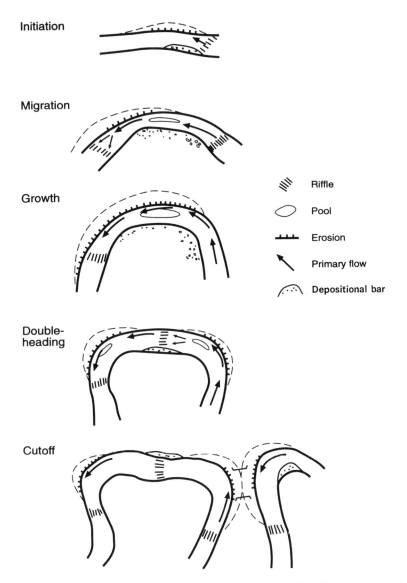

Figure 5.8 Qualitative model of sequence of changes in meander form typical of active gravel-bed rivers

measured (Figure 5.10). Analysis of the sequences on very many reaches of active meandering streams in Britain reveals consistent trends in the evolution. A qualitative model of the sequence of development of individual meander bends is presented in Figure 5.8. The sequence is as follows:

1. New bends—these develop from straight reaches; the current becomes directed against the bank and the bend develops rapidly.

2. Migrating bend—after a certain stage, a moderately curved bend is formed. It usually has a riffle symmetrically across the channel at the entrance to the bend and flow is direct against the bank just downstream of the apex.
3. Growth bends—many exhibit a later stage of growth. These bends are often smoothly curved and the main current hugs the outer bank for much of its length (Figure 5.9).
4. Double-headed bends—the flow becomes deflected against the bank upstream of the original apex and erosion is initiated there, leading to development of a lobe. The bends become too long for sediment to be carried right through the pool and a second riffle is deposited in the central zone.
5. Cutoff—this occurs by continuation of erosion in the lobes such that neighbouring lobes intersect.

Hooke and Harvey (1983) showed how, associated with the compound development, is the production of an extra riffle, and this has been reproduced experimentally (Whiting and Dietrich, 1993a and b). The applicability of this model is exemplified by one bend on the River Dane (Figure 5.9). The phase of rapid growth can be seen from the historical evidence. In 1983, as a result of the analysis of the relationship between bend form and types of behaviour (Hooke and Harvey, 1983), it was predicted that this bend would shortly begin to become compound and a riffle develop in the apex region. In 1987 this started to become evident, with suddenly increased erosion on the upstream side of the apex. This continued over the next few years and by 1992 a shallowing in the centre of the apex was visible. This has now developed into a distinct riffle, erosion is continuing on the upstream and downstream sides, but the bank in the centre part is stabilising.

Hickin and Nanson (1984) and later Hooke (1987a) and Biedenharn et al. (1989) have demonstrated that rates of migration or erosion increase rapidly once a certain curvature is reached. Why, when the bend is extending in length and thus dissipating energy over a greater distance, such acceleration should occur is still not satisfactorily explained but the model now appears to be of such wide applicability that it is suggested that the sequence is inherent in meander development. The exact form of the meanders and the rate of development are affected by the composition and resistance of the banks but the systematic changes emerge above these random effects.

The results of the work on unstable meanders indicate that such meanders exhibit no signs of attaining a stable form, an assumption used in equilibrium relations (e.g. Ackers and Charlton 1970); rather the development is continuous. Hooke and Redmond (1992) have suggested that cutoffs are part of this same sequence of meander development and that, therefore, a decrease in sinuosity should be incorporated; different explanations for such behaviour of channels need not be invoked. What is curious is that different streams or reaches of streams in Britain seem to be at different stages of this sequence and can be seen to be exhibiting an overall trend in behaviour. It would be expected that even if individual members show this sequence, meanders would be at different stages so that overall there would be statistical stability of characteristics. The Dane has shown a net increase in sinuosity of 10% in the period 1840–1984 but since 1984 several cutoffs have taken place (Figure 5.10a). On the neighbouring River Bollin recent evidence suggests sinuosity and bend develop-

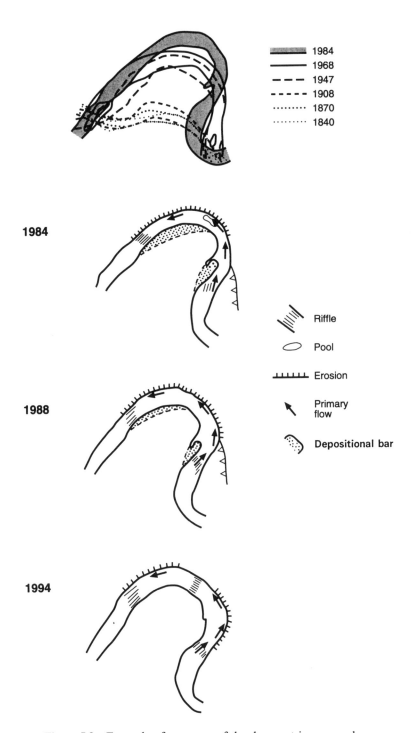

Figure 5.9 Example of sequence of development in a meander

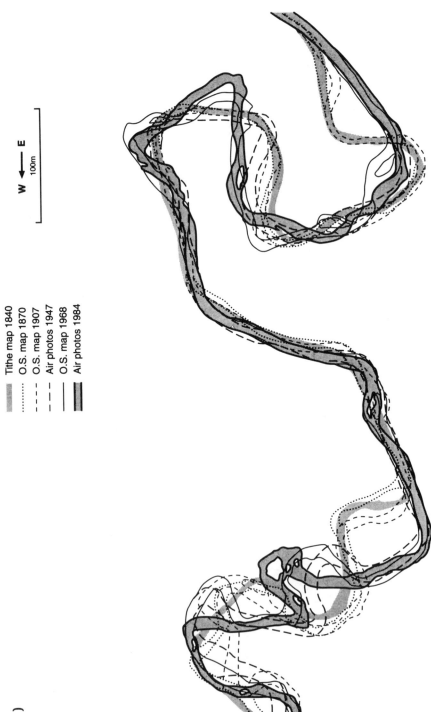

Figure 5.10 (a) Historical sequence of change for a section of the River Dane, Cheshire illustrating a range of stability, morphology and type of change

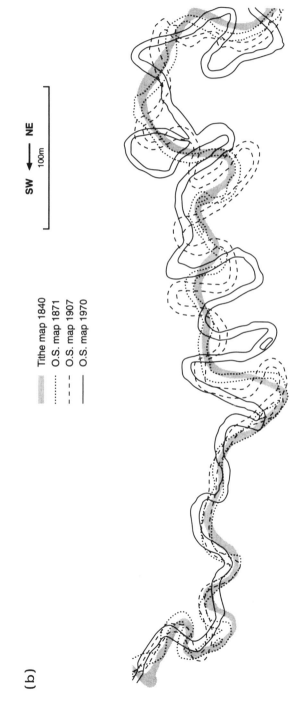

Figure 5.10 (b) Historical sequence of change, River Bollin, near Wilmslow, Cheshire

(b)

Tithe map 1840
O.S. map 1871
O.S. map 1907
O.S. map 1970

SW ◄──── NE
100m

ment can be seen to have preceded the recent cutoffs (Figure 5.10b). Many of the bends on Devon streams studied by Hooke (1977a and b) showed an increase in amplitude and curvature over the historical period. At these sites various possible causes of change, both anthropogenic and natural, were investigated to explain this increase in erosion and migration rates (Hooke and Kain, 1982) but no satisfactory explanation was found. On another reach of the Bollin from that illustrated (Mosley, 1975), and at many other sites (e.g. Thompson, 1984), changes in meander behaviour have been attributed to human alterations of the flow regime. The sensitivity of such reaches to changes in input have already been discussed but the possibility now arises that the degree of autogenesis in these meander changes is much greater than formerly suggested. Such autogenic and markedly non-linear behaviour may be suitably analysed within the context of chaos theory (Furbish, 1991; Hooke, 1991; Montgomery, 1993). Hooke (1991) has suggested that the time-scale for development of the bend sequence from initiation to cutoff, and for whole reaches from increase sinuosity to decrease and then fluctuation, depends on the rates of erosion and level of activity. The River Teme in the Welsh Borderland can be seen to have reached the stage of cut off more quickly than the Bollin, which in turn is eroding relative to its size, and therefore dimensions of meanders, more rapidly than the Dane.

The fact that change in river meanders appears to have such a high degree of consistency means that it should be possible to predict movement and construct viable predictive models. Various predictive models have been formulated and are becoming very sophisticated (Furbish, 1991; Howard, 1992). Most of these are based on theoretical assumptions of flow principles. Experimentation and numerical analysis is indicating increasingly that fundamental instability of flow is the underlying cause of meandering (Rhoads and Welford, 1991; Whiting and Dietrich 1993b). A simple model of rate of erosion in relation to curvature can replicate actual changes quite well and varying resistance to erosion can be incorporated which results in modification and skewing of the patterns, as in reality. What is still difficult to model is the behaviour of a sequence of bends with the interactions between them. It has been shown that curvature of one bend affects another but also that reaches can behave almost independently of one another. Rapidly eroding bends have a push factor and cutoffs can be seen to have a pull factor as well as propagating slope steepening upstream. The challenge now is to quantify the effects of site conditions and superimpose these on the overall models to give us more realistic predictions.

CONCLUSIONS

In the 18 years since the seminal volume *River Channel Changes* (Gregory, 1977) was published, significant progress has been made in understanding and explaining changes in river channel patterns. The dynamics, spatial occurrence and mechanisms of bank erosion have received much attention. Processes of fluvial entrainment and mass failure are dominant in erosion on the most laterally mobile rivers and the erosion is influenced, but not completely controlled, by the characteristics of flow; materials, bank geometry and soil moisture regime are also significant. Where rates of erosion are lower, i.e. where materials are more resistant or stream power is lower, other processes gain in importance either in their preparatory contribution to erosion or directly. The influence of vegetation is also much greater. Depositional morphology resulting from sediment transport processes has been a focus of study, more particularly in braided rivers, but the contributions of different styles of bars still require greater assessment and quantification. This applies similarly to chute flows, cutoffs and avulsions. Most attempts at modelling overall changes, at least in meandering patterns, have come from one of two directions. The first has been that of theoretical models and experimental simulations, based on principles of flow and assumed bed and bank conditions. The second is the more inductive approach of generalisation from documented changes of rivers. There is a certain convergence of results but what is now needed is a more effective combination of all these components and testing of specific conditions. If the stimuli to channel change, the sources and supply of sediment and the time-scale and position of storage could be more closely linked, then it should be possible to develop predictive models that not only aid geomorphological interpretation but also become useful for management purposes.

REFERENCES

Ackers, P. and Charlton, F.G. (1970) Meander geometry arising from varying flows. *Journal of Hydrology* 11, 230–252.

Amarsinghe, I. (1993) Effects of bank vegetation in waterways with special reference to erosion, variation of shear strength, root-density and hydraulics. PhD thesis, Open University.

Andrews, E. (1982) Bank stability and channel width adjustment, East Fork River, Wyoming. *Water Resources Research* 18, 1184–1192.

Arkell, B., Leeks, G.L., Newson, M.D. and Oldfield, F. (1983) Trapping and tracing: some recent observations of supply and transport of coarse sediment from upland Wales. *Special Publication International Association Sedimentologists* 6, 107–119.

Ashmore, P. (1991) How do gravel-bed rivers braid? *Canadian Journal of Earth Sciences* 28(3), 326–341.

Babinski, Z. (1992) The present-day fluvial processes of the Lower Vistula River. *Polish Geographical Studies* No. 157, 171pp.

Bathurst, J.C. (1979) Distribution of boundary shear stress in rivers. In: Rhodes, D.D. and Williams, G.P. (eds), *Adjustments in the Fluvial System*, Kendall/Hunt, Iowa.

Bathurst, J.C., Thorne, C.R. and Hey, R.D. (1977) Direct measurements of secondary currents in river bends. *Nature* 269, 504–506.

Best, J.L. (1986) The morphology of river channel confluences. *Progress in Physical Geography* 10, 157–174.

Best, J.L. and Bristow, C.S. (eds) (1993) *Braided Rivers*. Geological Society Special Publication No 75, 416pp.

Biedenharn, D.S., Combs, P.G., Hill, G.J., Pinkard, C.F. and Pinkston, C.B. (1989) Relationship between channel migration and radius of curvature on the Red River. In: Wong, S.S. (ed.), *Sediment Transport Modelling*, ASCE, New York, 536–541.

Bridge, J.S. and Jarvis, J. (1982) The dynamics of a river bend: a study in flow and sedimentary processes. *Sedimentology* 29, 499–541.

Burkham, D.E. (1972) *Channel Changes of the Gila River in Safford Valley, Arizona, 1846–1970*. US Geological Survey Professional Paper 665-G.

Carson, M.A. (1986) Characteristics of high-energy 'meandering' rivers: The Canterbury Plains, New Zealand. *Bulletin of the Geological Society of America* 97, 886–895.

Carson, M.A. and Griffiths, G.A. (1987) Bedload transport in gravel channels. *Journal of Hydrology (NZ)* 26, 1–151.

Church, M. and Jones, D. (1982) Channel bars in gravel-bed rivers. In: Hey, R.D., Bathurst, J.C. and Thorne, C.R. (eds), *Gravel-bed Rivers*, Wiley, Chichester, 291–338.

Church, M. and Miles, M.J. (1982) Discussion of 'Processes and mechanisms of bank erosion'. In: Hey, R.D., Bathurst, J.C. and Thorne, C.R. (eds), *Gravel-bed Rivers*, Wiley, Chichester, 259–268.

Dietrich, W.E. (1987) Mechanics of flow and sediment transport in river bends. In: Richards, K.S. (ed.) *River Channels: Environment and Processes*, Blackwell, Oxford, 179–227.

Ellis, L. (1993) River bank erosion and different hydrologic regimes. M.Phil. thesis, University of Nottingham.

Erskine, W., McFadden, C. and Bishop, P. (1992) Alluvial cutoffs as indicators of former channel conditions. *Earth Surface Processes and Landforms* **17**, 23–38.

Ferguson, R.I. (1993) Understanding braiding processes in gravel-bed rivers: progress and unsolved problems. In: Best, J.L. and Bristow, C.S. (eds), *Braided Rivers*, Geological Society Special Publication No. 75, 73–87.

Ferguson, R.I., Ashmore, P.E., Ashworth, P.J., Paola, C. and Prestegaard, K.L. (1992) Measurements in a braided river chute and lobe. I flow flow pattern, sediment transport and channel change. *Water Resources Research* **28**, 1877–1886.

Friedkin, J.F. (1945) *A Laboratory Study of the Meandering of Alluvial Rivers*. US Waterways Experiment Station Report, Vicksburg, Mississippi.

Friend, P.F. (1993) Control of river morphology by the grain-size of sediment supplied. *Sedimentary Geology* **85**, 171–177.

Furbish, D.J. (1991) Spatial autoregressive structure in meander evolution. *Bulletin of the Geological Society of America* **103**, 1576–1589.

Goff, J.R. and Ashmore, P. (1994) Gravel transport and morphological change in braided Sunwapta River, Alberta, Canada. *Earth Surface Processes and Landforms* **19**(3), 195–212.

Graf, W.L. (1983) Downstream changes in stream power in the Henry Mountains, Utah. *Annals Association of American Geographers* **73**, 373–387.

Gregory, K.J. (ed.) (1977) *River Channel Changes*. Wiley, Chichester.

Gregory, K.J., Davis, R.J. and Downs, P.W. (1992) Identification of river channel change due to urbanization. *Applied Geography* **12**, 299–318.

Harvey, A.M. (1986) Geomorphic effects of a 100 year storm in the Howgill Fells, Northwest England. *Zeitschrift fur Geomorphologie* **30**, 71–91.

Harvey, A.M. (1991) The influences of sediment supply on the channel morphology of upland streams: Howgill Fells, Northwest England. *Earth Surface Processes and Landforms* **16**, 675–684.

Hasegawa, K. (1989) Studies on qualitative and quantitative prediction of meander channel shift. In: Ikeda, S. and Parker, G. (eds) *River Meandering*, American Geophysical Union, Washington, 485pp.

Hey, R.D. (1982) Gravel-bed rivers: form and processes. In: Hey, R.D., Bathurst, J.C. and Thorne, C.R. (eds), *Gravel-bed Rivers*, Wiley, Chichester, 5–13.

Hey, R.D. and Thorne, C.R. (1975) Secondary flows in river channels. *Area* **7**, 191–195.

Hickin, E.J. (1983) River channel changes: retrospect and prospect. In: Collinson, J.D. and Lewin, J. (eds), *Modern and Ancient Fluvial Systems*, Blackwell, Oxford, 61–83.

Hickin, E.J. and Nanson, G. (1984) Lateral migration rates of river bends. *ASCE Journal of Hydraulic Engineering* **110**, 1557–1567.

Higgs, G. and Petts, G. (1988) Hydrological changes and river regulation in the UK. *Regulated Rivers: Research and Management* **2**, 349–368.

Hoey, T. (1992) Temporal variations in bedload transport rates and sediment storage in gravel-bed rivers. *Progress in Physical Geography* **16**(3), 319–338.

Hooke, J.M. (1977a) An analysis of changes in river channel pattern. Unpublished PhD Thesis, University of Exeter.

Hooke, J.M. (1977b) The distribution and nature of changes in river channel patterns. In: Gregory, K.J. (ed.), *River Channel Changes*, Wiley, Chichester, 265–280.

Hooke, J.M. (1979) An analysis of the processes of river bank erosion. *Journal of Hydrology* **42**, 39–62.

Hooke, J.M. (1980) Magnitude and distribution of rates of river bank erosion. *Earth Surface Processes and Landforms* **5**, 143–157.

Hooke, J.M. (1984) Changes in river meanders: a review of techniques and results of analysis. *Progress in Physical Geography* **8**, 473–508.

Hooke, J.M. (1986) The significance of mid-channel bars in an active meandering river. *Sedimentology* **33**, 839–850.

Hooke, J.M. (1987a) Changes in meander morphology. In: Gardiner, V. (ed.), *International Geomorphology*, Wiley, Chichester, 591–609.

Hooke, J.M. (1987b) Commentary on paper by M.D. Newson and G.J. Leeks. In: Thorne, C.R., Bathurst, J.C. and Hey, R.D. (eds), *Sediment Transport in Gravel-bed Rivers*, Wiley, Chichester, 219–222.

Hooke, J.M. (1990) The linkages between bank erosion and meander behaviour in gravel-bed rivers. *Portsmouth Polytechnic, Department of Geography Working Papers*, No. 14, 20pp.

Hooke, J.M. (1991) Non-linearity in river meander development: 'chaos' theory and its implications. *Portsmouth Polytechnic, Department of Geography Working Papers*, No. 19, 23pp.

Hooke, J.M. and Harvey, A.M. (1983) Meander changes in relation to bend morphology and secondary flows. In: Collinson, J.D. and Lewin, J. (eds), *Modern and Ancient Fluvial Systems*, Blackwell, Oxford, 121–132.

Hooke, J.M. and Kain, R.J.P. (1982) *Historical Change in the Physical Environment: A Guide to Sources and Techniques*. Butterworths, Sevenoaks.

Hooke, J.M. and Redmond, C.E. (1989a) Use of cartographic sources for analysis of river channel change in Britain. In: Petts, G.E. (ed.), *Historical Changes on Large Alluvial European Rivers*, Wiley, Chichester, 79–93.

Hooke, J.M. and Redmond, C.E. (1989b) River channel changes in England and Wales. *Journal of the Institution of Water and Environmental Management* **3**, 328–335.

Hooke, J.M. and Redmond, C.E. (1992) Causes and nature of river planform change. In: Billi, P. *et al.* (eds), *Dynamics of Gravel-bed Rivers*, Wiley, Chichester, 549–563.

Hooke, J.M., Bayliss, D.H. and Clifford, N.J. (1991) *Bank Erosion on Navigable Waterways.* Report to National Rivers Authority, University of Portsmouth, 68pp.

Hooke, J.M., Horton, B.E., Moore, J. and Taylor, M.P. (1994) *Upper River Severn (Caersws) Channel Study*. University of Portsmouth, RACER, Report to Countryside Council for Wales.

Howard, A.D. (1992) Modelling channel migration and floodplain sedimentation in meandering streams. In: Carling, P.A. and Petts, G.E. (eds), *Lowland Floodplain Rivers: Geomorphological Perspectives*, Wiley, Chichester, 1–41.

Klimek, K. (1989) Flood plains activity during floods in small mountain valleys, the Bieszczady Mountains, the Carpathians, Poland. Quaestiones Geographicae **2**, 93–100.

Lapointe, M.F. and Carson, M.A. (1986) Migration patterns of an asymmetric meandering river: The Rouge River, Quebec. *Water Resources Research* **22**, 731–743.

Lawler, D.M. (1986) River bank erosion and the influence of frost: a statistical examination. *Transactions Institute British Geographers* **11**, 227–242.

Lawler, D.M. (1987) Bank erosion and frost action: an example from South Wales. In: Gardiner, V. (ed.) *International Geomorphology 1986*, Part 1, Wiley, Chichester, 575–590.

Lawler, D.M. (1992) Process dominance in bank erosion systems. In Carling, P.A. and Petts, G.E. (eds), *Lowland Floodplain Rivers, Geomorphological Perspectives*, Wiley, Chichester, 117–144.

Leopold, L.B. (1992) Sediment size that determines channel morphology. In Billi, P. *et al.* (eds), *Dynamics of Gravel-Bed Rivers*, Wiley, Chichester, 297–312.

Lewin, J. (1976) Initiation of bedforms and meanders in coarse-grained sediment. *Bulletin of the Geological Society of America* **87**, 281–285.

Lewin, J. (1978) Floodplain morphology. *Progress in Physical Geography* **2**, 408–437.

Lewin, J. (1983) Changes of channel patterns and floodplains. In: Gregory, K.J. (ed.), *Background to Palaeohydrology*, Wiley, Chichester, 303–319.

Lewin, J. (1987) Historical river channel changes. In Gregory, K.J., Lewin, J. and Thornes, J.B. (eds), *Palaeohydrology in Practice—A River Basin Analysis*, Wiley, Chichester, 161–176.

Lewis, G.W. and Lewin, J. (1983) Alluvial cutoffs in Wales and the Borderlands. *Special Publication International Association of Sedimentologists* 6, 145–154.

Masterman, R. and Thorne, C.R. (1994) Analytical approach to flow resistance in gravel-bed channels with vegetated banks. In: Kirkby, M.J. (ed.) *Process Models and Theoretical Geomorphology*, Wiley, Chichester, 201–218.

Matthes, G.H. (1948) Mississippi River cutoffs. *Transactions of the American Society of Civil Engineers* 113, 1–15.

Montgomery, K. (1993) Non-linear dynamics and river meandering. *Area* 25, 97–108.

Mosley, M.P. (1975) Channel changes on the River Bollin, Cheshire, 1872–1973. *East Midland Geographer* 6, 185–199.

Murgatroyd, A.L. and Ternan, J.L. (1983) Impact of afforestation on stream bank erosion and channel form. *Earth Surface Processes and Landforms* 8, 357–369.

Nanson, G.C. and Hickin, E.J. (1986) A statistical analysis of bank erosion and channel migration in western Canada. *Bulletin of the Geological Society of America* 97, 497–504.

Nanson, G.C. and Young, R.W. (1981) Downstream reduction of rural channel size with contrasting urban effects in small coastal streams in south-eastern Australia. *Journal of Hydrology* 52, 239–255.

Nanson, G.C., Krusenstierna, A. and Bryant, E.A. (1994) Experimental measurements of river-bank erosion caused by boat-generated waves on the Gordon River, Tasmania. *Regulated Rivers: Research and Management* 9, 1–14.

Newson, M. (1981) Mountain streams. In: Lewin, J. (ed.), *British Rivers*, Allen and Unwin, London, 59–89.

Newson, M.D. and Leeks, G.J. (1987) Transport processes at the catchment scale. In: Thorne, C.R., Bathurst, J.C. and Hey, R.D. (eds), *Sediment Transport in Gravel-bed Rivers*, Wiley, Chichester, 187–223.

Newson, M.D. and Macklin, M.G. (1990) The geomorphologically-effective flood and vertical instability in river channels—a feedback mechanism in the flood series for gravel-bed rivers. In: White, W.R. (ed.), *International Conference on River Flood Hydraulics*, Wiley, Chichester, 123–140.

Page, K. and Nanson, G.C. (1982) Concave bank benches and associated floodplain formation. *Earth Surface Processes and Landforms* 7, 529–542.

Parker, G. and Johannesson, H. (1989) Observations on several recent theories of resonance and overdeepening in meandering channels. In: Ikeda, S. and Parker, G. (eds), *River Meandering*, Water Resources Monograph Series, Vol. 12, AGU, Washington DC, 379–416.

Parker, G. Diplas, P. Akiyama, J. (1983) Meander bends of high amplitude. *Journal of Hydraulic Engineering* 109, 1323–1327.

Passmore, D.G., Macklin, M.G., Brewer, P.A., Lewin, J., Rumsby, B.T. and Newson, M.D. (1993) Variability of late Holocene braiding in Britain. In: Best, J.L. and Bristow, C. (eds), *Braided Rivers*, Geological Society Special Publication **75, 205–229.**

Pearthree, M.S. and Baker, V. (1987) *Channel Change along the Rillito Creek System of Southeastern Arizona, 1941 through 1983.* Arizona Bureau of Geology and Mineral Technology Special Paper No. 6, 58pp.

Pizzuto, J.E. (1984) Equilibrium bank geometry and the width of shallow sand-bed streams. *Earth Surface Processes and Landforms* 9, 199–208.

Rhoads, R.L. and Welford, M.R. (1991) Initiation of river meandering. *Progress in Physical Geography* 15, 127–156.

Schumm, S.A. (1981) Evolution and response of the fluvial system, sedimentological implications. *Society of Economic Palaeontologists and Mineralogists Special Publication* 31, 19–29.

Thompson, A. (1984) Long and short term channel change in gravel-bed rivers. PhD thesis, University of Liverpool, 492pp.

Thompson, A. (1986) Secondary flows and the pool–riffle unit: a case study of the processes of meander development. *Earth Surface Processes and Landforms* 11, 631–641.

Thorne, C.R. (1982) Processes and mechanisms of river bank erosion. In: Hey, R.D, Bathurst, J.C. and Thorne, C.R. (eds), *Gravel-bed Rivers*, Wiley, Chichester, 227–259.

Thorne, C.R. (1990) Effects of vegetation on riverbank erosion and stability. In: Thornes, J. (ed.), *Vegetation and Erosion*, Wiley, Chichester, 125–144.

Thorne, C.R. and Lewin, J. (1979) Bank processes, bed material movement and planform development in a meandering river. In: Rhodes, D.D. and Williams, G.P. (eds), *Adjustments of the Fluvial System*, Allen and Unwin, London, 117–137.

Thorne, C.R. and Osman, A.M. (1988) Riverbank stability analysis. II. Applications. *Journal of Hydraulic Engineering* **114**(2), 151–172.

Thorne, C.R. and Tovey, N.K. (1981) Stability of composite river banks. *Earth Surface Processes and Landforms* **6**, 469–484.

Trimble, S.,W. (1994) Erosional effects of cattle on streambanks in Tennessee, USA. *Earth Surface Processes and Landforms* **19**, 451–464.

Whiting, P.J. and Dietrich, W.E. (1993a) Experimental studies of bed topography and flow patterns in large-amplitude meanders—1. Observations. *Water Resources Research* **29**(11), 3605–3614.

Whiting, P.J. and Dietrich, W.E. (1993b) Experimental studies of bed topography and flow patterns in large-amplitude meanders—2. Mechanisms. *Water Resources Research* **29**(11), 3615–3622.

Wolman, M.G. (1959) Factors influencing erosion of a cohesive river bank. *American Journal of Science* **257**, 204–216.

Wolman, M.G. and Gerson, R. (1978) Relative scales of time and effectiveness of climate in watershed geomorphology. *Earth Surface Processes and Landforms* **3**, 189–208.

6 Channel Cross-sectional Change

CHRIS C. PARK

Department of Geography, University of Lancaster, UK

INTRODUCTION

This chapter focuses on recent developments in the study and understanding of cross-sectional change in river channels. The emphasis is on developments since about 1980, for two main reasons. The first is that the evolution of studies before the late 1970s is well recorded in a number of books—including Gregory's (1977) *River Channel Changes*, Lewin's (1981) *British Rivers*, and Morisawa's (1985) *Rivers; Form and Process*—so it need not be described again here. By the late 1970s the study of river channels was well developed, and included theoretical and empirical work on equilibrium channel morphology, regime theory and hydraulic geometry, complex channel responses, anthropogenic channel adjustments, and channel changes during the Holocene and Pleistocene (Hickin, 1983).

The second reason for the post-1980 focus is the quite marked shift in emphasis from studies of river processes and fluvial landforms during the 1960s and 1970s—noted in reviews by Gregory (1978, 1979b, 1980)—towards studies of channel adjustments and responses to external factors during the 1980s and early 1990s—noted in more recent reviews by Gregory (1982b, 1983b, 1984) and by Richards (1987a and b). This emerging applied emphasis will doubtless continue to grow in importance, and Rhoads (1992a) has argued that one of the greatest challenges for fluvial geomorphology over the next decade is the need to predict the short- and long-term responses of river systems to anthropogenic changes in global climate.

The main developments in studies of channel cross-sectional change since about 1980 can be reviewed under four themes—equilibrium, geometry, adjustment and applications—which provide a convenient framework for the rest of this chapter.

EQUILIBRIUM

A hallmark of much geomorphological research, particularly since the 1960s, is an emphasis on equilibrium. The definition and use of equilibrium concepts in geomorphology is reviewed by Montgomery (1989), who highlights the problems of incorporating evolutionary change and historical constraints into working definitions of equilibrium and emphasises the difficulties of establishing causality at different spatial and temporal scales. J.D. Phillips (1988) echoes the need to focus on specific

Changing River Channels. Edited by Angela Gurnell and Geoffrey Petts.
© 1995 John Wiley & Sons Ltd.

spatial scales, and argues that relationships which operate over spatial scales an order of magnitude different are effectively independent of each other. A number of studies of channel morphology since the early 1970s have adopted ergodic reasoning based on space–time substitution, some seeking evidence of equilibrium via characteristic form and others seeking evidence of progressive change and relaxation times (Paine, 1985).

Fluvial equilibrium

The problems of defining and delimiting equilibrium apply equally well to studies of fluvial equilibrium, despite promising advances in approaches to the description and analysis of channel geometry (Richards, 1981) and in experimental studies of river hydraulics (Alexander, 1979). Underpinning such development is a basic assumption that natural physical laws are space- and time-invariant, thus allowing the study of channel adjustments to environmental change and the reconstruction of past processes from preserved evidence of former channel features (Ferguson, 1981; Richards and Ferguson, 1987). Most such studies have used statistical (particularly power-function) models to describe relationships between channel geometry and hydrological, hydraulic, sedimentological and biotic controls, although Rhoads (1992b) and others have proposed a wider use of more advanced models including simultaneous-equation models, continuously-varying parameter models, and distributed lag models.

Field studies in many different environments—including supraglacial streams (e.g. Knighton 1981a, 1985)—have shown how channel form is controlled by discharge, sediment load, bed and bank material composition, and valley slope (Knighton, 1987b), which all vary both along and between rivers. Miller's (1991a) study of river channel adjustment at over 250 sites in the Missouri River Basin indicates mutual adjustments between channel cross-sections, slope and sediment size.

Instability and change

Central to any understanding of river equilibrium is the notion of stability, although this is rarely defined in any tangible way. Most studies illustrate instability rather than stability. Some, such as Zeng Qinghua and Zhou Wenhao's (1991) study of the lower Yellow River in China, show natural channel adjustments to sediment transport and deposition. Process studies, such as Neller's (1988) analysis of erosion pin data in Armidale, New South Wales, can also reveal the location, timing and magnitude of river instability.

Repeated surveys of river cross-sections can also reveal temporal change. Recent studies of short-term (decadal) changes have focused on the Mississippi (Murphey and Grissinger, 1985) and on rivers in New Brunswick (Bray, 1987), and longer-term (100-year) studies are represented by remeasurement of representative cross-sections in south-central Kansas (Martin and Johnson, 1987). Hamlett et al. (1983) have associated increasing channel width and depth in east-central Iowa with increasingly intensive agricultural land use. In contrast, some repeat surveys—such as Bennett's (1989) study of the Suquehanna River in eastern New York State—show remarkable channel stability and persistence through time, even during periods of significant change in upstream drainage basins. A third type of repeat-survey study, illustrated by Lisle's (1982) monitoring of channel changes in north-west California, shows initial flood-related channel disturbance followed by progressive readjustment and relaxation over a period of years towards a new stable morphology.

Instability is widely revealed in studies of channel disturbance associated with catastrophic floods. Pitlick (1993) has shown how the failure of an earth dam triggered large-scale channel changes along part of Roaring River in the Rocky Mountains National Park, Colorado, and Erskine and Melville (1983) recorded the impacts of moderate floods on channel form and stability in the lower Macdonald River in New South Wales, Australia. Beschta (1983b) attributed significant increases in channel width on the Kowai River in New Zealand to a major storm in 1951, and Anderson and Calver (1982) have documented significant channel changes on central Exmoor in south-west England caused by the passage of the August 1952 flood. Not all large floods cause detectable channel changes, however. For example, Rhoads and Miller (1991) discovered relatively little impact of an estimated 100-year flood on part of the Des Plaines River in Illinois, which they attributed to low stream power, low hydrological variability, fine bed materials and cohesive banks along that reach of the river.

One interesting approach to the definition of river stability, illustrated by Graf (1983a, 1984), is to define probability functions for the likelihood of channel migration and erosional collapse of near-channel land which can then be used for economic evaluations of engineering channel control schemes. An alternative is to try to define and measure river channel sensitivity (Downs and Gregory, 1993; see also Chapter 16), again for use in formulating river management strategies.

Complex response

A number of recent studies have focused on the identification, controls and implications of discontinuities (thresholds) in the fluvial system which give rise to distinctive patterns of response to internal and external controlling factors (Howard, 1980).

The significance of such thresholds is well illustrated by Petts' (1980a and b, 1982) studies of river adjustment below reservoirs. These reveal complex readjustment of channel morphology and time-lags in patterns of response, which can reflect complex interconnections between processes, forms and vegetation but can inhibit or mask the establishment of new morphological and ecological equilibria. Xu's (1990a and b) analysis of complex response to reservoir construction of the Weihe River in China (Figure 6.1) revealed three stages of adjustment, starting with a decrease in channel sinuosity, followed by an increase, and eventually the establishment of constant sinuosity in equilibrium with the new hydrological regime.

Complex response is also apparent in river adjustments to other external changes. For example, Gellis et al. (1991) have documented complex response of arroyo development in the Colorado River Basin, since the late 19th century, in relation to peak discharges, sediment loads and storage, and vegetation colonisation. Variable sediment supply and transport often triggers complex response in river systems (Lane et al., 1982).

120

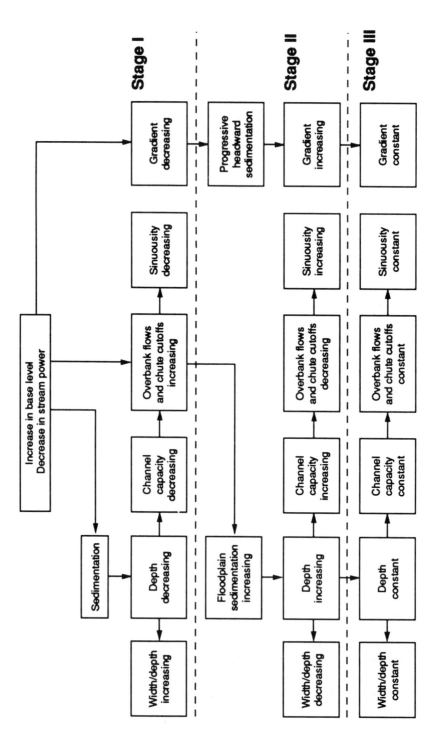

Figure 6.1 Complex response of river channel geometry upstream from a reservoir. (After Xu, 1990b, and Rhoads, 1992a, reproduced by permission of Edward Arnold)

GEOMETRY

The initial development of different approaches to the study of channel cross-sectional geometry, mainly during the 1950s and 1960s, was followed by a period of consolidation during the 1970s, when established techniques were applied to a wider range of environments to broaden our understanding of the variations and controls of channel geometry. This second phase saw the emergence of statistically-based models of fluvial equilibrium, experimental and simulation studies of channel form and change, and detailed field observations of the dynamics and processes of channel change and adjustment. Since the early 1980s the field has been further developed to embrace more critical explorations and more creative extensions of established approaches to the study of channel geometry, particularly in terms of hydraulic geometry and channel morphometry. Two other important developments during the 1980s were the emerging focus on channel adjustments to direct and indirect stimuli and the search for relevant applied uses of the emerging understanding of channel geometry, which are discussed in the following sections.

Hydraulic geometry

Leopold and Maddock's (1953) paper on hydraulic geometry—which defined the term, formulated the basic approach and has been a catalyst for a great deal of field research over the following decades—remains a classic. Indeed, the very fact that it has stood the test of time and has inspired such a wealth of follow-up work, speaks volumes for its importance to the post-war development of fluvial geomorphology.

Recent hydraulic geometric studies have sought to complement and extend Leopold and Maddock (Ferguson, 1986). One way in which this has been approached has been to further expand the range of river types on which the hydraulic geometry model has been applied. In this category, for example, fall studies by Griffiths (1981) of gravel-bed rivers in New Zealand, Rice's (1982) study of the hydraulic geometry of a proglacial river in Jasper National Park, Alberta, Ergenzinger's (1987) analysis of braided *torrente* channels in Italy, and the simple field study of some first order streams in the Lower Great Kei Basin, South Africa, reported by Frauenstein (1986–1987).

A limited number of attempts have been made to extend the hydraulic geometry model even further. The model has been applied to special cases such as supra-glacial rivers (Park, 1981a) or cave passages (White and Deike, 1989), in both of which the hydraulic geometry adjustments appear to be of a similar form to that typical of alluvial rivers, despite the obvious environmental contrasts. Bhowmik (1984) has stretched the model yet further, in his attempt to apply the basic principles of hydraulic geometry to floodplains on major rivers in the United States (relating floodplain width, cross-sectional area, surface area, depth, sinuosity and incision to stream order as a surrogate for discharge).

The four decades since Leopold and Maddock's initial paper have seen the publication of a great many hydraulic geometry studies, which have inspired some interesting comparative analyses of at-a-station and downstream exponents. The two earliest, carried out independently by Park (1977a) and Rhodes (1977, 1978), both used triaxial diagrams to examine the simultaneous variation of hydraulic geometry

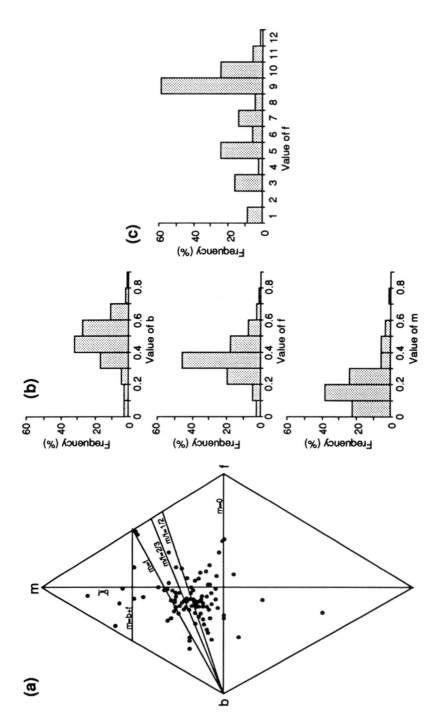

Figure 6.2 Summary of variations in downstream hydraulic geometry exponents from 91 rivers in different environments. The figure shows (a) simultaneous variations in the three exponents, (b) the frequency distribution of the exponents, and (c) the frequency distribution of different sets of responses according to channel type. (After Rhodes, 1987, reproduced by permission of *Geografiska Annaler*)

exponents, and both sought to identify possible controlling factors. Park focused mainly on environmental factors (such as climate and river type), while Rhodes concentrated on hydraulic factors. In his 1977 paper Rhodes confined the analysis to at-a-station variations, but he redressed the balance 10 years later (Rhodes, 1987) in a detailed analysis of downstream hydraulic geometry exponents (Figure 6.2) which emphasised 12 common sets of hydraulic responses to discharge and sediment load.

Many recent studies have focused on relationships between hydraulic geometry parameters and a variety of environmental factors and controls. Themes explored include flow variability (Bofu Yu and Wolman, 1987), controlled discharges (Mosley, 1982), bank stability and vegetation (Andrews, 1982), bed material properties (Rice, 1982), changes in bed slope (Ferguson and Ashworth, 1991), and fish habitat (Hogan and Church, 1989). Figure 6.3a shows typical at-a-station variations in the behaviour of hydraulic geometry variables during the rising and falling limbs of a flood hydrograph.

Whilst there have been many advances in understanding stream behaviour and channel response using the hydraulic geometry model, not all studies endorse Leopold and Maddock's conclusion that this is a rational or even a good way of describing cross-sectional channel adjustment. Caution is urged by Phillips and Harlin (1984), whose field analysis of a sub-alpine stream in a relatively homogeneous environment showed that hydraulic exponents are not stable over space. They argue that stream behaviour may be so influenced by local soil and subsurface conditions that it can make little sense to use single values of the b, f and m exponents to describe at-a-station variations in river systems.

There is, furthermore, mounting debate over whether the log-linear model of hydraulic geometry is either appropriate or meaningful, and a growing number of studies are proposing alternative or more sophisticated numerical models or approaches. Richards (1973, 1976) and Knighton (1975, 1979) were amongst the first to point out that many at-a-station hydraulic variations appeared to be distinctly non-linear (largely because of changing roughness and resistance with changing discharge), and to recommend the use of log-quadratic rather than log-linear equations. More recent analysis by Knighton (1987a) of the at-a-station hydraulic geometry of some Tasmanian streams further develops the theme of discontinuous adjustment, here related to sub-bankfull channelisation and the attainment of overbank stage. More detailed studies of velocity–discharge relationships (Knighton and Cryer, 1990) found that the traditional log-linear model applies fairly well in many cases and that the log-quadratic model is often only moderately successful. Williams (1987) tested a 'unit hydraulic geometry' approach, in which the coefficients and exponents of hydraulic geometry relations are based on unit discharge, and he concluded that this approach is more useful than conventional analysis in defining channel changes.

Interest in redefining the basic statistical ingredients of the hydraulic geometry model appears to have been rekindled in the early 1990s. Bates (1990), for example, explores a piecewise linear model which helps in the identification of thresholds in at-a-station hydraulic variations. Phillips (1990) concludes that a satisfactory generally applicable hydraulic geometry model has yet to be found, although he proposes an equation system based on the Darcy–Weisbach flow resistance equation. A different solution is preferred by Rhoads (1991a), who favours a continuously vary-

124

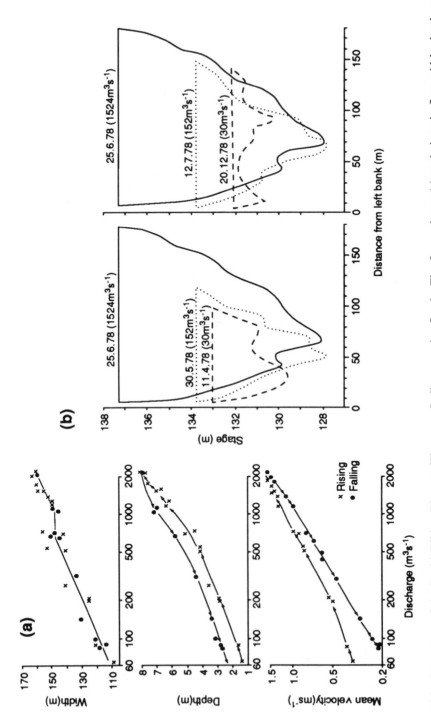

Figure 6.3 Responses of the Burhi Dihing River at Khowang, India, to major floods. The figure shows (a) variations in flow width, depth and mean velocity with discharge during the rising and falling stages of a flood in 1973, and (b) the sequence of scour and fill during a 1978 flood. (After Sarma and Basumallick, 1986, reproduced by permission of *Geografiska Annaler*)

ing parameter model of downstream hydraulic geometry. Ridenour and Giardino (1991) explore the usefulness of compositional data analysis in the statistical study of hydraulic geometry.

A recurrent theme in recent hydraulic geometry studies is the association between channel hydraulics and behaviour, and energy conservation and dissipation (Williams, 1978). Yang et al. (1981) and Yang and Song (1982) suggest that a river system constantly adjusts itself in response to varying constraints in such a manner that the rate of energy dissipation approaches a minimum value, and thus moves towards an equilibrium. Miller (1991b) pursues this idea further, seeing minimum variance as a reflection of equable change in at-a-station hydraulic geometry in response to changing discharge. Chang (1985b) explains adjustment of channel width during scour and fill (Figure 6.3b) in terms of a stream's tendency to establish equal power expenditure along the channel. Griffiths (1984) is more critical of extremal hypotheses (including minimum stream power, minimum unit stream power, minimum energy dissipation and maximum sediment transport rate) because they are often incompatible with field observations. Phillips (1991) considers the relevance of multiple modes of adjustment to extremal hypotheses for alluvial channel behaviour, and emphasises the need for a better understanding of the relative response time of hydraulic variables under different flow conditions.

Finally, attempts have recently been made to integrate hydraulic geometry concepts into a broader framework for the description and analysis of channel equilibrium. Ferguson and Ashworth (1991), for example, describe progressive downstream changes in channel character along small rivers in the Scottish Highlands in response to rapid changes in bed slope and bed sediment. Pizzuto (1992) combined hydraulic geometry equations with a watershed model to quantify downstream trends in equilibrium channel morphology in gravel-bed streams, and concluded that the study channels in Pennsylvania are best described as systems with multiple time-scales and multiple rates of response.

Channel morphometry

Whilst hydraulic geometry studies reveal much about the internal adjustment of channel form and size to changing discharge, both at-a-station and downstream, they reveal little about spatial variations or about interrelationships between channel form and possible environmental controls. For this reason a second major 'channel geometry school' has evolved, particularly since the mid-1970s, based on the morphometric analysis of relationships between channel cross-section geometry and either drainage area or distance downstream. Gregory and Park's (1974) analysis of channel changes downstream from a reservoir in south-west England is an early example, and Park (1978) described such changes in terms of allometric analysis.

This morphometric approach has developed in a number of important and fruitful ways since the early 1980s. One has been the application of existing approaches to new environments. Caine and Mool (1981), for example, showed that variations in channel geometry in tropical, high-relief streams in Nepal conform to the patterns established in temperate, low-relief studies. Kale (1990) reached a similar conclusion from extensive fieldwork in the western Deccan Trap Upland region of India, although the maximum monsoon discharge was found to strongly influence channel width and gradient.

A second important development has been the search for better or easier ways of describing channel geometry. Most early studies used simple measures such as channel width and depth, cross-sectional area, and width/depth ratio, based on field surveys of representative cross-sections. Surveying large numbers of complex cross-sections can be very time-consuming, and Robinson and Beschta (1989) have proposed a triangular procedure for estimating cross-sectional area. Little attention has been paid to cross-sectional channel asymmetry, despite the obvious downstream variations in asymmetry in meandering rivers. Knighton (1981b) and Milne (1983) have tried to develop meaningful measures of asymmetry, but few have followed their lead. Gregory's (1982a) proposed integrated index of river network power (which combines network volume and basin relief) is a promising and potentially very useful morphometric measure, but it too has yet to be fully developed or widely used. Graf (1983b) used similar ideas in his study of downstream changes in stream power in the Henry Mountains, Utah.

Development three has been the search for explanations (statistical and physical) of why channel geometry varies through both space and time. Mosley's (1979) analysis of channels of South Island, New Zealand, showed that nearly 70% of variation in channel morphology can be accounted for by differences in cross-sectional area, slope and cross-section shape, but only 53% of the morphological variability could be statistically 'explained' by the hydrological and sediment variables used. Causes of unexplained variability might include floodplain vegetation, variability of sediment character, boundary effects imposed by bedrock bluffs, and sequences of flood events (magnitude, frequency and precise timing), none of which are easily quantified. Other studies have adopted more empirical approaches to the identification of factors controlling channel form, including bedrock geology (Miller, 1991a), erodibility of cohesive bank materials (Pizzuto, 1984), critical bank height for mass failure (Thorne, 1991), thresholds of bedload movement (Inokuchi and Sasaki, 1985), coarse woody debris (Robinson and Beschta, 1990) and vegetation (Gregory and Gurnell, 1988; Ikeda and Izumi, 1990). Miller (1984) insists that traditional models fail to represent the mechanisms through which channel cross-sections adjust, and he proposes a multiple-equation model which incorporates the effect of mutual adjustment of channel properties.

Discontinuous response

Scale dependency and spatial threshold effects have been highlighted in a number of channel geometry studies. Klein (1981) was amongst the first to question whether channel geometry relations in small drainage basins can be extrapolated to larger basins, partly because of discontinuities in downstream changes in suspended sediment concentration (Klein, 1982). Carragher et al. (1983) identified a scale threshold at about 3 km^2 which they attributed to the changing contributions of rapid storm flow and to the effect of bed material size.

Spatial discontinuities associated with environmental controls have been identified in a number of studies. Sarma and Basumallick (1986) examined the impact of three knick-points on downstream changes in channel geometry along the Burhi Dihing

River in India, which appear to have influenced cross-sectional geometry more than meander geometry. Kale *et al.* (1986) account for discrete changes in downstream channel geometry in the Vashishti River, India, in terms of changing bed and bank material, channel slope and planform. Richards and Greenhalgh (1984) identified channel shrinkage downstream from a river diversion on the River Derwent in Yorkshire, England. Dury (1984) found abrupt variations in channel width along part of the River Severn, which appeared to be a function of local differences in bank strength.

This type of evidence of non-linear downstream adjustment of channel geometry has led Ebisemiju (1991) and others to urge caution in using spatial interpolation techniques to identify and quantify river channel changes caused by or associated with human activities. The extrapolation of upstream geometric relations to downstream reaches might not in all situations be either statistically appropriate or physically meaningful, and it might sometimes be more sensible to use segmented regional relationships rather than extrapolation of upstream relations.

Discontinuities in channel geometry are also evident at tributary junctions (Figure 6.4). Kennedy (1984) points out that most geomorphologists have long assumed that fluvial confluences will be morphologically accordant, despite much field evidence of discordant fluvial junctions and engineering practice which favours discordant confluences in non-erodible materials. Adjustments of channel width occur predominantly at junctions, so Richards (1980a) proposes a channel geometry model based on channel dimensions as power-functions of stream link magnitude. At the local scale, Rhoads (1987) has shown that changes in channel geometry at confluences are diverse, and they seem not to conform with predictions based on simple

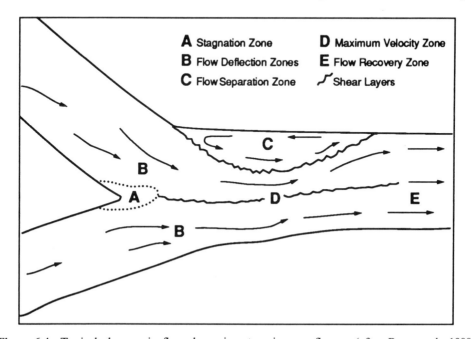

Figure 6.4 Typical changes in flow dynamics at a river confluence (after Roy *et al.*, 1988, reproduced by permission of John Wiley & Sons Ltd)

hydraulic geometry. Detailed field (Roy, 1983) and laboratory flume (Best, 1988) studies have shown the importance of confluence angle and ratio of discharge between the tributary and mainstream channels. The evidence suggests that channel cross-sectional area decreases below most confluences, implying an increase in flow velocity (Roy and Roy, 1988). This might be a function of many things, including changes in bed slope, reduction in grain roughness at the confluence, curvature of tributary and acute angle of entry (Roy *et al.*, 1988).

ADJUSTMENT

Most of the studies of channel geometry reviewed in the previous section seek to describe channel form as though it were fixed, although rivers are dynamic systems and are thus subject to change. In this section we examine some of the controls and processes of channel cross-sectional adjustment, then review recent research on adjustment triggered by human activities. Interest in actual and potential adjustments in the fluvial system is relatively new (Gregory, 1980; Park, 1981b), but understanding grew rapidly during the early 1980s (Rhodes and Williams, 1982).

Natural adjustments

Knighton (1988a) distinguishes between autogenic change, which is affected by channel migration (natural and artificial), and allogenic change which results from floods and climatic variation. Superimposed on these long-term natural adjustments of rivers are the short-term anthropogenic effects which often lead to channel instability, promote rapid and complex responses, and give rise to unwanted and costly environmental impacts.

Process–response models of channel change have shown the importance of mutual adjustments between form and process in streams in different environments (Rhoads, 1988). Perhaps not surprisingly a great deal of attention continues to be paid to the processes of channel change and the dynamics of adjustment. Conventional process studies, illustrated by Gardiner's (1983) field study of channel bank erosion on the River Lagan in Northern Ireland, are less common than they were in the 1970s. They seem to have been superseded by the more wide-ranging analyses of channel dynamics. Andrews (1982), for example, showed how the adjustment of channel width along the East Fork River in Wyoming during the 1974 snowmelt flood was controlled largely by scour and fill.

The role of high discharges in channel adjustment continues to attract attention. Neller (1980) questions the usefulness of the dominant discharge model on the basis of evidence that at least two sets of flows are responsible for channel geometry in his study areas near Sydney, Australia. The channel cross-sections there are modified by flows with one to three year recurrence intervals, but meander geometry appears to be adjusted to infrequent larger flows (at least 25 years recurrence interval). Carling (1988a) also casts doubt on the general applicability of the dominant discharge model, particularly in channels which are constrained by cohesive banks and/or compacted beds.

A number of recent studies demonstrate the importance of high-magnitude floods on stream channel response. Gupta (1983) argues that the impact of such events on channel form can be very persistent, particularly if they are associated with large supplies of coarse sediment, and smaller flows in the inter-flood period are unable to modify such forms. Rhoads (1990b) illustrates the impact of such catastrophic floods in a small desert mountain stream, where channel geometry was in dis-equilibrium with the short-term hydrologic regime.

Relatively limited attention has been paid to the impacts of recent changing weather patterns on channel form and adjustment. Channel changes associated with increased flooding on the Hunter River in Australia have been linked by Erskine and Bell (1982) with increases in annual rainfall, particularly the rainfall intensity of frequent storms. Miller *et al.* (1993) have documented fluvial responses in southern Illinois to a number of phases of climatic variation within the 20th century, includ-

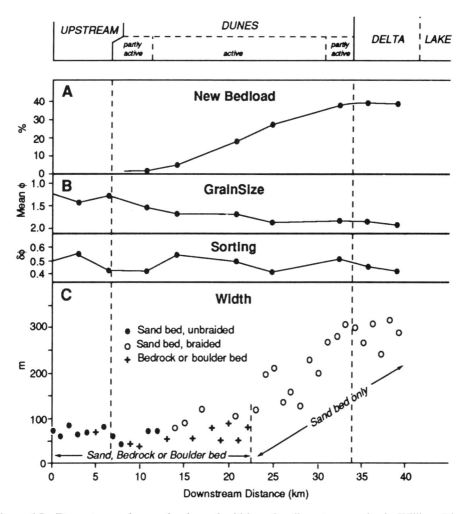

Figure 6.5 Downstream changes in channel width and sediment properties in William River, Canada, associated with the input of aeolian sediment from an extensive dune system. (After Smith and Smith, 1984, reproduced by permission of the Geological Society of America)

ing a phase of decreasing mean annual precipitation from 1904 to about 1945, a phase with many intense storms and high annual precipitation totals between 1945 and 1951, and a phase of increasing annual precipitation from 1952 to the early 1990s.

Sediment input and transport is obviously a major influence on channel adjustments. Both Rhoads (1990b) and Nolan and Marron (1985) highlight the importance to channel adjustment of sediment delivery during catastrophic storms, particularly where sediment delivery overwhelms transport capacities and promotes long-lasting changes in channel geometry. Recent studies have focused on channel cross-section changes caused by large inputs of sediment from a variety of sources including hillslope erosion (Beschta, 1983a; Brown, 1983), upstream gully development (Harvey, 1991) and landslides (Lyons and Beschta, 1983; C.J. Phillips, 1988). Sediment input from tributaries into regulated rivers (Petts, 1984) and massive sediment delivery associated with dam failure (Anthony and Harvey, 1991) also cause channel cross-sectional adjustments with minimal delay.

One of the most graphic illustrations of channel changes associated with sediment delivery is the William River in north-west Saskatchewan (Figure 6.5), described by Smith and Smith (1984). The river starts out as a relatively narrow, deep, single-channel stream, and then it picks up a 40-fold increase in bedload over a 27 km reach through a large sand dune field. Here it develops a braided pattern, increases in width five-fold, and increases in width/depth ratio 10-fold.

Adjustment to human impact

From the mid-1970s onwards a great deal of effort has been directed towards the identification, quantification and explanation of the impacts of human activities on channel morphology. From an early emphasis on the impacts of reservoir construction and urban development (Gregory, 1979c), interest has broadened to encompass land use changes, river flow regulation, deforestation and other direct and indirect impacts (Gregory, 1983a).

Rhoads' (1991b) analysis of anthropogenic factors that have altered channels in the Santa Cruz Valley, Arizona—which include upstream urbanisation, dam construction, artificial drainage diversion, obstruction of flow by transport routes, and stream channelisation—emphasises how varied and complex the human impact can be and underlines the need for more study of multiple impacts. A useful distinction is drawn by Warner and Bird (1988) between deliberate and inadvertent modifications of channels in New South Wales and Victoria, Australia. The deliberate modifications include channel regulation for water supply, power and navigation, as well as channelisation, vegetation management and urban drainage improvements. Inadvertent (non-deliberate) modifications include aggregate extraction and mining disturbance, flood mitigation, flow augmentation and flow diversion.

A variety of approaches have been used to determine where anthropogenic channel changes have taken place. Some studies use a space–time substitution approach, based on the relationships between channel geometry and drainage area or distance downstream (e.g. Figure 6.6a). Gregory et al. (1992) tested the usefulness of three additional approaches, based on the measurement of channel widths from large-scale topographic maps of different dates, the use of vegetation, structures and morpholo-

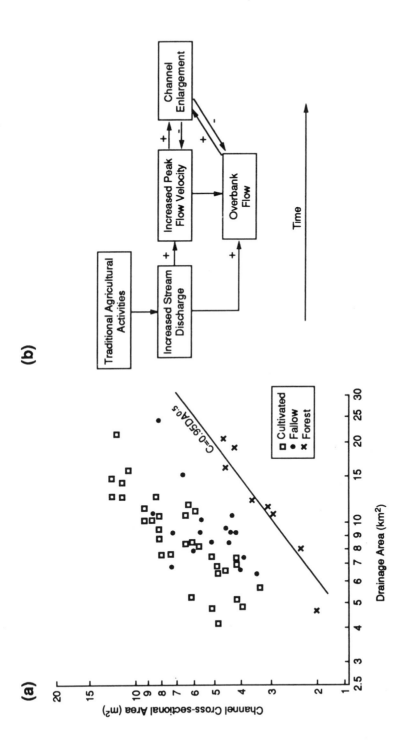

Figure 6.6 Channel cross-section changes associated with land use change in Nigeria. The figure shows (a) downstream variations in channel cross-sectional area for three different land uses, and (b) the sequence of channel adjustments to increasing discharge from traditional farming practices. (After Odemerho, 1984, reproduced by permission of *Geografiska Annaler*)

gical field indicators of channel change, and the field mapping of spatial variations in channel adjustments.

Induced channel changes

More effort has been devoted to induced channel changes than to natural channel changes since 1980, and there is now a massive literature on the subject. Because of space constraints it is only possible here to outline the variety of contexts in which induced changes have been identified, rather than offer a detailed review.

Urbanisation in a drainage basin promotes increased discharge and flow velocity, which in turn result in channel enlargement (Park, 1977b). Nanson and Young (1981) studied channel entrenchment and enlargement caused by urban development in south-east Australia. Morisawa and Lafleure (1982) detected a great increase in channel cross-sectional areas in basins with higher percentages of urbanisation in Pennsylvania and New York, although Neller (1989) failed to find such a relationship in New South Wales. Instead, he found greater channel enlargement on steeper headwater urban basins, and rapid gully development associated with disturbances such as road crossings. Whitlow and Gregory (1989) describe channel erosion and widening, coupled with increased drainage density, associated with urban development of parts of Harare in Zimbabwe. Other studies in the humid tropics, by Ebisemiju (1989a and b), reveal generally smaller channels than in the temperate zone, and a less regular channel adjustment to urban development. Odemerho (1992) found limited downstream response to urbanisation, which he attributed to storage of floodwaters and sediments on floodplains.

Reservoir construction tends to have the opposite effect, by decreasing discharge and flow velocity and causing channel shrinkage. Such impacts have been demonstrated on a number of rivers in England—including the River Ter in Essex (Petts and Pratts, 1983), the River Derwent in Derbyshire (Petts, 1987), and below Chew Valley Lake in Somerset (Petts and Thoms, 1986)—and elsewhere, including the Golan Heights in Israel (Inbar, 1990) and a range of alluvial channels in the United States (Williams and Wolman, 1984). Such impacts are not confined to major dams, because repeated surveys across sections of the Washita River in Oklahoma show channel adjustments to local floodwater-retaining structures (Schoof et al., 1987). Field studies in a variety of environments illustrate the complex response typically associated with reservoir construction (Petts, 1979), with some sections being affected by aggradation and others by degradation, and the immediate channel response post-construction being often different to final equilibrium conditions. For example, channel adjustments below Gardiner Dam in Canada included bed degradation and armouring, a significant decrease in channel width and a decrease in average channel bed slope (Rasid, 1979), but no marked change in channel pattern (Galay et al., 1985). Complex response is also evident in the Weihe River, China, downstream from the Sanmenxia Reservoir (Xu, 1990b). Carling (1988b) points out that each reservoir-drainage basin system is unique, so that singular responses are not to be expected and models of channel adjustment remain extremely general. Some recent studies have focused on the dynamics of channel change below reservoirs, and they highlight the importance of bar growth, in-channel bench construction, formation and bank attachment of longitudinal bars, and vegetation colonisation (Knighton, 1988b; Sherrard and Erskin, 1991).

Similar channel responses to decreased peak flows have been observed in rivers affected by flow diversion (out) or flow reduction. The abstraction for hydro-electric purposes of glacial meltwater in the Val d'Herens, Switzerland, for example, has promoted morphological responses downstream in terms of channel size, shape and pattern (Gurnell, 1983; Gurnell *et al.*, 1990). Similarly, channel shrinkage along parts of the Rio Grande have been attributed to upstream water development schemes which include water diversion and extraction (Everitt, 1993). The reverse situation, where peak discharges are increased by flow diversion (in), promotes corresponding channel adjustments. The diversion can be either natural, such as that associated with stream capture (Shepherd, 1982), or part of a water management strategy (Bradley and Smith, 1984).

Channels in some areas have also been affected by the changing sediment loads associated with mining activities. Sometimes the mining has taken place directly within the channel, with dramatic effect. For example, hydraulic gold mining of fluvial sediments along Bear River in California has triggered significant channel instability and change (James, 1989, 1991) as has historic tin streaming along some rivers on Dartmoor in south-west England (Park, 1979). Commercial gravel extraction has caused channel adjustment along the Manawatu River in New Zealand (Page and Heerdegen, 1985). In other instances the mining has taken place adjacent to the rivers, but still significantly increased sediment load. Graf (1979) describes some channel responses to gold and silver mining activities in the Central City District of Colorado, Touysinthiphonexay and Gardner (1984) study the response of small streams to surface coal mining in central Pennsylvania, Richards (1980b, 1982) outlines channel adjustment to china clay working in Cornwall, England, and Knighton (1989, 1991) documents channel bed and cross-section changes on Tasmania's Ringarooma River caused by local mining activities.

Channelisation also promotes river channel adjustments, mainly in terms of enlargement associated with increased flood flows, increased flow velocities and erosion (Brookes, 1985). Brookes' (1987) study of 46 channelised rivers in England and Wales found channel size increases of up to 146%. Detailed field study of the complex pattern of channel adjustment to channelisation in parts of Tennessee revealed the importance of bank collapse, in-channel flow deflection, recovery of stable bank form and the development of vegetated depositional surfaces along the banks (Simon and Hupp, 1986; Simon, 1989; Hupp and Simon, 1991; Chapter 11). Rhoads (1990a) has also studied adjustment to channelisation in south-central Arizona, where channel instability (involving both enlargement and degradation) accompanied the initial channel modification. Instability has been observed in channelised reaches in other areas, too. For example, Erskine (1992) attributed significant bed erosion along parts of the Hunter River in Australia to the impacts of channel straightening and reduced sediment supply by floodplain reworking. The impacts of river control works on sections of the Vistula River, Poland, have also been described by Babinski (1984) and Wyzga (1991).

Other aspects of induced channel changes have received much less attention, although some interesting studies have looked at the impacts on channel morphology of specific engineering schemes and land use changes. Illustrative of the former are studies of channel adjustments downstream from bridges, by Gregory and Brookes (1983) and Douglas (1985), which show increased widths and width/depth ratios. Land use changes are illustrated by Bird's (1982) study of channel incision along Eaglehawk Creek in Victoria, Australia, in response to land drainage and enclosure, and Odemerho's (1984) study of the relationships between channel size and basin land use in south-west Nigeria (Figure 6.6b). Channel adjustments have also been related to afforestation (Murgatroyd and Ternan, 1983), forest drainage (Tuckfield, 1980) and timber harvesting (Heede, 1985, 1991). The management implications for channel stability of wetland construction and operation have been explored in some detail by Rhoads and Miller (1990).

APPLICATIONS

Gregory (1979a) anticipated that a hallmark of channel studies in the 1980s would be a concern with applications, and this was indeed the case and continues to be so in the mid-1990s. As Gregory and Douglas (1982) point out, rivers are the most common land-forming agency, and whilst fluvial processes and adjustments are still not yet well understood, there is an emerging understanding of how change can be transmitted throughout a river network. Moreover, the application of knowledge of fluvial processes to the solution of practical environmental problems, and the reconstruction of past fluvial environments from preserved morphological sedimentological evidence are becoming more common (Gregory, 1985).

The three main applications briefly reviewed here—reflecting the emphasis in the literature—are the problems of designing stable channels, the use of channel geometry to estimate discharge at ungauged sites, and the reconstruction of past fluvial conditions and environments.

River engineering and stable channel design

Engineers and river managers have long been interested in the properties of stable natural channels, which provide invaluable clues in the design of stable artificial or regulated channels. This is true particularly where minimum engineering intervention is desired or acceptable, and where there is pressure to leave channels in as natural a state as possible (Newson, 1986). As Hey (1990) points out, geomorphological guidance in stable channel design—which reflects an understanding of the natural processes controlling channel shape and dimensions—is required to ensure that proposed schemes are feasible and sustainable. Such an approach must take into account the determinate nature of hydraulic geometry relationships (Hey, 1978) and the dynamic process–response system which gives rise to channel development and adjustment (Hey, 1979).

Most engineering solutions have been based on modelling the shape and size of steady-state channel cross-sections. A variety of approaches have been developed. Diplas and Vigilar (1992), for example, favour an approach based on numerically solving momentum balance equations for sediment particles at the point of impending motion, whereas Chang (1985a, 1986) uses an energy approach together with physical relationships of flow continuity, flow resistance and sediment transport.

Chang (1984) has also proposed a computer-based flood- and sediment-routing model which simulates river channel changes and might be usable for channel design purposes. Griffiths (1983) developed a theory of stable channel design based on temporal and spatial stability in the concentration of bed material load, whilst Ikeda and Izumi (1991) base their model on a balance of lateral bedload and lateral diffusive transport of suspended sediment. Barr *et al.* (1980) adopt a different approach again; they argue that the effects of bank slope and bed sediment load and size on cross-sectional geometry are relatively unimportant, and derive a functional relationship between channel width and depth. Some modelling exercises are inhibited by lack of real-world channel data either for comparative purposes or as basic inputs, although Brownlie's (1985) extensive compilation of laboratory and field data will doubtless prove invaluable in the future.

Estimating discharge from channel geometry

A variety of approaches have been developed for estimating peak discharge at ungauged sites on the basis of channel geometry (Wharton, 1992). Whilst the theme is explored in more detail in Chapter 15, it is appropriate to at least mention it in this overview because it is prominent in the channel cross-section literature.

Using channel morphology as a basis for predicting hydrological variables is not new, and Mosley's (1979) study of 73 rivers on South Island, New Zealand, builds on earlier work carried out mainly in the United States and Britain. Osterkamp and Hedman (1982) describe a detailed study based on 252 streamflow-gauging stations in the Missouri River Basin, which developed power-function equations relating various measures of channel geometry to various measures of discharge. Parrett and Cartier (1991) tested channel geometry-based methods of flow estimation against multiple regression equations relating streamflow to basin and climate variables, and also in comparison with concurrent flows at nearby gauging sites. They concluded that the channel geometry-based approaches are reliable, convenient and useful. Petoit and Daxhelet (1990) also estimated bankfull discharge from channel geometry on a number of rivers of different sizes in Belgium.

There are distinct advantages in using channel geometry as a basis for estimating peak discharges at ungauged sites. Hedman and Osterkamp (1982) stress that this approach is quick and inexpensive, and Wharton *et al.* (1989) show that such estimates are much more reliable than those obtained using relationships between discharge and drainage area.

Fluvial reconstruction and palaeohydrology

Palaeohydrology is based on using understanding of contemporary fluvial processes and morphology to reconstruct past fluvial conditions and environments, and the field has developed a great deal since the early 1980s. Collinson and Lewin (1983) offer a comprehensive overview of the early work.

More long-term reconstructions often involve less complete evidence, make more assumptions about relationships between form and process in contemporary channels, and infer past processes using morphological and sedimentological evidence. Kozarski's (1983) study of palaeochannel geometry along the Warta River in Poland also used palaeobotanical analyses and radiocarbon measurement to date inferred channel changes, which were attributed to Late Vistulian and Holocene changes in basin vegetation, which strongly influenced runoff conditions and in turn promoted channel adjustment. Starkel (Chapter 2) and Brown (Chapter 3) illustrate other reconstructions of channel changes during the late Pleistocene and Holocene.

The palaeohydrology studies which seek to estimate former discharges on the basis of preserved morphological evidence—particularly former channel geometry— provide an interesting development of the theme outlined in the previous section. Dury (1985), however, is critical of recent attempts to retrodict palaeodischarge which, he argues, must remain subject to great uncertainty unless former roughness can be reconstructed. Gregory (1983a) points out that most approaches to palaeohydrology are based on what is effectively a two-dimensional approach, based on the analysis of preserved channel cross-sections and on what is known of the relationship between channel geometry and peak discharge in contemporary river systems. He calls for a more three-dimensional approach, based on drainage network volume and its relationship with flood discharge.

CONCLUSIONS

This overview of recent developments in the study and understanding of river channel cross-sections has deliberately taken a broad perspective because its objective has been to illustrate the rich variety of themes and approaches which have attracted attention since about the early 1980s, in the development of many of which Ken Gregory has been actively involved if not a pioneer.

The field is alive and well, and has capitalised on the process studies which began in earnest during the 1960s and 1970s, on the channel geometry and adjustment studies which rose to prominence particularly during the 1970s and 1980s, and on the applications which have attracted increasing attention since the early 1980s. What the future holds in store for this area within fluvial geomorphology is difficult to anticipate, but towards the close of the millennium it has reached maturity, proved its worth, and inspired a great deal of interesting and illuminating research.

REFERENCES

Alexander, D. (1979) Simulation of channel morphology; problems and prospects. *Progress in Physical Geography* 3(4), 544–572.

Anderson, M.G. and Calver, A. (1982) Exmoor channel patterns in relation to the flood of 1952. *Proceedings—Ussher Society* 5(3), 362–367.

Andrews, E.A. (1982) Bank stability and channel width adjustment, East Fork River, Wyoming. *Water Resources Research* 18(4), 1184–1192.

Anthony, D.J. and Harvey, M.D. (1991) Stage-dependent cross-section adjustments in a meandering reach of Fall River, Colorado. *Geomorphology* 4(3–4), 187–203.

Babinski, Z;. (1984) The effects of human activity on changes in the Lower Vistula channel. *Geographia Polonica* 50, 271–282.

Barr, D.I.H., Alam, M.K. and Nishat, A. (1980) A contribution of regime theory relating principally to channel geometry. *Proceedings, Institution of Civil Engineers, Part 2* 69, 651–670.

Bates, B.C. (1990) A statistical log piecewise linear model of at-a-station hydraulic geometry. *Water Resources Research* **26**(1), 109–118.

Bennett, S.J. (1989) An example of channel stability; the Susquehanna River in eastern New York State. *Northeastern Geology* **11**(1), 29–39.

Beschta, R.L. (1983a) Channel changes following storm-induced hillslope erosion in the Upper Kowai Basin, Torlesse Range, New Zealand. *Journal of Hydrology (New Zealand)* **22**(2), 93–111.

Beschta, R.L. (1983b) Long-term changes in channel widths of the Kowai River, Torlesse Range, New Zealand. *Journal of Hydrology (New Zealand)* **22**(2), 112–122.

Best, J.L. (1988) Sediment transport and bed morphology at river channel confluences. *Sedimentology* **35**(3), 181–198.

Bhowmik, N.K. (1984) Hydraulic geometry of floodplains. *Journal of Hydrology* **68**(1–4), 369–401.

Bird, J.F. (1982) Channel incision at Eaglehawk Creek, Gippsland, Victoria, Australia. *Proceedings, Royal Society of Victoria* **94**(1–2), 11–22.

Bofu Yu and M.G. Wolman (1987) Some dynamic aspects of river geometry. *Water Resources Research* **23**(3), 500–509.

Bradley, C. and Smith, D.W. (1984) Meandering channel response to altered flow regime; Milk River, Alberta and Montana. *Water Resources Research* **20**(12), 1913–1920.

Bray, D.I. (1987) A study of channel changes in a reach of the North Nashwaaksis Stream, New Brunswick, Canada. *Earth Surface Processes and Landforms* **12**(2), 151–165.

Brookes, A. (1985) River channelisation; traditional engineering methods, physical consequences and alternative practices. *Progress in Physical Geography* **9**(1), 44–73.

Brookes, A. (1987) River channel adjustments downstream from channelisation works in England and Wales. *Earth Surface Processes and Landforms* **12**(4), 337–351.

Brown, A.J. (1983) Channel changes in arid badlands, Borrego Springs, California. *Physical Geography* **4**(1), 82–102.

Brownlie, W.R. (1985) Computation of alluvial channel data. *Journal of Hydraulic Engineering—ASCE* **111**(7), 1115–1119.

Caine, N. and Mool, P.K. (1981) Channel geometry and flow estimates for two small mountain streams in the Middle Hills, Nepal. *Mountain Research and Development* **1**(3–4), 231–243.

Carling, P. (1988a) The concept of dominant discharge applied to two gravel-bed streams in relation to channel stability thresholds. *Earth Surface Processes and Landforms* **13**(4), 355–367.

Carling, P.A. (1988b) Channel change and sediment transport in regulated UK rivers. *Regulated Rivers: Research and Management* **2**(3), 369–387.

Carragher, M.J., Klein, M. and Petch, J.R. (1983) Channel width–drainage area relations in small basins. *Earth Surface Processes and Landforms* **8**(2), 177–181.

Chang, H.H. (1984) Modelling of river channel changes. *Journal of Hydraulic Engineering—ASCE* **110**(2), 157–172.

Chang, H.H. (1985a) River morphology and thresholds. *Journal of Hydraulic Engineering—ASCE* **111**(3), 503–519.

Chang, H.H. (1985b) Channel width adjustment during scour and fill. *Journal of Hydraulic Engineering—ASCE* **111**(10), 1368–1370.

Chang, H.H. (1986) River channel changes; adjustments of equilibrium. *Journal of Hydraulic Engineering—ASCE* **112**(1), 43–55.

Collinson, J.D. and Lewin, J. (eds) (1983) *Modern and Ancient Fluvial Systems.* Blackwell Scientific; International Association of Sedimentologists, Special Publication 6.

Diplas, P. and Vigilar, G. (1992) Hydraulic geometry of threshold channels. *Journal of Hydraulic Engineering—ASCE* **118**(4), 597-614.

Douglas, I. (1985) Hydrogeomorphology downstream of bridges; one mechanism of channel widening. *Applied Geography* **5**(2), 167–170.

Downs, P.W. and Gregory, K.J. (1993) The sensitivity of river channels in the landscape system. In: Thomas, D.S.G. and Allison, R.J. (eds), *Landscape Sensitivity*, Wiley, Chichester, 15–30.

Dury, G.H. (1984) Abrupt variation in channel width along part of the River Severn, near Shrewsbury, Shropshire, England. *Earth Surface Processes and Landforms* 9(5), 485–492.

Dury, G.H. (1985) Attainable standards of accuracy in the retrodiction of palaeodischarges from channel dimensions. *Earth Surface Processes and Landforms* 10(3), 205–213.

Ebisemiju, F.S. (1989a) The response of headwater stream channels to urbanisation in the humid tropics. *Hydrological Processes* 3(3), 237–253.

Ebisemiju, F.S. (1989b) Patterns of stream channel response to urbanisation in the humid tropics and their implications for urban land use planning; a case study from south-western Nigeria. *Applied Geography* 9(4), 273–286.

Ebisemiju, F.S. (1991) Some comments on the use of spatial interpolation techniques in studies of man-induced river channel changes. *Applied Geography* 11(1), 21–34.

Ergenzinger, P. (1987) Chaos and order; the channel geometry of gravel bed braided rivers. *Catena Supplement* 10, 85–98.

Erskine, W.D. (1992) Channel response to large-scale river training works; Hunter River, Australia. *Regulated Rivers: Research and Management* 7(3), 261–278.

Erskine, W. and Bell, F.C. (1982) Rainfall, floods and river channel changes in the upper Hunter. *Australian Geographical Studies* 20(2), 183–196.

Erskine, W. and Melville, M.D. (1983) Impacts of the 1978 floods on the channel and floodplain of the lower Macdonald River, N.S.W. *Australian Geographer* 15(5), 284–292.

Everitt, B. (1993) Channel responses to declining flow on the Rio Grande between Ft. Quitman and Presidio, Texas. *Geomorphology* 6(3), 225–242.

Ferguson, R.I. (1981) Channel form and channel changes. In: Lewin, J. (ed.), *British Rivers*, Allen & Unwin, London, 90–125.

Ferguson, R.I. (1986) Hydraulics and hydraulic geometry. *Progress in Physical Geography* 10(1), 1–31.

Ferguson, R. and Ashworth, P. (1991) Slope-induced changes in channel character along a gravel-bed stream; the Allt Dubhaig, Scotland. *Earth Surface Processes and Landforms* 16(1), 65–82.

Frauenstein, G. (1986–1987) The hydraulic geometry of a first order stream in the Lower Great Kei Basin. *South African Geographer* 14(1–2), 107–113.

Galay, V.J., Pentland, R.S. and Halliday, R.A. (1985) Degradation from the South Saskatchewan River below Gardiner Dam. *Canadian Journal of Civil Engineering* 12(4); 849–862.

Gardiner, T. (1983) Some factors promoting channel bank erosion, River Lagan, County down. *Journal of Earth Sciences—Royal Dublin Society* 5(2), 231–239.

Gellis, A., Hereford, R., Schumm, S.A. and Hayes, B.R. (1991) Channel evolution and hydrologic variations in the Colorado River Basin; factors influencing sediment and salt loads. *Journal of Hydrology* 124(3–4), 317–344.

Graf, W.L. (1979) Mining and channel response. *Annals Association of American Geographers* 69(2), 262–275.

Graf, W.L. (1983a) Flood-related channel change in an arid-region river. *Earth Surface Processes and Landforms* 8(2), 125–139.

Graf, W.L. (1983b) Downstream changes in stream power in the Henry Mountains, Utah. *Annals Association of American Geographers* 73, 373–387.

Graf, W.L. (1984) A probabilistic approach to the spatial assessment of river channel instability. *Water Resources Research* 20(7), 953–962.

Gregory, K.J. (ed.) (1977) *River Channel Changes*. Wiley, Chichester, 448pp.

Gregory, K.J. (1978) Fluvial geomorphology. *Progress in Physical Geography* 2(2), 346–354.

Gregory, K.J. (1979a) Hydrogeomorphology; how applied should we become? *Progress in Physical Geography* 3(1), 84–101.

Gregory, K.J. (1979b) Fluvial geomorphology. *Progress in Physical Geography* 3(2), 274–282.

Gregory, K.J. (1979c) River channels. In: Gregory, K.G. and Walling, D.E. (eds), *Man and Environmental Processes; A Physical Geography Perspective*, Dawson, Folkestone, 123–143.

Gregory, K.J. (1980) Fluvial geomorphology. *Progress in Physical Geography* 4(3), 421–430.

Gregory, K.J. (1982a) River power. In: Adlam, B.H. *et al.* (eds), *Papers in Earth Studies*, Geo Books, Norwich, 1–20.

Gregory, K.J. (1982b) Fluvial geomorphology; less uncertainty and more practical application? *Progress in Physical Geography* **6**(3), 427–438.

Gregory, K.J. (1983a) Human activity and palaeohydrology; a review. *Quaternary Studies in Poland* **4**, 73–80.

Gregory, K.J. (1983b) Fluvial geomorphology. *Progress in Physical Geography* **7**(3), 385–396.

Gregory, K.J. (1984) Fluvial geomorphology. *Progress in Physical Geography* **8**(3), 421–430.

Gregory, K.J. (1985) Fluvial geomorphology—process explicit and implicit? *Progress in Physical Geography* **9**(3), 414–424.

Gregory, K.J. and Brookes, A. (1983) Hydrogeomorphology downstream from bridges. *Applied Geography* **3**(2), 145–159.

Gregory, K.J. and Douglas, I. (1982) Rivers must be managed. *Geographical Magazine* **54**(2), 104–106.

Gregory, K.J. and Gurnell, A.M. (1988) Vegetation and river channel form and process. In: Viles, H.A. (ed.), *Biogeomorphology*, Blackwell, Oxford, 11–42.

Gregory, K.J. and Park, C.C. (1974) Adjustment of river channel capacity downstream from a reservoir. *Water Resources Research* **10**, 840–873.

Gregory, K.J., Davis, R.J. and Downs, P.W. (1992) Identification of river channel change due to urbanisation. *Applied Geography* **12**(4), 299–318.

Griffiths, G.A. (1981) Hydraulic geometry relationships of some New Zealand gravel bed rivers. *Journal of Hydrology (New Zealand)* **19**(2), 106–118.

Griffiths, G.A. (1983) Stable-channel design in alluvial rivers. *Journal of Hydrology* **65**(4), 259–270.

Griffiths, G.A. (1984) Extremal hypotheses for river regime; an illusion of progress. *Water Resources Research* **20**(1), 113–118.

Gupta, A. (1983) High-magnitude floods and stream channel response. In: Collinson, J.D. and Lewin, J. (eds), *Modern and Ancient Fluvial Systems*, Blackwell Scientific, International Association of Sedimentologists, Special Publication 6, 219–227.

Gurnell, A.M. (1983) Downstream channel adjustments in response to water abstraction from hydro-electric power generation from alpine glacial melt-water streams. *Geographical Journal* **149**(3), 342–354.

Gurnell, A.M., Clark, M.J. and Hill, C.T. (1990) The geomorphological impact of modified river discharge and sediment transport regimes downstream of hydropower scheme melt-water intake structures. In: Sinniger, R.O and Monbaron, M. (eds), *Hydrology in Mountainous Regions II*, IAHS Publication 194, 165–170.

Hamlett, J.M., Baker, J.L. and Johnson, H.P. (1983) Channel morphology changes and sediment yield for a small agricultural watershed in Iowa. *Transactions—American Society of Agricultural Engineers* **26**(5), 1390–1396.

Harvey, A.M. (1991) The influence of sediment supply on the channel morphology of upland streams; Howgill Fells, Northwest England. *Earth Surface Processes and Landforms* **16**(7), 675–684.

Hedman, E.R. and Osterkamp, W.R. (1982) Streamflow characteristics related to channel geometry of streams in western United States. *US Geological Survey Water-Supply Paper* 2193, 17pp.

Heede, B.H. (1985) Application of geomorphological concepts to evaluate timber harvest influences on a stream channel—a case study. *Zeitschrift fur Geomorphologie, Supplement-band* **55**, 121–130.

Heede, B.H. (1991) Response of a stream in disequilibrium to timber harvest. *Environmental Management* **15**(2), 251–255.

Hey, R.D. (1978) Determinate hydraulic geometry of river channels. *Proceedings of the American Society of Civil Engineers, Journal of the Hydraulics Division* **104** (HY6), 869–885.

Hey, R.D. (1979) Dynamic process-response model of river channel development. *Earth Surface Professes and Landforms* **4**, 59–72.

Hey, R.D. (1990) Environmental river engineering. *Journal—Institution of Water and Environmental Management* **4**(4), 335–340.

Hickin, E.J. (1983) River channel change; retrospect and prospect. In: Collinson, J.D. and

Lewin, J. (eds), *Modern and Ancient Fluvial Systems*, Blackwell Scientific, International Association of Sedimentologists, Special Publication 6, 61–83.

Hogan, D.L. and Church, M. (1989) Hydraulic geometry in small, coastal streams; progress towards quantification of salmonid habitat. *Canadian Journal of Fisheries and Aquatic Sciences* 46(5), 844–852.

Howard, A.D. (1980) Thresholds in river regimes, In: Coates, D.R. and Vitek, J.D. (eds), *Thresholds in Geomorphology*, George Allen & Unwin, London, 227–258.

Hupp, C.R. and Simon, A. (1991) Bank accretion and the development of vegetated depositional surfaces along modified alluvial channels. *Geomorphology* 4(2), 111–124.

Ikeda, S. and Izumi, N. (1990) Width and depth of self-formed straight gravel rivers with bank vegetation. *Water Resources Research* 26(10), 2353–2364.

Ikeda, S. and Izumi, N. (1991) Stable channel cross sections of straight sand rivers. *Water Resources Research* 27(9), 2429–2438.

Inbar, M. (1990) Effects of dams on mountainous bedrock rivers. *Physical Geography* 11(4), 305–319.

Inokuchu, M. and Sasaki, T. (1985) Channel morphology of the lower Ishikari River, Hokkaido, Japan. *Transactions—Japanese Geomorphological Union* 6(2), 87–100 (in Japanese).

James, L.A. (1989) Sustained storage and transport of hydraulic gold mining sediment in the Bear River, California. *Annals Association of American Geographers* 79(4), 570–592.

James, L.A. (1991) Incision and morphological evolution of an alluvial channel recovering from hydraulic mining sediment. *Bulletin, Geological Society of America* 103(6), 723–736.

Kale, V.S. (1990) Morphological and hydrological characteristics of some allochthonous river channels, western Deccan Trap Upland region, India. *Geomorphology* 3(1), 31–43.

Kale, V.S., Karlekar, S.N. and Deodhar, L.A. (1986) Channel morphology and hydraulic characteristics of Vashishti River, Maharashtra (India). *Transactions—Institute of Indian Geographers* 8(2), 113–126.

Kennedy, B.A. (1984) On Playfair's law of accordant junctions. *Earth Surface Processes and Landforms* 9(2), 153–173.

Klein, M. (1981) Drainage area and the variation of channel geometry downstream. *Earth Surface Processes and Landforms* 6(6), 589–593.

Klein, M. (1982) The relation between channel geometry and suspended sediment transport in the downstream direction. *Zeitschrift fur Geomorphologie* 26(4), 491–494.

Knighton, A.D. (1975) Variations in at-a-station hydraulic geometry. *American Journal of Science* 275, 186–218.

Knighton, A.D. (1979) Comments on log-quadratic relations in hydraulic geometry. *Earth Surface Processes* 4(3), 205–209.

Knighton, A.D. (1981a) Channel form and flow characteristics of supraglacial streams, Austre Okstinbreen, Norway. *Arctic and Alpine Research* 13(3), 295–306.

Knighton, A.D. (1981b) Asymmetry of river channel cross-sections; part I. Quantitative indices. *Earth Surface Processes and Landforms* 6(6), 581–588.

Knighton, A.D. (1985) Channel form adjustment in supraglacial streams, Austre Okstindbreen, Norway. *Arctic and Alpine Research* 17(4), 451–466.

Knighton, A.D. (1987a) Streamflow characteristics of north-eastern Tasmania: II. Hydraulic geometry. *Papers and Proceedings—Royal Society of Tasmania* 121, 125–135.

Knighton, A.D. (1987b) River channel adjustment—the downstream adjustment. In: Richards, K.S. (ed.), *River Channels*, Blackwell, Oxford, 95–128.

Knighton, A.D. (1988a) River channel adjustment. *Geography Review* 1(4), 9–14.

Knighton, A.D.(1988b) The impact of the Parangana Dam on the River Mersey, Tasmania. *Geomorphology* 1(3), 221–237.

Knighton, A.D. (1989) River adjustment to changes in sediment load; the effects of tin mining on the Ringarooma River, Tasmania, 1875-1984. *Earth Surface Processes and Landforms* 14(4), 333–359.

Knighton, A.D. (1991) Channel bed adjustment along mine-affected rivers of Northeast Tasmania. *Geomorphology* 4(3–4), 215–219.

Knighton, A.D. and Cryer, R. (1990) Velocity–discharge relationships in three lowland rivers. *Earth Surface Processes and Landforms* 15(6), 501–512.

Kozarski, S. (1983) River channel changes in the middle reach of the Warta Valley, Great Poland Lowland. *Quaternary Studies in Poland* **4**, 159–169.

Lane, L.J. *et al.* (1982) Relationships between morphology of small streams and sediment yield. *Journal of the Hydraulics Division, ASCE* **108**(HY11), 1328–1365.

Leopold, L.B. and Maddock, T. (1953) The hydraulic geometry of stream channels and some physiographic implications. *US Geological Survey Professional Paper 252*, 56pp.

Lewin, J. (ed.) (1981) *British Rivers. George Allen & Unwin, London*, 216pp.

Lisle, T.E. (1982) Effects of aggradation and degradation on riffle–pool morphology in natural gravel channels, north-western California. *Water Resources Research* **19**(2), 463–471.

Martin, C.W. and Johnson, W.C. (1987) Historical channel narrowing and riparian vegetation an expansion in the Medicine Lodge River basin, Kansas, 1871–1983. *Annals Association of American Geographers* **77**(3), 436–449.

Miller, S.O., Ritter, D.F., Kochel, R.C. and Miller, J.R. (1993) Fluvial responses to land-use changes and climatic variations within the Drury Creek watershed, southern Illinois. *Geomorphology* **6**(4), 309–329.

Miller, T.K. (1984) A system model of stream-channel shape and size. *Bulletin. Geological Society of America* **95**(2), 237–241.

Miller, T.K. (1991a) A model of stream channel adjustment; assessment of Rubey's hypothesis. *Journal of Geology* **99**(5), 699–710.

Miller, T.K. (1991b) An assessment of the equable change principle in at-a-station hydraulic geometry. *Water Resources Research* **27**(10, 2751–2758.

Milnes, J.A. (1983) Variation in cross-sectional asymmetry of coarse bedload river channels. *Earth Surface Processes and Landforms* **8**(5), 503–511.

Montgomery, K. (1989) Concepts of equilibrium and evolution in geomorphology; the model of branch systems. *Progress in Physical Geography* **13**(1), 47–66.

Morisawa, M. (1985) *Rivers; Form and Process*. Longman, Harlow.

Morisawa, M. and Lafleure, E. (1982) Hydraulic geometry, stream equilibrium and urbanisation. In: Rhodes, D.D. and Williams, G.P. (eds), *Adjustments of the Fluvial System*, Allen & Unwin, London, 333–350.

Mosley, M.P. (1979) Prediction of hydrological variables from channel morphology, South Island rivers. *Journal of Hydrology (New Zealand)* **18**(2), 109–120.

Mosley, M.P. (1982) Analysis of the effect of changing discharge on channel morphology and instream uses in a braided river, Ohau River, New Zealand. *Water Resources Research* **18**(4), 800–812.

Murgatroyd, A.L. and Ternan, J.L. (1983) The impact of afforestation on stream bank erosion and channel form. *Earth Surface Processes and Landforms* **8**(4), 357–369.

Murphey, J.B. and Grissinger, E.H. (1985) Channel cross-section changes in Mississippi's Goodwin Creek. *Journal of Soil and Water Conservation* **40**(1), 148–153.

Nanson, G.C. and Young, R.W. (1981) Downstream reduction of rural channel size with contrasting urban effects in small coastal streams of south-eastern Australia. *Journal of Hydrology* **52**(3–4), 239–255.

Neller, R.J. (1980) Channel changes on the Macquarie Rivulet, New South Wales. *Zeitschrift fur Geomorphologie* **24**(2), 168–179.

Neller, R.J. (1988) A comparison of channel erosion in small urban and rural catchments, Armidale, New South Wales. *Earth Surface Processes and Landforms* **13**(1), 1–7.

Neller, R.J. (1989) Induced channel enlargement in small urban catchments, Armidale, New South Wales. *Environmental Geology and Water Sciences* **14**(3), 167–171.

Newson, M.D. (1986) River basin engineering—fluvial geomorphology. *Journal—Institution of Water Engineers and Scientists* **40**(4), 307–324.

Nolan, K.M. and Marron, D.C. (1985) Contrast in stream-channel response to major storms in two mountainous areas of California. *Geology* **13**(2), 135–138.

Odemerho, F.O. (1984) The effects of shifting cultivation on stream channel size and hydraulic geometry in small headwater basins of south-western Nigeria. *Geografiska Annaler, Series A* **66A**(4), 327–340.

Odemerho, F.O. (1992) Limited downstream response of stream channel size to urbanisation in a humid tropical basin. *Professional Geographer* **44**(3), 332–339.

Osterkamp, W.R. and Hedman, E.R. (1982) Perennial-streamflow characteristics related to channel geometry and sediment in Missouri River basin. *US Geological Survey Professional Paper* 1242, 37pp.

Page, K.J. and Heerdegen, R.G. (1985) Channel change on the lower Manawatu River. *New Zealand Geographer* **41(1), 35–38.**

Paine, A.D.M. (1985) 'Ergodic' reasoning in geomorphology; time for a review of the term? *Progress in Physical Geography* 9(1), 1–15.

Park, C.C. (1977a) World-wide variations in hydraulic geometry exponents of stream channels—an analysis and some observations. *Journal of Hydrology* 33, 133–146.

Park, C.C. (1977b) Man-induced changes in stream channel capacity. In: Gregory, K.J., (ed.) *River Channel Changes*, Wiley, Chichester, 121–144.

Park, C.C. (1978) Allometric analysis and stream channel morphometry. *Geographical Analysis* 10, 211–228.

Park, C.C. (1979) Tin streaming and channel changes; some preliminary observations from Dartmoor, England. *Catena* 6(3–4), 235–244.

Park, C.C. (1981a) Hydraulic geometry of a supra-glacial stream; some observations from the Val d'Herens, Switzerland. *Revue de Geomorphologie Dynamique* 30(1), 1–9.

Park, C.C. (1981b) Man, river systems, and environmental impacts. *Progress in Physical Geography* 5, 1–31.

Parrett, C. and Cartier, K.D. (1991) Methods for estimating monthly streamflow characteristics at ungaged sites in western Montana *US Geological Survey Water Supply Paper* 2365, 30pp.

Petit, F. and Daxhelet, C. (1990) Evaluation of channel capacity at the bankfull stage and of the bankfull discharge recurrence interval for various rivers of mid- and upland Belgium. *Bulletin—Societe Geographique de Liege* 25, 69–84 (in French).

Petts, G.E. (1979) Complex response of river channel morphology subsequent to reservoir construction. *Progress in Physical Geography* 3(3), 329–362.

Petts, G.E. (1980a) Implications of the fluvial process–channel morphology interaction below British reservoirs for stream habitats. *Science of the Total Environment* 16(2), 149–163.

Petts, G.E. (1980b) Long-term consequences of upstream impoundment. *Environmental Conservation* 7(4), 325–332.

Petts, G.E. (1982) Channel changes in regulated rivers. In: Adlam, B.H. *et al.* (eds), *Papers in Earth Studies*, Geo Books, Norwich, 117–142.

Petts, G.E. (1984) Sedimentation within a regulated river. *Earth Surface Processes and Landforms* 9(2), 125–134.

Petts, G. (1987) The regulation of the Derbyshire Derwent. *East Midland Geographer* 10(2), 54–63.

Petts, G.E. and Pratts, J.D. (1983) Channel changes following reservoir construction on a lowland English river. *Catena* 10(1–2), 77–85.

Petts, G.E. and Thoms, M.C. (1986) Channel aggradation below Chew Valley Lake, Somerset. *Catena* 13(3), 305–320.

Phillips, C.J. (1988) Geomorphic effects of two storms on the upper Waitahaia River catchment, Raukumara Peninsula, New Zealand. *Journal of Hydrology (New Zealand)* **27(2), 99–112.**

Phillips, J.D. (1988) The role of spatial scale in geomorphic systems. *Geographical Analysis* 20(4), 308–317.

Phillips, J.D. (1990) The instability of hydraulic geometry. *Water Resources Research* 26(4), 739–744.

Phillips, J.D. (1991) Multiple modes of adjustment in unstable river channel cross-sections. *Journal of Hydrology* 123(1–2), 39–49.

Phillips, P.J. and Harlin, J.M. (1984) Spatial dependency of hydraulic geometry exponents in a sub-alpine stream. *Journal of Hydrology* **71(3–4), 277–283.**

Pitlick, J. (1993) Response and recovery of a subalpine stream following a catastrophic flood. *Bulletin, Geological Society of America* 105(5), 657–670.

Pizzuto, J.E. (1984) Bank erodibility of shallow sandbed streams. *Earth Surface Processes and Landforms* 9(2), 113–124.

Pizzuto, J.E. (1992) The morphology of graded gravel rivers; a network perspective. *Geomorphology* **5**(3–5), 457–474.

Rasid, H. (1979) The effects of regime regulation by the Gardiner Dam on downstream geomorphic processes in the South Saskatchewan River. *Canadian Geographer* **23**(2), 140–158.

Rhoads, B.L. (1987) Changes in stream channel characteristics at tributary junctions. *Physical Geography* **8**(4), 346–361.

Rhoads, B.L. (1988) Mutual adjustments between process and form in a desert mountain fluvial system. *Annals Association of American Geographers* **78**(2), 271–287.

Rhoads, B.L. (1990a) The impact of stream channelisation on the geomorphic stability of an arid-region river. *National Geographic Research* **6**(2), 157–177.

Rhoads, B.L. (1990b) Hydrologic characteristics of a small desert mountain stream; implications for short-term magnitude and frequency of bedload transport. *Journal of Arid Environments* **18**(2), 151–163.

Rhoads, B.L. (1991a) A continuously varying parameter model of downstream hydraulic geometry. *Water Resources Research* **27**(8), 1865–1872.

Rhoads, B.L. (1991b) Impact of agricultural development on regional drainage in the lower Santa Cruz Valley, Arizona, USA. *Environmental Geology and Water Sciences* **18**(2), 119–135.

Rhoads, B.L. (1992a) Fluvial geomorphology. *Progress in Physical Geography* **16**(4), 456–499.

Rhoads, B.L. (1992b) Statistical models of fluvial systems. *Geomorphology* **5**(3–5, 433–455.

Rhoads, B.L. and Miller, M.V. (1990) Impact of riverine wetland construction and operation on stream channel stability; conceptual framework for geomorphic assessment. *Environmental Management* **14**(6), 799–807.

Rhoads, B.L. and Miller, M.V. (1991) Impact of flow variability on the morphology of a low-energy meandering river. *Earth Surface Processes and Landforms* **16**(4), 357–367.

Rhodes, D.D. (1977) The b-f-m diagram; graphical representation and interpretation of at-a-station hydraulic geometry. *American Journal of Science* **277**, 73–96.

Rhodes, D.D. (1978) Discussion: World-wide variations in hydraulic geometry exponents of stream channels; an analysis and some observations (by Chris C. Park). *Journal of Hydrology* **39**(1–2), 193–197.

Rhodes, D.D. (1987) The b-f-m diagram for downstream hydraulic geometry. *Geografiska Annaler, Series A* **69A**(1), 147–161.

Rhodes, D.D. and Williams, G.P. (eds) (1982) *Adjustments of the Fluvial System.* Allen & Unwin, London.

Rice, R.J. (1982) The hydraulic geometry of the lower portion of the Sunwapta River valley train, Jasper National Park, Alberta. In: Davidson-Arnott, R, Nickling, W. and Fahey, B.D. (eds), *Research in Glacial, Glacio-fluvial, and Glacio-lacustrine Systems.* Geo Books, Norwich, 141–173.

Richards, K.S. (1973) Hydraulic geometry and channel roughness—a non-linear system. *American Journal of Science* **273**, 877–896.

Richards, K.S. (1976) Complex width–discharge relations in natural river systems. *Bulletin, Geological Society of America* **87**, 199–206.

Richards, K.S. (1980a) A note on changes in channel geometry at tributary junctions. *Water Resources Research* **16**(1), 241–244.

Richards, K.S. (1980b) Influence of suspended kaolinite on the channel form of a polluted stream in Cornwall. *Clays and Clay Minerals* **28**(2), 157–159.

Richards, K.S. (1981) River channel forms. In: Goudie, A. *et al.* (eds), *Geomorphological Techniques,* Allen & Unwin, London, 56–61.

Richards, K.S. (1982) Channel adjustment to sediment pollution by the china clay industry in Cornwall, England. In: Rhodes, D.D. and Williams, G.P. (eds), *Adjustments of the Fluvial System,* Allen & Unwin, London, 309–331.

Richards, K.S. (1987a) Rivers: environment, process and form. In: Richards, K.S. (ed.), *River Channels,* Blackwell, Oxford, 1–13.

Richards, K.S. (1987b) Fluvial geomorphology. *Process in Physical Geography* **11**(3), 432–457.

Richards, K.S. and Ferguson, R.I. (1987) River channel dynamics—introduction. In: Gardiner, V. (ed.), *International Geomorphology 1986, Proceedings of the 1st Conference, Volume 1*, Wiley, Chichester, 541–547.

Richards, K.S. and Greenhalgh, C. (1984) River channel change; problems of interpretation illustrated by the River Derwent, north Yorkshire. *Earth Surface Processes and Lanforms* **9**(2), 175–180.

Ridenour, G.S. and Giardino, J.R. (1991) The statistical study of hydraulic geometry; a new direction for compositional data analysis. *Mathematical Geology* **23**(3), 349–366.

Robinson, E.G. and Beschta, R.L. (1989) Estimating stream cross-sectional area from wetted width and thalweg depth. *Physical Geography* **10**(2), 190–198.

Robinson, E.G. and Beschta, R.L. (1990) Coarse woody debris and channel morphology interactions for undisturbed streams in Southeast Alaska, USA. *Earth Surface Processes and Landforms* **15**(2), 149–156.

Roy, A.G. (1983) Optimal angular geometry models of river branching. *Geographical Analysis* **15**(2), 87–96.

Roy, A.G. and Roy, R. (1988) Changes in channel size at river confluences with coarse-bed material. *Earth Surface Processes and Landforms* **13**(1), 77–84.

Roy, A.G., Roy, R. and Bergeron, N. (1988) Hydraulic geometry and changes in flow velocity at a river confluence with coarse bed material. *Earth Surface Processes and Landforms* **13**(7); 583–598.

Sarma, J.N. and Basumallick, S. (1986) Channel form and process of the Burhi Dihing River, India. *Geografiska Annaler, Series A* **68A**(4), 373–382.

Schoof, R.R., Gander, G.A. and Welch, N.H. (1987) Effect of floodwater-retarding reservoirs on selected channels in Oklahoma. *Journal of Soil and Water Conservation* **42**(2), 124–127.

Shepherd, R.G. (1982) River channel and sediment responses to bedrock lithology and stream capture. In: Rhodeds, D.D. and Williams, G.P. (eds), *Adjustments of the Fluvial System*, Allen & Unwin, London, 255–275.

Sherrard, J.J. and Erskine, W.D. (1991) Complex response of a sand-bed stream to upstream impoundment. *Regulated Rivers: Research and Management* **6**(1), 53–70.

Simon, A. (1989) A model of channel response in disturbed alluvial channels. *Earth Surface Processes and Landforms* **14**(1), 11–26.

Simon, A. and Hupp, C.R. (1986) Channel widening characteristics and bank slope development along a reach of Cane Creek, west Tennessee. *US Geological Survey Water Supply Paper* 2290, 113–126.

Smith, N.G. and Smith, D.G. (1984) William River; an outstanding example of channel widening and braiding caused by bedload addition. *Geology* **12**(2), 78–82.

Thorne, C.R. (1991) Analysis of channel instability due to catchment land-use change. *In: Peters, N.E. and Walling, D.e. (eds), Sediment and Stream Water Quality in a Changing Environment*, IAHS Publication 203, 111–122.

Touysinthiphonexay, K.C.N. and Gardner, T.W. (1984) Threshold response of small streams to surface coal mining, bituminous coal fields, central Pennsylvania. *Earth Surface Processes and Landforms* **9**(1), 43–58.

Tuckfield, C.G. (1980) Stream channel stability and forest drainage in the New Forest, Hampshire. *Earth Surface Processes and Landforms* **5**(4), 317–329.

Warner, R.F. and Bird, J.F. (1988) Human impacts on river channels in New South Wales and Victoria. In: Warner, R.F. (ed.), *Fluvial Geomorphology of Australia*, Academic Press, 343–363.

Wharton, G. (1992) Flood estimation from channel size; guidelines for using the channel-geometry method. *Applied Geography* **12**(4), 339–359.

Wharton, G., Arnell, N.W., Gregory, K.J. and Gurnell, A.M. (1989) River discharge estimated from channel dimensions. *Journal of Hydrology* **106**(3–4), 365–376.

White, W.B. and Deike, G.H. (1989) Hydraulic geometry of cave passages. In: White, W.B. and White, E.L. (eds), *Karst Hydrology; Concepts from the Mammoth Cave Area*, Van Nostrand Reinhold, 223–258.

Whitlow, J.R. and Gregory, K.J. (1989) Changes in urban stream channels in Zimbabwe. *Regulated Rivers: Research and Management* **4**(1), 27–42.

Williams, G.P. (1978) Hydraulic geometry of river cross-sections; theory of minimum variance. *US Geological Survey Professional paper* 1029, 47pp.

Williams, G.P. (1987) Unit hydraulic geometry—an indicator of channel changes. *US Geological Survey Water Supply Paper* **2330, 77–89.**

Williams, G.P. and Wolman, M.G. (1984) Downstream effects of dams on alluvial rivers, *US Geological Survey Professional Paper* 1286, 83pp.

Wyzga, B. (1991) Present-day downcutting of the Raba River channel (western Carpathians, Poland) and its environmental effects. *Catena* **18**(6), 551–566.

Xu, J. (1990a) An experimental study of complex response in the channel adjustment downstream from a reservoir. *Earth Surface Processes and Landforms* **15**(1), 43–54.

Xu, J. (1990b) Complex response in response in adjustment of the Weihe River channel to the construction of the Sanmenxia Reservoir. *Zeitschrift fur Geomorphologie* **34**(2), 233–245.

Yang, C.T. and Song, C.C.S. (1982) Dynamic adjustments of alluvial channels. In: Rhodes, D.D. and Williams, G.P. (eds), *Adjustments of the Fluvial System*. Allen & Unwin, London, 55–67.

Yang, C.T., Song, C.C.S and Woldenberg, M.J. (1981) Hydraulic geometry and minimum rate of energy dissipation. *Water Resources Research* **17**(4), 1014–1018.

Zeng Qinghua and Zhou Wenhao (1991) Some characteristics of sediment transport in the lower Yellow River. In: Peters, N.E. and Walling, D.E. (eds), *Sediment and Stream Water Quality in a Changing Environment*, IAHS Publication 203, 133–142.

Part II

PROCESSES OF CHANGE

7 Suspended Sediment Yields in a Changing Environment

D.E. WALLING

Department of Geography, University of Exeter, UK

THE CONTEXT

It was 35 years ago that Frédéric Fournier published his classic monograph entitled *Climat et Erosion*, which attempted to review existing information on the sediment yields of world rivers and to identify the primary controls on the global pattern of denudation (Fournier, 1960). Although more recent increases in data availability have inevitably caused some of Fournier's assumptions and conclusions to be questioned (cf. Walling and Webb, 1987), the potential for using information on the sediment loads of the world's rivers to provide a general overview of rates and patterns of global denudation undoubtedly remains. Jansson (1982, 1988), Milliman and Meade (1983), Walling and Webb (1983a), Dedkov and Mozzherin (1984) and Milliman and Syvitski (1992) have, for example, taken advantage of the order of magnitude or more increase in the number of world rivers for which sediment yield data are available to provide updated assessments of the yields and patterns involved. Previous uncertainties regarding the magnitude of the total annual suspended sediment transport from the land to the oceans have been largely resolved and a value of *c.* 15×10^9 tonnes is currently widely accepted for this flux. Maximum specific suspended sediment yields are now known to be of the order of $50\,000\,\mathrm{t\,km^{-2}\,year^{-1}}$, and the maps of the global distribution of suspended sediment yield produced by Walling and Webb (1983a), Dedkov and Mozzherin (1984) and Lvovitch *et al.* (1991) have established the major patterns and controls involved.

Current assessments of the global pattern of suspended sediment yield and the loads involved must, however, be seen as representing a snapshot, albeit a multi-exposure composite image, of the global denudational system, based on recent measurements. It is important to recognise that suspended sediment yields may be highly sensitive to changes in climate, land use and other environmental conditions and that the system is a dynamic one. Meade (1969) has emphasised that present-day sediment yields may not be representative of the geological past and there is growing evidence that rates of erosion and sediment yield have increased significantly in many areas of the world during the recent past and that such increases are continuing. Table 7.1, for example, contrasts typical rates of soil erosion documented under natural undisturbed conditions with those occurring in cultivated areas. In all cases there is an order of magnitude increase, and in several instances the increases are

Changing River Channels. Edited by Angela Gurnell and Geoffrey Petts.
© 1995 John Wiley & Sons Ltd.

Table 7.1 A comparison of soil erosion rates (in kg m^{-2} year^{-1}) under natural undisturbed conditions and under cultivation in selected areas of the world. Based on Morgan (1986)

Country	Natural	Cultivated
China	<0.20	15.00–20.00
USA	0.003–0.30	0.50–17.00
Ivory Coast	0.003–0.02	0.01– 9.00
Nigeria	0.05 –0.10	0.01– 3.50
India	0.05 –0.10	0.03– 2.00
Belgium	0.01 –0.05	0.30– 3.00
UK	0.01 –0.05	0.01– 0.30

even greater. The global implications of the data presented in Table 7.1 become clearer when it is recognised that the area of the earth's surface given over to crop production and livestock grazing has increased by more than five-fold over the past 200 years (cf. Buringh and Dudal, 1987) and that the recent ISRIC/UNEP global survey of human-induced soil degradation (Oldeman *et al.*, 1991) has shown that nearly 10% of the total land surface of the globe is currently adversely affected by water erosion.

A substantial proportion of the eroded material generated by any increase in either the incidence or intensity or water erosion is likely to find its way into rivers and there are reports of increased sediment yields in many areas of the world, and particularly in developing countries (cf. Douglas, 1967). Figure 7.1, based on the work of Abernethy (1990), provides an example of the magnitude of the increase documented for several reservoir catchments in Southeast Asia which have experienced substantial land clearance and intensification of land use. Data obtained from reservoir surveys at different times in the past have been used to demonstrate that sediment yields in these catchments have evidenced annual rates of increase of between 2.5 and 6.0%. Abernethy (1990) suggested that these increases closely paralleled the rate of population growth in the areas concerned (cf. Figure 7.1, inset), although he found the ratio of the rate of increase of sediment yield to that for the population to be greater than unity. Based on this evidence he suggested that annual suspended sediment yields in many developing countries were currently increasing at a rate equivalent to 1.6 times the rate of population increase and could therefore be expected to double in about 20 years. Contrasting situations will, however, exist in other areas of the world where reservoirs now trap a major proportion of the sediment formerly transported by the rivers. In the classic case of the River Nile, the construction of the Aswan High Dam has caused the annual suspended sediment yield at the mouth of this major river to decrease from *c.* 100 million tonnes to near zero.

With the current concern for global change and the impact of both climate change and human activity on the global system, promoted by international programmes such as the International Geosphere Biosphere Programme, there is clearly a need to consider changes in sediment yield as a key component of the system and to assess their sensitivity to environmental change. Knowledge concerning changing

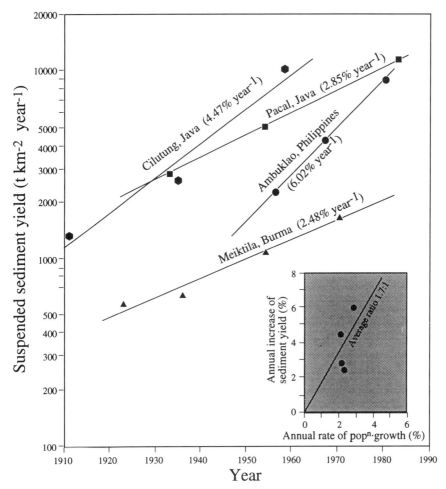

Figure 7.1 Trends of increasing sediment yields in selected reservoir catchments in Southeast Asia (based on Abernethy, 1990)

sediment fluxes is important in relation to studies of land degradation, terrestrial inputs to the oceans, and global element budgets, and such changes can also have wider environmental, economic and social implications relating to aquatic habitats, river management, loss of valuable reservoir storage and adverse effects on river water quality (cf. Clark *et al.*, 1985; Mahmood, 1987; Braune and Looser, 1989). This contribution attempts to review available evidence concerning the sensitivity of sediment yields to environmental change and the magnitude of the changes involved.

CHANGING SEDIMENT YIELDS

In comparison with other hydrological and hydrometeorological parameters, such as annual runoff, river floods and annual precipitation, information on the sensitivity of river basin sediment yields to environmental change is more difficult to assemble, due to the general absence of reliable long-term records of sediment yield in most areas of the world and the lack of obvious surrogates. Few, if any, records extend back beyond the present century and, in view of the many problems associated with obtaining accurate estimates of river loads (cf. Walling and Webb, 1981, 1987, 1988), the reliability of early records is frequently open to question. Additional sources of information must therefore also be exploited. These sources represent a variety of time-scales ranging from a long-term geological perspective, through the evidence afforded by lake sediments which extends back over time-scales of 10^2 to 10^3 years, to recent catchment experiments and attempts to use contemporary data to provide a longer-term perspective by means of space–time substitution. The evidence furnished by these various sources of information can be reviewed in turn.

RECONSTRUCTING PAST SEDIMENT YIELDS

The long-term geological perspective

Information on the present-day distribution of sedimentary rocks of different ages on the earth's surface and their associated mass can provide a basis for estimating the sedimentation rate at different periods in the past (cf. Gregor, 1985). Since, in very broad terms, these estimates of sedimentation rate can be equated with the global erosion rate, it is possible to derive estimates of global denudation rates in the past. These values will include both mechanical and chemical denudation, but, because the two are closely related (cf. Walling and Webb, 1983b), and the former dominates the total denudation rate, they can provide a useful indication of gross variations in global erosion over the past 500 million years. Coupled with estimates of the areal extent of the land area at different times in the geological past, these values can in turn be used to generate estimates of changing rates of specific sediment yield over this period. This approach has been used by Tardy et al. (1989) to reconstruct the temporal pattern of global specific sediment yield over the past 500 million years depicted in Figure 7.2A. Figure 7.2A indicates that global specific sediment yields have ranged between about 30 t km^{-2} year^{-1} and 70 t km^{-2} year^{-1} during the geological past. These variations reflect fluctuations in the global climate, and more particularly in runoff amounts, as well as in vegetation cover, the relief of the land masses, and tectonic activity. The Cambrian, Devonian, Cretaceous and Tertiary periods were thus characterised by relatively high sediment yields, whereas values in the Carboniferous, Permian and Triassic were substantially lower. The Devonian period (c. 350–400 $\times$ 10^6 years BP), which is marked by relatively high sediment yields, is known to have been a particularly wet period with high runoff rates.

Also of significance in Figure 7.2A is the trend of increasing sediment yields towards the present, which Tardy et al. (1989) have represented by an estimate of contemporary sediment yield (i.e. 108 t km^{-2} year^{-1}). This estimate is not strictly comparable to those based on the mass of accumulated sediments and it could also be seen as being too low, since it does not include the dissolved load associated with chemical denudation. In this respect a value closer to 125 t km^{-2} year^{-1} would be more appropriate. Nevertheless, if as seems reasonable, the overall trend is accepted

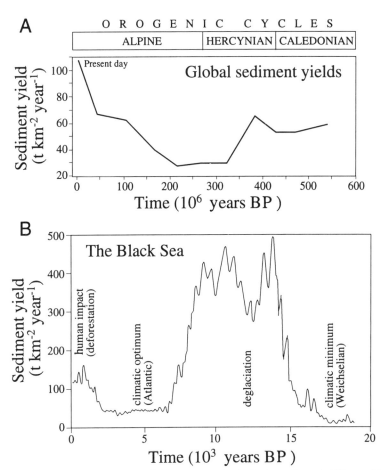

Figure 7.2 Variations of sediment yield over geological time. Based (A) on Tardy *et al.* (1989) and (B) on Degens *et al.* (1991)

as meaningful, at least part of the increase can be ascribed to the increasing rates of tectonic activity that have prevailed since the end of the Jurassic period (*c.* 130 × 10^6 years BP). However, the present-day estimate must also reflect the influence of human activity, and more particularly forest clearance and agricultural activity, in increasing sediment yields. The importance of these latter effects have been emphasised by Milliman *et al.* (1987) in their study of rates of Holocene sedimentation in the Yellow and East China Seas which receive sediment from the Yellow River. In this case the recent river input was estimated to be approaching an order of magnitude greater than that existing in the early and middle Holocene, as a result of land clearance and agricultural development in the loess region of the Middle Yellow River Basin.

The estimates of past global sediment yields presented in Figure 7.2A serve to emphasise the 'natural' variability of sediment yields in response to long-term environmental change and more particularly changes in climate and tectonic activity. However, they also highlight the important role of human activity in causing the total sediment yield from the land surface of the earth to increase approximately three-fold relative to the geological background. More detailed information regarding fluctuations in sediment yields over the recent geological past may be usefully introduced by considering sedimentary evidence from the Black Sea, where detailed analysis of sediment cores (cf. Ross and Degens, 1974) has been used by Degens *et al.* (1976, 1991) to reconstruct the record of sediment input from its *c.* 2.3 × 10^6 km^2 catchment area over the past 20 000 years (cf. Figure 7.2B). This record demonstrates that the sediment input to the Black Sea, and therefore sediment yields from its catchment area, were relatively low during the Weichselian glaciation. The sediment input increased dramatically during the subsequent period of deglaciation in response to the increased runoff, the abundant supply of sediment exposed by the retreating ice and the lack of vegetation cover, but slowly declined towards the Atlantic climatic optimum when a relatively dense vegetation cover would have existed. The significant increase in sediment inputs during the past 2000 years has been directly related to the impact of human activity, and more specifically deforestation and development of agriculture within the area, which has caused sediment yields to increase by a factor of about three. The data presented in Figure 7.2B thus again afford a useful means of demonstrating the impact of human activity on sediment yields in this region, but they also indicate that natural climatic variations, particularly those associated with periods of glaciation and deglaciation, may cause even greater changes in sediment yield.

Although it is inevitably fraught with uncertainties, particularly as the time-scale involved increases, reconstruction of temporal patterns of sediment yield during the geological past provides a useful means of demonstrating both the long-term 'natural' variability of sediment yields and the importance of recent human activity in perturbing the system. In the latter case, increases in sediment yield of the order of two to three times have been documented by several studies. In many situations such increases will exceed the long-term 'natural' variability of the system, but in the example of the Black Sea cited above, the natural variability can be seen to be as high as an order of magnitude and thus to be substantially greater than the more recent increases caused by human activity.

Evidence from lake sediments

Where lakes occur at the outlet of a drainage basin and trap a large proportion of the sediment output, detailed analysis of the sedimentary record can provide valuable evidence concerning fluctuations in sediment loads during the more recent past. The Black Sea example cited above demonstrates the value of this approach for reconstructing long-term trends, but lake sediments are also capable of providing more detailed information on both the magnitude and timing of changes in sediment inputs from the lake catchment over periods of 10^2 and 10^3 years. Furthermore, where the catchment draining to the lake is relatively small in size, it may be possible to relate the reconstructed record of sediment yield to documentary evidence regarding land use changes and other human impacts within the catchment (cf. Dearing *et al.*, 1990). The reliability of the reconstructed record of sediment yield will depend on the accuracy of the core dating techniques employed and the number of cores and on the precision of the core correlation procedures used to estimate the total *volumes* of sediment deposited during particular periods. Evidence of changing rates of deposition obtained from a single core can provide a basis for evaluating changes in the *relative magnitude* of sediment inputs through time, but multiple cores and core correlation techniques are an essential prerequisite for estimating the volumes of sediment involved and therefore the absolute values of sediment yield

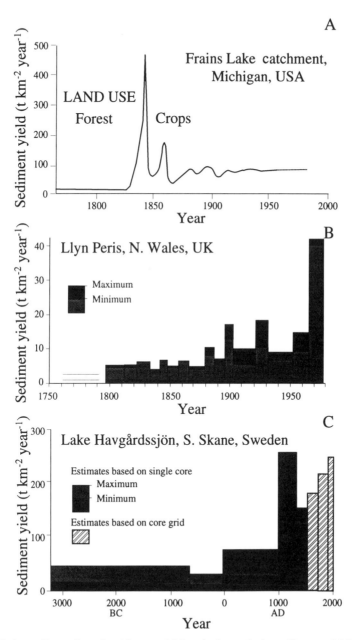

Figure 7.3 Lake-sediment-based evidence of historical trends in sediment yields. Based (A) on Davis (1976), (B) on Dearing *et al.* (1981) and (C) on Dearing *et al.* (1987)

(cf. Dearing and Foster, 1986). Because of the increased temporal resolution associated with dating techniques applicable to the recent past (e.g. lead-210 dating) this approach to reconstructing the sediment yield record has been most widely applied to the historical period and particularly to periods extending back several hundreds of years. Several examples of the evidence provided are presented in Figure 7.3 and these will be considered in turn.

The classic example of the Frains Lake catchment in Michigan, USA, illustrated in Figure 7.3A is based on the work of Davis (1976). This lake is located at the outlet of a small 0.18 km^2 drainage basin and a detailed programme of sediment coring, core analysis and dating enabled the record of sediment yield for the past 200 years to be reconstructed. The reconstructed sediment yield record shows low rates of sediment yield in pre-settlement times, rising by a factor of up to 70 with the onset of settlement and agricultural clearance after 1830, and stabilizing after 1900 at a rate about 10 times the pre-settlement level. Because of the greatly increased temporal resolution afforded by this example, it is possible to distinguish both the immediate short-term impact of land clearance and the longer-term sustained increase in sediment yields.

The reconstructed record of sediment yields from the 38 km^2 catchment of Llyn Peris in North Wales, UK, illustrated in Figure 7.3B, was produced by Dearing *et al.* (1981). Again substantial changes in sediment inputs to the lake are apparent, with sediment yields increasing about eight-fold, from about 5 t km^{-2} year^{-1} in the earliest period, to about 40 t km^{-2} year^{-1} in recent years. These increases have been ascribed to the impact of mining, quarrying, overgrazing and recent constructional activity in increasing sediment mobilisation from this upland catchment. Maximum and minimum estimates of sediment yield have been provided for the individual periods in order to take account of some of the uncertainties involved in the calculations.

An example of a reconstructed record of sediment yield extending back over a longer period of several thousand years is provided by Figure 7.3C which is based on the work of Dearing *et al.* (1987) on Lake Havgårdssjön, a small lake set in the hummocky moraine landscape of southern Skane, Sweden. In this study a grid of 47 cores was used to reconstruct the record of sediment yield for the period post-1550, and evidence from a single 4m core was used to extend the record on a more tentative basis back to 3050 BC. Estimates based on the single core have been ascribed precision limits in order to take account of uncertainties in sediment dating. Based on this record, it can be seen that sediment yields during the period 3000 BC to 50 BC were of the order of 25 t km^{-2} year^{-1}, a level which is consistent with an area of essentially undisturbed woodland. From about 50 BC sediment yields increased, rising to a peak of 86 to 250 t km^{-2} year^{-1} during the period AD 950–1300. This is again consistent with the known history of the area, since this was a period of forest clearance, village establishment, and agricultural expansion, following the introduction of the heavy-wheeled plough. The decrease in sediment yields in the subsequent period from AD 1300 to AD 1550 also coincides with the agrarian depression documented for many areas of north-west Europe and with the climatic deterioration of the early part of the 'Little Ice Age'. More recent increases in sediment yield in the period post-1550 again correspond closely with the historical records which point to an expansion in the area under cultivation and therefore susceptible to higher rates of surface runoff and erosion.

Table 7.2 Lake-sediment-based evidence of increases in sediment flux due to catchment disturbance by human activity from tropical and sub-humid environments

Lake	Location	Documented increase in sediment flux	Source
Lake Patzcuaro	Mexico	×7	O'Hara *et al.* (1993)
Lake Sacnao	Mexico	×35	Deevey *et al.* (1979)
Lake Ipea	Papua, New Guinea	×10	Oldfield *et al.* (1980)
Lac Azigza	Morocco	×5	Flower *et al.* (1989)

The examples of lake-sediment-based reconstructions of temporal variations in sediment yield presented above relate to temperate environments in Europe and North America. Similar investigations have been undertaken in other areas of the world where it has again proved possible to document changes in sediment yields reflecting anthropogenic disturbance. For example, O'Hara *et al.* (1993) report results obtained from Lake Patzcuaro in the severely degraded landscape of the volcanic highlands of central Mexico which reconstructed sediment inputs to the lake over the past 4000 years. The results indicate that sediment yields increased more than five-fold as a result of extensive land clearance and were at least as high under the land management of the indigenous population as after the Spanish conquest of the region. These findings and those from several other studies in tropical and sub-humid environments are summarised in Table 7.2. The trends evidenced in Figure 7.3 and Table 7.2 again demonstrate the sensitivity of sediment yields to land use change. This sensitivity is reflected in both the substantial gross increases evident from a comparison of sediment yields before and after human-induced land disturbance, and the variations reflecting particular phases of human activity.

THE EVIDENCE FROM LONG-TERM RECORDS

As noted in the introduction, there is a paucity of reliable long-term records of suspended sediment load which can be used to analyse long-term trends in sediment yields and their controlling factors. Changes in sampling equipment, in sampling frequency and protocols and in load calculation procedures can all introduce uncertainties into the consistency and continuity of long-term records. In addition, major changes in river behaviour, such as produced by reservoirs or river regulation, may introduce further discontinuities into the records such that it becomes increasingly difficult to interpret measured loads in terms of the sediment yields from the upstream catchment. More work is clearly required to collate available long-term records and to evaluate their potential for documenting long-term trends. Several examples based on available analyses may, however, be usefully introduced.

In most areas of the world, analysis of longer-term trends of increasing sediment loads will need to take account of both the effects of changes in land use and the impact of climatic change in modifying the precipitation and runoff regime. Long-term sediment load records now available for the former Soviet Union analysed by Bobrovitskaya (1994) serve to emphasise this point. Figures 7.4 and 7.5, based on

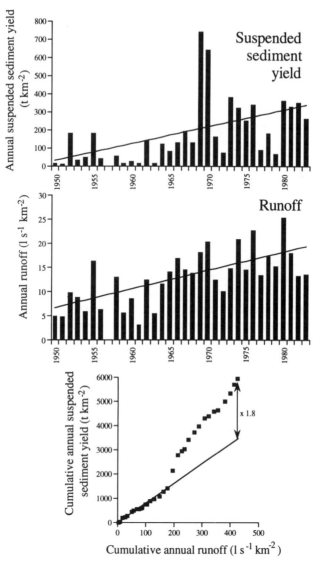

Figure 7.4 Longer-term trends in suspended sediment transport and runoff for the Dnestr River at Sambur, Ukraine, 1950–1983. (Based on data supplied by Dr N. Bobrovitskaya, State Hydrological Institute, St Petersburg, Russia)

data provided by Bobrovitskaya (personal communication), document recent trends in the suspended sediment loads transported by the Dnestr River at Sambur in the Ukraine (850 km^2) and the Kolyma River at Srednekansk in western Siberia (99 400 km^2), as represented by the least squares-fitted trend lines and double mass plots of cumulative annual sediment yield versus cumulative annual runoff. In the case of the Dnestr River (Figure 7.4), the trend line for the annual sediment load suggests that loads have increased by as much as five-fold since the early 1950s. This increase undoubtedly reflects the impact of forest clearance in the upstream catchment, but it

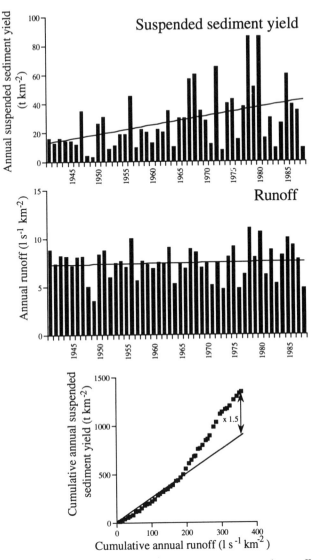

Figure 7.5 Longer-term trends in suspended sediment transport and runoff for the Kolyma River at Srednekansk in western Siberia, 1941–1988. (Based on data supplied by Dr N. Bobrovitskaya, State Hydrological Institute, St Petersburg, Russia)

must also reflect climatic change and the general increase in runoff amounts that has occurred over the period and more particularly since the late 1960s (cf. Figure 7.4). The double mass plot of cumulative sediment yield versus runoff affords a means of establishing the timing of the changes involved and of tentatively distinguishing the effects of land use and climate change. In this case, the break in the double mass plot around 1968 suggests that the impact of forest clearance was particularly felt after this date. Assuming that there is a linear relationship between runoff and sediment load and that the effects of increased runoff are therefore not reflected by a change in the slope of the double mass plot, this disturbance can be estimated to have caused a 1.8-fold increase in the sediment load of this river. In the case of the records from the 99 400 km^2 basin of the Kolyma River depicted in Figure 7.5, the trend line provides no evidence of changing runoff rates and therefore of a climatically-driven shift. In this river the increase in sediment loads of c. 1.5-fold dating from about 1964 can be related to the impact of gold mining activity within the upstream catchment.

The long-term records available for rivers in the USA also provide evidence of increases in sediment yield consequent upon land use change. Uri (1991), for example, reports the case of the Iowa River above Coralville Reservoir near Iowa City, where the suspended load has increased progressively over the period 1948–1985. This increase has in turn been related to the expansion of the cultivated area within the catchment and more particularly to the increasing dominance of corn and soybean cropping. The area planted to soybeans almost doubled between 1960 and 1985 and Uri (1991) has calculated that for each 1% increase in the area planted to corn and soybeans, the mean suspended sediment concentration increased by approximately 0.42%. Other US rivers have evidenced a marked reduction in sediment loads, rather than an increase, as a result of damming for reservoir development. Meade and Parker (1985) have, for example, documented how the construction of five major dams on the Missouri River between 1953 and 1963 caused a marked reduction in the sediment loads transported by this river, such that the load entering the Mississippi was reduced to only about 25% of its former value (cf. Figure 7.6). Since the Missouri River formerly represented the major supply of sediment to the Mississippi, the sediment load of that river has also declined, and the load at its mouth in 1984 was less than one-half of the value before 1953. Similarly, the same authors have used available data for the five major rivers of Georgia and the Carolinas from the early part of the century and from the early 1980s to demonstrate that the loads of these rivers are now only about one-third of those in 1910. Even greater reductions in suspended sediment transport have been documented by Meade and Parker (1985) for rivers draining the arid and semi-arid regions of the south-west of the USA, where reservoir development has again caused major changes in river behaviour. In the case of the Rio Grande, its annual sediment discharge to the Gulf of Mexico has declined from about 20 million tonnes in 1940 to something less than 1 million tonnes at present. Similarly, the Colorado River now discharges about 100 000 tonnes of sediment to the Gulf of California each year, whereas before about 1930 the load was more than three orders of magnitude greater and averaged 125–150 million tonnes.

It is clear from the above examples that reservoir construction on many of the world's major rivers will have caused marked reductions in sediment loads. The River Nile probably affords an even more extreme example, because the annual sediment load transported into the delta has decreased from c. 100 million tonnes to almost zero as a result of the closure of the Aswan Dam. Data compiled by UNESCO (1978) relating to the major reservoirs of the world indicate that these impoundments now control around 10% of the total runoff from the land to the oceans. Although it is reasonable to assume that the proportion of the total sediment flux from the land to the oceans which is trapped in these reservoirs will be similar, it is important to note that many of the major reservoirs of the world are

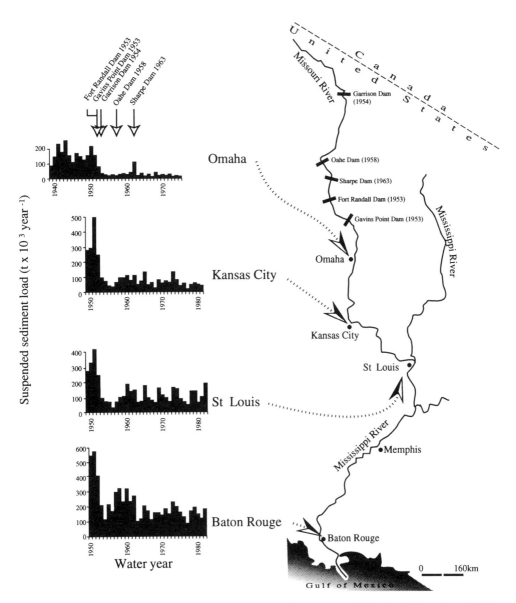

Figure 7.6 Changes in the annual suspended loads transported by the Missouri and Mississippi Rivers resulting from reservoir construction (based on Meade and Parker, 1985)

located in arid and semi-arid regions where sediment yields are relatively high. Furthermore, the important role of smaller dams and reservoirs in trapping sediment transported from upstream tributaries must also be considered and the overall reduction in sediment yield caused by river impoundments is thus likely to be substantially more than 10%. Indeed, it seems likely that in some areas of the globe increases in sediment transport caused by land clearance and land use change will have been balanced by reductions caused by reservoir development, to the extent that current sediment yields may be fairly close to those existing before the onset of widespread land disturbance by human activity.

EVIDENCE FROM OTHER CONTEMPORARY MEASUREMENTS

Catchment experiments

Most catchment experiments are, by design, concerned with relatively small areas and specific treatments, or land use changes applied to the entire catchment area in order to monitor their impact (cf. Ward, 1971; Walling, 1979). As such the results obtained cannot generally be directly extrapolated to larger heterogeneous drainage basins more representative of regional conditions, in order to examine longer-term trends in sediment yields. Available results do, however, provide valuable information on the likely *magnitude* of the changes in sediment response associated with particular types of catchment disturbance or land use practices. The examples of increases in sediment yield documented by experimental catchment studies listed in Table 7.3 provide information of this type and again emphasise the sensitivity of sediment yields to catchment disturbance. In these cases increases of up to more than two orders of magnitude have been documented.

Whilst most experimental catchment studies have focused on assessing the magnitude of *increases* in sediment yield caused by land disturbance and land use practices, it is also important to consider studies that have documented the impact of soil conservation and other catchment management practices aimed at *reducing* suspended sediment yields. Such results provide a contrasting perspective on the sensitivity of sediment yields to change and indicate the potential for reversing the detrimental effects of land use change and land use practices. Table 7.4 lists some results obtained from four small watersheds in the severely eroded loess region of the Middle Yellow River where a range of soil conservation measures, including both tree planting and construction of terraces and gully check dams, have been employed to achieve a marked reduction in sediment yields. In this case, sediment yields have been reduced by *c*. 90% or more. However, it is important to recognise

Table 7.3 Some results from experimental basin studies of the impact of land use change on sediment yield

Region	Land use change	Increase in sediment yield	Reference
Westland, New Zealand	Clearfelling	× 8	O'Loughlin *et al*. (1980)
Oregon, USA	Clearfelling	× 39	Fredriksen (1970)
Northern England	Afforestation (ditching and ploughing)	× 100	Painter *et al*. (1974)
Texas, USA	Forest clearance and cultivation	× 310	Chang *et al*. (1982)
Maryland, USA	Building construction	× 126–375	Wolman and Schick (1967)

Table 7.4 The impact of soil and water conservation measures on catchment sediment yields in the Loess Region of the Middle Yellow River, China

Catchment	Area (km^2	Area controlled (%)	Sediment yield reduction (%)	Reference
Wangmao Gully	5.97	68	89	Mou (1991)
Wangjia Gully	9.1	71	91	Mou (1991)
Nanxiaohe Gully	36.3	58	97	Mou (1991)
Yangjiagou Gully	0.87	40	93	Li (1992)

that the catchment areas involved were small and that there is a general lack of definitive evidence regarding the potential for such measures to effect a significant reduction in the sediment yield of much larger river basins.

Space–time substitution

In the absence of reliable long-term records of sediment yield capable of demonstrating trends associated with specific environmental changes, space–time substitution, or the ergodic hypothesis, can afford an effective means of demonstrating the likely trends involved. In essence the approach involves assembling information on spatial variations in sediment yield in response to, for example, variations in land use, and using these variations to examine the likely magnitude of changes in sediment yield associated with known temporal changes in catchment condition. Figure 7.7 provides several examples of the potential of this approach. In each case, by assuming that the low sediment yields associated with forested or natural areas afford a benchmark against which changes associated with alternative land uses can be assessed, increases in sediment yield of up to several orders of magnitude can be inferred. It is necessary to assume that other factors influencing the sediment yields remain constant and the ergodic hypothesis can clearly only be applied with rigour within a relatively small area. Thus, for example, in the case of the data presented for Kenya in Figure 7.7, which is based on the work of Dunne (1979), it may not be strictly correct to infer that sediment yields from the grazing areas were once as low as those in the forest areas, since the former occur primarily in semi-arid areas, whilst the latter are found mainly in the more humid regions.

The classic example of the potential of this approach for reconstructing the long-term record of sediment yield for an area is provided by the study undertaken by Wolman (1967) in the Piedmont region of Maryland, USA. In this case, data such as those shown in Figure 7.7 were combined with a generalised record of land use change in the region to synthesise the record of temporal changes in sediment yield depicted in Figure 7.8. In this region, sediment yields can be seen to have increased by an order of magnitude as a result of forest clearance by European settlers and the expansion of cropping, reaching a maximum around 1900. After this date sediment yields declined somewhat, in response to the introduction of conservation measures and a decline in the area under cultivation. Construction activities associated with urban and suburban expansion caused a further increase in sediment yields in the 1960s, but values subsequently declined to levels approaching those

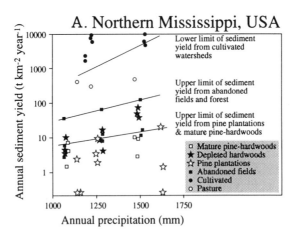

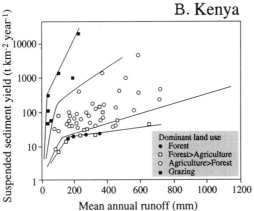

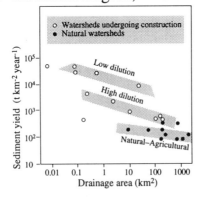

Figure 7.7 Examples of studies which have investigated the spatial variation of annual suspended sediment yields in response to land use activities. Based (A) on Ursic and Dendy (1965), (B) on Dunne (1979) and (C) on Wolman and Schick (1967)

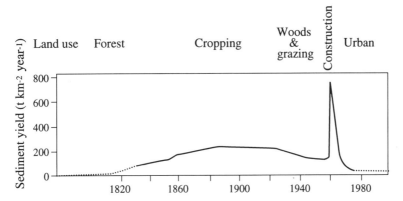

Figure 7.8 The reconstruction of the historical record of suspended sediment yields in the Piedmont region of Maryland, USA, proposed by Wolman (1967) on the basis of space–time substitution

associated with the natural forest land use as a result of the widespread existence of urban surfaces which tend to protect the soil from erosion. Similar results are presented by Phillips (1993) for the Lower Neuse basin in North Carolina, USA. Here, it is again suggested that suspended sediment yields increased by an order of magnitude, from less than 2 t km^{-2} $year^{-1}$ in the pre-colonial period to 10–17 t km^{-2} $year^{-1}$ in post-colonial times.

A further example of the information that can be obtained using the ergodic hypothesis, but this time at the global scale, is provided by the work of the Soviet scientists Dedkov and Mozzherin (1984). These workers gathered suspended sediment yield data from more than 3600 river basins in the mountainous and non-mountainous (plains) regions of the world. These basins were grouped into large (< 5000 km^2) and small (5000 km^2) basins and further classified into three groups, according to the degree of disturbance by agricultural activity. Category I basins were largely undisturbed and were characterised by either a forest cover $> 70\%$ or a ploughed area $< 30\%$. Category II basins exhibited an average degree of landscape change and were characterised by an area under forest or cultivation of between 30% and 70%. In category III basins, the degree of disturbance by land use activities increased still further, with an area under forest $< 30\%$ or a ploughed area $> 70\%$. By comparing the average sediment yield of category I basins with those of category III basins it was possible to derive an approximate measure of the magni-

Table 7.5 Increases in the sediment yields of world rivers resulting from catchment disturbance by land use activities, based on data assembled by Dedkov and Mozzherin (1984)

Group	Small basins	Large basins	All basins
Lowland rivers ($N = 1854$)	× 13.0	× 8.1	× 10.0
Mountain rivers ($N = 1811$)	× 2.2	× 3.8	× 2.8

tude of the increase in sediment yield associated with land disturbance by human activity (cf. Table 7.5). In the plains regions of the world the average increases were 8.1 and 13.0 for large and small river basins, respectively, whereas in the mountain areas the equivalent values were 3.8 and 2.2. Whilst these values are necessarily limited by the fact that they represent simple comparisons of mean values of sediment yield for drainage basins in different categories, and are therefore heavily dependent on the representativeness of the sample of river basins, they nevertheless provide a useful indication of the general magnitude of the increase in the sediment yields of world rivers resulting from catchment disturbance by land use activities. The values involved are closely in line with those presented in Figure 7.3 where lake sediments have provided the opportunity to reconstruct the detailed record of change for particular drainage basins.

SOME PROBLEMS OF INTERPRETATION

Linking upstream catchment behaviour to downstream sediment yield

In the preceding discussion of the various sources of evidence for the magnitude of changes in the suspended sediment loads of rivers in response to environmental change, a close and synchronous linkage between changing sediment loads and climate change, catchment disturbance, reservoir construction and other forcing variables has been implicitly assumed. It is, however, important to recognise the complexity of the drainage basin sediment delivery system which may attenuate the linkage between changing erosion rates and sediment output (cf. Robinson, 1977; Walling, 1983, 1989). A good example of this complexity is afforded by the work of Trimble (1976, 1981, 1983, Chapter 9) who investigated the long-term response of the sediment budget of Coon Creek in the Driftless Area of Wisconsin, USA, to land use change during the period 1853–1975. Using information from a range of field and documentary sources he was able to reconstruct the sediment budget of this 360 km^2 basin for two periods, namely, 1853–1938 and 1938–1975. The first period represented one of poor land management, which resulted in severe erosion, whereas the second was characterised by the introduction of conservation measures and substantially reduced erosion rates. The sediment yields estimated for the two periods were, however, almost identical at c. 110 and 115 t km^{-2} year^{-1}, respectively. During the first period, large volumes of soil were eroded from the slopes of the basin, but only a small proportion (c. 5%) of this was transported out of the basin. Most of the eroded material was stored on the lower slopes as colluvium and in the valley floors. During the latter period, when widespread conservation measures were introduced, rates of soil loss from upland sheet and rill erosion were reduced by about 25%, but the sediment yield at the basin outlet remained essentially the same, because sediment stored in the tributary valleys and upper main valley was remobilised. In this case, therefore, significant changes in land use and land management within the upstream basin were not reflected in equivalent changes in downstream sediment yield. A similar situation has been described by Meade (1982) for other rivers in the eastern USA.

A more specific examples of the potential significance of temporal discontinuities in sediment delivery associated with the storage and remobilisation of large volumes of sediment in the channel system is provided by the difficulties involved in attempting to interpret the long-term (>40 year) sediment yield records available for the Orange River in South Africa and the San Juan River, a tributary of the Colorado

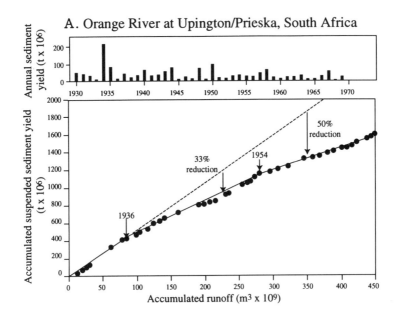

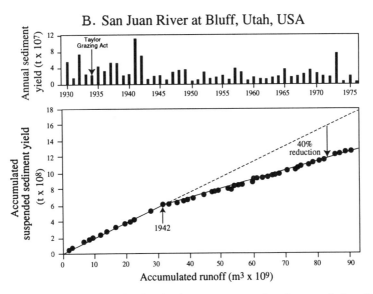

Figure 7.9 Temporal discontinuities in the long-term records of suspended sediment yield from the Orange River, South Africa, and the San Juan River, USA, demonstrated by double mass plotting. Based (A) on data contained in Rooseboom and Maas (1974) and (B) on data from US Geological Survey records

River, in south-west USA. Both drain large areas of semi-arid rangeland which were subjected to considerable grazing pressure during the latter part of the 19th century and the early part of the 20th century. This grazing pressure was subsequently reduced in an attempt to limit rates of soil loss and downstream sediment transport. Inspection of the long-term sediment yield records for these two rivers, which date from 1930 (Figure 7.9), provide some evidence of declining sediment yields which is confirmed by double mass plotting of accumulated sediment yield versus accumulated runoff. Hadley (1974) has suggested that the decline in the annual sediment yields of the San Juan River which was initiated in 1942 can be linked to the reduced grazing pressure and consequent improved vegetation cover and reduced soil loss which resulted from the Taylor Grazing Act of 1934. More detailed work by Graf (1985) has, however, cast doubt on this simple interpretation. Graf argues that the reduction in sediment loads in the Colorado and its tributaries which has occurred since the 1940s is due primarily to a dramatic reversal in the behaviour of the alluvial valleys and channels upstream, whereby erosional or throughput channel reductions were replaced by aggradation. Much of the reduction in sediment yield which occurred after 1940 can therefore be attributed to increased losses to channel storage and Graf (1985) indicated that the amount of sediment currently stored in the canyon floors of the Colorado and its tributaries is equal to about 200 times the present annual sediment yield.

The long-term sediment yield record for the Orange River presented in Figure 7.9 demonstrates a similar reduction in sediment yield since 1936 to that evidenced by the San Juan River. This reduction has been noted and discussed by Rooseboom and von Harmse (1979), who ascribed it to a progressive exhaustion of erodible sediments following a period of accelerated erosion initiated by human activity. This is consistent with the analysis of the alluvial valley fill deposits reported by Butzer (1971), who suggested that the intensive anthropogenic disturbance by burning and overgrazing which occurred during the period 1880–1930 and which corresponded with a run of dry years, caused increased surface runoff and a rapid dissection of the valley fill deposits. The decline in sediment yields post-1936 can therefore be tentatively ascribed to the completion of a 'cut' cycle, and the further reduction apparent since 1954 may represent the initial stages of a 'fill' cycle since Butzer (1971) refers to recent alluviation of the major valleys.

Both the examples cited above emphasise the potential importance of the storage of fine sediment within alluvial valley floors in controlling the downstream record of sediment yield and that this record may not be a direct reflection of the temporal variation of erosion rates within the upstream basin. In this context, the importance of scale to the interpretation of the impact of catchment changes on downstream sediment yield must also be recognised. Whereas such changes may be rapidly evident in small basins, the time-scale involved may be considerably greater in larger basins where large stores of sediment exist in alluvial valleys and channels to buffer the upstream change.

Response to climatic change

Most of the information and discussion presented in this contribution has been concerned primarily with the impact of land use change and catchment disturbance on sediment yields. However, in view of current interest and concern for the potential impact of climatic change it is important also to direct explicit attention to this component of global change. The general absence of reliable long-term records of river sediment loads highlighted previously has precluded detailed analysis of trends related to recent climatic fluctuations, but a variety of potential scenarios may be considered in relation to future climatic changes. In this context it is important to recognise the multiple controls on catchment sediment yields and their interrelationships. For example, to consider the case of reduced annual precipitation and therefore a shift towards a drier climate, it is possible to suggest that sediment yields would decline in response to reduced runoff. Conversely, if the interaction of precipitation amounts and vegetation cover density is considered, it is possible to argue that sediment yields might increase in response to reduced cover density, in line with the classic Langbein–Schumm rule (cf. Langbein and Schumm, 1958). Changes in the frequency and intensity of rainfall, as well as changes in land use consequent upon reduced moisture availability, could add further complexity to the ultimate scenario.

The rapidity of any climatic fluctuation, as well as the direction of change, could also prove significant, since this will influence the degree of adjustment between precipitation inputs and land cover density. This may be illustrated further by considering Figure 7.10, which provides a schematic representation of possible changes in sediment yield associated with increases and decreases in effective precipitation. Where precipitation increases, and the shift occurs rapidly, a significant short-term increase in sediment yield could result from the increased erosive energy impacting on catchment surfaces where the vegetation cover density has had insufficient time to adjust to the increased moisture availability and therefore reflects the previous drier conditions. As the vegetation cover density increases, in response to the increased moisture availability, sediment yields could decline to a level below those existing originally. At the end of this 'wet' period, the shift in effective precipitation back to the original conditions could have little or no impact on sediment yield, although yields could increase slightly once the vegetation cover density had reduced in response to the reduced moisture availability. In the case of a decrease in effective precipitation, sediment yields are shown in Figure 7.10A to increase slightly in response to the reduced vegetation cover density. However, at the end of this 'dry' period, the restoration of the precipitation back to its original level is marked by a rapid, but again short-lived, increase in sediment yield in view of the delay associated with the increase in vegetation cover. Thus, although in the longer term increases and decreases in effective precipitation are shown as giving rise to, respectively, a slight decrease and increase in sediment yield, the periods of transition to increased precipitation are marked by substantial short-lived increases in sediment yield.

The schematic trends portrayed in Figure 7.10A are aimed primarily at demonstrating the potential complexity of climatic change/sediment yield interactions, but some evidence for the occurrence of the responses suggested is available. Roberts and Barker (1993) cite the case of the response of the sediment yields of two Kenyan rivers documented by Walling (1984) to changing climatic conditions in this region (cf. Figures 7.10B and C). The period at the end of the 1950s and into the early 1960s was characterised by a shift to wetter conditions, consequent upon

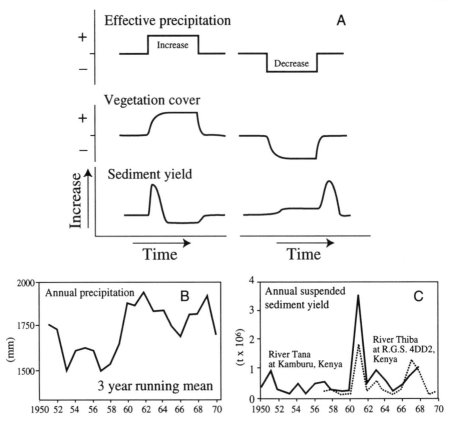

Figure 7.10 (A) A schematic representation of the impact of climatic change on sediment yield (based partly on Knox, 1972). (B) The trend of annual precipitation over Lake Victoria between 1950 and 1970 (based on Kite, 1981, after Roberts and Barker, 1993). (C) The variation of the sediment loads of the Tana and Thiba Rivers, Kenya, between 1950 and 1970 (based on Walling, 1984)

changes in atmospheric circulation (cf. Lamb, 1966) and this was reflected by a marked, but essentially temporary, increase in the sediment loads of the Tana and Thiba Rivers which corresponds closely to the schematic trends postulated in Figure 7.10A.

It is also important to recognise the potential interactions between land use change and changes in climate, both in terms of isolating their respective impacts and in relation to synergistic effects. The detailed studies of the changes in the sediment load of the Sanchaunhe River in China reported by Zhao *et al.* (1992) afford a useful example of the problems of separating the effects of catchment management and climate change on sediment yields. The Sanchuanhe River drains a 4161 km^2 basin in the gullied–hilly loess region of the Middle Yellow River basin and has been the object of extensive soil conservation and water conservancy activities during the 1980s. Thus, by the end of 1989, 267 km^2 of bench terrace had been constructed in the catchment, 703 km^2 had been planted with forest for soil conservation purposes and nine reservoirs had been commissioned. A comparison of the mean annual sediment loads for the periods 1957–1969 and 1980–1989 indicated

Table 7.6 The relative importance of climatic variation and watershed management in accounting for the recent reduction in the sediment yield of the Sanchaunhe River, China. Based on Zhao *et al.* (1992)

Annual sediment yield 1957–1969	36.81×10^4 t
Annual sediment yield 1980–1989	9.63×10^4 t
Total reduction	74%
Reduction due to reduced rainfall	33–38%
Reduction due to watershed management	36–41%

that sediment yields in the latter period had been reduced to only about 25% of those in the period 1957–1969. The soil and water conservation works undertaken in the river basin were partly responsible for this reduction, but it also reflected a shift to drier conditions, with the mean annual precipitation in 1980–1989 being only 509 mm as compared to 542 mm in 1957–1969. By using a range of data analysis techniques, Zhao *et al.* (1992) were able to demonstrate that in the case of the Sanchaunhe River basin, approximately 50% of the reduction in sediment yield noted for the period 1980–1989 could be attributed to the drier conditions and approximately 50% to the soil and water conservation measures (cf. Table 7.6).

In some situations it is likely that changes in catchment condition and changes in climate may operate synergistically to produce changing sediment yields. Thus it is possible to conceive of a situation where land use change causes general degradation within a catchment leading to instability, but a shift in climate towards wetter conditions is needed to trigger a major change in erosion and sediment yields. Climatic change could therefore provide an important forcing function for more general catchment disturbance.

PERSPECTIVE

The preceding review of sources of information regarding the response of suspended sediment yields to environmental change, and the magnitude of the changes involved, has highlighted the sensitivity of this parameter to global change and the need to view compilations of global sediment yield data as snapshots of a dynamic system. Evidence from the geological past emphasises that sediment yields may exhibit substantial temporal variations, even under natural conditions, in response to changes in climate and tectonic activity. In more recent times, increases in sediment loads of several-fold have been widely documented in river basins impacted by land use change and other forms of human-induced disturbance and it has been suggested that, at the global scale, the sediment yields of river basins in lowland areas that have been heavily influenced by agricultural activity could now be approaching an order of magnitude greater than they were under undisturbed conditions. Such increases have been particularly marked in many areas of Asia and Oceania which account for a major proportion of the total transport of sediment from the land to the oceans and therefore exert an important control on the magnitude of this flux (cf. Milliman and Meade, 1983). Furthermore, there is a growing body of evidence that recent climatic change has also caused significant increases in sediment yield in several areas of the world including eastern Africa and the former Soviet Union. However, it is important to recognise that there are also many instances where the sediment loads of rivers have declined, or at least remained stable, in the recent past. In the case of the Sanchuanhe River in China cited in Table 7.6, climate change, as reflected in reduced annual rainfall, has been shown to have caused reduced sediment yield and there are numerous examples of rivers where reservoir construction has resulted in reduced sediment loads. In an attempt to assess the trends in the sediment load records from rivers throughout the former Soviet Union, Bobrovitskaya (1994) has suggested that over a major proportion of this vast region more than 50% of the rivers in specific areas show evidence of significant non-stationarity in their sediment load records. These non-stationary records include approximately similar numbers of rivers with increasing and decreasing trends.

In view of the economic and environmental significance of increased sediment loads in rivers, it is important that their sensitivity to change is viewed as a key component of the phenomenon of global change which is currently attracting widespread attention. Lack of reliable long-term records has generally precluded detailed assessments of the problem, but there is clearly scope to analyse available data in greater detail and to exploit other sources of evidence such as those reviewed in this contribution. Further work is also undoubtedly required to provide a global-scale assessment of the response of river sediment fluxes to global change, to elucidate the complexities of relating changes in upstream erosion rates to changes in downstream sediment yields, to establish the likely impact of existing climatic change scenarios, to evaluate the potential effectiveness of catchment management techniques for reducing sediment yields, and to develop effective prediction procedures. There is also considerable scope to link such analysis to other aspects of global change which are currently attracting attention. The increasing rate of degradation of the global soil resource and changing material fluxes between the land and the oceans are, for example, two related issues that would clearly benefit from closer integration.

ACKNOWLEDGEMENTS

This contribution is based in part on a report produced by the author for Project H-1-2 of the 4th phase of the UNESCO International Hydrological Programme. The assistance of many individuals in providing the author with information on river sediment loads, and of Dr Phil Owens in data collation and processing, are gratefully acknowledged.

REFERENCES

Abernethy, C. (1990) The use of river and reservoir sediment data for the study of regional soil erosion rates and trends. Paper presented at the International Symposium on Water Erosion, Sedimentation and Resource Conservation, Dehradun, October, 1990.

Bobrovitskaya, N.N. (1994) Assessment of trends to sediment discharge variations in rivers of the former Soviet Union (FSU). In: *Proceedings of the International Symposium on East–West, North–South Encounter on the State-of-the-art in River Engineering Methods and Design Philosophies*, St Petersburg, Russia, May 1994, State Hydrological Institute, St Petersburg, 32–39.

Braune, E. and Looser, U. (1989) Cost impacts of sediments in South African rivers. In: *Sediment and the Environment* (Proceedings of the Baltimore Symposium, May 1989), IAHS Publication No. 184, 131–143.

Buringh, P. and Dudal, R. (1987) Agricultural land use in space and time. In: Wolman, M.G. and Fournier, F.G.A. (eds), *Land Transformation in Agriculture*, SCOPE Report no. 32, Wiley, Chichester, 9–43.

Butzer, K.W. (1971) Fine alluvial fills in the Orange and Vaal basins of South Africa. *Proceedings Association of American Geographers* 3, 41–48.

Chang, M., Roth, F.A. and Hunt, E.V. (1982) Sediment production under various forest-site conditions. In: *Recent Developments in the Explanation and Prediction of Erosion and Sediment Yield* (Proceedings of the Exeter Symposium, July 1982), IAHS Publication no. 137, 13–22.

Clark, E.H., Haverkamp, J.A. and Chapman, W. (1985) *Eroding Soils: The Off-Farm Impacts*. The Conservation Foundation, Washington, DC.

Davis, M.B. (1976) Erosion rates and landuse history in southern Michigan. *Environmental Conservation* 3, 139–148.

Dearing, J.A. and Foster, I.D.L. (1986) Lake sediments and palaeohydrological studies. In: Berglund, B.E. (ed.), *Handbook of Holocene Palaeocology and Palaeohydrology*, Wiley, Chichester, 67–90.

Dearing, J.A., Elner, J.K. and Happey-Wood, C.M. (1981) Recent sediment flux and erosional processes in a Welsh upland lake catchment based on magnetic susceptibility measurements. *Quaternary Research* 16, 356–372.

Dearing, J.A., Hakansson, H., Liedberg-Johnsson, B., Persson, A., Skansjo, S., Widholm, D. and El Daoushy, F. (1987) Lake sediments used to quantify the erosional response to land use change in southern Sweden. *Oikos* 50, 60–78.

Dearing, J.A., Alstrom, K., Bergman, A., Regnell, J. and Sandgren, P. (1990) Recent and long-term records of soil erosion from southern Sweden. In: Boardman, J., Foster, I.D.L. and Dearing, J.A. (eds), *Soil Erosion on Agricultural Land*, Wiley, Chichester, 173–191.

Dedkov, A.P. and Mozzherin, V.T. (1984) *Eroziya i Stock Nanosov na Zemle*. Izdatelstvo Kazanskogo Universiteta.

Deevey, E.S., Rice, D.S., Rice, P.M., Vaughan, H.H., Brenner, M. and Flannery, M.S. (1979) Mayan urbanism: Impact on a tropical karst environment. *Science* 206, 298–306.

Degens, E.T., Paluska, A. and Eriksson, E. (1976) Rates of soil erosion. In: Svensson, B.H. and Soderlund, R. (eds), *Nitrogen, Phosphorus and Sulphur—Global Cycles*, SCOPE Report no. 7, Ecol. Bull. (Stockholm) 22, 185–191.

Degens, E.T., Kempe, S. and Richey, J.E. (1991) Summary: Biogeochemistry of major world rivers. In: Degens, E.T., Kempe, S. and Richey, J.E. (eds), *Biogeochemistry of Major World Rivers*, Wiley, Chichester, 323–347.

Douglas, I. (1967) Man, vegetation and the sediment yields of rivers. *Nature* 215, 925–928.

Dunne, T. (1979) Sediment yield and land use in tropical catchments. *Journal of Hydrology* 42, 281–300.

Flower, R.J., Stevenson, A.C., Dearing, J.A., Foster, I.D.L., Airey, A., Rippey, B., Wilson, J.P.F. and Appleby, P.G. (1989) Catchment disturbance inferred from paleolimnological studies of three contrasted sub-humid environments in Morocco. *Journal of Paleolimnology* 1, 293–322.

Fournier, F. (1960) *Climat et Erosion*. PUF, Paris, 201pp.

Fredriksen, R.L. (1970) Erosion and sedimentation following road construction and timber harvest on unstable soils in three small western Oregon watersheds. *US Forest Service Research Paper*, PNW 104.

Graf, W. (1985) Channel processes and sediment yield in arid region drainage basins. In: Spencer, T. (ed.), *Abstracts of Papers for the First International Geomorphology Conference (Manchester, UK, Sept. 1985)*, p. 223.

Gregor, C.B. (1985) The mass–age distribution of Phanerozoic sediments. In: Snelling, N.J. *et al.* (eds), *The Geochronology of the Geological Record*, The Geological Society of London, 284–289.

Hadley, R.F. (1974) Sediment yield and land use in southwest United States. In: *Effects of*

Man on the Interface of the Hydrological Cycle with the Physical Environment (Proceedings of the Paris Symposium, September 1974), IAHS Publication no. 113, 96–98.

Jansson, M.B. (1982) *Land Erosion by Water in Different Climates*, UNGI Rapport no. 57. Department of Physical Geography, Uppsala University.

Jansson, M.B. (1988) A global survey of sediment yield. *Geografiska Annaler*, series A, **70**, 81–98.

Kite, G.W. (1981) Recent changes of level of Lake Victoria. *Hydrological Sciences Bulletin* **26**, 233–243.

Knox, J.C. (1972) Valley alluviation in southwestern Wisconsin. *Annals of the Association of American Geographers* **62**, 401–410.

Lamb, H.H. (1966) Climate in the 1960s: Change in the world's wind circulation reflected in prevailing temperatures, rainfall patterns and the levels of African lakes. *Geographical Journal* **132**, 183–212.

Langbein, W.B. and Schumm, S.A. (1958) Yield of sediment in relation to mean annual precipitation. *Transactions American Geophysical Union* **39**, 1076–1084.

Li, Z. (1992) The effects of forest in controlling gully erosion. In: *Erosion, Debris Flows and Environment in Mountain Regions* (Proceedings of the Chengdu Symposium, July, 1992), IAHS Publication no. 209, 429–437.

Lvovich, M.I., Karasik, G.Ya., Bratseva, N.L., Medvedeva, G.P. and Meleshko, A.V. (1991) *Contemporary Intensity of the World Land Intracontinental Erosion*, USSR Academy of Sciences, Moscow, 336p.

Mahmood, K. (1987) *Reservoir Sedimentation: Impact, Extent and Mitigation*, World Bank Technical paper no. 71.

Meade, R.H. (1969) Errors in using modern stream-load data to estimate natural rates of denudation. *Geological Society of America Bulletin* **80**, 1265–1274.

Meade, R.H. (1982) Sources, sinks and storage of river sediment in the Atlantic drainage of the United States. *Journal of Geology* **90**, 235–252.

Meade, R.H. and Parker, R.S. (1985) Sediment in rivers of the United States. In: *National Water Summary 1984, US Geological Survey Water Supply Paper* **2275**, 49–60.

Milliman, J.D. and Meade, R.H. (1983) World-wide delivery of river sediment to the oceans. *Journal of Geology* **91**, 1–21.

Milliman, J.d. and Syvitski, J.P.M. (1992) Geomorphic/tectonic control of sediment discharge to the ocean: The importance of small mountainous rivers. *Journal of Geology* **100**, 325–344.

Milliman, J.D., Qin, Y-S., Ren, M-E. and Saito, Y. (1987) Man's influence on the erosion and transport of sediment by Asian rivers: the Yellow River (Huanghe) example. *Journal of Geology* **95**, 751–762.

Morgan, R.P.C. (1986) *Soil Erosion and Conservation*. Longman, Harlow, 298pp.

Mou, J. (1991) The impact of environmental change and conservation measures on erosion and sediment loads in the Yellow River basin. In: *Sediment and Stream Water Quality in a Changing Environment* (Proceedings of the Vienna Symposium, August 1991) IAHS Publication no. 203, 47–52.

O'Hara, S.L., Street-Perrott, F.A. and Burt, T.P. (1993) Accelerated soil erosion around a Mexican highland lake caused by prehispanic agriculture. *Nature* **362**, 48–51.

O'Loughlin, C.L., Rowe, L.K. and Pearce, A.J. (1980) Sediment yield and water quality responses to clearfelling of evergreen mixed forests in western New Zealand. In: *The Influence of Man on the Hydrological Regime, with Special Reference to Representative and Experimental Basins*, IAHS Publication no. 130, 285–292.

Oldeman, L.R., Hakkeling, R.T.A. and Sombroek, W.G. (1991) *World Map of the Status of Human-Induced Soil Degradation: An Explanatory Note*. ISRIC, Wageningen.

Oldfield, F., Appleby, P.G. and Thompson, R. (1980) Palaeoecological studies of lakes in the Highlands of Papua New Guinea. I. The chronology of sedimentation. *Journal of Ecology* **68**, 457–477.

Painter, R.B., Blyth, K., Mosedale, J.C. and Kelly, M. (1974) The effect of afforestation on erosion processes and sediment yield. In: *Effects of Man on the Interface of the Hydrological Cycle with the Physical Environment*, IAHS Publication no. 113, 150–157.

Phillips, J.D. (1993) Pre- and post-colonial sediment sources and storage in the Lower Neuse basin, North Carolina. *Physical Geography* **14**, 272–284.

Roberts, N. and Barker, P. (1993) Landscape stability and biogeomorphic response to past and future climatic shifts in intertropical Africa. In: Thomas, D.S.G. and Allison, R.J. (eds), *Landscape Sensitivity*, Wiley, Chichester, 65–82.

Robinson, A.R. (1977) Relationship between soil erosion and sediment delivery. In: *Erosion and Solid Matter Transport in Inland Waters* (Proceedings of the Paris Symposium, July 1977), IAHS Publication no. 122, 159–167.

Roseboom, A. and Harmse, H.J. von K. (1979) Changes in the sediment load of the Orange River during the period 1939–1969. In: *The Hydrology of Areas of Low Precipitation* (Proceedings of the Canberra Symposium, Dec. 1979), IAHS Publication no. 128, 459–470.

Roseboom, A. and Maas, N.F. (1974) Sedimentafvoergegewens vir die Orange-Tugalaen Pongolariviere, Tech. Report no. 59, Dept. of Water Affairs, Pretoria, South Africa.

Ross, D.A. and Degens, E.T. (1974) Recent sediments of the Black Sea. In: *The Black Sea— Geology, Chemistry and Biology*, American Association Petroleum Geologists Memoir no. 20, 249–278.

Tardy, Y., N'Kounkou, R. and Probst, J.L. (1989) The global water cycle and continental erosion during Phanerozoic time. *American Journal of Science* **289**, 455–483.

Trimble, S.W. (1976) Sedimentation in Coon Valley Creek, Wisconsin. In: *Proceedings of the Third Federal Interagency Sedimentation Conference*, US Water Resources Council, Washington, DC, 5-100–5-112.

Trimble, S.W. (1981) Changes in sediment storage in Coon Creek Basin, Driftless Area, Wisconsin, 1853–1975. *Science* **214**, 181–183.

Trimble, S.W. (1983) A sediment budget for Coon Creek basin in the Driftless Area, Wisconsin, 1853–1977. *American Journal of Science* **283**, 454–474.

UNESCO (1978) *World Water Balance and Water Resources of the Earth*. UNESCO, Paris.

Uri, N.D. (1991) Detecting a trend in water quality. *Research Journal WPCF* **63**, 869–872.

Ursic, S.J. and Dendy, F.E. (1965) Sediment yields from small watersheds under various land uses and forest covers. *Proceedings Federal Interagency Sedimentation Conference*, USDA Miscellaneous Publication 970, 47–52.

Walling, D.W. (1979) Hydrological processes. In: Gregory, K.J. and Walling, D.E. (eds), *Man and Environmental Processes*, Dawson, Folkestone, 57–81.

Walling, D.E. (1983) The sediment delivery problem. *Journal of Hydrology* **65**, 209–237.

Walling, D.E. (1984) The sediment yields of African rivers. In: *challenges in African Hydrology* (Proceedings of the Harare Symposium, July 1984), IAHS Publication no. 144, 265–283.

Walling, D.E. (1989) Linking erosion and sediment yield: some problems of interpretation. *International Journal of Sediment Research* **4**, 13–26.

Walling, D.E. and Webb, B.W. (1981) The reliability of suspended sediment load data. In: *Erosion and Sediment Transport Measurement* (Proceedings of the Florence Symposium, June, 1981), IAHS Publication no. 133, 79–88.

Walling, D.E. and Webb, B.W. (1983a) Patterns of sediment yield. In: Gregory, K.J. (ed.), *Background to Palaeohydrology*, Wiley, Chichester, 69–100.

Walling, D.E. and Webb, B.W. (1983b) The dissolved loads of rivers: a global overview. In: *Dissolved Loads of Rivers and Surface Water Quantity/Quality Relationships* (Proceedings of the Hamburg Symposium, August 1983), IAHS Publication no. 141, 3–20.

Walling, D.E. and Webb, B.W. (1985) Estimating the discharge of contaminants to coastal waters by rivers: some cautionary comments. *Marine Pollution Bulletin* **16**, 488–492.

Walling, D.E. and Webb, B.W. (1987) Material transport by the world's rivers. In: *Water for the Future: Hydrology in Perspective* (Proceedings of the Rome Symposium, April 1987), IAHS Publication no. 164, 313–329.

Walling, D.E. and Webb, B.W. (1988) The reliability of rating curve estimates of suspended sediment yield; some further comments. In: *Sediment Budgets* (Proceedings Porto Alegre Symposium, December 1988), IAHS Publication no. 174, 337–350.

Ward, R.C. (1971) *Small Watershed Experiments: An Appraisal of Concepts and Research Developments*, University of Hull Occasional Papers in Geography, no. 18.

Wolman, M.G. (1967) A cycle of erosion and sedimentation in urban river channels. *Geografiska Annaler* **49A**, 385–395.

Wolman, M.G. and Schick, A.P. (1967) Effects of construction on fluvial sediment: urban and suburban areas of Maryland. *Water Resources Research* **6**, 1312–1326.

Ye, Q., Jing, K., Yung, Y., Chen, Y. and Zhang, Y. (1983) Changes of the river course of the Lower Yellow River with respect to erosion of Loess Plateau. In: *Proceedings of the Second International Symposium on River Sedimentation*, Nanjing, china, October 1983, 597–607.

Zhao, W., Jiao, E., Wang, G. and Meng, X. (1992) Analysis on the variation of sediment yield in Sanchuanhe River basin in 1980s. *International Journal of Sediment Research* **7**, 1–19.

8 Bedload Transport and Changing Grain Size Distributions

BASIL GOMEZ

Department of Geography and Geology, Indiana State University, USA

INTRODUCTION

The absolute size of bedload is typically characterised by a central value of its grain size distribution, usually the median grain size (D_{50}). The convention detracts from an obvious property of bedload, namely that it is composed of a range of particle sizes. It has long been appreciated that, all other factors being equal, variations in the absolute size of bedload give rise to different transport relations (see Leopold and Emmet (1976) for a graphical example), but the effect of relative size on the transport rate of different size fractions is commonly ignored. Relative particle size is often expressed as D_i/D_{50m} (where D_i and D_{50m} are the geometric mean size of the *i*th fraction and the median grain size of the subsurface bed material). Einstein (1950) incorporated a hiding factor in his bedload function to correct for differences in the magnitude of the lift force acting on particles smaller than the characteristic bed material size that are sheltered by larger grains. Parker *et al.* (1982) and Parker (1990) developed relations that explicitly seek to represent the effects that hiding and differences in the particle size distribution of the surficial and subsurface bed material have on the transport of heterogeneous bedload.

Harrison (1950) used the term 'normal transport' to describe the condition when all particles are in motion, and the percentage of particles of a given size exposed at the surface at a given time is the same as the percentage by weight of that size in the total bed mixture. Parker *et al.* (1982) incorporated the analogous concept of 'equal mobility' in their model. It was not offered as a physical necessity but as an approximation. Equal mobility arises as a consequence of the shielding of small grains from the flow and the preferential exposure of large particles, coupled with their relative abundance on the bed surface. The adjustments combine to counteract the absolute size effect of particle weight by making coarse particles more available to the flow, thereby enhancing their probability for entrainment. As Kuhnle (1992) noted, there are two facets to equal mobility. 'Equal entrainment mobility' is defined as the case when all particle sizes comprising the bed material begin to move at the same flow strength. 'Equal transport mobility' refers to the situation where all particle sizes are transported according to their relative proportions in the bed material, so that the bedload and bed material grain size distributions are identical. Departures from these conditions give rise to differences in the transport rate of individual

Changing River Channels. Edited by Angela Gurnell and Geoffrey Petts.
© 1995 John Wiley & Sons Ltd.

size fractions, which in channels with heterogeneous bed materials are likely as con-
spicuous as temporal variations in the total transport rate, and to hydraulic sorting.
Hydraulic sorting is known to occur during the entrainment, transport and deposition of heterogeneous bedload; it is important because of its links to channel
armouring and downstream fining (Gomez, 1983a; Ashworth and Ferguson, 1989;
Kuhnle, 1992; Paola et al., 1992).

There has been considerable discussion about the extent to which the condition
of equal mobility or departures from it dominate in natural streams (Komar,
1987a; Ashworth and Ferguson, 1989; Komar and Shih, 1992). However, the
debate has been constrained by a lack of familiarity with a range of field conditions, despite the fact that variations in the particle size distribution of bedload are
well documented in diverse settings (although complete bedload and bed material
particle size distributions are often excluded from summary reports). The purpose
of this paper is to summarise the factors that influence the movement of heterogeneous sediment, and to review the role that absolute and relative size effects play
in regulating the particle size distribution of the bedload. To this end the following
sections consider the manner in which bed surface texture influences the initiation
of particle motion in heterogeneous sediment, and the way that fractional bedload
transport rates in natural channels vary with the flow conditions once motion has
been established.

INITIATION OF MOTION

The notion of size-selective entrainment arises intuitively as a consequence of the
difference in absolute size and correspondingly greater weight of large particles compared to small grains (Hjulström, 1935; Shields, 1936; Rubey, 1938). It is inextricably linked with the evaluation of flow competence; the estimation of hydraulic
parameters from the (maximum) size of particles transported by a given flow (Nevin,
1948; Komar, 1987b; Wilcock, 1992a). Many empirical relations that purport to
describe the inception of motion of heterogeneous bed materials are modifications of
the Shields curve which applies to uniform sediment (Miller et al., 1977). Scatter
about these relations is attributable to the inherent difficulty of defining initial
motion (Kramer, 1935; Neill, 1968; Gessler, 1971; Paintal, 1971; Helland-Hansen et
al., 1974; Lavelle and Mofjeld, 1987; Wilcock, 1988). The shear stress at which particle motion is initiated in heterogeneous sediment may be approximated by Shields'
relation if the median grain size is used to characterise the entire sediment mixture
(Day, 1980; Wiberg and Smith, 1987; Wilcock, 1992b; Kuhnle, 1993). However, even
in the absence of variations in particle shape or bed microtopography and neglecting
differences in density, the critical shear stress for individual size fractions in heterogeneous bed material is known to deviate from that associated with the motion of
uniform sediment (Laronne and Carson, 1976; Brayshaw et al., 1983; Komar and
Li, 1986; Hassan, 1992; James, 1993). The departure from the Shields curve is primarily attributed to topographic irregularities that cause bed pocket geometry to
vary. Pocket geometry influences particle mobility by modifying factors such as particle protrusion and friction angle that determine the critical boundary shear stress
for individual particles (Wiberg and Smith, 1987).

Bed–grain interactions

Particle protrusion is a measure of a grain's exposure to the flow. Kirchner *et al.* (1990) considered particle protrusion to be an amalgamation of two parameters; projection, the level of the top of a grain with respect to the local mean bed elevation, and exposure, the elevation of a grain above the local upstream maximum elevation. Fenton and Abbot (1977) investigated the influence of particle protrusion on the boundary shear stress required to initiate motion for the simplified case of a co-planar bed composed of uniform or well-sorted grains. Their data show that the critical Shields stress (τ_{*c}) is strongly dependent on particle protrusion (where $\tau_{*c} = \tau_c/(s-1)\rho g D_{50m}$, τ_c is the critical value of $\tau = \rho g S Y$, ρ is the density of water, g is the acceleration due to gravity, S is the energy slope, s is the specific gravity of sediment, and Y is the flow depth). In fully turbulent flow relative masking of a grain by $0.2D_n$ increased the critical Shields stress from 0.03 to 0.06 (where D_n is the nominal particle diameter). For the optimal case of a spherical particle resting in a triangular pocket on a co-planar bed composed of like particles ($p/D_n = 0.82$) the critical Shields stress was found to be 0.01 (where p is the protrusion of a particle above the surrounding particles). Carling (1983) and Andrews (1994) determined that critical Shields stress declined with increasing relative protrusion of gravel particles on natural streambeds in a manner similar to that reported by Fenton and Abbot (1977).

The friction angle (also termed the pivoting angle or angle of repose) is a measure of the difficulty of moving a grain out of the pocket in which it rests. For particles entrained by drag forces, critical boundary shear stress is proportional to the tangent of the friction angle (Komar and Li, 1986; Wiberg and Smith, 1987). Friction angle decreases with increasing grain size relative to the median bed grain size (Eagleson and Dean, 1961; Miller and Byrne, 1966; Li and Komar, 1986). Water-worked beds exhibit a wide range of friction angles (Kirchner *et al.*, 1990; Buffington *et al.*, 1992). Lower friction angles are associated with poorer sorting because infilling by fines permits fewer deep pockets to be preserved on the bed surface. Large grains exhibit a tendency to move at low friction angles whereas small grains move over a broader range of friction angles.

Egiazaroff (1965) developed a model for the critical boundary shear stress of individual size fractions in heterogeneous sediment that accounted for the effect of relative grain size on the exposure of individual size fractions. He found that the critical shear stress was size-independent when $D_i < D_{50m}$. Wiberg and Smith (1987) derived an expression for the critical boundary shear stress of non-cohesive sediment from the balance of forces acting on individual particles that assimilated effects due to particle protrusion and friction angle. They found that for heterogeneous bed particles the range of boundary shear stresses required to initiate motion of particles with diameters ranging from 0.2 to 4.0 times the scale size of the bed was not as large as that implied by the Shields curve. Their analysis also provided theoretical justification for equal entrainment mobility. Calculated values of τ_{*c} for D_{max}/D_{50s} (where D_{max} is the diameter of the largest mobile particle and D_{50s} is the median grain size of the surficial bed material) were in good agreement with Andrews and Erman's (1986) data for Sagehen Creek, which indicated that particles between about 0.5 and 2.0 times the median diameter of the surficial bed material were entrained at approximately the same boundary shear stress.

Grass (1970) expressed variations in particle mobility in terms of a probability distribution of critical boundary shear stress. Kirchner *et al.* (1990) and Buffington *et al.* (1992) applied a simplified version of Wiberg and Smith's (1987) model to individual test grains resting on a variety of water-worked beds. Although the influence of factors such as grain packing and burial were ignored, their analyses demonstrated that, for a given particle size, variations in particle protrusion and friction angle combined to produce a probability distribution of critical boundary shear stress, so that the likelihood for motion of any grain depended on its exact location on the bed. Commensurate with the size differential that permits small grains to reside in a wider range of bed pockets than large particles, the probability distributions became broader with decreasing grain size and increasing bed roughness. For a given bed surface the minimum critical boundary shear stress was approximately the same for all grain sizes. This helped to explain why all size fractions became mobile at nearly the same shear stress. However, the probability distributions of critical boundary shear stress diverged from a common origin so that large particles exhibited a narrow low range of critical shear stress, whereas small particles had a very broad range of mobilizing boundary shear stress. The latter effect is a product of the selective trapping of small grains in the bed (Laronne and Carson, 1976). It implies that the mobility of much of the finer material on the bed surface may be constrained by the movement of larger particles (cf. the rapid rise in bedload transport rate with increasing boundary shear stress), and is consistent with the observation that the travel distance of individual particles smaller than the median size of the superficial bed material is size-independent (Church and Hassan, 1992; Hassan and Church, 1992).

Reference shear stress

The relative proportion of each size fraction in a mixture is also known to influence the range of shear stresses required to initiate motion of heterogeneous sediments. Wilcock (1992b, 1993) and Kuhnle (1993) compared the critical boundary shear stress of individual size fractions (τ_{ci}) in well-sorted unimodal, weakly bimodal and strongly bimodal sediment mixtures from laboratory and field settings. The conditions for initial motion were defined by approximating τ_{ci} as the reference shear stress (τ_{ri}) that generated a small, arbitrarily defined reference bedload transport rate for each size fraction (where τ_{ri} produces a dimensionless fractional transport rate (W_{*i}) of 0.002, $W_{*i} = [(s-1)gq_{bi}]/[(f_i(\tau/\rho)^{3/2}]$, q_{bi} is the bedload transport rate of fraction i per unit width, and f_i is the proportion of fraction i in the bed material). This was accomplished by measuring bedload transport rates over a wide range of conditions and determining the shear stress that corresponded to the reference rate from a fitted relation between the dimensionless shear stress and dimensionless transport rate for each size fraction.

Relations between the reference shear stress and particle size for bed materials with unimodal and bimodal particle size distributions are summarised in Figure 8.1. Laboratory and field data suggest that all particles in well-sorted, unimodal bed materials begin moving, at an appreciable rate, at approximately the same reference shear stress. Parker and Klingeman (1982) argued that equal mobility is a state towards which equilibrium transport systems tend to evolve. The development of a

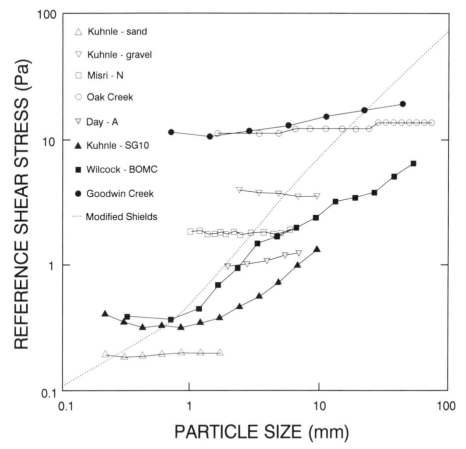

Figure 8.1 Variation of reference bed shear stress with particle size for unimodal sediments (open symbols) and bimodal sediments (solid symbols), after Kuhnle (1993) and Wilcock (1993). Field data are shown by circles. Weakly bimodal sediments are differentiated by the stippled symbol. The modified Shields curve was developed by Miller *et al.* (1977)

coarse surface layer facilitates this adjustment. The critical shear stress for coarse particles is reduced relative to that for uniform sediment; the mobility of coarse particles is enhanced by the presence of fine grains, and the preferential exposure and relative abundance of coarse particles on the bed surface increases their probability for motion by making them more available to the flow (Iseya and Ikeda, 1987; Wilcock and Southard, 1988; Kirchner *et al.*, 1990). The critical shear stress for the finer size fractions in the mixture correspondingly increases; the presence of coarse grains reduces the mobility of fine particles, and hiding decreases their probability for motion by making them less available to the flow.

The behaviour of bimodal sediments is more complex (Figure 8.1). Incipient motion is influenced by five factors: the absolute size of the sediment; the relative size of each fraction; the separation in size between the two modes; the proportion in each mode; and the fraction of sediment contained in the sizes between the two modes (Wilcock, 1992b; Kuhnle, 1993). Weakly bimodal bed material apparently behaves in a similar manner to unimodal sediment; strongly bimodal sediment behaves somewhat differently (Figure 8.1). The reference shear stress generally increases with increasing particle size. Reference shear stresses for the finer fractions in a strongly bimodal bed material are slightly larger than those for uniform sediments. Reference shear stresses for the coarser size fractions are less than those for uniform material defined by the Shields curve, but they are larger than those required to move well-sorted bed material. The reason for the differential is incompletely understood, but it is postulated that the relatively large proportion of sand in strongly bimodal bed material and the presence of partial transport in the coarse size fractions at high flows inhibit the coarsening process (Kuhnle, 1993; Parker and Wilcock, 1993; Wilcock, 1993). When the bedload is predominantly composed of sand, local size sorting separates the fully mobile fine size fractions into relatively homogeneous zones (McLean, 1981; Ferguson *et al.*, 1989). Within these zones the finer size fractions behave like a well-sorted sediment. The remainder of the bed surface coarsens, but if the coarse size fractions are not fully mobile the degree of coarsening may be insufficient to completely compensate for the inherent differences in the critical shear stress of the individual fractions that make up the bed material. In Goodwin Creek, Mississippi, the bed surface is somewhat coarser than the subsurface bed material, all fractions of the bed material are fully mobile at high flows and over a period of years the weighted mean particle size distribution of the bedload is the same as that of the subsurface bed material (Kuhnle, 1992; Kuhnle and Willis, 1992). Thus, as in Oak Creek, Oregon, the reference shear stress changes very little with particle size (Figure 8.1).

Armour development

The bed surface is coarser than the subsurface bed material in many channels. Surface coarsening is often associated with the development of a stable or mobile armour (Andrews and Parker, 1987; Sutherland, 1987; Parker and Sutherland, 1990). A stable armour forms when heterogeneous sediment is subjected to a flow that is of insufficient magnitude to move the coarsest fraction(s) of the bed material, in combination with a vanishing or near-vanishing supply of sediment from upstream (Sutherland, 1987). Stable armour development typically involves degradation and wholesale rearrangement of the surficial bed material (Harrison, 1950; Gomez, 1994). It is a product of size-selective erosion and the bedload is finer than the subsurface bed material, although all particle sizes are consistently present in the bedload and on the bed surface. Given the constraints of a fixed D_0 and D_{100}, the particle size distribution of the surficial bed material typically becomes more positively skewed and better sorted as a stable armour develops (Proffitt, 1980). This reflects a tendency for the upper portion of the grading curve to steepen and the lower portion to flatten relative to the original curve as the relative percentage of coarse particles on the bed increases at the expense of the finer fractions. The effect is accentuated in well-sorted sediments, in which a large proportion of the grain size distribution occupies a narrow size range, and is correspondingly less pronounced in poorly sorted bed materials. The particle size distribution of a stable armour is largely conditioned by the most recent flow that was capable of setting the surficial bed material in motion (Proffitt, 1980; Gomez, 1983b). It is unaffected by flows of a lesser magnitude than the constructing flow, but if the flow is increased the surface will coarsen, to the point where the size distributions of the bedload and subsurface bed material are identical and the armour is washed out. Reactivation of a stable armour involves concurrent motion of coarse and fine size fractions in the surficial bed material (Gessler, 1965; Günter, 1971), but does not preclude size-dependent transport (cf. Proffitt, 1980; Gomez, 1994).

The formation of a mobile armour is associated with a small but persistent rate of bedload transport. It provides the source for the bedload and facilitates the exchange of particles between the sediment bed and the flow over it by compensating for the disparity in mobility between coarse and fine particles (Parker et al., 1982; Parker and Klingeman, 1982; Andrews and Parker, 1987; Parker and Sutherland, 1990). Under the constraint that the particle size distributions of the bedload and the sub-surface bed material are identical, the bed surface progressively coarsens with decreasing transport rate; a limiting state of mobile armour development is defined by the formation of a stable armour (Dietrich et al., 1989; Parker and Sutherland, 1990; Lisle et al., 1993). At the other extreme, as transport rates increase the mobile armour becomes progressively finer and ultimately disappears once the particle size of the surficial and subsurface bed material coincide (cf. Kuhnle, 1989).

Parker et al. (1982), Parker and Klingeman (1982) and Andrews and Parker (1987) considered the effect of a mobile armour on critical boundary shear stress in terms of the particle size distribution of the subsurface bed material. Their analysis suggested that the difference in mobility between coarse and fine particles in a variety of gravel-bed rivers was quite small. The critical Shields stress is correlated with the grain size ratio such that $\tau_{*ci} = \alpha(D_i/D_{50m})^\beta$ (where τ_{*ci} is the critical Shields stress for fraction i). The exponent (β) should be zero if the critical boundary shear stress is proportional to particle size as the Shields curve implies. For precise equal entrainment mobility the dimensionless critical boundary shear stress and the grain size ratio are inversely proportional and β should have a value of -1.0. Several data sets have generated values of β in analogous relations that approach -1.0 (Andrews, 1983; Andrews and Erman, 1986; Wilcock and Southard, 1988). There is a degree of scatter about these relations, but they are interpreted as upholding the concept of equal entrainment mobility. Other data perhaps suggest a looser agreement of β with the value of -1.0, implying that size-selection persists despite the existence of strong relative size effects (Ashworth and Ferguson, 1989; Ferguson et al., 1989; Ashworth et al., 1992; Kuhnle, 1992). Parker (1990) considered the relation of critical boundary shear stress to relative grain size for Oak Creek in the context of a surface-based transport relation and obtained a value of β of approximately -0.9 rather than -1.0. He took this to imply that the surficial particles did not precisely satisfy the condition of equal entrainment mobility even when the transport relation was identical for all particle sizes, so that at low flows the bed-load is finer than the surficial bed material. That is, a mobile armour helps to establish a common threshold for the entrainment of all sizes present on the bed surface, by eliminating the intrinsic differences in the mobility of coarse and fine particles on the bed surface. It does not impede size-selective transport, even though the flow may be capable of mobilising most particle sizes present on the bed surface (cf. Diplas, 1987; Komar and Shih, 1992), nor is the presence of a mobile armour inconsistent with variations in the particle size distribution of the bedload that are driven by changing flow conditions (Parker *et al.*, 1982; Parker 1990).

FRACTIONAL BEDLOAD TRANSPORT RATES

At the lower limit of active transport, where rolling is the dominant transport mode, bed pocket geometry determines which particle sizes are mobile (Kirchner *et al.*, 1990; Andrews and Smith, 1992; Andrews, 1994). Figure 8.2A is a plot of fractional transport rate against relative particle size for discharges that approximate the bankfull discharge of 2.0 m^3 s^{-1} and discharges of about 1.5 and 0.75 times the bankfull discharge in Sagehen Creek, California (Andrews, 1994). The fractional transport rate (q_{bi}) has been scaled by the proportion of each size fraction in the surficial bed material ($D_{50s} = 58$ mm). Relations for two different transport conditions are presented. The degree to which a relation deviates from the horizontal indicates how much the particle size distribution of the bedload departs from that of the bed material and from equal transport mobility. Despite the fact that, for the most part, all size fractions in the bed material regularly appear in the bedload, size-dependent transport predominates. There is little change in the proportion of the total bedload size distribution that is accounted for by a given size fraction with increasing discharge, the armour remains intact at high flows and there is no general tendency towards equal transport mobility. In Sagehen Creek, the dearth of fine grains on the bed surface ($<-2\%$ of the surficial bed material is finer than 11 mm, compared to *c.* 32% in the subsurface bed material), coupled with their generally higher probability for entrainment relative to coarse particles, probably accounts for the manifestation of size-invariant transport of particles in the 5.6–11 mm size range (cf. Andrews and Erman, 1986; Kuhnle and Southard, 1988; Ashworth and Ferguson, 1989; Wilcock and McArdell, 1993). That is, these factors outweigh constraints on the mobility of fine grains imposed by the lack of movement of coarse particles (cf. Kirchner *et al.*, 1990; Buffington *et al.*, 1992). The systematic decrease in transport rate with increasing grain size for particles in the 11–64 mm size range is a product of the decline in the frequency of entrainment of coarse particles with increasing grain size (cf. Andrews and Erman, 1986). In Sagehen Creek about 40% of the surficial bed material is composed of material coarser than 64 mm in diameter. These particles are entrained very rarely or not at all but, because they have a relatively great mass compared to that of the total bedload sample, the frequency with which they appear in the bedload profoundly influences the total transport rate. Thus, for a given discharge, the bedload transport rate in Sagehen Creek typically varies by a factor of five depending on the presence or absence of particles in the 64–120 mm size range.

For comparison, in Figure 8.2B the fractional transport rate has been scaled by the proportion of each size fraction in the subsurface bed material ($D_{50m} = 30$ mm). The subsurface bed material contains significantly more fine material than the surficial bed material and the difference in the form of the two relations reflects the degree to which the particle size distribution of the bed surface differs from that of the subsurface bed material. The dearth of both fine and coarse fractions in the bedload ($D_{50m} = 26$ mm) is emphasised by these relations, but in all other respects the fractional transport rates display the same trends as the surface-based data.

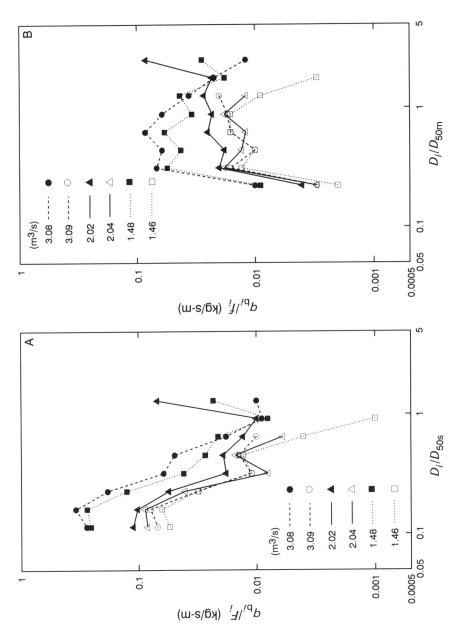

Figure 8.2 Average fractional bedload transport rate as a function of relative size for different flow and sediment transport conditions in Sagehen Creek (data from Andrews, 1994). The transport rate is the average for 8 or 12 samples collected during two or three traverses of the river with a 150 mm wide Helley-Smith sampler. The sampling duration was four minutes. Open symbols denote a mean transport rate of 0.0223–0.0255 kg/s-m and solid symbols a mean rate of 0.0048–0.0058 kg/s-m. In Figure 8.2A the fractional transport rate has been scaled by the proportion of each fraction in the surficial bed material, and in Figure 8.2B by the proportion of each size fraction in the subsurface bed material

Figure 8.3A is a plot of average fractional transport rate against relative particle size for selected groups of bedload samples derived from Milhous' (1973) Oak Creek data for the winter of 1971. The data are shown in the context bedload transport rate–stream power relation in Figure 8.3B. The fractional transport rate has been scaled by the proportion of each size fraction in the subsurface bed material ($D_{50m} = 20$ mm) and are averages for the indicated groups of bedload samples obtained over a relatively narrow range of stream powers (0.76–0.90, 4.2–4.8 and 6.9–9.2 kg/s-m, respectively). Averages are employed solely to provide a more stable estimate of the transport rate. Average fractional transport rates at the higher stream powers were derived from groups of three or four consecutive samples

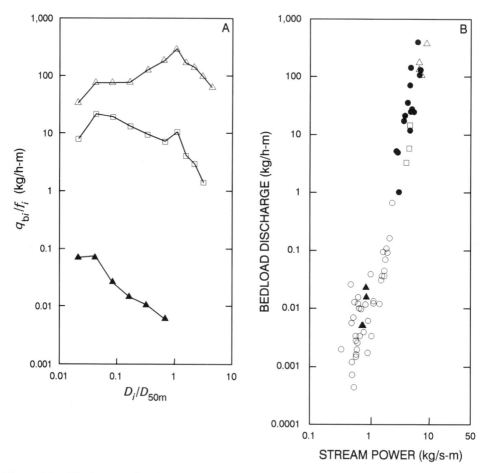

Figure 8.3 (A) Average fractional bedload transport rate as a function of relative size for different flow conditions in Oak Creek (data from Milhous, 1973). The transport rate is the average for consecutive samples or groups of samples, obtained under similar flow conditions in January 1971 using a stream-wide vortex trap. (B) Individual samples are shown in the context of the bedload discharge–stream power relation for the complete winter 1971 data set. The solid circles indicate the data that Parker *et al.* (1982) employed in the derivation of their bedload transport formula

obtained under comparable flow conditions, and at times when a similar range of particle sizes was in motion. The data for the lowest range of stream power are an amalgam of two groups of two consecutive samples with similar particle size distributions (excluding single large particles). All the samples were collected in January 1971.

In Oak Creek, fractional bedload transport rates at low stream powers exhibit the same size-dependence that is manifest in the Sagehen Creek data. Large particles (20–30 mm in diameter) exhibit a small probability for movement but the armour remains intact and bedload transport rates are low. Sand and fine gravel constitute the majority of the bedload and the transport rate of these size fractions is determined by their availability within the surficial bed material (Milhous, 1973). In the case of the smallest particles this is comparable with their availability within the subsurface bed material. As the stream power increases coarser particles begin to be dislodged from the armour with increasing regularity, but the transport rate of the finest size fraction declines because a proportion of these particles are selectively trapped in the bed (cf. Laronne and Carson, 1976; Church and Hassan, 1992). Some of the finer grains may also move into suspension. At still higher stream powers a significant proportion (about one-third) of the particles comprising the armour are mobile. There is an increasing tendency towards equal transport mobility but the coarse and fine fractions of the subsurface bed material remain underrepresented in the bedload.

Partial transport

Partial transport arises from differences in grain mobility. The proportion of fully mobilised size fractions varies between the limits of no motion and perfect equal transport mobility. The separation between the two modes, the proportion of sediment in each mode, and the fraction of sediment contained in the sizes between the two modes combine to determine whether the transition from partial to fully mobilised transport in a bimodal sediment will be smooth or abrupt (Wilcock, 1992b; Kuhnle, 1993). In Wilcock and McArdell's (1993) experiments the finest size fractions in the bed material were the first to achieve equal transport mobility. Any size fraction that was less than fully mobile had a diminished presence in the bedload and the coarsest fully mobile size fraction increased with increasing boundary shear stress. The condition where the proportion of each size fraction on the bed surface depends directly on its proportion in transport is a consequence of the manner in which equilibrium conditions are attained in a recirculating flume (Parker and Wilcock, 1993); the final state depends on the initial conditions and neither the equilibrium size distribution of the bedload nor the surficial bed material can be specified in advance. Partial transport cannot be maintained under steady-state sediment feed conditions since the material that passes through the flume must be identical to that entering it. The conditions in a recirculating flume may locally be approximated in rivers that transport a high throughput load or exhibit a strongly bimodal bed material so that the low flow transport is dominated by the finer size fractions. The latter situation prevails, for example, in Goodwin Creek. The bed material is strongly bimodal, and there is little difference between the surficial ($D_{50} = 11.7$ mm) and subsurface ($D_{50} = 8.3$ mm) bed material size distributions (Kuhnle, 1992). The fine (sand) size fractions move at much lower boundary shear stresses than the coarser (gravel) particles. At low flows the bedload is nearly all sand, but increasingly larger particles are transported as the boundary shear stress increases. Wilcock and McArdell (1993) employed a subset of Milhous' (1973) winter 1971 data to argue for the existence of partial transport in Oak Creek. They amalgamated bedload samples with transport rates falling within a factor of 1.5 without regard to their sequencing. This ignores variations in the particle size distribution and transport rate of bedload during and between flood events documented by Milhous (1973). Averaged fractional transport rates for groups of consecutive samples collected under similar flow conditions do not show a consistent decrease with increasing grain size, nor is there a consistent increase in the grain size separating fully and partially mobilized transport (Figure 8.3A). This suggests that the effects Wilcock and McArdell (1993) noted represent a time-averaged condition but, in the sense that only a portion of the sizes present in the bed material are in motion, partial transport prevails in the short-term.

Komar and Carling (1991) commented on the tendency for the coarser size fractions (e.g. D_{90}) of the bedload in Oak Creek to shift towards larger sizes more slowly than the D_{50} with increasing boundary shear stress. In other rivers the D_{50} has been found to remain almost constant and the D_{90} to increase with increasing shear stress (Ashworth et al., 1992), or to increase the same rate as the D_{90} (Komar and Carling, 1991). Reference to Figures 8.2B and 8.3A suggests that the response of the intermediate and coarser size fractions in the bedload to increasing boundary shear stress reflects the degree of similarity that exists between the particle size distribution of the surficial and subsurface bed material. In rivers where the intermediate size fractions of the subsurface bed material ($D_i/D_{50} = 1.0$) are well represented on the bed surface, plots of fractional transport rates against relative particle size exhibit a characteristically convex-up form (cf. Figure 8.2B: the plot for the highest stream powers at which bedload was sampled in Oak Creek also displays this form). The D_{50} of the bedload remains constant or almost constant with increasing shear stress while the D_{90} increases as increasingly coarser particles are entrained by the flow until all particle sizes become fully mobile. When the intermediate size fractions in the subsurface bed material mixture are underrepresented on the bed surface (as generally appears to be the case in Oak Creek) then mobilisation of these size fractions is dependent on the movement of coarser grains ($D_i/D_{50} > 1.0$). If, as Kirchner et al. (1990) suggest, large particles exhibit a narrow, low range of critical boundary shear stress, the number and size of coarse particles that are in motion at a given time determine the particle size distribution of the bedload and the transport rate. The D_{50} of the bedload increases as the probability for motion of large particles increases, but because large particles are in motion at relatively low shear stresses the D_{90} of the bedload increases at a slower rate than does the D_{50}.

In rivers where the bed materials are compacted and there is no armour, bed microtopography plays an increasing role in determining the mobility of a given size fraction (cf. Laronne and Carson, 1976; Komar and Carling, 1991). Fine particles infilling void spaces exhibit a tendency to move before the coarse framework gravels and the D_{50} and D_{90} increase with increasing boundary shear stress as the probability for motion of a greater range of particle sizes increases. Komar and Carling (1991) concluded that contrasts between the mobility of the intermediate and coarse size fractions in Oak Creek and Great Eggleshope Beck, UK, were in part a product of different bedload sampling techniques, but that they were generally consistent with differences in the bed material particle size distribution in the two rivers. A highly skewed bed material particle size distribution may contribute to the different behaviour of Oak Creek. However, the manner in which individual components of the bedload size distribution respond to increasing boundary shear stress primarily appears to be a function of their availability on the bed surface. It reflects the degree to which the particle size distribution of the surficial bed material deviates from that of the subsurface bed material, and is likely conditioned by the range and sequence of flows that the bed material is subject to (cf. Gomez, 1983b; Andrews, 1984). This observation is in agreement with the premise that the particle size distribution of the bedload is related to that of the subsurface bed material through the medium of the bed surface. It is supported by the results of studies of equilibrium bedload transport of mixed sized fractions in recirculating flumes (Wilcock and

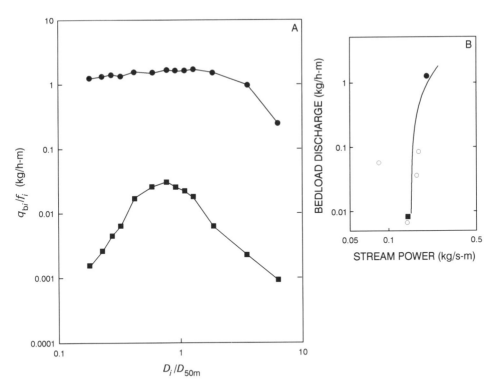

Figure 8.4 (A) Average fractional bedload transport rate as a function of relative size for different flow conditions in the Narrator Brook. The transport rate is the average for an entire flood event determined with a pit trap (Gomez, 1979). (B) The bedload discharge–stream power relation for the autumn 1976 data. The solid line indicates the regression relation for events following the initial rise when the transport rate was enhanced by the greater availability of fine sediment

Southard, 1989; Wilcock and McArdell, 1993), and is consistent with the behaviour of the fine size fractions in heterogeneous bedload.

In many cases the finest size fractions ($D_i/D_{50} < 1.0$) in the subsurface bed material are underrepresented on the bed surface because they are selectively trapped in the bed. Their presence in the bedload is dependent on the motion of coarse particles. Small grains exhibit higher median critical shear stresses and lower erodibility than do large and intermediate size particles (Kirchner *et al.*, 1990). The proportion of finer fractions in the bedload increases once the boundary shear stress increases to the point where coarse particles are mobilised. Differences in size and mass mean that, while the movement of coarse particles mobilises the finer size fractions, the erosion of fine grains inhibits the entrainment of coarse particles. Thus the proportion of the finer size fractions in the bedload may increase more rapidly than the proportion of coarser grains. If all but the largest particles are mobilised within a relatively narrow range of boundary shear stresses, a small additional increase in shear stress may be sufficient to mobilise almost the entire range of particle sizes present on the bed. Kirchner *et al.* (1990) also speculated that mobilisation of an increasing number of more resistant particles gave rise to the rapid increase in bedload transport rate with increasing boundary shear stress. Both effects are illustrated in Figure 8.4, using data from the Narrator Brook, UK (Gomez, 1979). Plots of average fractional transport rate against relative particle size are shown for two flood events, as well as the bedload transport rate–stream power relation for the complete autumn 1976 data. The fractional transport rate has been scaled by the proportion of each size fraction in the subsurface bed material ($D_{50m} = 2.0$ mm), and the bedload transport is integrated over an entire flood event.

Influence of sediment supply

Figure 8.5A is a plot of average fractional transport rate against relative particle size for selected groups of bedload samples derived from Jones and Seitz's (1980) Snake River, Idaho, data set for the spring of 1973 and 1974. The samples are shown in the context bedload transport rate–stream power relation for the complete data set in Figure 8.5B. The fractional transport rate has been scaled by the proportion of each size fraction in the subsurface bed material ($D_{50m} = 35$ mm), and the data are averages for those sampling periods when the bedload exhibited a similar particle size distribution, prior to, during and after disturbance of the armour. The Snake River has a high sand throughput load, and this accounts for the persistent presence of the finest size fractions in the bedload. These size fractions always appear to be almost fully mobile (Church *et al.* (1991) found a similar situation prevailed in Harris Creek, British Columbia). Although they account for *c.* 20% of the subsurface bed material in the Snake River, particles > 90 mm in diameter rarely move, even at the highest stream powers. Their stability limits movement of the intermediate size fractions, which typically are underrepresented in the bedload. These factors combine to give the fractional transport rate–relative size relations for the Snake River their pronounced concave-down form, which persists despite an apparent tendency towards fully mobilised transport at higher transport rates.

Very high transport rates have been observed in rivers where the availability of bedload apparently is unconstrained by sediment supply. Figure 8.6 is a plot of

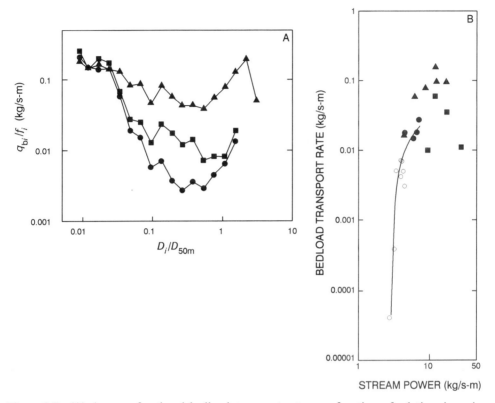

Figure 8.5 (A) Average fractional bedload transport rate as a function of relative size prior to (open and solid circles), during (solid triangles) and after (solid squares) disturbance of the armour layer in the Snake River, 1973–1974 (data from Jones and Seitz, 1980). The transport rate is the average for all samples pertaining to the specified bed condition. On each occasion 40 samples were collected during two traverses of the river with a 150 mm wide Helley-Smith sampler. The sampling duration was 30 or 60 seconds. (B) The bedload transport rate–stream power relation for the individual 1973–1974 data. The solid line indicates the regression relation for the period prior to disruption of the armour

average fractional transport rate against relative size for a range of flow conditions in the North Fork Toutle River, Washington. The fractional transport rate has been scaled by the proportion of each size fraction in the subsurface bed material ($D_{50m} = 10$ mm). Following the eruption of Mount St Helens in May 1980, the North Fork Toutle River was inundated with volcanic debris and mudflow deposits. The high sediment transport rates are a product of the erosion of unconsolidated avalanche deposits. Bedload transport rates (1.19–2.90 kg/s-m) are an order of magnitude greater than transport rates recorded in Oak Creek, over a similar range of stream power (6–12 kg/s-m). Transport rates and transport efficiencies (11–22%) compare well with Laronne and Reid's (1993) data for an ephemeral desert stream in which the bed lacks vertical structure. In the North Fork Toutle River the D_{50} of the bedload approximates that of the bed material when the discharge exceeds 100 m^3 s^{-1}, and Dinehart (1989) observed active gravel dunes in the reach immediately

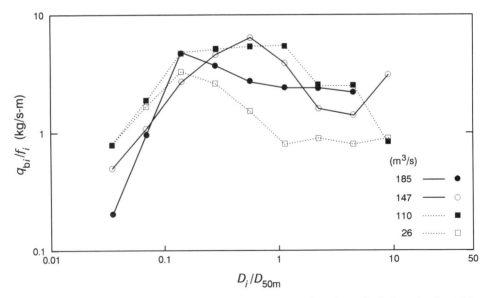

Figure 8.6 Average fractional bedload transport rate as a function of relative size for different flow conditions in the North Fork Toutle River at Kid Valley: 7 and 8 June 1985, and 13 and 26 February 1986 (data from Childers *et al.*, 1988; Williams and Rosgen, 1989). The transport rate is the average for 11 to 20 samples obtained during two to four traverses across the active bed with 150 or 300 mm wide Helley-Smith type samplers. The sampling duration varied between 5 and 30 seconds depending on the transport conditions

upstream from the sampling section. There is a tendency towards equal transport mobility at high flows, but both the fine and coarse size fractions are consistently underrepresented in the bedload (no sediment in the 128–256 mm size range was present in the bedload). Size-selective transport is preserved in the absence of a well-developed armour, during high flows and at high bedload transport rates. Thus, as Paola *et al.* (1992) observed, size-selective transport appears to be a natural consequence of the transport and deposition of heterogeneous sediment, and equal transport mobility may not be realised even if surface coarsening is inhibited because the local sediment supply exceeds the transport rate (cf. Dietrich *et al.*, 1989; Parker, 1990).

CONCLUDING DISCUSSION

Correct interpretation of the influence that surface coarsening has on the mobility of individual size fractions is crucial to understanding the behaviour of heterogeneous bedload. A coarse surface layer has been observed to develop in both feed and recirculating flumes. In the former case it is a requirement for equilibrium transport. It is not necessary in the latter case, so that an armour forms at some equilibrium transport rates but not others (Wilcock and Southard, 1989; Parker and Wilcock, 1993; Wilcock and McArdell, 1993). In many gravel-bed rivers the bed surface is also coarser than the underlying bed material (Milhous, 1973; Kuhnle, 1992) and Parker (1990) demonstrated that a coarse bed surface can be maintained over a wide range of flow and sediment transport conditions. The persistence or absence of active bedload transport determines whether the coarse surface layer constitutes a mobile or stable armour (Dietrich *et al.*, 1989; Parker and Sutherland, 1990; Lisle *et al.*, 1993). All particle sizes present in the underlying bed material are present in the armour and few particles are completely immobile (Harrison, 1950; Gessler, 1965; Milhous, 1973).

Armour development attempts to compensate for the intrinsically lower mobility of coarse particles relative to that of fine grains and render all particle sizes on the bed surface equally mobile (Parker and Klingeman, 1982; Andrews and Parker, 1987). Support for equal entrainment mobility has come from theoretical and empirical analyses of the critical boundary shear stress of individual size fractions in heterogeneous sediment, and from laboratory and field investigations (Andrews and Erman, 1986; Wiberg and Smith, 1987; Kirchner *et al.*, 1990; Buffington *et al.*, 1992; Kuhnle, 1993; Wilcock, 1993). Calculations of the boundary shear stress required to initiate motion of heterogeneous bed material suggest that the critical shear stress for each grain size in a heterogeneous sediment is characterised by a probability distribution that diverges from a common origin. Bed surface topography determines the shape of the probability distribution (Kirchner *et al.*, 1990; Buffington *et al.*, 1992). The presence of fine grains mobilising coarse particles and of coarse particles inhibiting the entrainment of fine grains exert fundamental controls on the mobility of a given bed surface. All particle sizes in well-sorted bed material exhibit a tendency to move at the same reference shear stress because the development of a coarse surface layer compensates for the inherent difference in the shear stress required to move the coarser and finer fractions of the bed material (Parker and Klingeman, 1982; Wilcock, 1993). In bimodal bed material the presence of sand and the persistence of partial transport at high flows may inhibit the coarsening process, so that the reference shear stress of individual size fractions tends towards size dependency (Kuhnle, 1993). Particle shape exerts an important additional control on the mobility of different size fractions in heterogeneous bed material (Gomez, 1994), and in the absence of a coarse surface layer local variations in bed microtopography may have the same effect on the incipient motion of heterogeneous sediment (Komar and Carling, 1991).

A vanishing or near-vanishing sediment supply motivates the development of a static armour; a mobile armour develops during active gravel transport. The development of an effective armour appears to involve wholesale rearrangement of the bed surface; this necessitates that all particle sizes on the bed are potentially mobile and thus appear in the bedload (Harrison, 1950; Gessler, 1965; Milhous, 1973; Gomez, 1993, 1994). This requirement may not be met if only a portion of the sizes present in the bed material are mobile. The sand fraction in strongly bimodal bed material may also act to inhibit the coarsening process and encourage partial transport. In such cases size-selective entrainment may predominate (Kuhnle, 1993; Wilcock, 1993; Wilcock and McArdell, 1993). Equal entrainment mobility is approximated in rivers with bimodal sand–gravel bed material where nearly all particle sizes are fully mobile at high flows (Kuhnle, 1993). Nevertheless, armour development may be insufficient to compensate for all of the differences in the critical shear stresses of the fractions present in the surficial bed material, so that size-selective transport persists (Parker, 1990).

Although the bed surface does not necessarily act to perfectly regulate the mobility of different size fractions, the particle size distribution of the bedload is related to that of the subsurface bed material through the medium of the bed surface and the textural changes that occur as the imposed flow and sediment transport conditions vary. At high boundary shear stresses and high bedload transport rates a mobile armour gradually disappears as the particle size distribution of the bedload approaches that of the subsurface bed material and all particle sizes in the bed become equally mobile (Parker and Klingeman, 1982; Parker et al., 1982; Andrews and Parker, 1987; Parker and Sutherland, 1990). Laboratory experiments provide empirical support for the existence of equal transport mobility at high boundary shear stresses ($\tau < 2\tau_c$; Kuhnle, 1989; Wilcock and Southard, 1989; Wilcock, 1992b; where τ_c is the critical boundary shear stress for the bulk mixture).

In some rivers relative size effects act to constrain the transport of fine grains and the transport of coarse grains is conditioned by their absolute size, even though their probability for motion may have been increased because of their increased abundance on the bed surface. Partial transport, in the sense that only a portion of the sizes present in the bed material are in motion, may be a common condition (Leopold, 1992), but many alluvial rivers appear able to transport all particle sizes present on the bed within the prevailing range of flows. This may or may not involve disruption of a mobile or static armour at high flows and the attainment of equal transport mobility with respect to the subsurface bed material (Milhous, 1973; Andrews, 1984). The particle size distribution of the bedload approaches that of the subsurface bed material as the imposed boundary shear stress increases beyond that required to generate a threshold (late-stage) stable armour (Günter, 1971; Proffit, 1980). Stable armours are known to be disrupted by high flows (Gomez, 1983b), and there is a tendency towards full mobility when the flow conditions are only slightly below those needed to establish a threshold armour (Gomez, 1994). The relation of the boundary shear stress at bankfull discharge to the threshold value for bed motion likely determines which transport regime prevails.

In self-formed rivers a stable channel geometry and active bedload transport may be maintained if $\tau/\tau_c < 1.2$, so that the threshold value is exceeded only on the central portion of the bed (Parker, 1978). If $\tau/\tau_c > 1.2$ the threshold value is exceeded around the entire perimeter and the banks erode laterally to a point where the reduction in shear stress again confines sediment transport to the central portion of the channel. Riparian vegetation and cohesive bank materials may permit natural channels to maintain a stable geometry at slightly higher boundary shear stresses ($\tau/\tau_c \approx 2.0$; Andrews, 1984). General motion of the bed is rare in these channels, bedload rates are low and the armour remains in place even though nearly all particle sizes present on the bed may be mobile (cf. Andrews, 1994). In channels which cannot adjust freely or are adjusting to a changing sediment load, conditions where $\tau/\tau_c \gg 2.0$ are possible (Kuhnle, 1992; Pitlick, 1992). At discharges approaching bankfull, bedload transport rates are high, general motion of the bed material occasions the disruption of any armour, and there is a tendency towards equal transport mobility.

During low and intermediate flows the bedload is typically finer than the bed material, even though all size fractions present on the bed surface may be mobile. At low flows the fractional transport rate of fine and intermediate particle sizes is also determined by their availability on the bed surface. If this corresponds to their availability in the subsurface bed material, these size fractions appear fully mobile when the fractional bedload transport rate is scaled by the particle size distribution of the subsurface bed material. As the flow increases, increasingly coarser particle sizes are transported. In rivers which transport a large amount of fine bedload the proportion of the finer size fractions in transport may remain close to their proportion in the bed. In other rivers the movement of coarse particles increases the likelihood that the fine particles will be trapped by the bed and the proportion of the fine size fractions in transport declines. Moving a portion of the finer size fractions into suspension has the same effect. If the boundary shear stress approaches the threshold value for bed motion the armour may be disrupted during high flows, the proportion of coarse particles in transport then approaches their proportion in the bed and all size fractions exhibit a tendency to be fully mobile.

Thus, the manner in which the transport rate of a given size fraction varies with the flow is conditioned by the imposed supply regime (which helps to determine sediment availability) and the interaction of the coarse and fine size fractions (which helps to determine their mobility). If the relation of the boundary shear stress at bankfull discharge to the threshold value for bed motion determines whether or not an armour persists or will be disrupted during high flows, the frequency of disruption may be low in many rivers, even though the flow may always be capable of mobilising most size fractions present on the bed surface. This is because an armour can be sustained by a small proportion of partially mobile particles (Harrison, 1950). The tendency towards size-selective transport is reenforced and it may dominate even at high flows. Size-selective transport also appears to be preserved in rivers where surface coarsening may be inhibited by a high sediment supply.

REFERENCES

Andrews, E.D. (1983) Entrainment of gravel from naturally sorted riverbed material. *Geological Society of American Bulletin* **94**, 1225–1231.

Andrews, E.D. (1984) Bed material entrainment and hydraulic geometry of gravel-bed rivers in Colorado. *Geological Society of America Bulletin* **95**, 371–378.

Andrews, E.D. (1994) Marginal bed load transport in a gravel bed stream, Sagehen Creek, California. *Water Resources Research* **30**, 2241–2250.

Andrews, E.D. and Erman, D.D. (1986) Persistence in the size distribution of surficial bed material during an extreme snowmelt flood. *Water Resources Research* **22**, 191–197.

Andrews, E.D. and Parkers, G. (1987) Formation of a coarse surface layer as the response to gravel mobility. In: *Thorne, C.R., Bathurst, J.C. and Hey, R.D. (eds), Sediment Transport in Gravel-bed Rivers*, Wiley, Chichester, 269–300.

Andrews, E.D. and Smith, J.D. (1992) A theoretical model for calculating marginal bedload transport rates of gravel. In: Billi, P., Hey, R.D., Thorne, C.R. and Tacconi, P. (eds), *Dynamics of Gravel-bed Rivers*, Wiley, Chichester, 41–48.

Ashworth, P.J. and Ferguson, R.I. (1989) Size-selective entrainment of bed load in gravel bed streams. *Water Resources Research* **25**, 627–634.

Ashworth, P.J., Ferguson, R.I., Ashmore, P.E., Paola, C., Powell, D.M. and Prestegaard, K.L. (1992) Measurements in a braided river chute and lobe 2. Sorting of bed load during entrainment, transport and deposition. *Water Resources Research* **28**, 1887–1896.

Brayshaw, A.C., Frostick, L.F. and Reid, I. (1983) The hydrodynamics of particle clusters and sediment entrainment in coarse alluvial channels. *Sedimentology* **30**, 137–143.

Buffington, J.M., Dietrich, W.E. and Kirchner, J.W. (1992) Friction angle measurements on a naturally formed gravel streambed: implications for critical boundary shear stress. *Water Resources Research* **28**, 411–425.

Carling, P.A. (1983) Threshold of coarse sediment transport in broad and narrow natural streams. *Earth Surface Processes and Landforms* **8**, 1–18.

Childers, D., Hammond, S.E. and Johnson, W.P. (1988) Hydrological data for computation of sediment discharge, Toutle and North Fork Toutle Rivers near Mount St. Helens, Washington, water years 1980–84, *US Geological. Survey Open-file Rept*, 87–548.

Church, M. and Hassan, M.A. (1992) Size and distance of travel of unconstrained clasts on a streambed. *Water Resources Research* **28**, 299–303.

Church, M, Wolcott, J.F. and Fletcher, W.K. (1991) A test of equal mobility in fluvial sediment transport: behaviour of the sand fraction. *Water Resources Research* **27**, 2941–2951.

Day, T.J. (1980) A study of initial motion characteristics of particles in graded bed material. *Geol, Survey Canada Pap.* **80-1A**, 281–286.

Dietrich, W.E., Kirchner, J.W., Ikeda, H. and Iseya, F. (1989) Sediment supply and the development of the coarse surface layer in gravel-bedded rivers. *Nature* **340**, 215–217.

Dinehart, R.L. (1989) Dune migration in a steep, coarse-bedded stream. *Water Resources Research* **25**, 911–923.

Diplas, P. (1987) Bedload transport in gravel-bed streams. *Journal of Hydraulic Engineering* **113**, 277–292.

Eagleson, P.S. and Dean, R.G. (1961) Wave-induced motion of bottom sediment particles. *Trans. Am. Soc. Civil Engr* **126**, 1162–1186.

Egiazaroff, I.V. (1965) Calculation of nonuniform sediment concentrations. *Journal of the Hydraulics Division, American Society of Civil Engineers* **91**, 225–247.

Einstein, H.A. (1950) The bedload function for sediment transportation in open channels. *US Department of Agriculture, Soil Conservation Service, Technical Bulletin,* **1026**, 78pp.

Fenton, J.D. and Abbott, J.E. (1977) Initial movement of grains on a stream bed: the effect of relative protrusion. *Proc. R. Soc. Lond.* **253A**, 523–537.

Ferguson, R.I., Prestegaard, K.L. and Ashworth, P.J. (1989) Influence of sand on hydraulics and gravel transport in a braided gravel river. *Water Resources Research* **25**, 635–643.

Gessler, J. (1965) Der Geschiebetriebbeginn bei Mischungen untersucht an natuerlichen Abpflaesterungserscheinungen in Kanaelen. *Mitteilung der Versuchsantstalt fur Wasserbau und Erdbau, Zurich*, No. 69. [The beginning of bedload movement of mixtures investigating as natural armoring in channels, English translation: W.M. Keck Laboratory of Hydraulics and Water Resources, California Institute of Technology, **Translation T-5**, 1967.]

Gessler, J. (1971) Critical shear stress for sediment mixtures. *International Association of Hydraulic Research, Proc. 17th Congr.* **3**, C1.1–C1.8.

Gomez, B. (1979) Bedload discharge in a small Devon stream. *Report of the Transactions of the Devonshire Association for the Advancement of Science* **111**, 31–48.

Gomez, B. (1983a) Temporal variations in bedload transport rates: the effect of progressive bed armouring. *Earth Surface Processes and Landforms* **8**, 41–54.

Gomez, B. (1983b) Temporal variations in the particle size distribution of surficial bed material: the effect of progressive bed armouring. *Geografiska Annaler* **65A**, 183–192.

Gomez, B. (1993) Roughness of stable, armored gravel beds. *Water Resources Research* **29**, 3631–3642.

Gomez, B. (1994) Effects of particle shape and mobility on stable armor development. *Water Resources Research* **30**, 2229–2239.

Grass, A.J. (1970) Initial instability of fine sand beds. *Journal of the Hydraulics Division, American Society of Civil Engineers* **96**, 619–632.

Günter, A. (1971) Die Kritische Mittlere Schlenschubspannung bei Geschiebemischungen Unter Berucksichtigung der Deckschichtbildung und der Terbulenzbedingten Sohlenschubspannungsschwankungen. *Mitteilungen der Versuchsantalt fur Wasserbau, Hydrologie und Glaziologie, Zurich*, No. 3. [The critical mean bed shear stress in the case of a mixed bedload—taking into account the formation of an armour coat and turbulence-induced fluctuations in the shear stress, English translation: Geological Survey of Canada, Ottawa, No. **751589**, 1976.]

Harrison, A.S. (1950) Report on special investigation of bed sediment segregation in a degrading bed. Inst. Engr. Res., Univ. California, Berkerley, Berkeley, California, **33-1**, 205pp.

Hassan, M.A. (1992) Structural controls of the mobility of coarse material in gravel-bed channels. *Israel Journal of Earth Sciences* **41**, 105–122.

Hassan, M.A. and Church, M. (1992) The movement of individual grains on the streambed. In: Billi, P., Hey, R.D., Thorne, C.R. and Tacconi, P. (eds), *Dynamics of Gravel-bed Rivers*, Wiley, Chichester, 159–173.

Helland-Hansen, E., Milhous, R.T. and Klingeman, P.C. (1974) Sediment transport at low Shields-parameter values. *Journal of the Hydraulics Division, American Society of Civil Engineers* **100**, 216–265.

Hjulström, F. (1935) Studies of the morphological activity of rivers. *Univ. Uppsala Geol. Inst. Bull.* **25**, 221–527.

Iseya, F. and Ikeda, H. (1987) Pulsations in bedload transport rates induced by a longitudinal sediment sorting: a flume study using sand and gravel mixtures. *Geografiska. Annaler* **69A**, 15–27.

James, C.S. (1993) Entrainment of spheres: an experimental study of relative size and clustering effects. *Spec. Publ. Internat. Assoc. Sediment.* **24**, 3–10.

Jones, M.L. and Seitz, H.R. (1980) Sediment transport in the Snake and Clearwater Rivers in the vicinity of Lewiston, Idaho. *US Geological Survey Open-File Rept*, 80-69, 179pp.

Kirchner, J.W., Dietrich, W.E., Iseya, F. and Ikeda, H. (1990) The variability of critical shear stress, friction angle, and grain protrusion in water-worked sediments. *Sedimentology* **37**, 647–672.

Komar, P.D. (1987a) Selective grain entrainment by a current from a bed of mixed sizes: a reanalysis. *Journal of Sedimentary Petrology* **57**, 203–211.

Komar, P.D. (1987b) Selective gravel entrainment and the empirical evaluation of flow competence. *Sedimentology* **34**, 1165–1176.

Komar, P.D. and Carling, P.A. (1991) Grain sorting in gravel-bed streams and the choice of particle sizes for flow-competence evaluations. *Sedimentology* **38**, 489–502.

Komar, P.D. and Li, Z. (1986) Pivoting analyses of the selective entrainment of sediments by shape and size with application to gravel threshold. *Sedimentology* **33**, 425–436.

Komar, P.d. and Shih, S-M. (1992) Equal mobility versus changing bedload grain sizes in gravel-bed streams. In: Billi, P., Hey, R.D., Throne, C.R. and Tacconi, P. (eds), *Dynamics of Gravel-bed Rivers*, Wiley, Chichester, 73–93.

Kramer, H. (1935) Sand mixtures and sand movement in fluvial models. *Transactions of the American Society of Civil Engineers* **100**, 798–838.

Kuhnle, R.A. (1989) Bed-surface size changes in gravel-bed channel. *Journal of Hydraulic Engineers* **115**, 731–743.

Kuhnle, R.A. (1992) Fractional transport rates of bedload on Goodwin Creek. In: Billi, P., Hey, R.D., Throne, C.R. and Tacconi, P. (eds), *Dynamics of Gravel-bed Rivers*, Wiley, Chichester, 141–155.

Kuhnle, R.A. (1993) Fluvial transport of sand and gravel mixtures with bimodal size distributions. *Sedimentary Geology* **85**, 17–24.

Kuhnle, R.A. and Southard, J.B. (1988) Bed load transport fluctuations in a gravel bed laboratory channel. *Water Resources Research* **24**, 247–260.

Kuhnle, R.A. and Willis, J.C. (1992) Mean size distribution of bed load on Goodwin Creek. *Journal of Hydraulic Engineering* **118**, 1443–1446.

Laronne, J.B. and Carson, M.A. (1976) Interrelationships between bed morphology and bed material transport for a small gravel-bed channel. *Sedimentology* **23**, 67–85.

Laronne, J.B. and Reid, I. (1993) Very high rates of bedload sediment transport by ephemeral desert rivers. *Nature* **366**, 148–150.

Lavelle, J.W. and Mofjeld, H.O. (1987) Do critical shear stresses for incipient motion really exist? *Journal of Hydraulic Engineering* **113**, 370–385.

Leopold, L.B. (1992) Sediment size that determines channel morphology. In: Billi, P., Hey,

R.D., Throne, C.R. and Tacconi, P. (eds), *Dynamics of Gravel-bed Rivers*, Wiley, Chichester, 297–307.

Leopold, L.B. and Emmett, W.W. (1976) Bedload measurements, East Fork River, Wyoming, *Proceedings of the National Academy of Sciences* 73, 1000–1004.

Li, Z. and Komar, P.D. (1986) Laboratory measurements of pivoting angles for applications to selective entrainment. *Sedimentology* 33, 413–423.

Lisle, T.E., Iseya, F. and Ikeda, H. (1993) Response of a channel with alternate bars to a decrease in supply of mixed-size bed load: a flume experiment. *Water Resources Research* 29, 3623–3269.

McLean, S.R. (1981) The role of non-uniform roughness in the formation of sand ribbons. *Marine Geology* 42, 49–74.

Milhous, R.T. (1973) Sediment transport in a gravel-bottomed stream. *Unpublished Ph.D. thesis, Oregon State University, Corvallis, Oregon, 232pp.*

Miller, M.C., McCave, I.N. and Komar, P.D. (1977) Threshold of sediment motion under unidirectional currents. *Sedimentology* 24, 507–527.

Miller, R.L. and Byrne, R.J. (1966) The angle of repose for a single grain on a fixed rough bed. *Sedimentology* 6, 303–314.

Neill, C.R. (1968) A re-examination of the beginning of movement of coarse granular bed materials. *Hydraul. Res. Stat., Wallingford, Rept,* INT 68, 37pp.

Nevin, C. (1948) Competency of moving water to transport debris. *Geological Society of America Bulletin* 57, 651–674.

Paintal, A.S. (1971) Concept of critical shear stress in loose boundary open channels. *Journal of Hydraulics Research* 9, 91–113.

Paola, C., Parker, G., Seal, R., Sinha, S.K., Southard, J.B. and Wilcock, P.R. (1992) Downstream fining by selective deposition in a laboratory flume. *Science* 258, 1757–1760.

Parker, G. (1978) Self-formed rivers with equilibrium banks and mobile bed, Part II, the gravel river. *Journal of Fluid Mechanics* 89, 127–148.

Parker, G. (1990) Surface-based bedload transport relation for gravel rivers. *Journal of Hydraulics Research* 28, 417–436.

Parker, G. and Klingeman, P.C. (1982) On why gravel-bed streams are paved. *Water Resources Research* 18, 1409–1423.

Parker, G. and Sutherland, A.J. (1990) Fluvial armor. *Journal of Hydraulics Research* 28, 529–544.

Parker, G. and Wilcock, P.R. (1993) Sediment feed and recirculating flumes: a fundamental difference. *Journal of Hydraulic Engineering* 119, 1192–1204.

Parker, G., Klingeman, P.C. and McLean, D.G. (1982) Bedload and size distribution in paved gravel-bed streams. *Journal of the Hydraulics Division, American Society of Civil Engineers* 108, 544–571.

Pitlick, J. (1992) Flow resistance under conditions of intense gravel transport. *Water Resources Research* 28, 891–903.

Proffitt, G.T. (1980) *Selective Transport and Armouring of Non-uniform Alluvial Sediments.* Res. Rept. Dept. Civil Engr., Univ. Canterbury, Christchurch, New Zealand, 80/22, 203pp.

Rubey, W.W. (1938) The force required to move particles on a stream bed. *US Geological Survey Prof. Paper* 189-E, 121–140.

Shields, A. (1936) Anwendung der Aehnlichkeitsmechanik und der Turbulenzforschung auf die Geschiebebewegung. *Mitteilungen Preussischen Versuchsantalt fur Wasserbau Schiffbau, Berlin,* 26. [Application of similarity principles and turbulence research to bedload movement, English translation: W.M. Keck Laboratory of Hydraulics and Water Resources, California Institute of Technology, **Rept 167**.

Sutherland, A.J. (1987) Static armour layers by selective erosion. In: Thorne, C.R., Bathurst, J.C. and Hey, R.D. (eds), *Sediment Transport in Gravel-bed Rivers*, Wiley, Chichester, 243–260.

Wiberg, P.L. and smith, J.D. (1987) Calculations of the critical shear stress for motion of uniform and heterogeneous sediments. *Water Resources Research* 23, 1471–1480.

Wilcock, P.R. (1988) Methods for estimating the critical shear stress of individual fractions in mixed-size sediment. *Water Resources Research* **24**, 1127–1135.

Wilcock, P.R. (1992a) Flow competence: a criticism of a classic concept. *Earth Surface Processes and Landforms* **17**, 289–298.

Wilcock, P.R. (1992b) Experimental investigation of the effects of mixture properties on transport dynamics. In: Billi, P., Hey, R.D., Thorne, C.R. and Tacconi, P. (eds), *Dynamics of Gravel-bed Rivers*, Wiley, Chichester, 109–131.

Wilcock, P.R. (1993) Critical shear stress of natural sediments. *Journal of Hydraulic Engineering* **119**, 491–505.

Wilcock, P.R. and McArdell, B.W. (1993) Surface-based fractional transport rates: mobilization thresholds and partial transport of a sand–gravel sediment. *Water Resources Research* **29**, 1297–1312.

Wilcock, P.R. and Southard, J.B. (1988) Experimental study of incipient motion in mixed sized sediment. *Water Resources Research* **4**, 1137–1151.

Wilcock, P.R. and Southard, J.b. (1989) Bed load transport of mixed sized sediment: fractional transport rates, bed forms, and the development of a coarse bed surface layer. *Water Resources Research* **25**, 1629–1641.

Williams, G.P. and Rosgen, D.C. (1989) Measured total sediment loads (suspended loads and bedloads) for 93 United States streams. *US Geological Survey Open-file Rept*, 89-67.

9 Catchment Sediment Budgets and Change

STANLEY W. TRIMBLE

Department of Geography, University of California, Los Angeles, USA

INTRODUCTION

Steady state rarely, if ever, occurs in streams. That is, there is almost always gain or loss of sediment storage so that input from the slopes rarely equals output as sediment yield (Walling, 1982). Therefore, sediment yields are not indicators of upland erosional rates unless storage fluxes are also taken into account. In most cases, it was merely assumed that there was no net change of sediment storage during the period of sediment yield measurement (Trimble, 1977). The purpose of this paper is to illuminate the magnitudes and time-scales of sediment storage fluxes in small streams under different environmental conditions.

Many sediment reservoirs exist in most basins, but these may be grouped into slope storage, or colluvium, and stream valley storage, or alluvium. If colluvium is transported farther by gravity or water it may remain as colluvium, become alluvium or become efflux. Alluvium can be eroded from the channel or floodplain and thence be redeposited as alluvium or become efflux. The facility with which sediment can move through the system is affected by many variables, some of which are given in Table 9.1. These factors may apply at time scales of hours to millennia.

This study will consider only alluvial or valley storage. In general, the most important process in valley storage gain appears to be vertical accretion from overbank streamflow. Other important storage zones are alluvial fans from small tributaries adjacent to the floodplain and colluvial deposits from adjacent slopes which interfinger into the vertical accretion and in some places may be more important than vertical accretion. In areas of steep, unstable slopes adjacent to the floodplain, mass movement may be an important source of valley storage gain.

Although steady state is rare, it is still a useful concept in analysing sediment budgets. One may describe the concept by saying that input equals output, upland erosion equals sediment yield, storage gain equals storage loss, or the net storage flux is zero. Although not the same, the concept has important similarities to regime theory (ASCE, 1982). Since the ideal rarely exists, it is necessary to speak of disequilibria. If the gain of storage during a given period is greater than the loss of storage, then a positive equilibrium exists, characterised in this study by a net storage gain and consequent aggradation of channels and/or floodplain. If storage loss is greater than storage gain, then a negative disequilibrium exists whereby erosion of channels and floodplains is dominant and a consequent net sediment loss occurs. These transitions were systematically described by Trimble (1983) and given a more

Changing River Channels. Edited by Angela Gurnell and Geoffrey Petts.

Table 9.1 Examples of morphologic and process factors that affect rates of sediment delivery. From Trimble and Lund (1982)

Factors	Range in rate	
	Low	High
Morphologic		
Slope	Gentle	Steep
Sediment texture	Coarse	Fine
Areas of erosion	Small, discontinuous	Large, continuous
Floodplain width	Wide	Narrow
Hydraulic roughness, channel and floodplain	High	Low
Process		
Type of erosion	Sheet	Channel
Stream regime	Equitable	Flashy

rigorous treatment by Phillips (1987). With this framework in mind, I shall identify five conceptual models of valley storage fluxes. I do not claim universality for these models, but most examples from the literature seem to fit one or more of them.

MODEL 1. HUMID REGION, QUASI STEADY STATE

A condition approaching a steady state is endemic to a humid region because only there can vegetation develop adequately to stabilise the landscape. We consider here a basin in which there is very little upland erosion and consequently little sediment load. The hydrologic regime and the longitudinal stream profile are both stable. Although there may occasionally be lateral erosion of stream banks, the opposite bank is built up a commensurate amount by lateral accretion so that there is no net sediment loss. During overbank flows, however, there may be a slow sediment gain because the well-vegetated floodplain would keep tractive force below critical values, not only protecting the surface from erosion, but also promoting deposition. Although sediment loads are low, such a process could accrete floodplains by several centimetres over millennia (Knox, 1972). Strictly speaking, this is not a steady state, but it may be the closest natural approximation. The concept is shown graphically in figure 9.1. In this and in the following figures steady state is shown by the horizontal line. Sediment gain (positive disequilibrium) is above the line and below is sediment loss (negative disequilibrium), a concept first proposed by Vita-Finzi (1969). In this model, there is a varying but low sediment gain attributable to vertical accretion (line A). Such a situation was obtained periodically in the eastern United States before European settlement. The general development of non-stratified Mollisols on floodplains there is strong evidence that both lateral stream migration and vertical accretion were very slow processes (Trimble and Lund, 1982). It seems reasonable to assume that much of the humid United States was in such a quasi steady state. A slight variant to the model above is when the stream is laterally

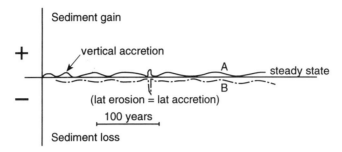

Figure 9.1   Model 1: quasi steady-state

eroding a bank higher than the floodplain. Because the higher surface, such as a terrace or hillside, is replaced by the lower-lying floodplain, storage loss is occurring
and the gain from vertical accretion can be offset to the point that net sediment loss
occurs (line B). The perturbation in the mid-line is a big discharge event which
would both erode the channel somewhat and deposit sediment on the floodplain.

MODEL 2. PERTURBATION OF HUMID AREA QUASI STEADY STATE

In this model (Figure 9.2), we consider a rupture of the system wherein quantities of
sediment are produced from the upland which are beyond the transport capacity of
the stream. Such a disturbance could and did come from strong climatic forces (e.g.
Knox, 1972), or from human alteration of the landscape. The excessive sediment
causes vertical accretion of floodplains and possibly aggradation of channels. A
decrease in the rate of gain might come about because of an amelioration or reversal of the conditions which caused the disturbance. If sediment supply remains low,
the stream will attempt to remove the accumulated sediment. If the channel has
been aggraded, vertical cutting will probably predominate at first and will be very
rapid. As a longitudinal profile of equilibrium is reached, vertical cutting will slow

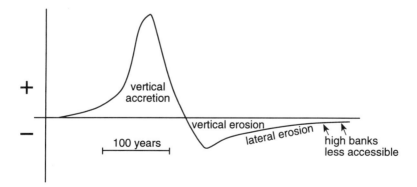

Figure 9.2   Model 2: great sediment influx with later amelioration

and lateral erosion will become more important. As this proceeds, the eroding high banks, now terrace escarpments, become increasingly less accessible to the stream so that the rate of sediment loss increases, but as long as terraces or other high banks are available, the sediment loss curve will be asymptotic to the steady state line (Trimble, 1984, 1993). As explained under Model 1, vertical accretion rates on the newly created floodplain may offset or even exceed removal rates, and when this obtains, Model 2 reverts to Model 1. Perhaps 'steady state' conveys the impression of geomorphic inactivity, but the foregoing shows that highly dynamic stream reaches can still qualify.

In comparing total removal with total accretion in Model 2, there is no necessary implication that the two volumes are equal, certainly not in the short run of 100–200 years. While the aggrading stream could, by means of vertical accretion, deposit sediment across the entire valley, including the channel, the eroding stream can generally erode only its channel and banks. In fact, measurements in the southern Piedmont and the Paleozoic Plateau show that a rapidly eroding stream can still be aggrading the old floodplain by vertical accretion, but only so long as overbank flows are possible (Trimble, 1975). Even extreme discharge events seldom develop enough tractive force to erode most vegetated floodplains. Wolman and Eiler (1958) found that a catastrophic event in New England only managed to erode as much as it deposited on the floodplain.

The removal phase may be not only of that accretion immediately previous, but the eroding stream may be removing sediment from several previous episodes (Duijsings, 1987). The order of sediment removal is complex; generally, the last sequence deposited is the first to be removed, but parts of the remainder of old sequences of up to 10 000 years may be removed at the same time (Dietrich et al., 1982). The rate of removal of any one sequence tends to be an exponential decay function of age, but rates of any one or more sequences are subject to considerable variation, including cessation. This concept of slowing sediment removal was first suggested by G.K. Gilbert (1917) who predicted that hydraulic mining debris in the South Yuba River of California would be largely removed within 50 years.

An excellent example of the above processes is given by the Southern Piedmont (Trimble, 1974). As the result of abusive agriculture there, channels and valleys filled with up to 6 m of sediment, mostly by vertical accretion, although fans and colluvium interfinger along the edge of floodplains. With the reversion of cropland to forest and the stabilisation of land by soil conservation measures, some streams have been eroding since the late 1940s. Most channel erosion had been vertical through the early 1970s, but a brief field inspection in late 1983 indicated that, despite vegetated banks, some streams had undergone considerable lateral erosion over the previous decade. In the northern part of the Piedmont, Jacobson and Coleman (1986) report lateral erosion with the creation of new, lower floodplains.

In the Paleozoic Plateau, the quasi-steady state was upset by agricultural erosion. Floodplains and channels were aggraded up to depths of 5 m, primarily by vertical accretion, although the role of colluvium and especially alluvial fans appears to be more important than on the Piedmont. Headwater streams have already passed through the erosional stage and are now mild sediment sinks (Trimble, 1993). Presently, the upper main valley is passing through the erosional stage. Lateral cutting has proceeded apace with maximum short-term rates of cut-bank retreat being of the order of 0.5 m yr^{-1}, but long-term average rates are only a fraction of that. Indications are that left unfettered by human attempts at stabilisation (considering the agricultural value of terrace land, this is unlikely), it would take streams several hundred years to remove all of the old, higher floodplain, now a terrace, and replace it with a new, lower floodplain adjusted to the present, milder hydrologic regime.

The debris from hydraulic mining in the California Sierras has behaved in almost perfect accordance with Model 2. In a sense, one may perceive the whole hydraulic-mining debacle as a gigantic flume or stream-table experiment which, despite its depredations on the landscape, gives many insights into sediment transport storage fluxes. By the use of high-pressure water jets, gold-bearing Tertiary gravels were removed from high terraces and routed into Sierra streams. Mountain valleys were buried to depths of over 25 m and debris was transported from the Sierras to the main tributaries of the Sacramento River flowing across the northern half of the Great Valley of California. One of these stream systems, the Yuba River, was studied by Gilbert in 1917 and by Adler in 1980. Aggradation, both of mountain tributaries and of the trunk river, was extremely rapid. After mining was abruptly halted in 1884, vertical stream incision of the headwater deposits began and the entrained sediment migrated downstream. That the peak of channel aggradation at Sacramento occurs before that upstream at Marysville may be a fluke, or it may be due to debris from other streams affecting the Sacramento River closer to the city of Sacramento, notably the American River (Gilbert, 1917). In any case, this mobile wave of sediment has been depicted as a downstream purging of the erosional debris, but, in fact, mainly the in-channel sediment was involved and floodplain deposits mostly remain. Adler's study shows that some channel elevations on the Yuba River upstream of Marysville did not crest until 1906 or 1912. This disparity of dates is especially interesting from a sediment budget viewpoint because it shows that at the same time channels were eroding, floodplains were still rapidly aggrading, a situation which also occurred in the Paleozoic Plateau (Trimble, 1975, 1983; Knox, 1987). The crest probably represents the initial downstream movement of material while the erosion since then is the vertical bed cutting of Model 2. Since then, lateral erosion has followed, but at slower rates than those predicted by Gilbert (Adler, 1980).

The hydraulic mining debris also demonstrates another important point: sediment budgets can differ greatly at various points along the same stream system, a concept termed 'out of phase' by Schumm (1973) and 'distributed' by Trimble (1993). In the Yuba River, for example, sediment accumulation in mountain valleys was probably at its peak no later than 1884 when mining ceased, and there was probably a net loss after that time. Downstream in the 30 km stretch between the mountains and Marysville, however, peak rates of sediment accumulation occurred about 1880–1900, a lag of several years (Figure 9.3). Profiles surveyed in 1899, 1906 and 1912 at the head of this reach show the highest level to have been 1906. If the sediment budget (Figure 9.3) is smoothed to give an approximate continuous value, the curve, almost identical with Model 2, shows net sediment loss beginning about 1905. The 20-year lag was due not only to the travel time of sediment from the initial aggradational phase but also to the downstream augmentation from upstream erosion. James (1989) working nearby in Bear Creek found similar processes there while Knighton (1989) found similar processes in Tasmania. Ordinarily, as this erosional

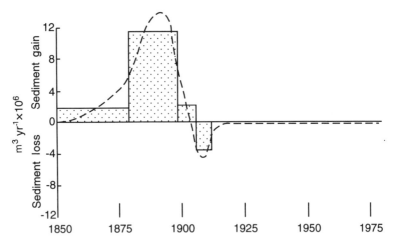

Figure 9.3 Sediment budget, lower Yuba River, California, 1850–1979. Modified from Adler (1980)

process and the eroded material move downstream, the in-stream effect would probably decrease because so much of the eroded material would go back into storage on floodplains. However, because the California streams now have high artificial levées, stream discharge is contained within the channel and stream elevations at both Marysville and Sacramento continue to drop.

Despite the stream channel erosion of the past 100 years, most of the mining debris is still in storage. This is clearly demonstrated by Figure 9.3 which shows that, out of 2.9×10^8 m^3 deposited in this 30 km length, only 0.3×10^8 m^3, or about 10%, has been transported away and the present rate appears to be very low. Gilbert (1917) predicted that by the mid-1960s, the volume stored in this stretch would be reduced to $108–170 \times 10^6$ m^3, but Adler measured about 261×10^6 m^3 still remaining in 1979. This demonstrates not only the slow progress of lateral erosion (Leopold et al., 1964; Meade, 1982; Trimble, 1983, 1993), but also the effectiveness of engineering works along the river designed to prevent lateral stream erosion. Thus, it would appear that most of the erosional debris is there to stay, perhaps over the next few hundred or even thousand years. A similar programme to prevent lateral erosion in the Paleozoic Plateau should give like results (Trimble, 1993).

The foregoing discussion has been based on human disruption of fluvial systems, but the model appears to apply well to the Holocene. As early as 1972, Church and Ryder suggested that much sediment from streams in British Columbia was coming from near-stream glacial deposits and, in 1989, Church and Slaymaker were much more definitive. Moreover, a recent review by Ashmore (1993) stunningly concludes that stream and valley-side sediment sources are dominant for all of Canada! Gordon (1979) has also suggested a similar phenomenon in New England, with duration of 10^5 years, which has furnished a constant sediment yield for 8000 years. On the other hand, artificial channel modifications in some parts of Britain have prevented the remobilisation of Holocene materials (Brown, 1987).

MODEL 3. VALLEY TRENCHING OR ARROYO CUTTING

Perturbations, whether climatic or human-induced, do not necessarily induce aggradation. If the perturbation causes an incremental increase in stream discharge greater than the incremental increase in sediment discharge, the result may be trenching or gullying (Cooke and Reeves, 1976; Schumm, 1977). The best-known and most-studied examples of this area are found in studies of 'arroyo cutting' in the semi-arid south-western United States. A useful compendium of these studies plus some important original work is found in the study by Cooke and Reeves (1977). The cultural explanation of this phenomenon is very straightforward: fragile glasslands were overgrazed, the vegetation was thinned drastically and the soil was severely compacted, resulting in greater stormflows, resulting in arroyo cutting. The climatic hypothesis is a bit more tenuous, but whether caused by overgrazing or climate or both, many valleys of the semi-arid south-west have been trenched. In recent years, many of these arroyos have been partially stabilised and even begun refilling (Emmett, 1974; Leopold, 1976). Whether this is due to amelioration of the original problem, or feedback from channel forms or vegetation, is uncertain, and may vary by location and time.

Not as well known is the arroyo cutting which occurred in the humid United States. In the Paleozoic Plateau and other regions it was known as 'valley trenching' and included headward extensions of tributaries sometimes extending from or through high, sandy Pleistocene terraces (Happ et al., 1940). Most such valleys were sediment sinks until after the turn of this century, some with vertical accretion up to 2 m thick. This would seem to indicate a period when the incremental supply of sediment was greater than the incremental supply of runoff. Such valleys became swampy and boggy during this period. Model 3 (Figure 9.4) features an initial period of sediment gain, and this often applied to the arid south-west (R.U. Cooke, personal communication). The runoff–sediment balance changed at some time, generally between 1890 and 1920, so that these former sediment sinks became prolific sediment sources to downstream floodplains. The cause appeared to be a combination of increased grazing and deteriorated hydrologic properties of soil caused by a half century of poor agricultural practices, but climatic change does not appear to be important (Trimble and Lund, 1982). By the 1930s, many tributary valleys were

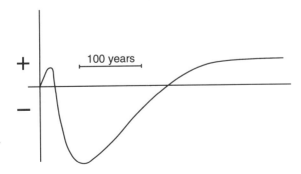

Figure 9.4 Model 3: arroyo cutting or valley trenching

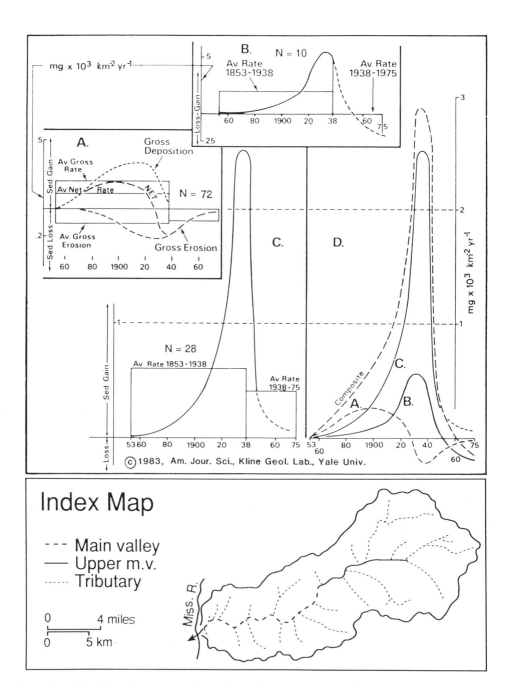

Figure 9.5 Distributed and composite sediment budgets, Coon Valley, Wisconsin, 1853–1975. (A) tributaries, (B) upper main valley, (C) lower main valley, (D) superimposition and composite. Redrawn from Trimble (1983)

trenched (Happ *et al.*, 1940). The early stages of this arroyo cutting were especially interesting from a sediment budget standpoint because at the same time that channels were rapidly enlarging, floodplains were still receiving vertical accretion. A sediment budget for the tributaries of Coon Creek, Wisconsin (Figure 9.5), shows that gross deposition (from vertical accretion) exceeded gross erosion of the channel until about 1935. The net budget showed the greatest sediment storage loss at about 1940 and that has now become asymptotic for reasons explained earlier. Additionally, many of the trenches have stabilised, presumably because the original problem was ameliorated. The thick vegetation in these stabilised gullies together with a more equitable hydrologic regime from improved land use has allowed some of these channels to become sediment sinks, thus according well with Model 3.

The longevity of valley sediment storage, its relatively slow rates of removal and the downstream migration of sediment are all well demonstrated by Coon Creek. As just described, many headward tributary valleys were sediment sinks for 30–60 years, then became strong downvalley sediment sources, and now may be near a steady state. Although the high banks of the old floodplain, now a terrace, are eroded on occasion, the new, lower floodplain created between the high bank is still aggrading and such stream reaches are now mild sediment sinks (Trimble, 1993).

Downstream in the upper main valley, rates of sediment accretion peaked in the 1930s. One reason for this is that the floodplain had built up by vertical accretion so that it was accessible to fewer overflows. For example, the banks at the village of Coon Valley were only 1.5–2 m high in 1912, but were about 4 m high by 1938. This greatly enhanced channel capacity is probably morphologic evidence of the deteriorated hydrologic regime already mentioned, wherein the magnitude of any given return frequency became larger during the first part of this century. Since that time, the sediment supply from downstream has been decreased and the hydrologic regime has ameliorated (Trimble, 1975; Trimble and Lund, 1982). The result is that (a) the old floodplain no longer receives significant sediment, and (b) as the high banks of the old floodplain retreat from erosion, the lateral accretion on the opposite bank is much lower, about 2 m above mean water level, which is presumably a bank height in adjustment with the present hydrologic regime (Trimble, 1983, 1993). Obviously, if a retreating high bank is being replaced by an advancing low bank, and there is no accretion on the old floodplain (now a terrace), this channel stretch is now a sediment source, so that Model 2 applies here. The locus of this process appears to be moving downstream to the lower main valley. There, constant accretion has occurred since European settlement, although the rates are only about 1 or 2% of the highs of the 1930s (Figure 9.5, Trimble, 1983, 1993). Assuming there are no renewed upstream sediment sources, it appears reasonable that the process of sediment loss will continue to move downstream to the lower main valley and that, in fact, has been occurring (Trimble, 1993). Thus, Model 2 applies here, but the sequence of events has not yet progressed to the sediment loss stage. Thus, sediment storage gain and loss fluxes are highly variable with space and time within a basin and no one sediment budget model characterises the entire basin, although a composite can be constructed (Figure 9.5D). Presumably, the complexity of sediment budgets generally increases with basin size, but this is a hypothesis which should be tested by further research. Indeed, I believe that documenting the complex fluxes and storage of sediment through space and

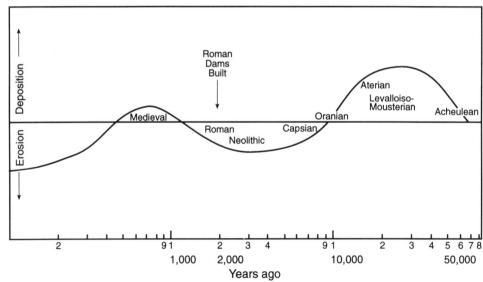

Figure 9.6 Chronology of erosion and deposition in the Mediterranean area. A logarithmic scale has been used to compress the time base. Redrawn from Vita-Finzi (1969)

time along with establishing the interrelationships is the most pressing need in sediment budget studies.

Perhaps the best long-term record of valley sediment storage fluxes is provided by Vita-Finzi (1969) in his comprehensive study of the Mediterrean region. Drawing from many published sources and his own work, his 70 000 year record shows two full cycles of Model 3 (Figure 9.6, note log scale for time). Indications are that Mediterrean valleys had generally aggraded during the last 60 000 years of the last glaciation. This fill developed calcareous crusts at various stages. By 10 000 years BP, streams were eroding but vertical cutting was impeded by the crusts. Later, the Romans built dams to retain moisture, but their effect was to curtail valley trenching. By 1100 BP, the valleys were again aggrading, a condition which lasted about 600 years. Finally the streams have been trenching for the past 400–500 years.

MODEL 4. URBAN STREAMS

The generally accepted model of urban sediment production is that of Wolman (1967). It shows a brief flare of erosion while urban construction is under way, but once that is stabilised, there is very little erosion of the mature urban landscape. Often ignored is the second part of Wolman's study: at the same time that upland areas are being stabilised, they are also becoming increasingly impervious so storm runoff increases while erosion decreases. At this stage, channel erosion, or arroyo cutting, begins (Douglas, 1974). Urban storm discharge peaks may be many times those of the rural countryside and the result, as expected, is the same as in arroyo cutting (Model 3), only usually more severe. Such channel erosion becomes a down-

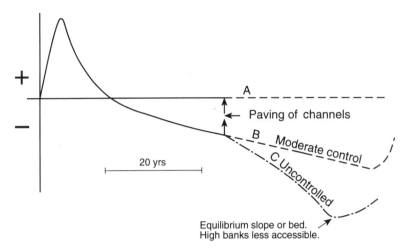

Figure 9.7   Model 4: urban streams

stream sediment source and it is easy to perceive downstream sediment yields becoming larger with time. However, the saving grace is that the land consumed by urban arroyos is expensive real estate so that there is usually an attempt to stabilise the channels, often by paving them. This model, then, is a severe case of Model 3, arroyo cutting, but the resolution of channel erosion tends to be more definite and thus Model 4 is presented (Figure 9.7). Like the arroyo model, there is an initial period of storage gain, but thereafter a sharp period of storage loss. What happens during this period becomes problematic because channels are rarely left to erode. Indeed, there is usually every effort to control them. If the channels are paved, they are also straightened and given a slope and cross-section which usually keeps them from gaining sediment, so the result is an immediate and permanent steady state by design (line A). If channels are given only moderate controls, such as sills, drop structures, vegetated banks, and wire or brush revetments, erosion may continue, but probably at a slower rate (line B). Eventually, it appears probable that uncontrolled channel slopes, widths and channel morphologies will adjust so there may be decreased sediment loss. Likewise, the uncontrolled channel (line C) will erode at increasing rates for a while and the erosion will probably subside eventually.

A somewhat unusual case of arroyo cutting in an urbanising area is occurring in the San Diego Creek on the coastal plain south of Los Angeles. A former basin of interior drainage where sediment has been accumulating for centuries, it has been channelised and the flow directed into Newport Bay. The greatly lowered base level has set off rapid channel cutting which is accelerated by increasing urbanisation (Trimble, 1981). Many artificial sills tend to retard additional vertical cutting, but lateral erosion is rampant in many places. Hoag (1983) estimates that storage loss from channel erosion accounts for at least 28% of the stream sediment yield, but the proportion may well be larger.

Vita-Finzi (1969) records a situation in ancient Jordan analogous with that of San Diego Creek but with tectonism as the control. During the Quaternary, terrestrial and lacustrine sediments were deposited in the interior drainage of the Dead Sea

Rift. Further downfaulting and the consequent degradation of the Jordan River dropped the local base level, causing widespread downcutting of tributaries.

MODEL 5. HIGH-ENERGY INSTABILITY, MOUNTAIN AND ARID STREAMS

A sediment budget for a stream with a very narrow floodplain and very steep valley sides is much more episodic than cyclic in nature, as the others have been. Mountain valleys tend to be relict to the last big event or events. Net storage gain can come from vertical accretion, much coming as fans from small lateral tributaries, but there is also a high probability of mass movements from valley sides (Dietrich and Dunne, 1978; Kelsey, 1980; Lehre, 1981; Dietrich *et al.*, 1982; Pearce and Watson, 1986; Smith and Swanson, 1987; Cain and Swanson, 1989; Benda, 1990; Slaymaker, 1993). Sediment loss tends to come with large events which 'flush' (Schumm, 1977; Nanson, 1986) sediment downstream. Unlike floodplains of most humid areas, mountain floodplains are steep, usually of non-cohesive materials reflecting the high-energy environment, are poorly anchored, and are narrow, so that high flows can lavish much more tractive force per unit area. The result is that such floodplains are often mobile, especially in very high magnitude events (Nanson and Croke, 1992). Work by Schick (1977), Graf (1983), Boison and Patton (1985), Patton and Boison (1986) and Schick and Lekach (1993) indicates that some arid basins also share this instability and variability.

Model 5 (Figure 9.8) shows a period of accretion, followed by a flushing event. This is followed by a massive mass movement event and then followed by another flushing event. This is followed by a period of vertical accretion and lateral fans with episodic gains from mass movements. Despite the variability endemic to mountain streams, Madej (1984) was able to construct a 33 year sediment budget for Redwood Creek, California, which shows a tripartite distributed sediment budget similar in many ways to that of Trimble (1983) and Knox (1989).

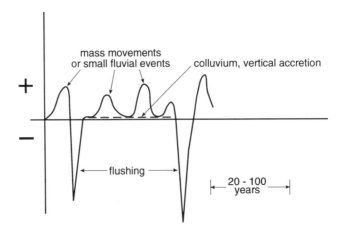

Figure 9.8 Model 5: high-energy instability, mountain and arid streams

CONCLUSIONS

Work on sediment budgets has greatly enhanced our understanding of fluvial processes. To help illuminate the magnitudes and time-scales of sediment storage fluxes under different environmental conditions, five sediment budget models have been presented. For this perspective, they generalise much of the sediment budget work done during the past quarter of a century. I conclude that the most pressing problem is to now identify the various sub-budgets which can operate within the overall basin budget and to establish the causal interrelationships among these sub-budgets. Research on such sub-budgets or 'distributed' budgets will call for close controls on time and space variations within basins. The presented models are not meant to be the final word but rather are presented with an Hegelian sense of provoking insights and revisions.

ACKNOWLEDGEMENTS

R.H. Meade and R.U. Cooke critiqued an earlier version of this paper but any faults remain mine. I thank Alesia Haymon for typing, Alex Mendel for research, and Chase Langford for the diagrams.

REFERENCES

Adler, L.L. (1980) Adjustment of the Yuba River, California, to the influx of hydraulic mining debris, 1849–1979. MA thesis, University of California, Los Angeles.

ASCE Sedimentation Committee (1982) Relationships between morphology of small streams and sediment yield. *Journal of the Hydraulics Division, American Society of Civil Engineers* **108**, 1328–1365.

Ashmore, P. (1993) Contemporary erosion of the Canadian landscape. *Progress in Physical Geography* **17**, 190–204.

Benda, L. (1990) The influence of debris flows on channels and valley floors in the Oregon Coast Range, U.S.A. *Earth Surface Processes and Landforms* **15**, 457–466.

Boison, P.J. and Patton, P.c. (1985) Sediment storage and terrace formation in Coyote Gulch basin, south-central Utah. *Geology* **13**, 31–34.

Brown, A.G. (1987) Long-term sediment storage in the Severn and Wye catchments. In: Gregory, K.J. and Thornes, J.B. (eds), *Paleohydrologyin Practice*, Wiley, Chichester.

Caine, N. and Swanson, F.J. (1989) Geomorphic coupling of hillslope and channel systems in two small mountain basins. *Zeitschrift fur Geomorphologie* **33**, 189–203.

Church, M. and Ryder, J.M. (1972) Paraglacial sedimentation: a consideration of fluvial processes conditioned by glaciation. *Geological Society of America Bulletin* **83**, 3059–3072.

Church, M. and Slaymaker, O. (1989) Disequilibrium of Holocene sediment yield in glaciated British Columbia. *Nature* **337**, 352–454.

Cooke, R.U. and Reeves, R.W. (1977) *Arroyos and Environmental Change in the American South-West*. Clarendon Press, Oxford.

Dietrich, W.E. and Dunne, T. (1978) Sediment budget for a small catchment in mountainous terrain. *Zeitschrift fur Geomorphologie* Supplementband **29**, 191–206.

Dietrich, W.E., Dunne, T., Humphrey, N. and Reid, L.M. (1982) Construction of sediment budgets for drainage basins. In: Swanson, F.J, Janda, R.J., Dunne, T. and Swanson, D.N. (eds), *Sediment Budgets and Routing in Forested Drainage Basins*, USDA Forest Service General Technical Report PNW-141, 5–23.

Douglas, I. (1974) The impact of urbanisation on river systems. In: *Proceedings of the Inter-national Geographical Union Regional Conference*, New Zealand Geographical Society, 307–317.

Duijsings, J.J. (1987) A sediment budget for a forested catchment in Luxembourg and its implications for channel development. *Earth Surface Processes and Landforms* 12, 173–184.

Emmett, W.W. (1974) Channel aggradation in western United States as indicated by observa-tion at Vigil Network sites. *Zeitschrift für Geomorphologie* 21, 52–62.

Gilbert, G.K. (1917) Hydraulic mining debris in the Sierra Nevada. *US Geological Survey Professional Paper* 105.

Gordon, R.B. (1979) Denudation rate of central New England determined from estuarine sedimentation. *American Journal of Science* 279, 1997–2006.

Graf, W.L. (1983) Variability of sediment removal in a semiarid watershed. *Water Resources Research* 19, 643–652.

Happ, S.C., Rittenhouse, G. and Dobson, G.C. (1940) Some principles of accelerated stream and valley sedimentation. *US Department of Agriculture Technical Bulletin* **695.**

Hoag, B.L. (1983) A geographic analysis of urban channel erosion in Orange County, Cali-fornia, MA thesis, Department of Geography, University of California, Los Angeles.

Jacobson, R.B. and Coleman, D.J. (1986) Stratigraphy and recent evolution of Maryland Piedmont flood plains. *American Journal of Science* **286**, 617–637.

James, L.A. (1989) Sustained storage and transport of hydraulic gold mining sediment in the Bear River, California. *Annals of the Association of American Geographers* 37, 570–592.

Kelsey, H.M. (1980) A sediment budget and an analysis of geomorphic process in the Van Duzen River basin, north coastal California, 1941–1975): Summary. *Geological Society of America Bulletin* **91**, 190–195.

Knighton, A.D. (1989) River adjustment to changes in sediment load: the effects of tin mining on the Ringarooma River, Tasmania, 1875–1984. *Earth Surfaces Processes and Landforms* **14**, 333–359.

Knox, J.C. (1972) Valley alluviation in southwestern Wisconsin. *Annals of the Association of American Geographers* 62, 401–410.

Knox, J.C. (1987) Historical valley floor sedimentation in the upper Mississippi valley. *Annals of the Association of American Geographers* 77, 224–244.

Knox, J.C. (1989) Long and short-term episodic storage and removal of sediment in water-sheds of southwestern Wisconsin and northwestern Illinois. In: *Sediment and the Environ-ment*, International Association of Hydrological Sciences Publication 184, 157–164.

Lehre, A. (1981) Sediment budget of a small California Coast Range drainage near San Fran-cisco. *International Association of Hydrological Sciences Publication* **132**, 123–139.

Leopold, L.B. (1976) Reversal of erosion cycle and climatic change. *Quaternary Research* 6, 557–562.

Leopold, L.B., Wolman, M.G. and Miller, J.P. (1964) *Fluvial Processes in Geomorphology*. W.H. Freeman & Co, San Francisco.

Madej, M.A. (1984) Recent changes in channel-stored sediment, Redwood Creek, California. *Redwood National Park Technical Report* No. 11. National Park Service, Arcata, California.

Meade, R.H. (1982) Sources, sinks and storage of river sediment in the Atlantic drainage of the United States. *Journal of Geology* 90, 235–252.

Nanson, G.C. (1986) Episodes of vertical accretion and catastrophic stripping: a model of dis-equilibrium flood-plain development. *Geological Society of American Bulletin* 97, 1467–1475.

Nanson, G.C. and Croke, J.C. (1992) A genetic classification of floodplains. *Geomorphology* 4, 459–486.

Patton, P.C. and Boison, P.J. (1986) Processes and rates of formation of Holocene alluvial terraces in Harris Wash, Escalante River basin, south-central Utah. *Geological Society of America Bulletin* **97**, 369–378.

Pearce, A.J. and Watson, A.J. (1986) Effects of earthquake-induced landslides on sediment budget and transport over a 50 year period. *Geology* **14**, 52–55.

Phillips, J.D. (1987) Sediment budget stability in the Tar River Basin, North Carolina. *Amer-ican Journal of Science* **287**, 780–794.

Schick, A.P. (1977) A tentative sediment budget for an extremely arid watershed in the southern Negev. In Doehring, D.O. (ed.), *Geomorphology in Arid Regions*, State University of New York, Binghampton, Publications in Geomorphology, 139–163.

Schick, A.P. and Lekach, J. (1993) An evaluation of two ten-year sediment budgets, Nahal Yael, Israel. *Physical Geography* **14**, 225–238.

Schumm, S.A. (1973) Geomorphic thresholds and complex response of drainage systems. In: Morisawa, M. (ed.) *Fluvial Geomorphology*, State University of New York, Binghampton, Publications in Geomorphology, 299–310.

Schumm, S.A. (1977) *The Fluvial System*. Wiley-Interscience, New York.

Slaymaker, O. (1993) The sediment budget of the Lillooet River Basin, British Columbia. *Physical Geography* **14**, 305–320.

Smith, R.D. and Swanson, F.J. (1987) Sediment routing in a small drainage basin in the blast zone of Mount St. Helens, Washington, USA. *Geomorphology* **1**, 1–13.

Trimble, S.W. (1974) *Man-induced Soil Erosion on the Southern Piedmont, 1700–1970*. Soil Conservation Society of America, Ankeny, Iowa.

Trimble, S.W. (1975) Response of Coon Creek, Wisconsin, to soil conservation measures. In: Zakrzewska-Borowiecki (ed.), *Landscapes of Wisconsin*, Association of American Geographers, Washington, DC, 24–29.

Trimble, S.W. (1977) The fallacy of stream equilibrium in contemporary denudation studies. *American Journal of Science* **277**, 876–887.

Trimble, S.W. (1981) Geomorphic analysis, Newport Bay Watershed, San Diego Creek comprehensive stormwater sedimentation control plan. *Technical Memorandum*. Boyle Engineering Corp., San Diego California.

Trimble, S.W. (1983) A sediment budget for Coon Creek basin in the Driftless Area, Wisconsin, 1853–1977. *American Journal of Science* **283**, 454–474.

Trimble, S.W. (1984) Storage of sediment in smaller streams at time scales of 10 to 10,000 years. Abstract in: *Symposium on Sediment Storage on River and Estuaries* 216, American Geophysical Union, Cincinnati, Ohio.

Trimble, S.W. (1993) The distributed sediment budget model and watershed management in the Paleozoic Plateau of the upper midwestern United States. *Physical Geography* **14**, 285–303.

Trimble, S.W. and Lund, S.W. (1982) Soil conservation and the reduction of erosion and sedimentation in the Coon Creek basin, Wisconsin. *US Geological Survey Professional Paper* 1234.

Vita-Finzi, C. (1969) *The Mediterranean Valleys*. Cambridge University Press, Cambridge.

Walling, D.E. (1983) The sediment delivery problem. *Journal of Hydrology* **65**, 209–237.

Wolman, M.G. (1967) A cycle of sedimentation and erosion in urban river channels. *Geografiska Annaler* **49A**, 385–395.

Wolman, M.G. and Eiler, J.P. (1958) Reconnaissance study of erosion and deposition produced by the flood of August, 1955, in Connecticut. *American Geophysical Union Transactions* **39**, 1–14.

10 River Channel Change: The Role of Large Woody Debris

E.A. KELLER[1] AND ANNE MACDONALD[2]

[1]*Environmental Studies and Geological Sciences, University of California, Santa Barbara, USA, and* [2]*PTI Environmental Services, Bellvue, Washington, USA*

INTRODUCTION

The role of large woody debris (LWD) in contributing to change of river channels in the United States has been long recognised, as illustrated by Lobeck's (1939) discussion of a huge debris jam on the Red River of north-western Louisiana. That debris jam—or the 'great raft', as it was known—was a 300-km-long accumulation of woody debris and sediment that prevented steamboat navigation access to the upper two-thirds of the river basin. President Jefferson sent an expedition in 1906 to the river and it was concluded that the great raft, which was thought to be several hundred years old, could never be removed. In the early 1800s, Henry Shreve designed and ran steamboats on the Mississippi River system. He also designed special steamboats to remove large woody debris (snags). Up to this time it was difficult to safely navigate rivers by steamboat because of fear of collision with snags that might damage the large boats (containing several decks, staterooms and a considerable degree of luxury) and inconvenience passengers. Shreve's snagboat was very successful and was used to clear snags from hundreds of kilometres of river channel, some of which were reportedly 20 m long, over a metre in diameter, and embedded up to 6 m into the riverbed sediments (McCall, 1988). It took Shreve and his snagboat from 1832 to 1839 to clear the great raft of the Red River. Approximately 10 years after Lobeck's description of the clearing of the 'great raft', Bevan (1948–49) (on the Middle Fork of the Willamete River in Oregon) concluded that LWD was responsible for more channel changes than any other factor.

Because rivers used for navigation that flow through forested areas have for decades been cleared of most LWD, we speculate that under natural conditions, much more woody debris was in stream and river channels than under present conditions, and that accumulations of LWD affected stream channel form and process more in the past than they do today (Triska, 1984). The nature of this influence is illustrated by several integrated studies of streams rich in woody debris located in north western California (Sedell and Luchessa, 1981; Sedell and Froggat, 1984).

The purposes of this chapter are to:

- Briefly review the role of LWD in influencing channel form and process. Discussion will focus on the study of seven channel reaches of Prairie Creek (Figure 10.1).

Changing River Channels. Edited by Angela Gurnell and Geoffrey Petts.
© 1995 John Wiley & Sons Ltd.

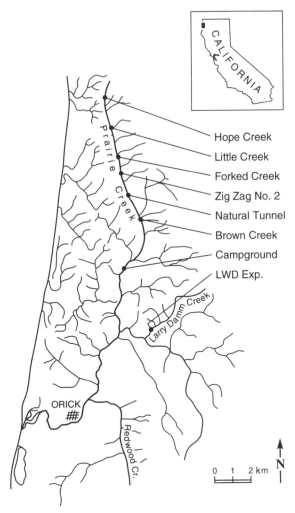

Figure 10.1 Map showing locations of study reaches in Prairie Creek and Larry Damm Creek, tributaries of Redwood Creek, north-western California

- Discuss channel changes resulting from LWD by presenting results of debris removal experiments in Larry Damm Creek (Figure 10:1).
- Discuss the hydraulics of LWD steps by considering changes in stream power with discharge over a step in Larry Damm Creek

CHANNEL FORM AND PROCESS

Dynamics of woody debris

Large woody debris consists of logs, limbs or root wads greater than 10 cm in diameter (Keller and Swanson, 1979). Study of the role of LWD on channel form and

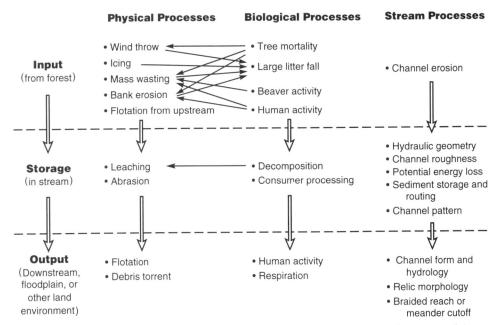

Figure 10.2 Diagram illustrating relationship between input, storage and output of large woody debris as it relates to physical, biological and stream processes

process is best approached by considering a budget approach that includes input, storage and output of LWD related to physical, biological and stream processes (Figure 10.2). Of particular importance are potential linkages between physical and biological processes related to input and storage of LWD in stream and river channels. Most significant processes are shown on Figure 10.2 and include:

- Tree mortality related to disease and damage by insects that render trees more vulnerable to wind throw, icing and mass wasting.
- Decomposition and consumer processing of LWD that aids in the process of leaching woody debris.
- Human activity (particularly timber harvesting) that may initiate mass wasting and bank erosion.
- Human activity (management and removal of LWD following timber harvesting) that changes the amount of LWD that may be removed from the system by flotation or that may accumulate only for short time periods.

Of particular importance in discussing the dynamics of LWD is the role that the debris plays in influencing channel form and process:

- Input of LWD from the forest may affect the stream channel by way of channel erosion. This results when a stream-side tree falls to the forest floor or into the stream channel. An upthrown root wad exposes the channel banks to erosion by causing a bowl-shaped zone of bank erosion.
- When LWD is stored in stream channels it locally modifies hydraulic geometry, increases channel roughness, concentrates potential energy loss, regulates sediment

storage and routing, and often alters channel pattern (Heede, 1972; Swanson *et al.*, 1976; Swanson and Leinkaemper, 1978; Keller and Swanson, 1979; Keller and Tally, 1979; Mosley, 1981; Marston, 1982; Gregory *et al.*, 1985, 1993; Gregory and Davis, 1992; Lisle, 1986; Keller *et al.*, 1995).

Prairie Creek: a case history

Prairie Creek is a tributary to Redwood Creek near Orick, California (Figure 10.1). At the time of evaluation the seven study reaches flowed through relatively undisturbed, old-growth redwood forest. Basic morphologic data for the study reaches are shown on Table 10.1. Figure 10.3 shows a morphological map for part of the Brown Creek reach, considered typical of upper Prairie Creek. Data on Table 10.1 and Figure 10.3 support the following conclusions concerning the role of LWD on stream channel form and process:

- LWD exerts a major control on channel form and process in streams draining old-growth redwood forest.
- In the headwater areas of Prairie Creek nearly all pools are either directly formed by, or are significantly influenced by LWD. As drainage basin area increases, and the size of the stream increases, fewer pools are directly formed by LWD, but it still plays an important role in influencing pool morphology (particularly pool depth).
- LWD in headward portions of drainage basins controls a significant amount of the total drop in elevation of the channel. Further downstream (as drainage basin area and streamflow volume increases) this generally decreases.
- LWD produces numerous areas where sediment is temporarily stored in the stream channel. These sediment storage sites provide an important sediment buffer system that modulates the routing of sediments through the stream system (Keller and Tally, 1979).
- LWD that consists of large trunks of redwood may reside in the stream channel for more than 200 years; therefore, LWD may be considered a semi-permanent part of the stream system.
- LWD produces a variety of channel forms and hydrologic variability important for fish habitat. Of particular importance are pools defended by LWD that provide undercut stream banks beneath LWD and root mats of living trees.

DEBRIS REMOVAL EXPERIMENTS

An important approach to understanding how LWD modifies channel form and process is through the use of carefully planned experiments that remove LWD from a stream channel and directly observe and measure change. Unfortunately, there have only been a few such experiments performed (Smith *et al.*, 1993a). For example, Shields and Smith (1992) studied changes in physical habitat diversity resulting from selective removal of LWD from a sand-bed system, the South Fork of the Obion River in western Tennessee; and Smith *et al.* (1993a and b) report effects of

Table 10.1 Drainage basin, channel and large woody debris characteristics for the seven study reaches of Prairie Creek shown on Figure 10.1

Study Reach	Hope Creek	Little Creek	Forked Creek	Zig Zag No. 2	Natural Tunnel	Brown Creek	Campground
Upstream basin area (km²)	0.7	3.5	6.6	8.2	11.2	16.7	27.2
Stream order	2	2	2	2	2	3	4
Slope	0.02	0.014	0.012	0.009	0.01	0.01	0.005
Debris loading (m³ m⁻²)	0.436	0.025	0.026	0.043	0.212	0.170	0.039
Pool to pool spacing (in channel widths)	6.2[c]	4.7[c]	2.6	6.6	2.7	6.0	4.0
% channel area pool[a]	49	34	46	36	41	26	25
% channel area riffle[a]	21	46	49	20	15	18	25
% channel in debris-stored sediment[a]	30	18	30	15	21	29	13
% channel area undercut banks[a]	1	4	3	4	1	<1	1
% pool morphology influenced by debris	86	71	87	50	80	67	50
% debris controlled drop in elevation of the channel (%)[b]	43	27	34	8	<1	18	<1

[a]Total percentages in stream environments may be less or greater than 100% due to overlaps such as pools that contain debris-stored sediment or existence or other environments not listed

[b]Ratio of cumulative loss of channel elevation associated with large organic debris to total fall of the stream reach

[c]Spacing controlled by organic debris

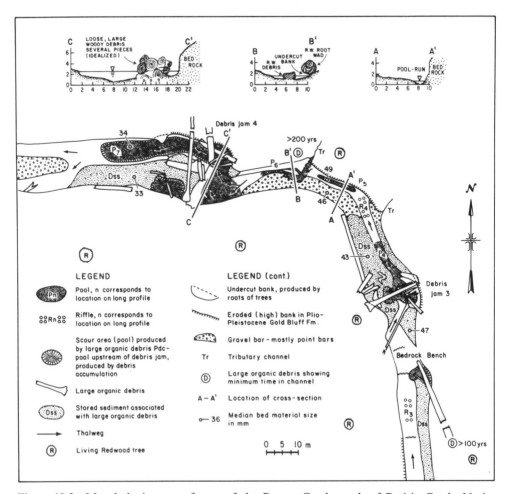

Figure 10.3 Morphologic map of part of the Brown Creek reach of Prairie Creek. Notice that the large woody debris at pool 6 has been in that location for more than 200 years, as determined from dendrochronology. Large debris accumulations (debris jams 3 and 4) greatly influence channel form and process by producing morphologic and hydrologic variability important for aquatic habitats (from Keller and Tally, 1979)

removal of LWD on channel morphology and bedload transport from a gravel-bed stream (Bambi Creek) in south-eastern Alaska.

Experimental removal of LWD from Larry Damm Creek, north-western California (Figure 10.1) was completed from 1979 to 1982 and briefly reported in Mac-Donald and Keller (1987). What follows below is a detailed discussion of the debris removal experiments in the Larry Damm Creek and the changes that occurred.

Larry Damm Creek: debris removal experiments

Larry Damm Creek is a tributary in the Redwood Creek watershed (Figure 10.1). The drainage basin area is 3.9 km², the average channel slope is moderate at 0.014,

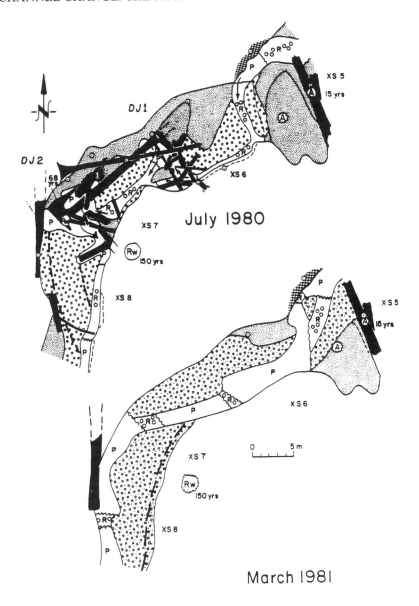

EXPLANATION

P   POOL

R   RIFFLE

▬   LARGE ORGANIC DEBRIS

⬭   MOSTLY SAND

⬭   MOSTLY GRAVEL

Ⓡⓦ   LIVING REDWOOD TREE

Ⓐ   LIVING ALDER TREE

⟶   DIRECTION OF FLOW

XSN   LOCATION OF A CROSS-SECTION

DJN   DEBRIS JAM

Figure 10.4 Morphologic maps of the upstream portion of the 'pull' reach, Larry Damm Creek, before and after debris removal (from Keller *et al.*, 1995)

and the characteristic width of the study reach (channel area ÷ channel length) is 8.2 m. Bankfull discharge at the study reach is approximately 3.5 m³ s⁻¹. The streambed prior to disturbance was dominantly fine gravel (D_{50} 28 mm). The two debris jams removed in the experiment had been present in the channel for at least 68 years, based on the age of living trees growing on the debris jams. These jams were 'caulked' or infilled with smaller pieces of LWD introduced to the channel during or after timber harvest. The approximate time of timber harvesting in the Larry Damm drainage basin was from 1954 to 1968.

In order to assess effects of LWD on channel form and process, including sediment storage and routing, nearly all LWD was removed from a 100 m stretch of Larry Damm Creek in August 1980. Approximately 70 m³ of LWD was removed, most in two discrete debris jams at the head of the 100 m reach (DJ1, upstream, and DJ2, downstream, on Figure 10.4). During the debris removal operations, no sediment was removed from the channel. However, the removal of the debris did disturb the sediment within the channel.

Prior to the experimental removal of LWD, 31 cross-sections were established in a 300 m reach of Larry Damm Creek. The central 100 m reach was the 'pull' reach where 17 cross-sections were established. Six cross-sections were established in the 100 m reach above the 'pull' reach and another eight cross-sections were established in the 100 m reach below the 'pull' reach. In addition, the long profile of the entire channel was surveyed three times (September, 1979; January, 1980; and August, 1980), and the channel was mapped with tape and compass two times to establish magnitudes and variation in scour and fill, channel pattern, and substrate size distribution in the stream channel. The channel was also surveyed to estimate the amount of debris-stored sediment within the 'pull' reach. We estimate that 90 m³ of debris-stored sediment was present in the stream channel prior to the removal of the LWD. As suggested on Figure 10.4, most of the debris-stored sediment was located: against the right bank between the two debris jams; upstream of DJ1 in the vicinity of cross-section 6 against the right bank; upstream of DJ1 in the vicinity of cross-section 5 left bank; and in the gravel bar downstream of DJ2.

Following removal of LWD in August 1980, most local adjustment of sediment to the removal was complete by 1 November 1980, following peak discharge of approximately 12.5% of bankfull. Subsequent storms in early December, late January and mid-February of the winter of 1980–1981 caused the most significant changes in sediment storage patterns. These are demonstrated on Figures 10.4, 10.5 and 10.6 that compare pre- and post-debris removal morphologic maps, channel cross-sections, and channel longitudinal profiles. The most significant changes in storage of sediment were:

- Erosion along the left bank above DJ1 (Figure 10.4), observed as increased bank undercutting at cross-section 6 (Figure 10.5).
- Removal of a substantial part of the sandbar downstream from cross-section 5 (Figure 10.4), which had previously been stabilised in the backwater of the DJ1.
- Formation of a point bar between cross-sections 7 and 8 (Figures 10.4 and 10.5).
- Development of three relatively deep pools between cross-sections 5 and 8 (Figures 10.4 and 10.6).

Measurement of scour and fill following removal of LWD allowed for the devel-

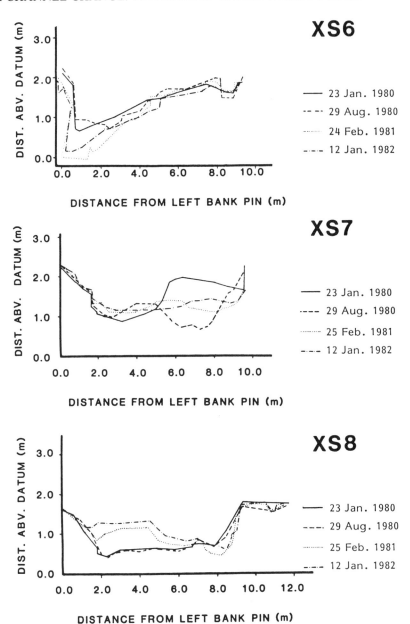

Figure 10.5 Cross-sections 6, 7 and 8 in the 'pull' reach of Larry Damm Creek. Locations of cross-sections are shown in Figure 10.4

opment of a mass balance sediment routing model to describe the movement of relatively fine-grained, debris-stored sediment (i.e. the sediment trapped behind debris accumulations as upstream slackwater deposits). Following removal of LWD, the dominate storage site in the 'pull' reach was a point bar located in the vicinity of cross-sections of 7 and 8 (Figure 10.4). A good deal of the sediment transported

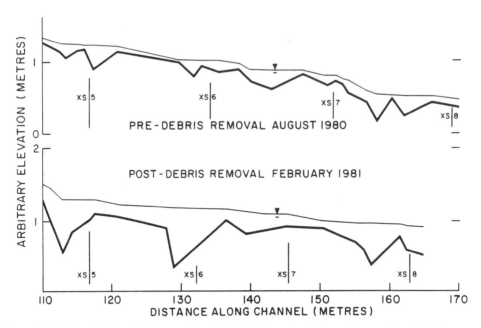

Figure 10.6 Longitudinal (bed) profile of the 'pull' reach of Larry Damm Creek, before and after debris removal. Cross-section numbers are those shown on Figures 10.4 and 10.5

into the 'pull' reach from upstream (50 m³ in the first year and 30 m³ in the second year, following removal) was deposited on this point bar. Finer sediment associated with the LWD was deposited lower in the 'pull' reach. Finally, approximately 30 m³ of debris-stored sediment was transported entirely out of the 'pull' reach and deposited above debris jams further downstream (MacDonald and Keller, 1987). This pattern of sediment transport related to channel variability is consistent with that reported by Mosley (1981), who recognised that when debris-stored sediment is mobilised it generally moves from one storage site to another in a complex manner.

The calibre of sediment within the 'pull' reach also changed following removal of LWD. This is best observed by examination of Figure 10.4. Much of the relatively finer bed material (mostly sand) present prior to debris removal was replaced following removal of the LWD by coarser sediment (mostly gravel).

The debris removal experiment of Larry Damm Creek also allowed direct documentation of the effect of organic debris on pools and riffles. Prior to debris removal, three of the five pools of the 'pull' reach were created or significantly enhanced by LWD, while the remaining two were associated with channel bends. Following debris removal, eight pools were identified in the 'pull' reach, and of these six were associated with channel bends and the remaining two occurred as chutes opposite lateral bars (Figure 10.4). The pool located between cross-sections 7 and 8 is defended by a large redwood bole that could not be removed with the other LWD because it was too deeply buried in the steep channel bank. Nevertheless, the removal of the LWD significantly reduced the effect of LWD on pool morphology. Pool to pool spacing within the 'pull' reach decreased from approximately 2.5 to 1.6 channel widths. This was due to the greater number of pools within the 'pull' reach following debris removal. The long profile through the upper part of the 'pull' reach before and following debris removal (Figure 10.6) shows that the post-removal pools are significantly deeper than the pools present prior to LWD removal. The pre-removal pools were related to LWD, and the presence of the debris encouraged deposition of fine sediment which restricted the depth of pools. Following removal of debris, flow convergence and development of point bars produced deeper pools. These pools are more similar to those found in alluvial stream reaches lacking LWD, or other large roughness elements. Pools in alluvial, gravel-bed rivers may be formed and maintained by flow processes related to convergence of flow that produces a shear stress, or velocity reversal (Keller, 1971; Lisle, 1979). On the other hand, pool formation and maintenance related to large roughness elements such as woody debris is probably best explained by a scour model related to turbulent velocity changes caused by interactions between the flowing water and obstructions to flow such as LWD (Beschta, 1983; Smith, 1990). Following removal of LWD in the

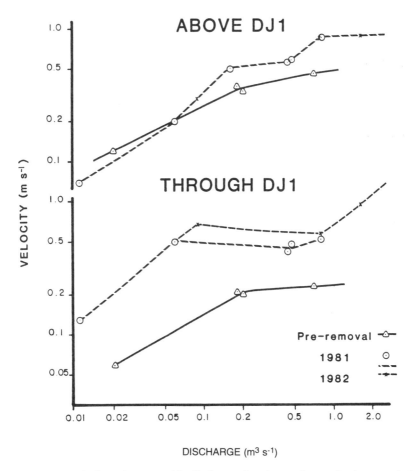

Figure 10.7 Mean velocity of water with discharge for the study reach above and through debris jam 1, Larry Damm Creek, before and after Debris removal. Location of debris jam 1 is shown on Figure 10.4 (from Keller *et al.*, 1995)

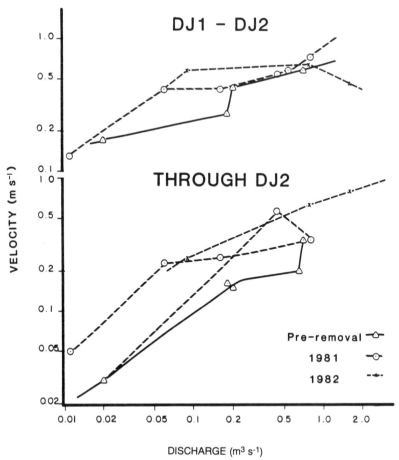

Figure 10.8 Mean velocity of water with discharge for reach above and through debris jam 2, Larry Damm Creek, before and after debris removal. Location of debris jam 2 is shown on Figure 10.4

'pull' reach of Larry Damm Creek, the pools formed due to scour as a result of flow convergence, rather than scour produced by large roughness elements.

Debris removal experiments in Larry Damm Creek allowed the evaluation of how accumulations of LWD affect the mean travel time of water through a particular reach of accumulation of LWD. Because channel slope was not altered by removing the LWD, changes in mean velocity through the 'pull' reach reflected adjustments in channel roughness. On the scale of sub-reaches, channel slope can also adjust to changes in roughness and both can account for changes in water velocity (Prestegaard, 1983). Salt tracing techniques, outlined by Calkins and Dunne (1970) and Church and Kellerhals (1970), were used to measure mean travel time of water through selected reaches before and after debris removal. A quantity of salt was introduced sufficiently upstream to be vertically and horizontally mixed within the desired reach. The passage of this salt-impregnated water was traced at reach end-points with conductivity meters. The travel time through the reach is the difference

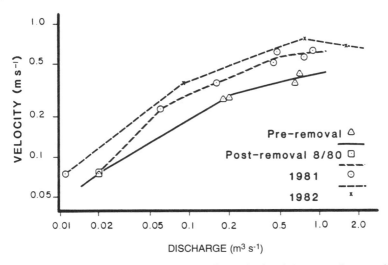

Figure 10.9 Mean water velocity with discharge from the head (upstream) part of the 'pull' reach through debris jam 2, Larry Damm Creek, before and after debris removal. Location is shown on Figure 10.4

in average arrival time to each endpoint. Mean velocity through a reach is calculated by dividing the centre line length of the wetted channel by the travel time. Results are shown on Figures 10.7, 10.8 and 10.9. Although complicated by changes in channel morphology as the 'pull' reach adjusted following debris removal, mean velocity is generally higher following debris removal over the range of discharges sampled. These relatively low discharges (less than 0.5 of bankfull) illustrated that velocity through a debris jam increased by up to 250% following removal of the LWD. If we were able to compare travel times and mean velocities at higher discharges (e.g. approximately 110% of bankfull or greater) when the debris jams are drowned out, we would expect that these differences would be much less, as reported by Gregory *et al.* (1985). Nevertheless, debris accumulations are very effective in retarding travel times at low and moderate flows that represent the dominant flow conditions experienced by aquatic biota.

LARGE WOODY DEBRIS STEPS: AN EXAMPLE FROM LARRY DAMM CREEK

Large woody debris plays an important role in the routing and storage of sediment. Debris accumulation such as woody debris steps produce storage compartments for sediments as shown in Figure 10.10. In steep mountain streams, LWD often acts as a local base level, producing stretches of the channel, where water surface slope, representing potential energy loss per unit length of channel, locally decreases above a debris jam or step and increases over the step relative to the average values. Generally, a pool and/or area of stored sediment is present above a jam and a plunge or scour pool is located below or downstream of a jam. This is a characteristic 'stepped bed' profile (Keller and Swanson, 1979; Heede, 1981; Marston, 1982), where a sig-

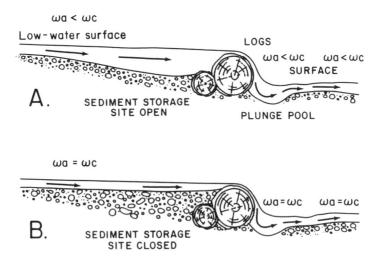

Figure 10.10 Idealised diagram of a simple step produced by large woody debris. Relationships between available (aw) and critical (cw) stream power (Bull, 1979) are shown in (A) for a sediment storage site that is open, and (B) for a sediment storage site that is closed. Notice that if the sediment storage site is full then available stream power is equal to critical stream power and sediment is transported through the reach. If the available stream power is less than the critical stream power then sediment will be deposited in storage sites (from Keller *et al.*, 1995)

nificant amount of a stream's potential energy may be dissipated at debris-created falls and cascades which occupy a relatively small percentage of the total stream channel length. Thus energy is expended at these locations rather than producing an incised channel with unstable, eroding channel banks. In the upper study reach of Prairie Creek, approximately 43% of the elevation loss within the reach is associated with LWD steps (Table 10.1). This debris control of elevation loss of the stream channel is most pronounced in headwater reaches and decreases downstream.

The simplest debris accumulation is a LWD step such as shown on Figure 10.11, which consists of a single tree step perpendicular to flow in the stream bed of Larry Damm Creek. The longitudinal profile shows debris-stored sediment above the step and a plunge pool below it. Because the pool upstream of the step in Figure 10.11 is associated with the bend, secondary circulation and convergence of flow probably will maintain the pool in this location. Therefore, the sediment storage compartment upstream at this step is full or nearly so.

A hypothetical comparison of unit stream power through LWD steps with filled vs. non-filled upstream storage sites is shown on Figure 10.10. Newly formed LWD steps produce open sediment storage sites that may accumulate sediment during relatively high flow events when bedload material is being transported. The storage site may eventually fill because the available stream power (stream power actually present) at the site is less than the critical stream power necessary to transport the load at that location (Figure 10.10A). As water flows over a LWD step the available stream power exceeds the critical steam power so scour and a plunge pool develops. Further downstream the available stream power may again be less than the critical

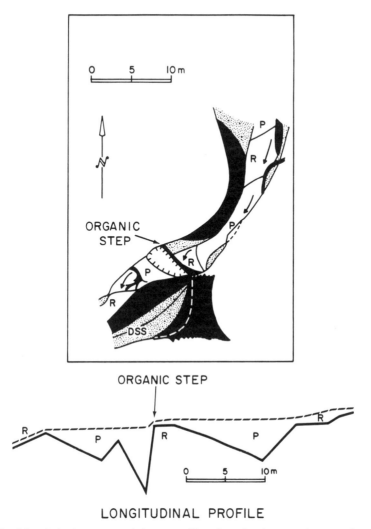

Figure 10.11 Morphologic map and long profile of a simple organic step, Larry Damm Creek: (from Keller *et al.*, 1995). P = pool, R = riffle and DSS = debris-stored sediment

power, and deposition may form a bar (Figure 10.10A). With time, the sediment storage site may become filled, and the available stream power eventually will become equal to the critical stream power, and sediment will be transported through the reach without net deposition (Figure 10.10B). At this point, an equilibrium of action is achieved, and the sediment is pumped through the system. This explanation of the dynamics of a LWD step utilises the concept of 'threshold of critical power', when the ratio of available to critical stream power is unity (Bull, 1979).

The hypothetical model for a filled storage compartment is supported by data collected for the step shown on Figure 10.11, and is shown in terms of stream power on Figure 10.12. The dashed line represents a discharge of 95% of bankfull. The location of the step on Figure 10.12 is coincident with the peak in stream power at a

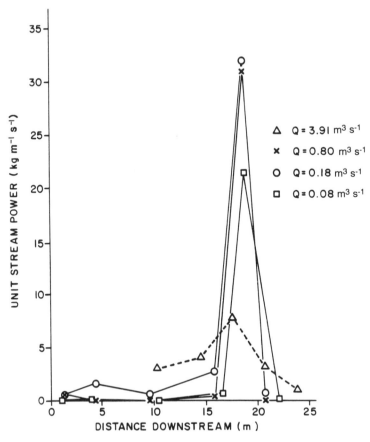

Figure 10.12 Distribution of unit stream power with discharge over the organic step shown in Figure 10.11. The location of the step is at the peak in stream power (distance about 18 m). From Keller *et al.* (1995)

distance of about 18 m. At this discharge, flow depth was observed to be approximately 0.7 m throughout the stream reach (except in the upstream pool, where depths exceeded 1.0 m). At this stage, predominantly sand-size sediment was being transported across the entire width of the active channel. Field observations at the LWD step in Larry Damm Creek (Figures 10.11 and 10.12) showed that with rising stage to about 95% bankfull, the plunge pool nearly filled with sediment. With falling stage it scoured, reproducing the LWD step and plunge pool environment. This example of unit stream power changes across the LWD step with discharge emphasises the complexities involved with interactions between flowing water, moving sediment and LWD. As the discharge increased, and the plunge pool below the LWD step filled with sediment, unit stream power decreased over the step itself, while increasing both upstream and downstream. The more uniform distribution of unit stream power allowed for the more uniform transport of sediment through the system, while the plunge pool filled during the rising stage.

DISCUSSION AND CONCLUSIONS

Evaluation of Prairie Creek in northwestern California clearly suggests that the role of LWD in forested streams greatly affects channel form and process. Large woody debris also has very important implications for aquatic habitats because it provides for hydrologic variability that produces a diversity of instream environments. Understanding of this principle has led to development of management strategies for streams flowing through forested areas that take into account the role of LWD. A recent summary of management strategies is provided by Gregory and Davis (1992). That summary concluded:

- Effects of LWD on stream channel management must consider the entire spectrum of ecological consequences, and, in particular, diversity of habitats.
- Development of management strategies of streams with LWD must consider the behaviour of debris accumulation under natural conditions. That is, the management strategy should strive to 'design with nature'.
- Processes that are likely to disturb stream channels such as timber harvesting should include strategies to protect aquatic resources such as utilisation of buffer zones.
- In specific instances, management plans may call for the introduction of new LWD into the stream system to provide for diversity of aquatic habitat.

Experiments that remove LWD in Larry Damm Creek are interesting because adjustments following the removal of the LWD produced a series of pools and riffles similar to those found in natural meandering gravel-bed channels. LWD removal experiments reported by Smith et al. (1993a and b) also emphasised post-removal adjustment that resulted in development of a series of pools and alternate bars. While these pools form valuable aquatic habitat, the total quality of habitats was reduced. This results because pools associated with LWD have a greater hydrologic complexity and the woody debris provides micro-habitats and cover for fish and other organisms living in the stream system (Smith et al., 1993a and b). Similarly, a study by Shields and Smith (1992) reported that diversity of physical habitat in the sand-bed stream they studied was strongly influenced by the amount of LWD in the system. They further report that selective removal of LWD resulted in lower quality aquatic habitat. This supports the conclusion by Keller et al. (1995) that overzealous removal of LWD can result in unnecessary damage to aquatic ecosystems. Therefore, a conservative practice concerning the removal of LWD should be adopted when managing streams and aquatic habitats.

ACKNOWLEDGEMENTS

Research on the role of LWD on stream channel form and process in the redwood forest of northern California was sponsored by the Water Resources Center, University of California, the US Forest Service Redwood Sciences Laboratory, Prairie Creek Redwood State Park and Redwood National Park.

REFERENCES

Beschta, R.L. (1983) The effects of large organic debris upon channel morphology: a flume study. Proceedings of the D.B. Simons Symposium on Erosion and Sedimentation. Simons Li, and Associates, Ft Collins, Colorado.

Bevan, A. (1948–1949) Floods and forestry. *University of Washington Forest Club Quarterly* **22**(2), 8pp.

Bull, W.B. (1979) Threshold of critical stream power. *Geological Society of America Bulletin* **90**, (I), 453–464.

Calkins, D. and Dunne, T. (1970) A salt tracing method for measuring channel velocities in small mountain streams. *Journal of Hydrology* **11**, 379–392.

Church, M. and Kellerhals, R. (1970) Stream gauging techniques for remote areas using portable equipment. Inland Waters Branch, Canadian Department of Energy, Mines and Resources, Techniqual Report 25, 89pp.

Gregory, K.J. and Davis, R.J. (1992) Coarse woody debris in stream channels in relation to river channel management in woodland areas. *Regulated Rivers: Research and Management* **7**, 117–136.

Gregory, K.J., Gurnell, A.M. and Hill, C.T. (1985) The permanence of debris dams related to river channel processes. *Hydro. Sci. J.* **30**, 371–381.

Gregory, K.J., Davis, R.J. and Tooth, S. (1993) Spatial distribution of coarse woody debris dams in the Lymington Basin, Hampshire, U.K. *Geomorphology* **6**, 207–224.

Heede, B.H. (1972) Influences of a forest on the hydraulic geometry of two mountain streams. *Water Research Bulletin* **8**(3), 523–530.

Heede, B. (1981) Dynamics of selected mountain streams in the western United States of America, *Zeitschrift fur Geomorphologie N.F.* **25**, 17–32.

Keller, E.A. (1971) Areal sorting of bedload material: the hypothesis of velocity reversal. *Geological Society of America Bulletin* **82**, 753–576.

Keller, E.A. and Swanson, F.J. (1979) Effects of large organic material on channel form and fluvial processes. *Earth Surface Processes* **4**, 361–380.

Keller, E.A. and Tally, T. (1979) Effects of large organic debris on channel form and process in the coastal redwood environment. In: Rhodes, D.D. and Williams, G.P. (eds), *Adjustment of the Fluvial System.* Annual Geomorphology Symposium, Binghampton, NY; Kendall Hunt Publications, Dubuque, Iowa, 169–198.

Keller, E.A., MacDonald, A., Tally, T. and Merrit, N.J. (1995) Effects of large organic debris on channel morphology and sediment storage in selected tributaries of Redwood Creek, northwestern California. *US Geological Survey Professional paper*, 1454-P (in press).

Lisle, T.E. (1979) A sorting mechanism for a riffle-pool sequence. *Geological Society of America Bulletin* **90**, 1142–1157.

Lisle, T.E. (1986) Stabilization of a gravel channel by large streamside obstructions and bedrock bends, Jacoby Creek, northwestern California. *Geological Society of America Bulletin* **97**, 999–1011.

Lobeck, A.K. (1939) *Geomorphology.* McGraw-Hill, New York, 428pp.

Macdonald, A. and Keller, E.A. (1987) Stream channel response to removal of large woody debris, Larry Damm Creek, northwestern California. In: Beschta, R.L., Blinn, T., Grant, G.E., Swanson, F.J. and Ice, G.G. (eds), *Erosion and Sedimentation in the Pacific Rim,* IAHS Publication no. 165, 405–406.

Marston, R.A. (1982) The geomorphic significance of log steps in forest streams. *Annals of the Association of American Geographers* **72**, 99–108.

McCall, E. (1988) The attack on the great raft. *Invention and Technology,* Winter, 10–16.

Mosley, M.P. (1981) The influence of organic debris on channel morphology and bedload transport in a New Zealand forest stream. *Earth Surface Processes and Landforms* **6**, 571–579.

Prestegaard, K.L. (1983) Variables influencing water-surface slopes in gravel-bed streams. *Geological Society of America Bulletin* **94**, 673–678.

Sedell, J.R. and Froggatt, J.L. (1984) Importance of streamside forests to large rivers: the isolation of the Willamette River, Oregon, U.S.A., from its floodplain by snagging and streamside forest removal. *Verh. Internat. Verlin. Limnol.* **22**, 1828–1834.

Sedell, J.R. and Luchessa, K.J. (1981) Using the historical record as an aid to salmonid habitat enhancement. In: Armentrout, N.B. (ed.), *Symposium on Acquisition and Utilization of Aquatic Habitat Inventory Information*, Western Division, America Fisheries Society, Portland, Oregon.

Shields, D.F. Jr and Smith, R.H. (1992) Effects of large woody debris removal on physical characteristics of a sand-bed river. *Aquatic Conservation: Marine and Freshwater Ecosystems* **2**, 145–163.

Smith, R.D. (1990) Streamflow and bedload transport in an obstruction-affected, gravel-bed stream. Unpublished PhD thesis, Oregon State University.

Smith, R.D., Sidle, R.C. and Porter, P.E. (1993a) Effects on bedload transport of experimental removal of woody debris from a forest gravel-bed stream. *Earth Surface Processes and Landforms* **18**, 455–468.

Smith, R.D., Sidle, R.C., Porter, P.E. and Noel, J.R. (1993b) Effects of experimental removal of woody debris on the channel morphology of a gravel-bed stream. *Journal of Hydrology* **152**, 153–178.

Swanson, F.J. and Leinkaemper, G.W. (1978) Physical consequences of large organic debris in Pacific Northwest streams. US Forest Service, GTR, PNW-69, 12pp.

Swanson, F.J., Leinkaemper, G.W. and Sedell, J.R. (1976) History, physical effects, and management inputations of large organic debris in western Oregon streams. USDA Forest Service, GTR, PNW-56, 15pp.

Triska, F.J. (1984) Role of wood debris in modifying channel morphology and riparian areas of a large lowland river under pristine conditions: a historical case study. *Verh. Int. Verlein. Limnol.* **22**, 1876–1892.

11 Vegetation Along River Corridors: Hydrogeomorphological Interactions

ANGELA M. GURNELL

School of Geography, University of Birmingham, UK

INTRODUCTION

This chapter focuses upon the interactions between hydrological and fluvial processes across river floodplains from the adjacent hillslopes to the river channel. In particular, it emphasises the way in which characteristics of the vegetation can be used to identify the nature and distribution of such processes. In 1984, Hickin observed (p. 123):

> Because we know so little about the ways in which vegetation influences river behaviour and fluvial geomorphology, any discussion of these matters is inevitably speculative, raising many more questions than it provides answers.

In the 1980s and 1990s, the river corridor environment has become a major theme for interdisciplinary research (Naiman and Décamps, 1990; Malanson, 1993). An interdisciplinary approach has been essential because river corridors, floodplains and riparian zones 'occur at interfaces—ecological interfaces between different ecosystems and conceptual interfaces between different disciplines' (Gregory *et al.*, 1989, p. 3). As a result of this research, it is now possible to identify more precisely both the direct influences of vegetation on fluvial processes and the information that the age, composition and vigour of vegetation gives us about the floodplain environment as a result of its sensitivity to physical environmental processes.

Increasing interest in interdisciplinary research on the river corridor environment has resulted from a realisation of the extent of the anthropogenic changes that have been imposed upon fluvial systems, particularly over the last 200 years (Petts, 1989, 1990a; Décamps and Fortuné, 1991). Many of these changes have directly impacted upon the vegetation of river corridors. Woody debris and vegetation clearance (Sedell and Froggatt, 1984), river channelisation (Zinke and Gutzweiler, 1990) and channel incision resulting from hydrological change have increasingly isolated rivers from their floodplains; flow regulation (Fenner *et al.*, 1985) and augmentation (Henszey *et al.*, 1991) have induced riparian vegetation change; and groundwater abstraction (Kondolf and Curry, 1984) and grazing pressure (Rucks, 1984) have reduced riparian vegetation biomass and so induced river bank erosion.

Sedell and Froggatt (1984, p. 1828) note that 'the relationship of floodplain and mainstem in large rivers in North America and Europe no longer exists and is

Changing River Channels. Edited by Angela Gurnell and Geoffrey Petts.
© 1995 John Wiley & Sons Ltd.

rapidly disappearing in Africa, South America and Asia'. Their description of the transformation of the Willamette River, Oregon, illustrates the isolation problem. The Willamette River corridor was once a series of multiple channels, sloughs and backwater areas, where fallen trees helped to create and maintain shoals, oxbow lakes and complex aquatic habitats. The river has been converted into a simple, single-channel, homogeneous habitat for aquatic vertebrates, with a four-fold decrease in river shoreline, as a result of 80 years of snag removal and riparian forest destruction.

In their natural state, riparian zones are one of the most dynamic components of the landscape, where frequent disturbances create complex mosaics of landforms and biological communities (Gregory *et al.*, 1991). In these environments, ecological refuges are dependent upon the extensive coupling of the channel with streamside vegetation, floodplain geomorphical features and groundwater (Sedell *et al.*, 1990). A fuller understanding of these complex environments is required to establish the rehabilitation needs of the many rivers that have been drastically altered by man's activities. The aim of this chapter is to explore the geomorphological and hydrological processes that generate the geomorphological surfaces which are the template for the development of riparian plant communities (Poff and Ward, 1990; Gregory *et al.*, 1991) and thus the degree to which the vegetation communities can reveal the nature and magnitude of these geomorphological and hydrological processes at a range of spatial scales. An increased understanding of such interactions is both of scientific interest and essential to the successful reinstatement of more natural riparian environments along heavily modified river corridors.

DEFINITIONS

Before considering the interactions between vegetation and physical environmental processes, the terms 'river corridor', 'riparian zone' and 'floodplain' as they are to be used in this chapter require definition.

A *river corridor* can be broadly defined as the river and river channel together with its associated wildlife and the adjacent riparian ecosystem (Angold *et al.*, 1994). Continuous interactions occur between the aquatic, riparian and hillslope ecosystems, so that a river corridor theoretically should be wide enough to cover the floodplain, both river banks, and an area of upland (Schaefer and Brown, 1992).

According to Gregory *et al.* (1991, p. 540) *'riparian zones* are not easily delineated but are comprised of mosaics of landforms, communities, and environments within the larger landscape'. Swanson *et al.* (1982) suggest that the riparian zone can be viewed at three scales. The most restricted scale confines the riparian zone to the water's edge. A broader definition includes the areas of streambed, banks and floodplain that are submerged for part of each year. This definition corresponds with that of Hupp (1992) who specifically excludes higher parts of the bottomland such as true floodplains and terraces. The most extensive definition includes the entire area of indirect interaction between aquatic and terrestrial environments. It, therefore, extends across the floodplain to the base of adjacent hillslopes. This last definition is adopted in the present discussion to represent both the river corridor and the riparian zone.

Bren (1993) provides a simple definition of a *floodplain* as a relatively level area of sediment deposited and periodically inundated by streams. The more complex definition of Nanson and Croke (192, p. 459) stresses the genetic association between rivers and the floodplains they construct (see also Chapter 9):

> An alluvial channel adjusts its hydraulic geometry and builds a surrounding floodplain in such a way as to produce a stable conduit for the transport of water and sediment. The active boundary of the channel is usually where the floodplain is being constructed or eroded, although during large floods this activity can extend across much of the floodplain surface.

As a result of this genetic association between rivers and their floodplains, Nanson and Croke develop a genetic classification of floodplains, which provides a framework for the discussion of floodplain vegetation in this chapter.

HILLSLOPE—FLOODPLAIN TYPES AND VEGETATION

Numerous studies have highlighted the relationship between hydrological and hydraulic factors, parent material and the overlying vegetation communities on both hillslopes (e.g. Gurnell and Gregory, 1987) and floodplains (e.g. Petts, 1990a). The association between vegetation species and communities and geomorphological features of the hillslope–floodplain system depend upon the sensitivity and tolerance of the vegetation to a variety of physical environmental characteristics including:

- Waterlogging and water quality and supply, and thus variability in the water table and soil moisture regime, which are dependent upon topographic position, the interaction of water from different source areas, and the hydraulic conductivity of the underlying soil and substrate.
- Flowing water and transported sediments, and thus the energy and frequency of flooding.
- Sedimentation, which may vary in calibre, organic content and rate according to position within the active channel or on the floodplain.

Certain properties of the vegetation species will also affect the degree to which they may be associated with particular geomorphological features of the hillslope–floodplain system:

- Seed supply, dispersal mechanisms (e.g. wind and/or water) and requirements for germination.
- Ability to survive rafting by the river from sites of erosion to sites of deposition.
- Ability to compete with other vegetation species under particular environmental conditions and stages of vegetation colonisation and succession.

Thus, an association between vegetation and floodplain geomorphic features can be anticipated and can provide a basis for furthering our understanding of floodplain environments. A particularly useful framework for assessing the geomorphological and hydrological information content of floodplain and hillslope vegetation communities is the genetic floodplain classification of Nanson and Croke (1992), since this identifies a range of typical geomorphological features of floodplains under different energy and substrate conditions. The classification is summarised in Table 11.1.

The different energy conditions that are incorporated into Nanson and Croke's classification indicate the degree to which areas of the river channel and floodplain are likely to be impacted by different processes which might influence the vegetation character. Nanson and Croke identify three classes of floodplain: class A—high-energy non-cohesive; class B—medium-energy non-cohesive; and class C—low-energy cohesive floodplains. High-energy floodplains would be expected to be characterised by vegetation that can survive damage by floodwaters and erosion, transport and deposition of sediments. This would be particularly true of confined floodplains (types A1 and A2—confined coarse-textured; confined vertical accretion). The less-confined semi-arid subclasses of type A would be expected to show patterns in vegetation which reflect tolerance of flood damage as well as specific water-supply requirements. Lower-energy floodplains (types B and C) will include a zone near the active river where the influence of direct flood and sediment damage and disturbance is important. However, vegetation on the broader floodplain, where inundation occurs less frequently and where flow velocities and thus sediment transport are lower during flood events, will reflect variability in other factors, particularly substrate character (including locally variable depositional environments) and soil water regime. Type B floodplains include types where fluvial features such as bars and islands (types B1 and B2; braided-river, and wandering gravel-bed river floodplains), and point bar, counterpoint bar and associated scroll bar and concave bench formation (Type B3, meandering, lateral-migration; and subtypes a—non-scrolled, b—scrolled, c—backswamp, d—counterpoint floodplains) provide characteristic environments with variable levels of disturbance for vegetation development. These floodplains consist of an inner zone of active floodplain development associated with channel migration, with overbank accretion becoming more important on floodplain zones that are more remote from the active channel. Type C is represented by either single thread or anastomosing channel systems which develop floodplains largely by overbank accretion of fine sediments. On these floodplains, the presence of abandoned channels and backswamps provides a variety of environments within which characteristic vegetation communities can develop.

The following two studies serve to illustrate how floodplain geomorphology and vegetation are related. Amoros et al. (1987) define four functional sectors (gorge, braided, meandering and anastomosed) in their ecological studies of the fluvial hydrosystems of the River Rhone, France, which they associate with variations in bed stability, planform dynamism, habitat diversity and expected biomass productivity of the floodplain. These functional sectors correspond to the type A1, type B1 and B2, type B3, and type C floodplains classified by Nanson and Croke. Furthermore, Harris (1988) identifies broad types of riparian vegetation which correspond with broad geomorphological valley and floodplain types in the Eastern Sierra Nevada, California. As a result of this correspondence, he produces a classification of geomorphic–vegetation units as a basis for identifying the potential sensitivity of stream systems and riparian zones to management.

The remainder of this chapter reviews the degree to which existing studies have highlighted particular physical environment influences on river corridor vegetation

Table 11.1 A genetic classification of floodplains (from Nanson and Croke, 1992, reproduced with permission of Elsevier Science)

Class A: High-energy non-cohesive floodplains
$\omega = > 300$ W m^{-2}. Disequilibrium floodplains which erode in response to extreme events, typically located in steep headwater areas where channel migration is prevented by valley confinements

Order/suborder	Type	Specific stream power, ω (W m^{-2})	Sediment	Erosional and depositional processes	Landforms	Channel planform	Environment
A1	Confined coarse-textured floodplains	>1000	Poorly sorted boulders and gravel; buried soils	Catastrophic floodplain erosion and overbank vertical accretion; abandoned-channel accretion; minor lateral accretion	Boulder levees; sand and gravel splays; back channels; abandoned channels and scour holes	Single-thread straight/irregular	Steep upland headwater valleys
A2	Confined vertical accretion floodplains	300–1000	Basal gravels and abundant sand with silty overburden	Catastrophic floodplain erosion and overbank vertical accretion	Large levées and deep back channels and scour holes	Single-thread straight/irregular	Upland headwater valleys
A3	Unconfined vertical accretion sandy floodplains	300–1000	Sandy-strata inter-bedded muds	Catastrophic channel widening; overbank vertical accretion; island deposition and abandoned-channel accretion. Minor lateral accretion	Flat floodplain surface	Single-thread wandering	Semi-arid open valleys
A4	Cut and fill floodplains	~300	Sands, silts and organics	Catastrophic gullying, overbank vertical accretion; abandoned-channel accretion	Flat floodplain surface; channel fills; swampy meadows	Straight/irregular	Upland dells and semi-arid alluvial-filled valleys

Table 11.1 (*continued*)

Class B: Medium-energy non-cohesive floodplains
ω = 10–300 W m^{-2}. Equilibrium floodplains formed by regular flow-events in relatively unconfined valleys

Order/suborder	Type	Specific stream power, ω (W m^{-2})	Sediment	Erosional and depositional processes	Landforms	Channel planform	Environment
B1	Braided-river floodplains	50–300	Gravels, sand and occasional silt	Braid-channel accretion and incision; overbank vertical accretion; minor lateral and abandoned-channel accretion	Undulating floodplain of abandoned channels and bars; backswamps	Braided	Abundant sediment load in tectonically and/or glacially active areas
B2	Wandering gravel-bed river floodplains	30–200(?)	Gravels, sands, silts and organics	As for braided and meandering rivers	Abandoned channels; sloughs; braid-bars; islands back channels	Braided meandering and anastomosing	Abundant sediment load; alternating sedimentation zones in tectonically and/or glacially active areas
B3	Meandering river, lateral-migration floodplains	10–60	Gravels, sands and silts	Cut-bank erosion; lateral point bar accretion; overbank vertical and abandoned channel accretion. Counterpoint accretion; minor oblique accretion	Flat to undulating floodplain surface; oxbows; backswamps	Meandering	Usually middle to lower valley reaches
B3a	Lateral-migration, non-scrolled-floodplains	ditto	ditto	ditto	ditto	ditto	ditto
B3b	Lateral-migration, scrolled floodplains	ditto	Sands and minor gravels	As for B3 with scroll-bar formation	Distinctly scrolled floodplains	ditto	ditto

		ω					
B3c	Later-migration, backswamp flood-plains	ditto	Sands, silts and organics	As for B3b	Central scrolled floodplain with flanking backswamps	ditto	ditto
B3d	Lateral-migration, counterpoint floodplains	ditto	Sands with abundant silts and organics	As B3 with pronounced counterpoint accretion	Concave benches with scrolled floodplains	Confined meandering	ditto

ClassC: Low-energy cohesive floodplains
ω = <10 W m^{-2}. Floodplains formed by regular flow-events along lateral stable single-thread or anastomosing low-gradient channels

		ω					
C1	Laterally stable, single-channel floodplains	<10	Abundant silts and clays with organics	Overbank vertical accretion	Flat floodplains with low levées; backswamps	Single-thread straight/meandering	Abundant fine sediment load middle–lower reaches
C2	Anastomosing-river floodplains	ditto	Gravel and sands with abundant silts and clays	Overbank vertical accretion; island deposition	Flat floodplains with extensive levees, islands and flood basins, crevasse-channels and splays	Anastomosing	Very low gradient with wide floodplains
C2a	Anastomosing-river, organic-rich floodplains	ditto	As for C2 with abundant organics and lacustrine deposits	As for C2 with peat formation, and lacustrine sedimentation	As for C2 with lakes and peat swamps	ditto	As for C2 in humid environments
C2b	Anastomosing-river, inorganic floodplains	ditto	As for C2 but with few or no organics	As for C2	As for C2	Anastomosing channels and co-existing flood-plain-surface braid channels	As for C2 in semi-arid environments

patterns. The discussion follows a sequence of influences from topographic–soil water regime effects; through physical impacts of flooding; river planform change; and finally the processes which affect the active river banks. Such a division of influences is artificial and thus there is some overlap between sections. However, in taking this sequence, the text is highlighting influences which provide increasingly local components in the vegetation mosaic. The text, therefore, follows a sequence in spatial scale. Temporal/successional elements are introduced into the discussion where appropriate as are references to river corridor environments in different climatic zones.

HYDROLOGICAL REGIME

Topography and the soil moisture and water table regime

The genetic classification of floodplains (Table 11.1) distinguishes between floodplain types according to the energy of the fluvial environment in which they occur, the degree of confinement of the floodplain, the types of sediment present (calibre and organic content), the channel planform and the floodplain landforms. The level of confinement dictates the lateral slope angle from hillslope to river; the channel planform and floodplain landforms dictate the local slope; and the sediments dictate the hydraulic conductivity of the materials from which the floodplain is constructed. The spatial distribution and variability of the soil moisture and water table regime reflect the general and local topography and the local hydraulic conductivity of the floodplain, which in turn influence the distribution of vegetation communities.

In headwater areas, where hillslopes often impinge directly on the river system, floodplains are either non-existent or narrow (e.g. types A1 and A2), and strong soil moisture and water table gradients exist close to river banks. Gurnell (1981) and Gurnell and Gregory (1986, 1987) illustrate how the vegetation communities on hillslopes in headwater heathland catchments closely reflect spatial variations in the water table regime and the saturated hydraulic conductivity of superficial deposits. However, as a result of the confinement of headwater floodplains, disturbances by flooding and hillslope geomorphological processes are often more important in defining the character of the riparian vegetation than simple topographically-induced moisture gradients (e.g. Hupp, 1982; Gecy and Wilson, 1990).

On more extensive floodplains (e.g. classes B and C), the association between topography, sediment calibre and vegetation that is driven by differential soil water drainage is more clearly defined. Thus Barnes (1978) noted an association between elevation, soil texture, carbon content and available water capacity, and the presence or absence of many herbaceous species in a Wisconsin river bottomland forest. Menges (1986) found elevation to be the best explanatory variable for herbaceous species within Wisconsin floodplain forests, and Shelford (1954) reconstructed floodplain vegetation communities in the Lower Mississippi Valley on the basis of age of deposits and their elevation above mean low water level. The differential interactions between river levels and water table levels close to the river, and between the rainfall regime and water table levels on adjacent hillslopes and on sections of floodplain remote from the river, is described by Bell and Johnson (1974). Such complexities may provide unexpected associations between topography, water table level and vegetation communities under certain combinations of local climate, river flow regime, and river corridor morphology.

In arid environments the soil water gradient down hillslopes and across flood-plains is particularly strongly reflected in the vegetation. Bennett *et al.* (1989) describe three types of riparian ecosystem (hydroriparian, mesoriparian and xeroriparian) associated with the perennial, intermittent and ephemeral availability of water in arid areas. The floral species associated with these ecosystems are influenced not only by the water regime but by other site characteristics such as latitude, elevation, slope, soil type, exposure and water chemistry. Gurtz *et al.* (1988) describe a downstream transition in riparian vegetation along an intermittent prairie stream in Kansas from grasses to shrubs to gallery forest and Stromberg (1993), interpreting patterns in riparian vegetation in a semi-arid catchment in Arizona, concludes that flow volume, and the related attributes of water table recharge and floodplain soil wetting, are the primary factors regulating riparian vegetation abundance. In arctic environments the soil water regime is also associated with the ground temperature regime and this may be reflected in the vegetation pattern. For example, Van Hees (1990) describes a typical successional sequence in vegetation across a floodplain in Alaska, where bare silt near the river gives way to a zone of feltleaf willow (*Salix alaxensis*), followed by balsam poplar (*Populus balsamifera*), then white spruce (*Picea glauca*), and finally black spruce (*Picea mariana*). The soil freezes earliest and deepest in the earliest successional stages, but also thaws earlier and supports warmer soil temperatures during the growing season. The later successional stages show a more restricted soil temperature range and the development of permafrost which impedes drainage leading to the development of forest wetland. The presence or absence of permafrost and the depth, water content, and temperature regime of the active layer are in part a result of the insulating effect of the biomass of the vegetation and associated dead organic matter.

Flooding

The direct impact of flooding on vegetation is identifiable in further altitudinally-related patterns in vegetation across the riparian zone. Broad zonation in floodplain vegetation frequently reflects the transition from hydraulically-dominated to hydrologically-dominated vegetation types along transects from the river to the adjacent hillslopes. For example, Parodi and Freitas (1990) identify associations between physiography and vegetation type in Peruvian Amazonia which are a function of both hydrological and hydraulic influences. Distinctive forest types are associated with riverine (levée) locations, creeks, backwater areas and terraces.

The relationship between flood levels and frequency, and vegetation cover has been highlighted in many studies. Bren (1988a and b) and Bren and Gibbs (1986) overlay maps of flood extent along a section of the River Murray, Australia. They show that particular vegetation associations are characteristic of areas with particular flood frequencies. Areas of very high flood frequency support giant rush, areas of high flood frequency support moira grass (*Pseudoraphis spinescens*) and moira grass in association with red gum (*Eucalyptus camaldulensis*); areas of moderate flood frequency support red gum in association with a variety of sedges and grasses;

and areas of low flood frequency support red gum in association with yellow box and black box woodlands (mainly *Eucalyptus largiflorens* and *Eucalyptus microcarpa*). The highest red gum quality is associated with higher frequencies of flooding. Furthermore, Bedinger (1971) suggests that the relationship between forest species and flood frequency is so close on the Lower White River Valley, Arkansas, that an analysis of forest species composition can be used to define flood limits for three frequencies of flooding: water hickory (*Carya aquatica*) and overcup oak (*Quercus lyrata*) dominate sites that are flooded 29–40% of the time; a more varied flora including nuttall oak (*Quercus nuttallii*), willow oak (*Quercus phellos*), sweetgum (*Liquidambar styraciflua*), southern hackberry (*Celtis laevigata*) and American elm (*Ulmus americana*) represent sites flooded 10–21% of the time; and southern red oak (*Quercus falcata*), shagbark hickory (*Carya ovata*) and black gum (*Nyssa sylvatica*) characterise sites which flood at intervals from two to eight years. Similarly, Teversham and Slaymaker (1976) identify indicator species for zones of different flood frequency on the Lilloet River, British Columbia.

Bedinger (1978) attributes the flood-associated distribution of species to their differing environment during early plant development. This is supported by the observations of Balslev *et al.* (1987) in Amazonian Ecuador, where the floodplain forest contains fewer species, but larger trees, than the adjacent unflooded forests. Bell (1980) suggests that the association between tree species and flood frequency may be the result of the combined effects of physiological factors and competition between tree species. Flood timing as well as magnitude may also affect the distribution of floodplain vegetation. For example, Fenner *et al.* (1985) illustrate the critical role of flood timing on the seed dispersal and germination of Fremont cottonwood (*Populus fremontii*) in the south-western United States. Fremont cottonwood produces seed which is dispersed by water and which has only a short period of viability. It requires flooding at the time of seed production (spring) to disperse the seeds, followed by lower flows (in summer) to expose damp areas of floodplain where the seeds can germinate. The spatial extent of this species, therefore, indicates both flood timing and extent. Similar observations are made by Thebaud and Debussche (1991), who explore the invasion of the Herault River system, France, by an introduced species, *Fraxinus ornus*, and relate its spatial extent to seed dispersal by water and its ability to establish and survive in flood-disturbed areas. The colonisation of the floodplains of the Salt and Gila Rivers, Arizona (Graf, 1982), by the artificially-introduced tamarisk (*Tamarix chinensis*), is also dependent on flooding, since seeds germinate while floating and can then colonise almost any type of unconsolidated material. Once established, tamarisk is susceptible to groundwater levels, which in the arid environment are significantly affected by recharge from flood events.

In addition to the influence of flood frequency; sediment erosion, transport and deposition associated with flooding also affect floodplain vegetation. For example, Hardin and Wistendahl (1983) describe scouring downstream and litter accumulation upstream of individual trees during floods on the Hocking River floodplain, Ohio. This results in an increase in herb density away from trees, particularly in an upstream direction. A more extreme example of the impact of sedimentation and flooding is described by Featherly (1940) as the result of inundation of an area of the floodplain of Deep Fork, Oklahoma, behind a large log jam. As a result of inundation and sedimentation, oak (*Quercus* spp.) and hickory (*Carya* spp.) trees, which require good drainage, died and *Fraxinus lanceolata*, *Populus deltoides* and *Salix nigra* colonised the floodplain along a gradient of increasing soil water content and waterlogging. The survival of trees on the wettest sites is as much a function of their ability to send out adventitious roots to keep pace with sedimentation, as it is a function of their ability to withstand waterlogging. The combined effects of such flood damage and sedimentation is used in the classic studies of Sigafoos (1964) and Sigafoos and Sigafoos (1966) which employ evidence from scar tissue, sprouts from damaged and decapitated trees, and adventitious roots that develop as a result of sedimentation, to reconstruct the flooding and sedimentation history of the flood-plain of the Potomac River. More recently, Hupp (1987, 1988) has combined flood damage and sedimentation information derived from vegetation in a manner similar to that of Sigafoos and Sigafoos (1966) with information on vegetation species composition and floodplain/channel morphological components (see below) to extend flood frequency records.

The association between hillslope drainage, river flood frequency, sediment erosion, transport and deposition, and associated damage to the riparian vegetation, is most clearly seen in the high-energy, confined environment of headwater rivers and their floodplains (e.g. types A1 and A2). The work of Hupp and Osterkamp provides information on the vegetation response within this type of environment at a range of spatial scales. For example, Hupp (1982) compares the composition and growth form of riparian forest along two adjacent sections of the Passage Creek, Virginia: a gentler section on non-resistant shales and a steeper gorge section on resistant sandstone. In the confined, steeper gorge section, the higher stream power intensifies flood flow velocities and increases the calibre of the sediment transported during flood events. This is reflected in stronger zonation in the vegetation parallel to the stream and a more diverse forest on the floodplain than that which occurs on the gentler section. Hupp (1983) shows how up to four vegetation zones, which are present on bars and islands along the Passage Creek, reflect the frequency and severity of flooding and flood damage. In protected areas of the channel, a first zone, the channel edge transition zone, contains patches of *Salix nigra*, *Orontium aquatica* and *Justicia americana* rooted below the low water line. Close to the channel there is a herbaceous zone, the edge zone, which contains no woody species. Next comes a shrub zone (*Alnus serrulata*, *Physocarpus opulifolius*, *Viburnum dentatum* and *Hamemalis virginiana*) where the shrubs show a history of flood damage since most stems are adventitious sprouts from old stumps. In this zone, herbaceous vegetation mainly occurs at the base of the shrubs. Finally, a transition zone between the shrub and riparian woods contains species from both zones. These four zones represent the transition from the zone of perennial water to the true flood-plain and have a combined width of 3 m where they are very compact or incomplete, to 30 m on the upstream ends of islands. Hupp (1990) and Osterkamp and Hupp (1984) define cross-valley and down-valley sequences of fluvial landforms which support distinct vegetation communities along the Passage Creek, South Fork Quantico Creek and Potomac River. Typical valley–floodplain cross-sections include the channel bed (under water at mean discharge), channel depositional bars which support herbaceous plants (at the level of about 40% flow duration), the channel shelf covered by riparian shrubs (best developed along steeper sections and corresponding to a 5–25% flow duration), the floodplain supporting floodplain forest (flood frequency of one to three years), and terraces with typical terrace forest assemblages, which represent former floodplains. In the upstream direction, the channel shelf persists well beyond the upper limits of the floodplain. In particular most first and second order streams are too steep and confined to develop a floodplain. Osterkamp and Hupp (1984, p. 1100) conclude that 'In the absence of discharge data to relate surfaces to flow frequencies, plants provide a convenient method of identifying geomorphic levels with relative confidence. Similarly, vegetative associations can aid in the mapping of geomorphic surfaces'.

In arid and semi-arid areas (e.g. floodplain types A3 and A4), the severity of flash floods coupled with their high sediment transport leads to strong associations between vegetation, its water requirements and resistance to flood damage; and zones of varying flood frequency. Hanes et al. (1989) describe the alluvial scrub which is adapted to the outwash environment of California. It occurs on outwash fans and riverine deposits where the lack of a reliable supply of water prevents the development of riparian woodlands. It is adapted to survival on poor fertility, porous substrates in locations of intense periodic flooding and erosion. Distinct phases (pioneer, intermediate and mature) of the scrub are identifiable by variations in both biomass and indicator species and are representative of the amount of time that has elapsed since the most recent flood. Similarly, Warren and Anderson (1985) identify four species assemblages within xeroriparian wash sites which dissect the bajada plain of the Sonoran Desert, Arizona. The first class is associated with watershed areas of less than 0.05 km^2 and represents areas of unreliable flow that cannot support preferential or obligate riparian species. The second class represents larger drainage areas which probably flow every year and are dominated by preferential and obligate riparian shrubs which are faster-growing and so replace the non-riparian species. The third class is associated with larger drainage areas of 2.5 to 120 km^2 where trees replace the shade-intolerant riparian shrubs, and shade-tolerant shrubs appear. The fourth class consists of tree-dominated banks and floodplain and shrub-dominated channel. Here the primary factors controlling the vegetation are the resistance of the channel shrubs to scouring by flood flows and the shade tolerance of the bank and floodplain species. Thus the vegetation in these desert washes result from the interaction between frequency and amount of runoff, shading, and channel scouring in channels draining catchments of different size.

A final example of the integrated effects of floodplain hydrology and the hydraulic impact of flooding on vegetation succession is given by Pautou and Décamps (1985). This case study concerns the lower-energy, less-confined environment of the upper Rhone between Geneva and Lyon, France. An increase in the number of woody species with altitude is associated with variation in the water table regime and the soil type, and a gradient in the degree of disturbance by flooding. The floodplain is sufficiently wide for the range in riparian forest to occur across individual floodplain transects. Patou and Décamps (1985) describe three altitudinal belts in the Rhone riparian forest (lower belt—pioneer communities on recently deposited alluvium; intermediate belt—willow communities on areas flooded during high flows; upper belt—hardwoods on more elevated areas) and suggest that similar belts exist on other lower floodplains in Central and Western Europe. Pautou et al. (1992) identify three groups of aquatic components of the river–floodplain system:

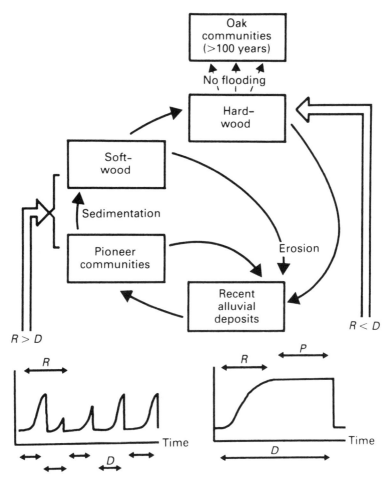

Figure 11.1   The influence of the flood disturbance regime on the succession of riparian plant communities along the Garonne and Ardour Rivers, France. Unstable systems are characterised by a recovery phase (*R*) greater than the time interval between floods causing disturbance to the vegetation (*D*). In stable systems *D* exceeds *R* leading to persistence (*P*). Where *R* > *D*, pioneer communities and softwood dominate. Where *R* < *D*, hardwood communities dominate. (Reproduced with permission of Blackwell Scientific Publications from Décamps and Tabacchi, 1994)

the aquatic components (active channels, sloughs and lakes); the semi-aquatic components (flooded for a lengthy period every year); and the terrestrial components (briefly and irregularly flooded, and linked to the presence of a water table) which may form a mosaic according to the local topography of the floodplain, but which at a coarse spatial scale support the broad altitudinal zones of riparian woodland described above. The interaction between these vegetation zones and hydrological/ flood disturbance is schematically represented by Décamps and Tabacchi (1994) (Figure 11.1).

FLUVIAL GEOMORPHOLOGICAL DISTURBANCES

The impact of flood disturbance described in the previous section leads to broad zonation in floodplain vegetation, but local variations in fluvial geomorphological processes lead to the development of a mosaic of disturbance patches within these broad zones, which are often associated with characteristic vegetation species and communities.

River planform change: associations between vegetation and floodplain landforms

Many studies employ patterns in vegetation to identify particular landforms on the floodplain and thus to establish the mode of formation of sections of floodplain.For example, Colonnello (1990) describes differences in tree species and species diversity between levée areas and intervening depressions on the Orinoco floodplain, Venezuela, as a result of differences in soil type and drainage, and inundation frequency. The pattern in the forest cover highlights the sequence of scroll bars which describe the migration of the river and the development of a scrolled (type B3b or c) floodplain. Similar vegetation variations are associated with scrolls on the Ucayali floodplain, Peru (Lamotte, 1990), but here there are also changes in vegetation with elevation which are associated with differences in inundation frequency and which may highlight components of the floodplain of different age. A three-fold classification of the Amazon floodplain from vegetation texture on remotely sensed (Landsat) imagery is proposed by Salo et al. (1986). Forests on the active floodplain exhibit clear scroll patterns in the primary forest succession. On some areas of floodplain a mosaic pattern occurs in the forest where the sequential successional forest has been cut from different angles so that a patchwork of successional stages is created. On former floodplain areas, the repetitive texture in the forest is not present . Although depressions and abandoned channels exist, much of the systematic undulation in the topography has been degraded and secondary forest succession, created by gaps opening in the canopy, adds to the heterogeneity of the forest cover. Page and Nanson (1982) clearly illustrate the differences in forest texture induced by the scrolled topography associated with concave-bench (concave downstream delineation) and point bar (convex downstream delineation) components of the floodplain on the Fort Nelson River, British Columbia. These patterns in the vegetation (Figure 11.2) indicate the presence of sections of scrolled counterpoint (type B3d) floodplain with sections of scrolled (type B3b) floodplain (Nanson and Croke, 1992). Thus, a range of scrolled floodplain types is revealed by patterns in the forest cover.

On braided and anastomosing rivers, other characteristics of the floodplain can be revealed by the vegetation cover. Channel planform change often results from channel abandonment rather than migration. The mode (from upstream or downstream) and rate (gradual or sudden) of abandonment is often reflected in the character and rate of sedimentation, and in the water regime within the abandoned stretch. The development of vegetation in abandoned channels can further enhance sedimentation rates and induce successions in vegetation which are representative of particular particle size, nutrient and drainage conditions (Amoros et al., 1987). Thus the pattern of vegetation on type C floodplains can be indicative of the sequence and the nature of evolution of the floodplain.

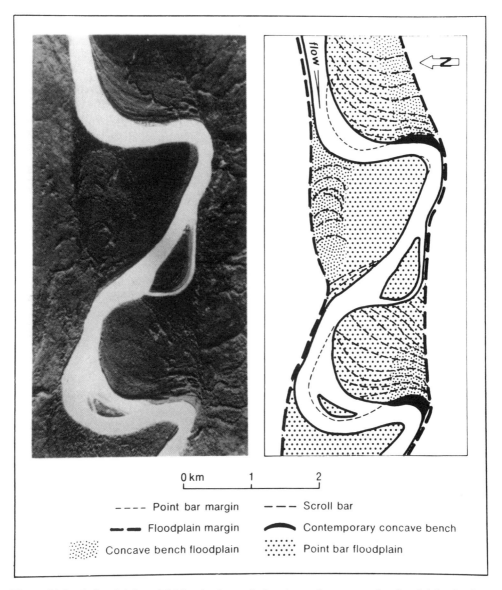

Figure 11.2 A floodplain exhibiting both scrolled point and counterpoint floodplain development (reproduced with permission of John Wiley and Sons from Page and Nanson, 1982)

River planform change and vegetative succession

Channel migration and abandonment, and the development of mid-channel and marginal bars exposes fresh areas for vegetation colonisation. The nature of the colonising vegetation associated with particular geomorphological features and the subsequent vegetative succession can help to reveal the age of particular features and their past and present environmental characteristics. Furthermore, the vegetation can directly influence the types of morphological feature that develop and the rate at which they evolve. For example, Hickin (1984) suggests that vegetation and debris rafting can have an important influence in accelerating the stabilisation of mid-channel bars through the influence of major roughness elements which can trap sediments. Woody debris accumulations can lead to the abandonment of channels and, through avulsion, the creation of new ones (Harwood and Brown, 1993; Nakamura and Swanson, 1993). The rate of development and morphology of scroll bars is also influenced by sedimentation resulting from the enhanced roughness of vegetation cover (Hickin, 1984).

A study by Bellah and Hulbert (1974) illustrates the interdependence between fluvial processes and vegetation colonisation and succession on exposed areas of alluvium on the Republican River floodplain, Kansas. Three pioneer tree species, sandbar willow (*Salix interior*), almondleaf willow (*Salix Amygdaloides*) and cottonwood (*Populus deltoides*), appear on alluvium within one or two years of exposure. The sandbar willow rarely persists for more than 10 years, almondleaf willow not more than 30 years, and cottonwood about a century. None of these three pioneer species can regenerate within established stands and so they gradually die out after different life spans for each species. The dense stands of vegetation stabilise the alluvium and trap sediments. Because of the rapid colonisation of alluvium by trees, the probability of removal of the sediment by a flood is inversely proportion to the time since exposure, as well as being directly proportional to flood severity. Within eight to ten years of initial colonisation, a variety of other tree and herb species start to appear on the stabilised alluvium so that after about 100 years dominant trees include American elm (*Ulmus americana*), common hackberry (*Celtis occidentalis*), green ash (*Fraxinus pennsylvanica*), red mulberry (*Morus rubra*) and boxelder (*Acer negundo*). Page and Nanson (1982) provide examples of the way in which vegetation colonisation and succession influences the evolution of bars and concave benches along the Murrumbidgee River, Australia. Three stages in the development of concave benches are described, starting with establishment of a bar through deposition from secondary currents near the concave bank. The bar is colonised by vegetation which becomes dominated by red gum (*Eucalyptus cameldulensis*) once vertical accretion has attained a level above that of frequent inundation. The largest trees develop on the highest point of the bar with younger trees representing the bankward and upstream extension of the bar to form a bench as the bend migrates. McBride and Strahan (1984) describe the interaction between vegetation colonisation and succession, and point bar development and stabilisation, in the formation of a scrolled floodplain along Dry Creek, California. Point bars can be divided into five environments: point bar bank, first ridge and swale, interior ridges and swales, lagoons, and the base of the floodplain terrace. Seedlings establish on the point bar bank as flows recede in spring with willows (*Salix* spp.) establishing on finer sediments than cottonwoods (*Populus* spp.). Seedling survival during the summer is dependent on water availability which is associated with depth to water table and particle size. As a result, surviving vegetation is usually distributed in strips which form the basis for ridge and swale development. McBride and Strahan interpret the distribution of riparian tree species on the ridges and swales as representing a successional sequence. Coarser sediments are initially trapped by mulefat (*Baccharis viminea*), willows and cottonwood to produce ridges of gravel on the outer edge of the point bar. The ridge provides a favourable environment for further cottonwood establishment. As the ridge builds, finer sediments are deposited on its surface and at the base of the floodplain terrace at the downstream end of the bar. New ridges form as the point bar extends. Growth of cottonwood and willows on the interior ridges traps more fine sediments which begin to fill the swales. The accumulating fine sediments support white alder (*Alnus rhombifolia*) in the interior swales and the interior downstream portions of the point bars.

The above examples illustrate how successional sequences in vegetation on fluvial depositional features provide an indication of their relative age and thus their evolu-

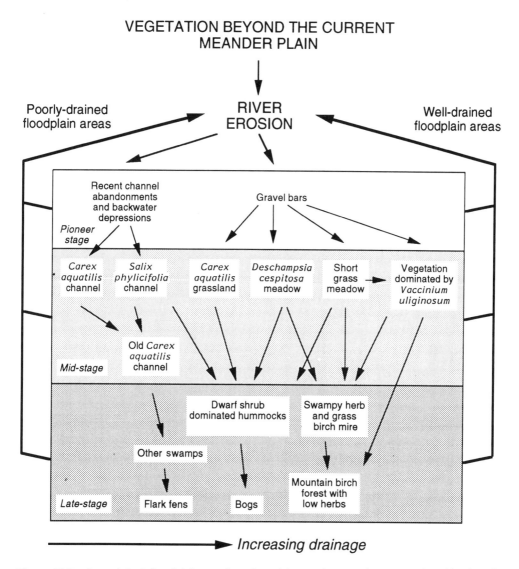

Figure 11.3 A model of floodplain erosion, deposition and vegetation succession (developed from Kalliola and Puhakka, 1988, reproduced with permission of Blackwell Science)

tionary sequence. A detailed study of the floodplain of the meandering River Kama-johka, Finland, by Kalliola and Puhakka (1988) generalises the use of vegetational successional sequences to aid understanding of floodplain evolution by developing a model of river dynamics and vegetation mosaicism. Here channel migration mainly determines the boundaries between vegetation groups and causes continuous site turnover. Figure 11.3 illustrates the model of erosion, deposition and vegetation succession that Kalliola and Puhakka propose. It makes the assumptions that the floodplain vegetation originates from riparian primary succession and that riparian vegetation can be destroyed by further river erosion. The heterogeneity of the floodplain vegetation is dependent on both spatial and temporal controls. The spatial pattern of the vegetation mosaic at any point in time reflects age differences and edaphic differences caused by variability in the geomorphology (particularly morphology which influences both drainage and inundation frequency, and sediment calibre) of the floodplain. Figure 11.3, therefore, illustrates differences in the succession of vegetation on gravel bars (coarse, well-drained) and in abandoned channels and backwaters (finer sediments and topographically lower, thus poorer drainage). The temporal aspect of change in floodplain vegetation relates to change in the location of sites of erosion and deposition, sedimentation patterns on the floodplain and in relict channels leading to a change in their elevation and thus their flood frequency and drainage characteristics, and their surface particle size distribution and organic content. Vegetation that colonises new areas of floodplain during the pioneer stage is resistant to the stresses caused by flooding and sedimentation. Figure 11.3 illustrates that particular species are associated with certain types of fluvial feature: *Deschampsia cespitosa* (gravel bars), *Salix phylicifolia* (fine-textured sediments in abandoned channels), *Carex aquatilis* (wettest parts of abandoned channels).

River bank form and process

At a local scale, riparian vegetation has a direct impact on river bank stability through its influence as a roughness element on marginal flow velocities and sedimentation, the impact of its roots in adding tensile strength to bank sediments, and through improved bank drainage via the network of macropores created by root development (Thorne, 1990). The interaction between vegetation cover and bank stability is well illustrated by a study of woody vegetation recovery along the banks of modified channels in Tennessee (Hupp, 1992). A model of bank recovery is associated with distinctive vegetation establishment patterns. Pioneer species (black willow, *Salix nigra*; river birch, *Betula nigra*; boxelder, *Acer negundo*; silver maple, *Acer saccharinum*; and cottonwood, *Populus deltoides*) establish in the middle stages of bank recovery, after initial mass movement and accretion have occurred. They establish low on the banks at about the 50% flow duration level where initial accretion occurs, and they survive because of their ability to tolerate medium to high levels of mass wasting and accretion. This establishment of species which are tolerant of accretion on the lower part of the bank leads to an interaction between vegetation growth and roughness, deposition of sediments, and a consequent reduction in the bank slope which is enhanced by mass movement processes. A second group of relatively stable site species (ironwood, *Carpinus caroliniana*; green ash, *Fraxinus pennsylvanica*; sweet gum, *Liquidambar styraciflua*; American elm, *Ulmus americana*; bald cypress, *Taxodium distichum*; and tupelo gum; *Nyssa aquatica*) establish in the later stages of recover when mass wasting and accretion rates have subsided. The final suite of bottomland oaks establishes after bank recovery is complete.

Hupp's (1992) model of bank recovery and vegetation establishment implies that the vegetation on river banks may be indicative of their stability and stage of evolution. Another aspect of the association between riparian trees and bank morphology and evolution is described by Rowntree (1991), who identifies benches at the toe of some river banks in South Africa, which are the direct result of tree root reinforcement of the bank. This important impact of tree roots on bank morphology is an extreme example of the way in which the roots of any vegetation cover influence bank stability. The reinforcement effect is only as deep as the rooting depth of the overlying vegetation and so the rooting depth as a proportion of the bank height will control whether the roots are able to resist bank failure or not (Thorne, 1990), and thus the dynamic form of the bank profile.

Other properties of bank vegetation can indicate the significance of geomorphological processes. For example, Gregory (1992) describes the way in which tree root exposure, the position of trees on river banks, and the presence of bent trunks can be used to identify eroding banks and, particularly, enlarging river channels. Lisle (1989) describes how riparian tree colonisation along the margins of low flow water levels in Californian streams defines an inner channel within the broad flood channels that are cut by flash floods. Sedimentation around these trees gradually establishes an inner channel profile between major flash flood events. Thus the morphology, sedimentation and riparian tree growth in these semi-arid channels provides an indication of the flow and channel evolution history of the site. More subtle relationships between in-channel flow characteristics and riparian vegetation have also been identified including variations in the number of species present along a stream velocity gradient (Nilsson, 1987) and variations in the distribution of species up the river banks in response to inundation, erosion and deposition gradients (Merry *et al.*, 1981).

INTEGRATION AND APPLICATION

This chapter has illustrated the hydrological and geomorphological information content of the vegetation species and communities within river corridors at a variety of spatial and temporal scales. This has reinforced the view that 'understanding the role of vegetation ... in contemporary river channels facilitates our understanding of the present and improves interpretation of the past' (Gregory and Gurnell, 1988, p. 35). Furthermore, an understanding of interactions between vegetation, hydrological and geomorphological processes across hillslopes and floodplains in semi-natural situations is essential to the successful management of riparian zones that have been heavily impacted by anthropogenic influences. River corridor management can, therefore, benefit from improved methods of recording (e.g. Petersen, 1992), classifying (e.g. Naiman *et al.*, 1992) and interpreting the potential (e.g. Kovalchik and Chitwood, 1990) of river corridor geomorphology and vegetation. An improved knowledge of how river corridors function at a variety of space and time-scales (Décamps and Fortuné, 1991) provides a basis for forecasting the impacts of specific developments such as hydro-power development (Pautou *et al.*, 1992), river regulation (Powers *et al.*, 1994) and flow diversion (Pearlstine *et al.*, 1985) and also provides a rationale for restoring rivers and their floodplains to a more natural state (Gregory *et al.*, 1995; Peterken and Hughes, 1995). The development of 'soft' engineering approaches to the design and restoration of river channels (e.g. Chapter 17) and advances in the appreciation of how vegetation may be incorporated into such designs (e.g. Hemphill and Bramley, 1989; Copping and Richards, 1990) can be extended, from a sound scientific basis, to the design and restoration of floodplains (e.g. Petts, 1990b) in the context of a variety of physical, ecological, economic and social constraints.

ACKNOWLEDGEMENTS

The author wishes to express her sincere thanks to Ken Gregory, Geoff Petts and Andrew Brookes for their critical reading of an early draft of this manuscript. At the time that he read this, Ken Gregory had no idea of its purpose, but he willingly and enthusiastically gave advice and encouragement as he has always done over the last 28 years.

REFERENCES

Amoros, C., Roux, A.L., Reygrobellet, J.L., Bravard, J.P. and Patou, G. (1987) A method for applied ecological studies of fluvial hydrosystems. *Regulated Rivers: Research and Management* **1**, 17–36.

Angold, P., Edwards, P.J. and Gurnell, A.M. (1994) *A Context for Developing Methodologies for Optimising the Value of River Corridor Survey Data*. National Rivers Authority R&D Note 273, National Rivers Authority, Bristol.

Balslev, H.G., Lutyen, J., Ollgaard, B. and Holm-Nielsen, L.B. (1987) Composition and structure of adjacent unflooded and floodplain forest in Amazonian Ecuador. *Opera Botanica* **92**, 37–57.

Barnes, W.J. (1978) The distribution of flood plain herbs as influenced by annual flood elevation. *Wisconsin Academy of Sciences, Arts and Letters, Transactions* **66**, 254–266.

Bedinger, M.S. (1971) *Forest Species as Indicators of Flooding in the Lower White River Valley, Arkansas*. US Geological Survey Professional Paper 750-C, C248–C253.

Bedinger, M.S. (1978) Relation between forest species and flooding. In: Greeson, P.E., Clark, J.R. and Clark, J.E. (eds), *Wetland Functions and Values: The State of Our Understanding*, American Water Resources Association, Minneapolis, 427–435.

Bell, D.T. (1980) Gradient trends in the streamside forest of central Illinois. *Bulletin of the Torrey Botanical Club* **107**, 172–180.

Bell, D.T. and Johnson, F.L. (1974) Ground-water level in the floodplain and adjacent uplands of the Sangamon River. *Transactions Illinois State Academy of Science* **67**, 376–383.

Bellah, R.G. and Hulbert, L.C. (1974) Forest succession on the Republican River floodplain in Clay County, Kansas. *The Southwestern Naturalist* **19**, 155–166.

Bennett, P.S., Kunzman, M.R. and Johnson, R.R. (1989) Relative nature of wetlands: riparian and vegetational considerations. In: *Proceedings of the California Riparian Systems Conference*, 22–24 September 1988, Davis, California, USDA Forest Service General Technical Report PSW-110, 140–142.

Bren, L.J. (1988a) Flooding characteristics of a riparian red gum forest. *Australian Forestry* **51**, 57–62.

Bren, L.J. (1988b) Effects of river regulation on flooding of a riparian red gum forest on the River Murray, Australia. *Regulated Rivers: Research and Management* **2**, 65–77.

Bren, L.J. (1993) Riparian zone, stream and floodplain issues: a review. *Journal of Hydrology* **150**, 277–299.

Bren, L.J. and Gibbs, N.L. (1986) Relationships between flood frequency, vegetation and topography in a river red gum forest. *Australian Forest Research* **16**, 357–370.

Colonnello, G. (1990) A Venezuelan floodplain study on the Orinoco River. *Forest Ecology and Management* **33/34**, 103–124.

Copping, N.J. and Richards, I.G. (eds) (1990) *Use of Vegetation in Civil Engineering*. Construction Industry Research and Information Association/Butterworths, London, 292pp.

Décamps, H. and Fortuné, M. (1991) Long-term ecological research and fluvial landscapes. In: Risser, P.G. (ed.), *Long-term Ecological Research*, Wiley, Chichester, 135–151.

Décamps, H., Fortuné, M., Gazelle, F. and Pautou, G. (1988) Historical influence of man on the riparian dynamics of a fluvial landscape. *Landscape Ecology* **1**, 163–173.

Décamps, H. and Tabacchi, E. (1994) Species richness in vegetation along river margins. In: Giller, P.S., Hildrew, A.G. and Raffaelli, D.G. (eds), *Aquatic Ecology: Scale, Pattern and Process*, Blackwell Scientific Publications, Oxford, 1–20.

Featherley, H.I. (1940) Silting and forest succession on Deep Fork in Southwestern Creek County, Oklahoma. *Proceedings of the Oklahoma Academy of Science* **20**, 63–64.

Fenner, P., Brady, W.W. and Patton, D.R. (1985) Effects of regulated water flows on regeneration of Fremont Cottonwood. *Journal of Range Management* **38**, 135–138.

Gecy, J.L. and Wilson, M.V. (1990) Initial establishment of riparian vegetation after disturbance by debris flows in Oregon. *American Midland Naturalist* **123**, 282–291.

Graf, W.L. (1982) Tamarisk and river channel management. *Environmental Management* **6**, 283–296.

Gregory, K.J. (1992) Vegetation and river channel processes. In: Boon, P.J., Calow, P. and Petts, G.E. (eds) *River Conservation and Management*, Wiley, Chichester, 255–269.

Gregory, K.J. and Gurnell, A.M. (1988) Vegetation and river channel form and process. In: Viles, H. (ed.), *Biogeomorphology*, Blackwell, Oxford, 11–42.

Gregory, K.J., Gurnell, A.M. and Petts, G.E. (1995) The role of dead wood in aquatic habitats in forests. In: *Forests and Water*, Proceedings of the 1994 Discussion Meeting of the Institution of Chartered Foresters, ICF, Edinburgh (in press).

Gregory, S.V., Lamberti, G.A. and Moore, K.M.S. (1989) Influence of valley floor landforms on stream ecosystems. In: *Proceedings of the California Riparian Systems Conference*, 22–24 September 1988, Davis, California, USDA Forest Service General Technical Report PSW-110, 3–8.

Gregory, S.V., Swanson, F.J., McKee, W.A. and Cummins, K.W. (1991) An ecosystem perspective of riparian zones. *BioScience* **41**, 540–551.

Gurnell, A.M. (1981) Heathland vegetation, soil moisture and dynamic contributing area. *Earth Surface Processes and Landforms* **6**, 553–570.

Gurnell, A.M. and Gregory, K.J. (1986) Water table level and contributing area: the generation of runoff in a heathland catchment. In: *Conjunctive Water Use*, International Association of Hydrological Sciences Publication 156, IAHS Press, Wallingford, 87–95.

Gurnell, A.M. and Gregory, K.J. (1987) Vegetation characteristics and the prediction of runoff: analysis of an experiment in the New Forest, Hampshire. *Hydrological Processes* **1**, **125–142.**

Gurtz, M.E., Marzolf, G.R., Killingbeck, K.T., Smith, D.L. and McArthur, J.V. (1988) Hydrologic and riparian influences on the import and storage of coarse particulate organic matter in a prairie stream. *Canadian Journal of Fisheries and Aquatic Sciences* **45**, 655–665.

Hanes, T.L., Friesen, R.D. and Keane, K. (1989) Alluvial scrub vegetation in coastal southern California. In: *Proceedings of the California Riparian Systems Conference*, 22–24 September 1988, Davis, California, USDA Forest Service General Technical Report PSW-110, 187–193.

Hardin, E.D. and Wistendahl, W.A. (1983) The effects of floodplain trees on herbaceous vegetation patterns, microtopography and litter. *Bulletin of the Torrey Botanical Club* **110**, 23–30.

Harris, R.R. (1988) Associations between stream valley geomorphology and riparian vegetation as a basis for landscape analysis in the Eastern Sierra Nevada, California, USA. *Environmental Management* **12**, 219–228.

Harwood, K. and Brown, A.G. (1993) Fluvial processes in a forested anastomosing river: flood partitioning and changing flow patterns. *Earth Surface Processes and Landforms* **18**, 741–748.

Hemphill, R.W. and Bramely, M.E. (eds) (1989) *Protection of River and Canal Banks*. Construction Industry Research and Information Association/Butterworths, London, 200pp.

Henszey, R.J., Skinner, Q.D. and Wesche, T.A. (1991) Response of montane meadow vegetation after two years of streamflow augmentation. *Regulated Rivers: Research and Management* **6**, 29–38.

Hickin, E.J. (1984) Vegetation and river channel dynamics. *Canadian Geographer* **28**, 111–126.

Hupp, C.R. (1982) Stream-grade variation and riparian-forest ecology along Passage Creek, Virginia. *Bulletin of the Torrey Botanical Club* **109**, 488–499.

Hupp, C.R. (1983) Vegetation pattern on channel features in the Passage Creek Gorge, Virginia. *Castanea* **48**, 62–72.

Hupp, C.R. (1987) Botanical evidence of floods and palaeoflood history. In: Singh, V.P. (ed.) *Regional Flood Frequency Analysis*, D. Reidel Publ. Co., 355–369.

Hupp, C.R. (1988) Plant ecological aspects of flood geomorphology and palaeoflood history. In: Baker, V.R., Kochel, R.C. and Patton, P.C. *Flood Geomorphology*, Wiley, Chichester, 335–356.

Hupp, C.R. (1990) Vegetation patterns in relation to basin hydrogeomorphology. In: Thornes, J.B. (ed.) *Vegetation and Erosion*, Wiley, Chichester, 217–237.

Hupp, C.R. (1992) Riparian vegetation recovery patterns following stream channelization: a geomorphic perspective. *Ecology* **73**, 1209–1226.

Kalliola, R. and Puhakka, M. (1988) River dynamics and vegetation mosaicism: a case study of the River Kamajohka, northernmost Finland. *Journal of Biogeography* **15**, 703–719.

Kondolf, G.M. and Curry, R.R. (1984) The role of vegetation in channel bank stability: Carmel River, California. In: Warner, R.E. and Hendrix, K.M. (eds), *California Riparian Systems*, University of California Press, Berkeley and Los Angeles, 124–133.

Kovalchik, B.L. and Chitwood, L.A. (1990) Use of geomorphology in the classification of riparian plant associations in mountainous landscapes of central Oregon, USA. *Forest Ecology and Management* **33/34**, 405–418.

Lamotte, S. (1990) Fluvial dynamics and succession in the Lower Ucayali River basin, Peruvian Amazonia. *Forest Ecology and Management* **33/34**, 141–155.

Lisle, T.E. (1989) Channel-dynamic control on the establishment of riparian trees after large floods in Northwestern California. In: *Proceedings of the California Riparian Systems Conference*, 22–24 September 1988, Davis, California, USDA Forest Service General Technical Report PSW-110, 8–13.

Malanson, G.P. (1993) *Riparian Landscapes*. Cambridge University Press, Cambridge.

McBride, J.R. and Strahan, J. (1984) Fluvial processes and woodland succession along Dry Creek, Sonoma County, California. In: Warner, R.E. and Hendrix, K.M. (eds), *California Riparian Systems*, University of California Press, Berkeley and Los Angeles, 110–119.

Menges, E.S. (1986) Environmental correlates of herb species composition in five southern Wisconsin floodplain forests. *The American Midland Naturalist* **115**, 106–117.

Merry, D.G., Slater, F.M. and Randerson, P.F. (1981) The riparian and aquatic vegetation of the river Wye. *Journal of Biogeography* **8**, 313–327.

Naiman, R.J. and Décamps, H. (eds) (1990) *The Ecology and Management of Aquatic–Terrestrial Ecotones*. Man and the Biosphere Series No. 4, Unesco, Paris, and Parthenon, Carnforth, UK, 316pp.

Naiman, R.J., Lonzarich, D., Beechie, T.J. and Ralph, S.C. (1992) General principles of classification and the assessment of conservation potential in rivers. In: Boon, P.J., Calow, P. and Petts, G.E. (eds), *River Conservation and Management*, Wiley, Chichester, 93–123.

Nakamura, F. and Swanson, F.J. (1993) Effects of coarse woody debris on morphology and sediment storage of a mountain stream system in western Oregon. *Earth Surface Processes and Landforms* **18**, 43–61.

Nanson, G.C. and Croke, J.C. (1992) A genetic classification of floodplains. *Geomorphology* **4**, 459–486.

Nilsson, C. (1987) Distribution of stream-edge vegetation along a gradient of current velocity. *Journal of Ecology* **75**, 513–522.

Osterkamp, W.R. and Hupp, C.R. (1984) Geomorphic and vegetative characteristics along three northern Virginia streams. *Geological Society of America Bulletin* **95**, 1093–1101.

Page, K. and Nanson, G. (1982) Concave-bank benches and associated floodplain formation. *Earth Surface Processes and Landforms* **7**, 529–543.

Parodi, J.L. and Freitas, D. (1990) Geographical aspects of forested wetlands in the Lower Ucayali, Peruvian Amazonia. *Forest Ecology and Management* **33/34**, 157–168.

Pautou, G. and Décamps, H. (1985) Ecological interactions between the alluvial forests and hydrology of the Upper Rhone. *Archiv für Hydrobiologie* **104**, 13–37.

Pautou, G., Girel, J. and Borel, J-L. (1992) Initial repercussions and hydroelectric developments in the French Upper Rhone Valley: a lesson for predictive scenarios propositions. *Environmental Management* **16**, 231–242.

Pearlstine, L., McKellar, H. and Kitchens, W. (1985) Modelling the impacts of a river diversion on bottomland forest communities in the Santee River floodplain, South Carolina. *Ecological Modelling* **29**, 283–302.

Peterken, G.F. and Hughes, F.M.R. (1995) Restoration of floodplain forests. In: *Forests and Water*, Proceedings of the 1994 Discussion Meeting of the Institution of Chartered Foresters, ICF, Edinburgh (in press).

Petersen, R.C. (1992) The RCE: a riparian, channel and environmental inventory for small streams in the agricultural landscape. *Freshwater Biology* **27**, 295–306.

Petts, G.E. (1989) Historical analysis of fluvial hydrosystems. In: Petts, G.E., Moller, H. and Roux, A.L. (eds), *Historical Change of Large Alluvial Rivers: Western Europe*, Wiley, Chichester, 1–18.

Petts, G.E. (1990a) Forested river corridors: a lost resource. In: Cosgrove, D. and Petts, G.E. (eds), *Water, Engineering and Landscape: Water Control and Landscape Transformation in the Modern Period*, Belhaven, London, 12–34.

Petts, G.E. (1990b) The role of ecotones in aquatic landscape management. In: Naiman, R.J. and Décamps, H. (eds), *The Ecology and Management of Aquatic–Terrestrial Ecotones*, Man and the Biosphere Series No. 4, Unesco, Paris, and Parthenon, Carnforth, UK, 227–261.

Poff, N.L. and Ward, J.V. (1990) Physical habitat template of lotic systems: recover in the context of historical pattern of spatiotemporal heterogeneity. *Environmental Management* **14**, 629–645.

Powers, A., Wright, P., Pucherelli, M. and Wegner, D. (1994) GIS efforts target long-term resource monitoring. *GIS World* **7**, 36–39.

Rowntree, K. (1991) An assessment of the potential impact of alien invasive vegetation on the geomorphology of river channels in South Africa. *South African Journal of Aquatic Science* **17**, 28–43.

Rucks, M.G. (1984) Composition and trend of riparian vegetation on five perennial streams in Southeastern Arizona. In: Warner, R.E. and Hendrix, K.M. (eds), *California Riparian Systems*, University of California Press, Berkeley and Los Angeles, 97–107.

Salo, J., Kalliola, R., Hakkinen, I., Makinen, Y., Niemala, P., Puhakka, M. and Coley, P.D. (1986) River dynamics and the diversity of Amazon lowland forest. *Nature* **322**, 254–258.

Schaefer, J.M. and Brown, M.T. (1992) Designing and protecting river corridors for wildlife. *Rivers* **3**, 14–26.

Sedell, J.R. and Froggatt, J.L. (1984) Importance of streamside forests to large rivers: the isolation of the Willamette River, Oregon, USA, from its floodplain by snagging and streamside forest removal. *Verh. Internat. Verein. Limnol.* **22**, 1828–1834.

Sedell, J.R., Reeves, G.H., Hauer, F.R., Stanford, J.A. and Hawkins, C.P. (1990) Role of refugia in recovery from disturbances: modern fragmented and disconnected river systems. *Environmental Management* **14**, 711–724.

Shelford, V.E. (1954) Some Lower Mississippi Valley flood plain biotic communities: their age and elevation. *Ecology* **35**, 126–142.

Sigafoos, R.S. (1964) *Botanical Evidence of Floods and Flood-plain Deposition*. US Geological Survey Professional paper 485-A, A1–A35.

Sigafoos, R.S. and Sigafoos, M.D. (1966) Flood history told by tree growth. *Natural History* **75**, 50–55.

Stromberg, J.C. (1993) Instream flow models for mixed deciduous riparian vegetation within a semiarid region. *Regulated Rivers: Research and Management* **8**, 225–235.

Swanson, F.J., Gregory, S.V., Sedell, J.R. and Campbell, A.G. (1982) Land–water interactions: the riparian zone. In: Edmonds, R.L. (ed.), *Analysis of Coniferous Ecosystems in the Western United States*, Hutchinson Ross, Stroudsberg, Pennsylvania, 267–291.

Teversham, J.M. and Slaymaker, O. (1976) Vegetation composition in relation to flood frequency in Lillooet River valley, British Columbia. *Catena* **3**, 191–201.

Thebaud, C. and Debussche, M. (1991) Rapid invasion of *Fraxinus ornus* L. along the Herault River system in southern France: the importance of seed dispersal by water. *Journal of Biogeography* **18**, 7–12.

Thorne, C.R. (1990) Effects of vegetation on riverbank erosion and stability. In: Thornes, J.B. (ed.) *Vegetation and Erosion*, Wiley, Chichester, 123–144.

Van Hees, W.W.S. (1990) Boreal forest wetlands—what and where in Alaska. *Forest Ecology and Management* **33/34**, 425–438.

Warren, P.L. and Anderson, L.S. (1985) Gradient analysis of a Sonoran desert wash. In: Johnson, R.R., Ziebell, C.D., Patton, D.R., Ffolliott, P.F. and Hamre, R.H. (eds), *Riparian Ecosystems and their Management: Reconciling Conflicting Uses*, US Forest Service General Technical Report RM-120, 150–155.

Zinke, A. and Gutzweiler, K.-A. (1990) Possibilities for regeneration of floodplain forests within the framework of the flood-protection measures on the Upper Rhine, West Germany. *Forest Ecology and Management* **33/34** 13-20.

Part III

INFORMATION FOR THE
MANAGEMENT OF CHANGE

12 Information Flow for Channel Management

MICHAEL J. CLARK

Department of Geography, University of Southampton, UK

LINKING INFORMATION WITH ACTION

River channel management is not, and should never become, an exclusively academic domain. The scientific (often academic) community has contributed enormously to the better understanding of the river system, and to the definition of appropriate and achievable targets for river managers. In addition, scientific methods have proved highly effective in monitoring and analysing river processes, as well as in designing and evaluating the tools of management. But the heart of the management process itself lies outside the scientific sphere. Its practitioners, though often scientifically or technologically trained, exercise their skills in a different context and by different rules. Since, in effect, rivers are usually managed by agencies, the management process is a corporate process, and the role of information as a component in river management has to be viewed against the role of information in corporate functions—a task which is introduced towards the end of the chapter. Furthermore, the inevitable embedding of corporate structures in economic and political infrastructures requires an even greater effort to look outwards to issues and imperatives beyond the scientific disciplines, rather than seeking comfort and satisfaction by looking inwards at that which is familiar.

All of this can be profoundly disturbing and depressing to the average practitioner of the river environment. Geomorphologists, ecologists, planners and engineers yearn for practical and robust guidelines for action, supported by adequate funding and implemented through efficient but flexible procedures. To such operational staff, United Nations environmental conferences, corporate information technology strategies and political debates on relative merits of such issues as inter-agency and single-agency management can appear at best to be diversions, and at worst to be barriers to effective practice.

For such reasons, it is fashionable to set the practitioners of any field apart from the theorists and philosophers, and to assume that their worlds are essentially in contention. This chapter shares with Chapter 18 the quest to demonstrate that in river management the three worlds intersect significantly, and that their interaction can be both helpful and (much more reassuring) controllable. The chapter concludes that information, and information systems, can contribute immediately and significantly to river management, increasing effectiveness as well as efficiency.

Changing River Channels. Edited by Angela Gurnell and Geoffrey Petts.
© 1995 John Wiley & Sons Ltd.

It demonstrates that developments in information handling are more pragmatic in implementation than some of their more evangelical adherents might suggest—but indicates that significant shifts in the emphasis of both river science and river management may in fact be taking place without the majority of those involved being conscious of these trends. However, in reaching this position, the argument has to confront counter views and explore some unfamiliar byways. In particular, it must start by considering the possibility that the whole information revolution, driven in equal measure by positivist science and technological society, might be fundamentally flawed. By thus challenging ourselves with the unthinkable, we can strengthen rather than weaken the positive role of information in channel and catchment management.

THE INFORMATION DILEMMA

The world is awash with information. Even by 1989, the UNESCO International Hydrological Programme was able to list 52 online water-related publicly available databases (Nieuwenhuysen, 1989). The enlightening trickle of information that built up over the 18th and 19th centuries became a machine-fed and machine-fuelling torrent by the middle of the 20th century, and, as we approach the end of the century, the torrent threatens to become an all-consuming flood. Suffocation competes with enlightenment. Information technologies, classifications and typologies are increasingly devoted to compressing and managing information. Information strategies become so complex that they often fail to move beyond the drawing board or the working party. Archives are constantly enlarged to store the flood, and information superhighways are promised to channel it. Our desktop computers act as sluices, aiming to nourish our work or our understanding by releasing just the right amount of information where and when we need it most—a kind of intellectual irrigation!

For the sceptic, the opportunities to denigrate this love affair with information are widespread and increasing. It is paradoxical to say the least that from the global perspective, significant sectors of the population continue to be denied either sufficient water or sufficiently high quality water, and that the waterways themselves continue to deteriorate in both environmental quality and their ability to perform the essential task of transmitting water to the sea without unacceptable detrimental impact on surrounding people. Even in the developed countries, the gulf between those areas where channel quality is notably and consistently improving and those many other areas where low standards prevail remains stark. Equally disturbing in its own way is the fact that few of the really important advances in channel management appear to have been driven by databanks and numerical models. For example, the move towards catchment-based holistic approaches, the trend away from hard engineering towards soft engineering, and the slow but progressive recognition that notions such as environmental priorities and the quest for sustainable development are more than just political bandwagons, were all born of scientific and philosophical reflection rather than the mere agglomeration or transformation of information.

Such a sober commentary on the role of the information revolution in river management clearly demands a creative response from those who retain a faith that information is power. In seeking such a response, it is necessary to consider the meaning of information and power, and then to assess their symbiosis in the quest for enhanced channel management. There is no conventional blueprint for such a re-assessment, so the journey is best attempted by seeking a few signposts from which to construct a vision of a future informatic channel management approach.

THE PROCESS OF INFORMING

Without dwelling on the semantics involved, it is useful to adopt as a starting point the assumption that, for present purposes, *informing* can be regarded as the timely communication of something which the recipient will find to be of interest and of value. It is a process of telling, and *information* is that which is told (and which also should be of interest and of value). In this sense, information is significantly different to data. Again for present purposes, data are descriptors (and often measures) of attributes, and to become genuinely *informative* such data need to pass the test of being of interest, of value and not only communicable but actually communicated. A database is successful if it stores data and is open to inputs and outputs, but an information system is successful only if it actively transmits a flow of information. The analogy of a lake and a river is not inappropriate, with the former being domi-nated by structure and storage, while the latter is dominated by adjustment to the need to create flow.

In our search to define the role of information in channel management, we are thus led to question what it is that renders data interesting, valuable and communic-able to channel managers. This is a search which in part will demonstrate a response to the evolution of channel management policies and strategies, and in part will counter our earlier scepticism by suggesting that this evolution was itself partly generated by the supply of information and of information-handling capabilities. As in so many branches of science and management, the roles of master and of servant are both played by information and the information processor.

MEETING THE CRITERION OF INTEREST

Data and information

It is entirely possible for a river management agency to suffer simultaneously from a glut of data and a deficit of information. A familiar but convincing example is the provision of detailed species lists for the benthic fauna of river sampling sites to a river engineer or planner. The data clearly have relevance, but not meaning. They can in principle illuminate important aspects of water or habitat quality, but in practice they achieve this only for those people with sufficient disciplinary skill to be able to interpret the data and derive the information. Paradoxically, in such cases it is possible greatly to enhance the information content by degrading the data (e.g. classifying species according to their sensitivity to a given water quality parameter). Part of the information strategy of any agency should be a rigorous assessment of the extent to which data can effectively be interpreted and applied by their recipients.

The data in such contexts have to pass both a test of relevance and a test of appropriateness of form and expression. One of the worrying consequences of the otherwise welcome move towards holistic (discipline-integrative) approaches in water management agencies is that specialists are increasingly being expected to assimilate information from outside their own discipline. On reflection, it quickly becomes apparent that the information content of a data set is not a fixed property. Rather, it is highly dependent upon the skills and assumptions of the recipient. This is hardly an innovative observation, but its implications are magnified as we adopt ever more automated processes of data communication. Technologies such as Geographic Information Systems (GIS) have the functional flexibility to draw upon a single data set (such as a species list), but present quite different versions of this to different users depending on their needs and abilities: one source informs in many different ways. In practice, this functionality is rarely used. Either the practitioners have concluded that a monothetic view of the world is preferable or, far more likely, they have simply never considered the advantages and feasibility of a polythetic presentation.

An outcome of this debate is that surrogate information and diagnostic indices can both be considered as entirely viable components in a long-term information strategy, rather than being viewed as sufficing but suboptimal alternatives to the proper data sets that would ideally be substituted.

Exploration and analysis

As the flood of data mounts, the challenge of extracting useful information from it also multiplies. The response of the scientific and management communities has been to bend with the flow, using the data to feed conventional numerical analyses and models, but turning increasingly to the need to adopt a more creative stance. Since most models represent, and many analyses seek, a formally structured rendering of the environment (the channel and the catchment being two effective contexts), they are both effectively *a posteriori* approaches: the pattern or structure is given, and the model or analysis seeks to substantiate or calibrate it. Increasingly, however, it has been appreciated that information that is currently unknown often resides within existing databases. An *a priori* approach to extracting this information is often best rooted in exploratory pattern seeking, which requires a very different attitude and an equally different set of numerical tools.

The creative exploratory approach is well suited to holistic (integrative) information structures, and though 'data mining' is founded on database science, it is no surprise that its spatial counterpart has emerged within the GIS community. River channels and catchments perfectly represent the fusion of space, time and attribute in effectively describing events, whether univariate of multivariate. Openshaw (1994) has suggested that pattern-spotting exploratory approaches ideally require a tri-space mode, searching the data for informative patterns which combine time, place and attribute dimensions. If such approaches (let alone their associated technologies) are to take a place in catchment management planning, there are two barriers to overcome—the technical development of the necessary tools for operational use, and the attitudinal acceptance by the end-user that the patterns concerned though currently unknown do exist, are accessible to automated analysis, and are of sufficient strength and value to justify the effort. Both the challenge and the prize are significant:

> The necessary tools have to be clever enough to filter out the rubbish, to ignore data redundancy, overlook the insignificant, and yet powerful enough to identify possible pattern against a background of uncertainty and noise. A new breed of database analysers are needed that are able to seek out systematically patterns that are in some ways unusual amongst masses of spatial information, some of which is probably meaningful and much of which is potentially junk. (Openshaw, 1994)

COMMUNICATING INFORMATION

As the familar cliché points out, while in theory there may be no difference between theory and practice, in practice there is! Thus the pundits of information strategy have spent much of the last decade extolling the theoretical virtues of enterprise-wide corporate databases. In practice, this concept may place such overheads on an organisation, and induce such problems of access and of mismatch between required and provided data format or structure, that the more limited option of the 'data warehouse' may be preferred (Bachmann, 1994). A data warehouse holds information relating to a designated subject, so that while it may still be enterprise-wide, it greatly limits the likely problems of access and structure. Once again, we are reminded that data may validly reside in a database (this is, after all, the core function of a data archive), but *information* only exists if a process of communication permits it to inform someone or some application.

Critical to the task of ensuring that information is actively communicated is the difficult challenge of assessing the appropriate balance between understanding and decision support. It is characteristic of science specialists of whatever discipline that they believe that their own commitment to the pivotal role of understanding is shared by operational managers. It is not, and it is difficult to provide a convincing argument that it should be. In science, understanding is both a worthy end in its own right, and an effective guide to both cause and consequence. In such a context, understanding provides a basis upon which to guide action. The remit of an operational manager is rarely discipline-focused, and in practice is increasingly holistic. While understanding can offer a valuable element of generalisation to the manager's approach, it is a luxury that will often be available at a relatively superficial level only.

Competent management requires a clear and effective link between characteristics of the present situation, and alternatives or preferences for appropriate action. Provided that the definition of input and output are both clear, the link can be achieved without the necessity for a high level of understanding. The intuitive judgement of experienced operatives derives its power from the strength of this link between observed present characteristics and preferred response. Can information flow provide an effective surrogate for such experience? Can information-based technology outperform experience? In the right circumstances, the answer to both questions may be yes, but in order to render the circumstances sufficient it is necessary to engineer the flow of information so as to maximise expertise (intelligent systems) rather than merely to maximise understanding. In this context, unusually, we may have identified a genuine distinction between the theory and practitioner domains.

POWER FROM INFORMATION

Empowering the organisation

The opening supposition of this chapter was that, despite all of their inherent challenges, information processing and communication were potential sources of power, which could be employed profitably in the service of river management. To achieve such ends, however, requires considerably more than simply making available to the manager information of relevance and interest. Positive steps have to be taken to ensure that communication is followed by action, and that action is maximised in terms of its value.

In examining the ways in which information can be put to use in an organisation, it is helpful to concentrate on the IT systems which carry out these functions. In river management, these range from conventional databases and spreadsheets, through complex modelling to the integrative technology of GIS (Browne, 1995). Since we have already argued that GIS emphasises the forward-looking holistic use of information, this can serve as a useful focus for our consideration of the power that is generated from information. In many ways, the primary aim is to increase the range of contribution which GIS makes to the work of an agency. Every new role that is found 'recycles' existing data and derives further benefit from the GIS investment that has been made. If an analogy with staff development can be used, then we can suggest that it is important to raise the level of *'responsibility'* that is granted to a GIS or any other IT system. Staff with responsible roles offer greater benefit to an organisation than staff in menial roles: similarly, GIS with a responsible role has a greater opportunity for yielding advantage to a management agency than GIS in an exclusively routine setting. As GIS thus comes to assume responsibility for some tasks, agency *confidence* in GIS increases and it becomes easier to delegate tasks to the GIS.

Such an evolution in the system's role has to take place step by step. Neither the corporate psychology of the agency nor the logic of systems development would support the notion of moving instantly from GIS as a routine data integration tool to GIS in a core decision-making role. Taken sequentially, however, this progression can be safe, convincing and corporately acceptable. The key lies in empowering the information system to mature in its contribution to the organisation, while simultaneously helping the organisation to mature in its use of information systems. The history of the adoption of GIS by the National Rivers Authority (NRA) of England and Wales admirably demonstrates this evolution.

Empowering the system

In essence, the strategy that is being suggested lies in a progressive migration of GIS from an input role through one of support to full management functions. The strategy applies equally well in a decision context (decision input, decision support, decision making) and in an operational context (operational input, operational support, operational management)—both of which are core components in river management agencies. As an *input*, GIS handles data acquisition, integration, storage and access. It is a basis for query and for visualisation. But it plays no creative role: the user takes the input and employs it in the service of the organisation. An effective example would be the multitude of maps and data sets that are assembled to provide a basis for catchment management planning. The data integration process is invaluable, and the catchment manager uses the maps and selective database queries to build reports (operational) and to provide input to decisions.

At the *support* level, GIS plays a creative role to manipulate information into a form which has a direct contribution to make to an operational or decision process. But while the role is genuinely creative, it does not carry responsibility—the user always intervenes to make the decision or to trigger the operational action. A 'support' example would be the catchment- or channel-related indices which can be calculated to prioritise actions. These might concern the identification of flood-prone housing in relation to a purchase or mortgage enquiry, or the use of GIS to identify channel ecological characteristics of management relevance (Gurnell *et al.*, 1995). In such tasks, GIS can assemble a variety of information and synthesise it into a single diagnostic index to identify and map areas of particular value or vulnerability. This synthesis is creative: it creates a spatial zonation that exists only because of the GIS processes. But the user retains all of the responsibility, decides what response procedures to put into operation and where these should be prioritised.

It is still relatively rare for GIS as such to be given a fully responsible role, though spatial data handling is actually embedded in many fully automated systems—an aircraft autopilot, an automated river flood response system and the water intake management system of a regional hydro-power scheme are just three cases. At the management level, GIS integrates and manipulates data against a background of 'rules' (of which one form is an expert system) from which action is decided and in some cases actually triggered (by control engineering, or by phoning the necessary human operatives).

As an information system strategy, this potential progression from input through support to management represents an attractive route to value-add, but also poses a number of challenges. As the sophistication of the task increases, so too must the robustness of the application. Even more fundamental, the first task is actually to identify within any individual organisation just what migration routes are available and feasible, which itself might be a component in more fundamental corporate reengineering, as is discussed later. It can thus be seen that the provision of information to the manager through the intermediary of a GIS model has significant potential, but it would be mistaken to assume that this is the only context in which 'power' can be achieved from information.

The roots of power

King and Kraemer (1993) explore the role of information provision through models from a very different perspective. Since their focus is on model support for policy makers rather than for operational decision making, they see the model as offering a form of 'estimated truth' which in practice fuels what they term 'datawars'—"Put simply, the models were used because they were effective weapons in ideological, partisan, and bureaucratic warfare over fundamental issues of public policy" (p.354). The suggestion that the larger-scale and more comprehensive the model, the more likely it is to be used as a device in policy debate rather than operational use does not, however, reduce the importance of three other powerful influences of models.

First, models clarify the issues being debated, and help to make the assumptions, content and constraints explicit. Secondly, modelling imposes disciplined analysis and discussion, drawing strength and consistency from the fact that while recalibration can offer scenarios and sensitivity, blatant tinkering with the structure for policy ends would usually destabilise the model. Thirdly, models are often extremely effective indicators of what not to do and just how bad the results of a truly problematic policy might be. Placed in this perspective, information processed through a model can be seen to offer very different opportunities to those usually sought from an operational numerical simulation. While such models are unreliable adjuncts to major policy shift, they can provide yet another real and significant role for information at times of paradigm quiescence.

It is tempting to suggest that in approaching the problem from a quite different direction, river managers from a number of major agencies have recently converged on much the same set of conclusions as those reached by King and Kraemer (1993). As the range and depth of information to be incorporated in decision making increases (with environmental and economic attributes being added to the more familiar physical and chemical parameters), managers move subtly from only using calibrated numerical simulations such as hydrological or hydraulic models, to adding more policy-relevant models such as a catchment management GIS. In so doing, the managers shift their stance from using the (quantitative) model to predict the most likely outcomes, to using the more holistic catchment GIS model to select preferred options largely on the basis of the rejection of the least acceptable outcomes. This is a substantial and significant change of emphasis. The fact that managers may well have implemented this polarised use of models without consciously assessing the reasons for so doing, merely serves to indicate that there must be some inherent virtue in the approach if several distinct communities have evolved it independently. Other possibly related shifts in philosophy will emerge later in the chapter.

THE LIMITS OF INFORMATION

Whatever the conceptual and structural context in which information is used, a variety of limitations must be confronted in order to link data with action. These difficulties lie at the core of the information flow for channel management, and are so wide ranging that they defy easy summary. Instead, it may suffice to break the challenge into three components—problems of data acquisition, data handling and information application—and to consider just one example under each of these headings.

Deficiencies in data acquisition

The concept of 'data quality' encompasses a variety of attributes which influence the value of information, including accuracy (the extent to which the recorded attributes faithfully represent the variable of interest), precision (the resolution of the measurement process) and reliability (subsuming associated properties such as objectivity) (Clark, 1995). In practice, deficiencies in quality are usually depicted as error, which itself is a complex concept involving both precision (in measurement, recording and processing) and accuracy (in sampling, the data model, the data structure and processing). Error sources and propagation have been well covered in the literature, and are often associated with notions of uncertainty in data. A helpful typology of data uncertainty with clear implications for data error is provided by Geertman and Ruddijs (1994):

- Completeness uncertainty.
- Spatial (locational/topographical) error.
- Temporal (relative and absolute time) error.
- Attribute error (sampling, measurement accuracy and precision, discretisation, generalisation, classification, *a priori* attributes).
- Logical consistency and conceptual error (semantics, data model, data structure).
- Data lineage uncertainty (recording, transfer, archiving, processing, presentation).
- Meta-uncertainty (uncertainty about the extent to which other uncertainties and errors are known to the user).

Each of these categories has immediate application to the world of channel management, and the typology is of particular value in alerting river managers to the breadth of accreditation that they should demand of their data. On the other hand, it is important to accept error as unavoidable—indeed, as an integral part of the process of recording observations from the real world. Consequently, operational emphasis should be placed upon designing systems sufficiently robust and flexible to cope with error, rather than on efforts to eliminate error. Interestingly, although this conclusion has been derived independently through the study of error (see, for example, Maffini *et al.*, 1989), it converges strongly on the recommended approach to handling uncertainty (see below).

Data quality assessment is used by both operational and decision-making staff to indicate how much reliance can be placed in-house on the results of any analysis using the data, and also to provide a foundation for public accountability (Clark, 1993). In practice, metadata (information about the properties of a data set which may be appended to that data set) increasingly provide a standard first-order response to error propagation.

Problems in data handling

A technical review of the challenges of data handling is beyond the scope of this chapter, but the essential links between technical considerations and management implications are well made by Browne (1995) with respect to GIS. It is sufficient here to emphasise the extent to which apparently technical matters are, in fact, often closely associated with methodological and conceptual issues of great management significance.

For example, one of the best-recognised (though not best-solved) problems of spatial data handling lies in what Openshaw (1984) called the modifiable areal unit problem—rooted in the lack of spatial coincidence between the areal units used to collect or record different data sets. With such disparate frameworks of data recording, the critically important tasks of data aggregation or analysis of interrelationship between spatially non-contiguous variables are rendered highly problematic. After more than a decade of attack, this problem remains significant despite the development of a panoply of response techniques based on spatial reallocation or interpolation of data using increasingly 'intelligent' approaches (Flowerdew and Green, 1992).

River management displays its own particular linear version of the modifiable unit area problem in the form of the mismatch between the lengths and locations of the reaches used to record and analyse river data (Gurnell *et al.*, 1995). The outcome is not only technically disruptive, but also introduces elements of scale dependency which make it possible for misleading management perceptions to derive from the results of analyses which do not take account of the reach mismatch. The problem appears exclusively technical, the route towards solution is highly conceptual, and the outcome is entirely practical. Again, the worlds of the theorist and practitioner converge constructively.

Uncertainties in information application

There is a tendency to assume that information is the illumination that will ultimately banish the shadows of uncertainty in our thinking and action. Indeed, so strong is the belief in the ultimate supremacy of certainty that shortfalls in the attainment of this target are conventionally attributed to technical deficiencies (model simplicity, limited data availability, inadequate measurement resolution), and the solution is sought through equally technical devices (error margins, confidence limits, factors of safety). More recently, it has become apparent that uncertainty may be both a deep-rooted and inherent part of natural and management systems, and the emphasis has moved from attempting to remove uncertainty to endeavouring to live with it (Clark and Gardiner, 1994), a clear parallel to the earlier discussion of response to data error. With the focus thus turning towards such mechanisms as investment phasing and a prioritisation of 'soft' (maintenance) engineering, the role of information in the river management structure has begun to change. Rather than being largely an input to scientific or technical modelling for design purposes, it is increasingly seen as the key to scenario building, preferred option choice and impact assessment. In a world of acknowledged uncertainty, risk analysis and sensitivity analysis join the conventional modelling packages in the information-handling portfolio (HM Treasury, 1991). The shift is inclusive rather than exclusive, but does represent another significant reshaping of the fluvial information machine.

Not surprisingly, the uncertain definition of the river (channel or basin) system encourages a management response that is adaptive and flexible, since this permits retuning of the response as further information becomes available—altering, though not necessarily reducing, the scope of the uncertainty (Department of the Environment, 1991; Clark and Gardiner, 1994). This transformation of emphasis is well advanced in a number of river basin management agencies, including the National Rivers Authority (NRA) of England and Wales. Not only are 'soft' approaches increasingly frequently adopted, but there is a slow though critically important increase in the appreciation of the role of Post Project Appraisal (PPA) and retrospective review of monitoring results. It is the feedback from a properly constituted PPA system that will provide the project steering necessary to accomplish a fully adaptive form of management, and it is the new forms of integrated (holistic) information handling—including integrative analytical GIS—that provide a technology capable of offering the manager a genuine interaction with this information.

MANAGING THE UNMANAGEABLE

This paper was introduced with the suggestion that institutional river management processes required assessment against the background of an understanding of the role of information in corporate structures and processes. Since the time of Adam Smith and Henry Ford, Western economies have developed and honed the approach of achieving 'corporate productivity'—whether producing cars or managing rivers—by fragmenting the processes (tasks) involved, and by creating specialised structures (departments) to undertake them. It need be no surprise that this approach has been as successful in river management agencies as it has on the factory production line. There is both logic and organisational appeal in the division of river management into tightly defined functional areas (departments) such as water quality, flood defence, fisheries, navigation and conservation.

It should also be no surprise that as the weaknesses in this traditional business model begin to emerge in industrial settings, particularly in terms of the dramatically negative influence of process fragmentation on the time, energy and procedure necessary to weld the whole operation together, the same weaknesses are surfacing in the river management institutions. The moves to recombine activities within the powerful structures such as the Catchment Management Plan (or its coastal-zone equivalent, the Shoreline Management Plan), and to emphasise area-based arrangements rather than specialist roles in the organisation of management agencies, both reflect this realignment of approach.

Recasting the organisation

The most extreme (but some would argue, the most therapeutic) form of reorganisation is currently heralded by the rise of corporate reengineering (Hamner and Champy, 1993), which offers to service and management agencies as well as to industrial companies an opportunity to revolutionise the organisation of communal actions. Interestingly, advances in information handling lie at the heart of corporate reengineering, and may provide a valuable spark through which to reignite the use of information in river management and engineering. In brief, corporate reengineering can be seen as an attack on the traditional process of fragmenting complex tasks. Indeed, it moves away from task-based thinking altogether, and concentrates on reconceptualising both the targets of the agency and the processes through which these can best be achieved. For example, the exploration of sustainable management sets revolutionary new targets for river managers, while the important debate concerning 'hard' versus 'soft' engineering concentrates on the relative merits of achieving an agreed target. Information permits reengineering by helping to define targets and fuel their control systems, but most effectively by facilitating the integration of sub-concepts and sub-tasks. For the river manager, the lesson of corporate reengineering is that massive (not just marginal) increase in effectiveness and efficiency can probably be achieved by focusing on task-integrative and investment-rationalising processes such as holistic catchment planning, and on skill-integrative processes such as the building of case teams with an area management remit.

While it is both satisfying and invigorating to appreciate that river managers are participating in a global shift of corporate organisation (albeit without most of them realising it!), the picture would be seriously incomplete if we failed to return to the roots both of information handling and of social and economic management, of which river management is but one small component. We end, as we started, with notions so deep and expansive that they sit uncomfortably with the day-to-day or even decade-to-decade aims of the river manager.

Recasting the scientific underpinning

One of the most provoking of the challenges to scientific orthodoxy that have emerged in the last decade has been that of Saul (1992). This is a broad-brush demolition of 400 years of government and management which, it is argued, have fatally betrayed Voltaire's hopes of replacing a sinister élitist autocracy with a well-meaning and openly efficient democracy. Power has passed from the élite (of Church and State) to the organisation, where it resides in, and hides behind, the twin evils of structure and rationalism. Since we have seen that information-handling is fundamental to the new and more powerful corporate/agency structures, and that the numerical process model is an advanced reflection of scientific rationalisation, it is difficult to avoid assuming that Saul's scorn would continue to be directed towards some of our present views of river management, no matter how reengineered. Information is a fickle partner, yielding elucidation and entrapment in equal measure, and it is tempting to hark back to the 'datawars' of King and Kraemer (1993) in this respect.

Without abandoning the aspirations of the Age of Reason, with its commitment to science and information, Saul's (1992) critique suggests that it is healthy to face the possibility that a new order is being fashioned in Western societies, and on the basis of the above discussion we can propose that environmental managers are already (perhaps unwittingly) an inherent part of this process. He challenges us with a rejection of '... that bizarre method which consists of using massive quantities of information to create confusion which in turn creates ignorance and thus removes power from those who receive the information.' In this model, government and management are both organisations based on suffocatingly complex fragmentation of remit and responsibility, which neuters managers whose lives are in principle dominated by consultation and coordination rather than by action. To counter this approach, both science and management need to inform rather than to confuse, to empower rather than to overwhelm, and to enfranchise rather than to exclude. The successes of information science and of operational development of information-based management techniques for the river environment over the past decade suggests that this is possible, though not easy. Nevertheless, in linking the new fluvial science to the new river management, these are the challenges that we must confront and overcome.

VISION AND REVISION: A ROLE FOR INFORMATION?

Notwithstanding any residual doubts about the effectiveness of current data acquisition, handling and dissemination systems, it appears that information remains rooted at the heart of scientific efforts to understand rivers, and practical efforts to manage them. Nevertheless, the nature of both the management process and the information processes which support it is undergoing significant change. Information provides us with a sharper and more complete vision of the systems which we wish to understand and manage, but society as a whole is transforming in terms of its expectations of science and its demands on managers. The flow of information continues, but its destination and exploitation are subject to great change.

This paper has argued that river management is experiencing a spectral shift which is apparent in at least three contexts, separately recognised but closely related in practice and in prospect. First, traditional formal data analyses are increasingly being supplemented with more creative and often less formal exploratory and pattern-spotting techniques. Secondly, conventional reductionist calibrated simulations are increasingly being complemented with holistic policy-relevant models, which differ greatly in both form and purpose. Thirdly, the current orthodoxy of river management organised within specific functional sectors is being challenged by a return to coherent task-oriented (generalist) approaches typified by catchment management planning. These three trends reflect underlying propensities within science and society, yet are outcomes of these shifts only in the most indirect sense. River managers are the product of their time, but are rarely conscious either of their paradigm or its consequences. The quest for sustainable river management eases us towards adaptive management which senses, and is sensitive to, the many signals emanating from the diverse components of the river environment. An acceptance that uncertainty is a freedom rather than a constraint urges us towards a more flexible and proactive catchment management than has been commonplace. And lying both within and between these imperatives is the flow of information which, when properly (or at least, adequately) managed, both nourishes our science and guides our management.

REFERENCES

Bachmann, C. (1994) Data warehouses allow information access. *GIS World* 7(4), 58.

Browne, T. (1995) The role of GIS in hydrology. In Foster, I.Dl., Gurnell, A.M. and Webb, B. (eds), *Sediment and Water Quality in River Catchments*, Wiley, Chichester (in press).

Clark, M.J. (1992) Data constraints on GIS application development for water resource management. In: Kovar, K. and Nachtnebel, H.P. (eds), *Application of Geographic Information Systems in Hydrology and Water Resources Management*. IAHS Publication No. 211, Vienna, 451–463.

Clark, M.J. (1995) Spatial data for long-term ecological modelling. *Proceedings, UNESCO/ONR Workshop on Long-term Ecological Monitoring, Galway, Ireland, 1994* (in press).

Clark, M.J. and Gardiner, J.L. (1994) Strategies for handling uncertainty in integrated river basin planning. In: Kirby, C. and White, W.R. (eds), *Integrated River Basin Development*, Wiley, Chichester, 437–445.

Department of the Environment (1991) *Policy Appraisal and the Environment*, London, HMSO, Chapter 2.

Flowerdew, R. and Green, M. (1992) Developments in areal interpolation methods and GIS. *Annals of Regional Science* 26, 67–78.

Geertman, S. and Ruddijs, S. (1994) Geographical sensitivity analysis: some procedures for generating meta-information. *EGIS/MARI '94: Proceedings of the Fifth European Conference on GIS*, vol. 1, 151–160.

Gurnell, A.M., Angold, P. and Gregory, K.J. (1995) Classification of river corridors: issues to be addressed in achieving an operational methodology. *Aquatic Conservation* 4, 219–231.

Hamner, M. and Champy, J. (1993) *Reengineering the Corporation*. Harper Business, New York, 223pp.

HM Treasury (1991) *Economic Appraisal in Central Government: a Technical Guide for Government Departments*. London, HMSO, pp. 63–64.

King, J.L. and Kraemer, K.L. (1993) Models, facts and policy process: the political ecology of estimated truth. In: Goodchild, M.F., Parks, B.O. and Steyaert, L.T. (eds), *Environmental Modeling with GIS*, Oxford University Press, Oxford, 353–360.

Maffini, G., Arno, M. and Bitterlich, W. (1989) Observations and comments on the generation and treatment of error in digital GIS data. In: Goodchild, M., and Gopal, S. (eds), *The Accuracy of Spatial Databases*, Taylor and Francis, London, 55–69.

Nieuwenhuysen, P. (1989) *Scientific and Technical Water-related Documentary and Information Systems*. UNESCO, Paris, 52pp.

Openshaw, S. (1984) *The Modifiable Areal Unit Problem*, Concepts and Techniques in Modern Geography 38, Geo Books, Norwich.

Openshaw, S. (1994) Exploratory space-time-attribute pattern analysers. In: Fotheringham, S. and Rogerson, P. (eds), *Spatial Analysis and GIS*, Taylor and Francis, London, 83–104.

Saul, J.R. (1992) *Voltaire's Bastards: the Dictatorship of Reason in the West*. Penguin Books, 640pp.

13 Investigating Change in Fluvial Systems Using Remotely Sensed Data

E.J. MILTON[1], D.J. GILVEAR[2] and I.D. HOOPER[3]

[1]*Department of Geography, University of Southampton, UK;* [2]*Department of Environmental Science, University of Stirling, UK; and* [3]*School of Applied Sciences, University of Wolverhampton, UK*

INTRODUCTION

Many of the world's major rivers, such as the Amazon, Yukon and Ob, have their headwater tributaries in remote and inaccessible areas, whilst in their lower reaches the size and multitude of channels create problems for field study. Smaller rivers in less inhospitable environments can also be difficult to study if the aim is to collect information in relation to a flood event, or from the whole of the catchment. Consequently, most information on fluvial systems is limited to a relatively small number of local studies undertaken over the last 30 or 40 years, primarily on a short stretch of river close to a centre of population. Remote sensing offers the possibility of observing the whole Earth at scales compatible with the fluvial systems of interest and provides a means of monitoring changing river channels by repetitive measurements from spaceborne and airborne systems.

Traditionally, research scientists have relied upon field investigations to obtain data on channel and floodplain characteristics, though these have been supplemented with data from aerial photographs since the 1940s and multispectral imagery since the 1960s. Field techniques vary among disciplines, but a common feature is that they are localised in extent. Geomorphologists have traditionally gained information on channel morphology using topographic survey at a number of cross-sections along the entire length of a river. Similarly, vegetation ecologists have typified floodplain vegetation using quadrat data for points along a transect from the valley-side to the river. When an indication of change is required repeat surveys are undertaken at the same localities, or devices are constructed that monitor change; for example erosion pins in the case of river bank erosion, either read manually at infrequent time intervals or logged automatically. Usually these studies are carried out for periods of less than five years. Although data prior to the commencement of the study would be advantageous, field data rarely exist for the locality in question.

In this chapter we focus on the application of remote sensing to the study of river channel change, but this is only one of a wide range of potential applications of the technique within hydrology as may be seen by the reviews by Salomonson *et al.* (1983), Hockey *et al.* (1990), Engman and Gurney (1991), and Muller *et al.* (1993). Remote sensing does not seek to replace the traditional field-based investigations

Changing River Channels. Edited by Angela Gurnell and Geoffrey Petts.

referred to above, but to complement them by allowing the results and interpretations of such studies to be extended across wider areas, from the cross-section to the reach and thence to the channel segment and the catchment scale. Furthermore, seeing the problem from a different perspective (literally) can provide new insights and suggest new hypotheses which can then be tested in the field.

PHYSICAL BASIS

The application of remote sensing to the study of river channel change is based on a number of assumptions. The first of these is that the changes of interest in river channels elicit measurable changes in the spatial or temporal pattern of electromagnetic radiation (emr). The amount of reflected or emitted emr at different wavelengths may be changed, or the way in which the surface scatters light or microwave energy may be changed so that a surface becomes more diffuse or more specular. Such changes may occur across space as when mapping the area of flood inundation, or through time as when monitoring vegetation growth in stable channels, or, most commonly may involve changes across space and through time. Fluvial systems are dynamic entities defined by the spatial pattern of hydrogeomorphological components through time; both dimensions are necessary to understand the complete picture.

The second assumption is that the remote sensing system used has a known level of radiometric precision, even if it has not been calibrated in an absolute sense. Measurements from remote sensing systems have errors associated with them just like those made at ground level, so for a measured change in digital number (dn) to be significant it must exceed the uncertainty level of the sensor. The precise wavelengths sensed can also be important, and when comparing data from different sensors it is important to bear this in mind, especially when studying features which show large changes in spectral response around the waveband in question. Modern airborne sensors are frequently returned to the laboratory to check their calibration; satellite sensors may have on-board calibration devices or may be calibrated with respect to areas of the Earth's surface thought to have near-constant reflectance.

Thirdly, the geometric distortions introduced by the sensor and platform are assumed to be known, so that real displacements of image features may be discriminated from those due to defects in the sensor, artefacts of the processing or the motion of the platform. Some of these distortions can be very subtle and can mimic the spatial patterns of features of interest; for example Boxall *et al.* (1993) described an image in which linear features were transformed into a sinusoidal 'meandering' pattern due to an electronic problem affecting an airborne imaging spectrometer. Similar effects are also commonly associated with data collected from airborne platforms if they are not roll-corrected. Furthermore, change detection and mapping generally assumes that geometric distortions may be corrected such that successive remotely sensed data sets may be superimposed and compared with related spatially-referenced data such as digitised topographic maps.

Townshend and Justice (1988) have reviewed those properties of remote sensing systems which control the accuracy of change detection procedures in the context of land cover transformations at regional and continental scales and have summarised

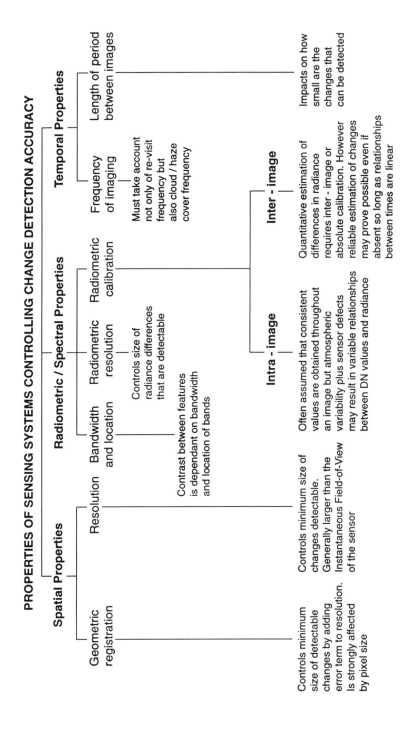

Figure 13.1 The main properties of remote sensing systems controlling the accuracy of change detection and analysis procedures. (After Townshend and Justice, 1988, reproduced by permission of Taylor & Francis Ltd and the authors)

the main factors in a diagram which is reproduced here (Figure 13.1). They emphasise the importance of matching the spectral, radiometric and spatial capabilities of the sensor with the properties of the surfaces being sensed. In addition, the timing and frequency of sensing must be considered, especially as affected by unpredictable barriers to sensing, such as cloud cover in the optical region.

Spectral properties of the fluvial environment

Figure 13.2 shows some representative spectral response curves from a number of surfaces found in the fluvial environment. These were measured by Hooper (1992) using a Daedalus AADS 1268 airborne multispectral scanner known as the ATM (Airborne Thematic Mapper), the spectral bands of which are shown in Table 13.1. Although in this example the individual bands have not been calibrated to radiance units it is apparent from the raw dn values that the spectral responses of these surfaces fall into three distinct classes: (i) water, shadow and aquatic vegetation, (ii) trees and herbs (green vegetation) and (iii) bare substrate (exposed bars and bare soil). These three classes may be thought of as the 'spectral end-members' of the fluvial environment. They may be represented in a co-spectral plot (Figure 13.3) based on two bands which show good discrimination between the spectral end-members whilst also being relatively insensitive to other factors such as absorption and scattering within the atmosphere. The first assumption above implies that a change in a hydrogeomorphological feature of interest causes the group of pixels delineating that

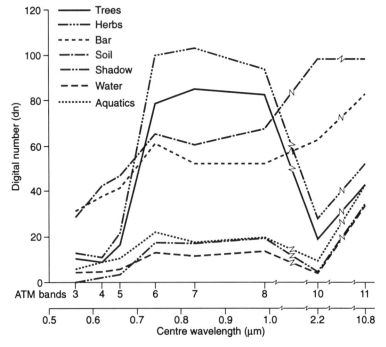

Figure 13.2 Representative spectral response curves from the River Teme (uncalibrated dn values). (Modified from Hooper, 1992)

Table 13.1 Daedalus ATM bands

1	0.42–0.45 μm	7	0.76–0.90 μm
2	0.45–0.52 μm	8	0.91–1.05 μm
3	0.52–0.60 μm	9	1.55–1.75 μm
4	0.605–0.625 μm	10	2.08–2.35 μm
5	0.63–0.69 μm	11	8.50–13.00 μm
6	0.695–0.75 μm		

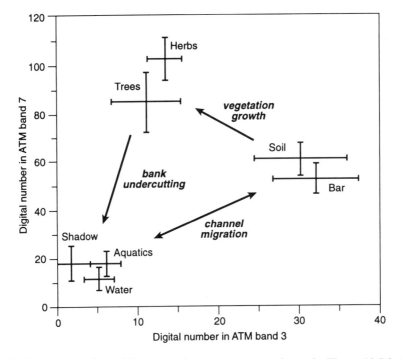

Figure 13.3 Co-spectral plots of the spectral response curves shown in Figure 13.2 indicating the lines of spectral mixing between pairs of spectral end-members. (Modified from Hooper, 1992)

feature to migrate between two or more of these spectral end-members. In a theoretical sense this is true: the spectral changes resulting from the change from water to a gravel bar as a channel migrates, or from green vegetation to water as a bank is undercut, are shown by the arrows on Figure 13.3; however, we must also take into account the spatial properties of the fluvial environment.

Spatial properties of the fluvial environment

Whereas the spectral dimension of the fluvial environment is well-suited to remote sensing, the spatial dimension is much less suitable. In part this is due to the technical limitations imposed by the scale of features in relation to the spatial resolution of the sensor, but it also has to do with the nature of classification. Some features, for example a river cliff, may be mapped unambiguously in the field and identified with similar precision in the spectral domain. The same is not true of a feature such as an in-channel bar which would be faithfully recorded by a remote sensing system as a fuzzy transition between bare substrate and water, no matter how fine the spatial resolution of the sensor. In the field a surveyor may choose to define the limit of the bar on hydrological or geomorphological criteria not represented by a spectral end-member and therefore inaccessible to remote sensing in the spectral domain.

The practical consequences of scale may be represented in terms of the size of the Instantaneous Working Area (IWA) (Clark, 1990) and the size of the effective Ground Resolution Element (GRE) (Wilson, 1988) in relation to the features of interest. The size of the IWA imposes a limitation which varies in its significance depending on the method of analysis chosen and the technical resources available. The significance of the GRE is more fundamental. Traditional methods of image classification were developed for large, uniform areas such as agricultural fields, and they allocate pixels to the 'most likely' or dominant class. They do not perform well when many of the GREs contain mixtures of scene components; a different 'scene model' is required (Strahler et al., 1986). In many fluvial environments, mixed pixels are the norm, whatever the size of the GRE. Furthermore, often the features of interest are themselves mixtures of spectral end-members. It is therefore necessary to develop methods to address this spatial complexity.

Several methods have been developed to deal with the complexities of unresolved scene components. The simplest method is to treat particular mixtures of end-members as valid classes in their own right, either by deliberately selecting training sites from representative areas of each or by introducing 'seed pixels' to force clusters to accrete around areas of the feature space known to be populated by mixed pixels (Christensen et al., 1988). The second method involves the use of a mixture model to estimate the proportion of each of the spectral end-members present within a single GRE. This approach has only recently been applied to the fluvial environment (Barton and Bathols, 1989; Mertes et al., 1993) but has considerable potential, particularly when applied to data from imaging spectrometers which record data in many spectral bands. Finally, the spatial context of GREs may be used to provide further information on their likely composition and significance. For example, a mixed pixel from an in-channel feature comprising unresolved elements of water, substrate and vegetation might have an identical spectral response to another class, such as riparian vegetation. However, by introducing an additional dimension into the classification process which represents the linear distance from the channel axis these two pixels may be separated and allocated to the 'most probable' class. In addition to being used to improve the accuracy of classification, contextual information may also be used to improve the definition of the fluvial features themselves. Mouchot (1991) demonstrated how a simple topological model could be applied to single-band Landsat TM data in order to improve the depiction of small channels in the Mackenzie delta and how the assessment of channel connectivity could then be used to study the flow of nutrients and sediment through the system.

Temporal properties of the fluvial environment

Change occurs over a very wide range of time-scales in the fluvial environment, and remote sensing has the potential to be applied to virtually all of them, from the observation and measurement of sediment pulses in a proglacial environment to the evolution of river systems over thousands of years (e.g. Jacobberger, 1988). Practical difficulties due to cloud cover, orbital limitations and unfavourable viewing geometry limit the operational use of spaceborne systems sensing in optical wavelengths for studying the fluvial environment (Hockey *et al.*, 1990), but results from the synthetic aperture radar on ERS-1 have been most encouraging (Blyth and Biggin, 1993; Rudant, 1994; Wagner, 1994). Airborne systems are inherently more flexible in timing and advanced sensors such as the Itres Instruments Compact Airborne Spectrographic Imager (CASI) (Babey and Anger, 1993) are able to produce high quality data when flown under cloudy skies.

ANALYTICAL METHODS

In addition to ensuring that the spectral, radiometric, spatial and temporal properties of the sensing system are compatible with the changes expected in the fluvial system, it is necessary to consider the influence of any image processing procedures applied to the data. In the present context the most important of these procedures is that of image registration. Change detection and mapping using remote sensing requires the spatial registration of multitemporal (or multi-date) data sets and this process is usually achieved using a least squares transformation of one image to another, or of one image to a map (see also Chapter 14). Three steps are involved, as described below.

Selection of Ground Control Points (GCPs)

The best GCPs are often artificial high contrast objects such as road junctions, runway intersections and fence intersections. Stream confluences can be used but these are clearly inappropriate in a study of river channel change.

Mathematical transformation of the image

Satellite platforms are highly stable and often a first-order transformation based on relatively few GCPs is sufficient to produce a satisfactory transformation. Airborne platforms are much less stable but techniques to geometrically correct photographic images from nadir and oblique-pointing cameras are well-developed (Li *et al.*, 1993). Airborne multispectral scanners pose far greater problems. Successful rectification of airborne scanner data may require the survey area to be split into many smaller sections (Christensen *et al.*, 1988) or the transformation to be localised (Devereux *et al.*, 1989), or use of a parametric correction procedure based on frequent measurements of aircraft attitude (Wilson, 1994), and may benefit from the incorporation of a digital elevation model (Cosandier *et al.*, 1994).

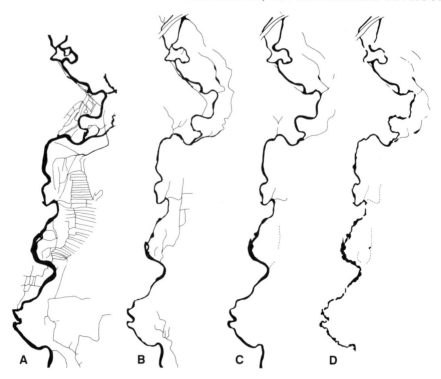

Figure 13.4 Visual interpretation of the channel part of the River Avon, Hampshire, from images subjected to different resampling regimes: (A) channel pattern shown on a 1:10 560 scale map of the area; (B) channel pattern interpreted from a simulated SPOT panchromatic image (10 metre resolution); the other maps show the channel pattern interpreted from simulated SPOT XS image (20 metre resolution) produced by bilinear interpolation (C), and nearest neighbour interpolation (D). (Interpretation by A.M. Gurnell and M.J. Clark)

Resampling to a new rectangular grid

The user is often faced with the choice between nearest-neighbour, bilinear, or cubic convolution resampling algorithms with perhaps the advice that 'cubic convolution is best but slow'. This important step deserves more careful consideration in the present context since there is a very real trade-off between radiometric fidelity and the preservation of edges in the image depending upon the algorithm used. Figure 13.4 shows the results of a manual photointerpretation of the river channel pattern in part of the River Avon in Hampshire based on airborne multispectral scanner data processed so as to simulate the SPOT XS and Pan sensors. Comparison with the 1:10 560 scale Ordnance Survey map (Figure 13.4A) shows the extent to which the small channels of an old water meadow system were not resolved either on the 10 metre resolution panchromatic image or the 20 metre resolution false-colour image, and also shows differences in the width and continuity of the main channel. Mismatch between the survey date of the map and the remotely sensed data means that the omission of small channels may not simply be due to the poorer spatial resolution of the imagery; the channels could now be dry or could be choked with vegetation and hence show little contrast with the surrounding land cover of this area. Similarly, observed differences in the width of the main channel could reflect real differences in river stage between the date of the overflight and that chosen for the mapped channel margin. However, the results also reveal significant differences in the channel pattern simply as a result of the resampling algorithm used. The blurring effect of bilinear interpolation (Figure 13.4C) caused the main channel width to be exaggerated compared with that shown on the 10 metre panchromatic image, but preserved its continuity. In contrast, the image produced by nearest neighbour interpolation (Figure 13.4D) shows a discontinuous channel pattern with a width similar to that shown on the panchromatic image. Nearest neighbour interpolation is often recommended for images which are to be subjected to automated classification or quantitative analysis because it is the only interpolation method to leave the dn values unchanged. The evidence of this example is that, for visual interpretation of river channel patterns, the choice of interpolant should depend upon whether the aim is simply to detect the presence of a channel or to measure its size and shape.

In this example, a human interpreter was able to overlook the fact that most of the channel pixels in the bilinear resampled image were mixtures of water and surroundings, and instead, concentrate upon their connectivity, thus readily defining the channel network. There is sometimes a tendency to think of the visual interpretation of remotely sensed images as being a stage in the evolution of the subject on the way to fully automated classification in which the human interpreter is redundant. This is not so; they are complementary methodologies and the assumptions made and opportunities offered by each should be clearly understood (Duggin and Robinove, 1990). For many applications in which textural and contextual information are important, and particularly those in which the spatial entities are small and arranged in complex patterns, interpretation by a skilled person is preferable to automated classification. In these cases the role of digital image processing is to provide an image whose geometry and information content has been optimised for the human eye-brain system. This is achieved by digital image processing techniques including contrast stretching, spatial frequency filtering, colour space rotation, and data compression.

Summary

It should be clear from the preceding discussion that there is no universally applicable remote sensing system for monitoring and mapping change in the fluvial environment. Effective use of remote sensing requires the use of a sensing system whose spectral, radiometric, spatial and temporal properties match those of the fluvial system of interest. This will now be demonstrated by a review of studies which have applied remote sensing to channel geomorphology, flood inundation mapping, floodplain geomorphology, and to vegetation in the channel and riparian zones.

CHANNEL GEOMORPHOLOGY

Variability and changes in channel planform

A knowledge of channel planform and changes in channel position through time is critical to many geomorphological and river management problems. Such change in

Table 13.2 Examples of studies using airborne and satellite data to study river planform and channel changes

Location	Sensor	Author(s)
F. Feshie, Scotland	B/W air photos	Ferguson and Werritty (1983)
Red Creek, Wyoming	B/W air photos	Schumann (1989)
R. Tay, Scotland	B/W air photos	Gilvear and Winterbottom (1992)
R. Ashley, New Zealand	B/W air photos	Warburton et al. (1993)
R. Ucayali and Amazon, Peru	Landsat MSS	Salo et al. (1986)
R. Brahmaputra, Bangladesh	Landsat MSS and SPOT	Thorne et al. (1993)
R. Ganga, Bangladesh	Landsat MSS and SPOTT	Thorne et al. (1993)
R. Yamuna	Landsat TM, Space Shuttle SIR	Ramasamy et al. (1991)
R. Niger	Landsat MSS	Brivio et al. (1988)
R. Niger	Landsat TM	Diakite et al. (1986)
R. Rapti, India	Landsat MSS and TM	Nagarajan et al. (1993)

channel planform and shifts in channel position have been identified by field-based measurements and ground-surveys. In a number of studies, field data have been complemented by air photo interpretation but sometimes only in a qualitative manner without rectification and digitisation of photographs (Table 13.2). For example, Ferguson and Werritty (1983) used aerial photographs, in conjunction with field studies, to map changes in the planform of a wandering gravel-bed river in Scotland over a 35 year period by direct visual comparison. Similarly, Schumann (1989) identified the 'parent' channel and the relative importance of anabranches in an anastomosing reach of Red Creek, Wyoming, using aerial photographs. The parent channel was darkest due to grasses and sagebrush flanking the channel where moisture availability was highest. Small anabranches were often lined with grass but did not have sagebrush growing on their banks and were lighter in tone. In contrast, Lewin and Weir (1977) determined the evolution of a meander loop on the River Rheidol by contouring the floodplain using photogrammetric techniques and aerial photographs for various dates.

More recently, satellite data have been used to enhance our knowledge of channel geomorphology, primarily the channel planform of large river systems (Table 13.2), for which the synoptic view of a spaceborne sensor has advantages over aerial photography. Thus, Salo et al. (1986) used multi-date Landsat MSS images of the meandering and anastomosing stretches of the Ucayali and Amazon in Peru, to quantify lateral migration rates of 2000 metres per year between 1979 and 1983. Landsat MSS and TM data of the Ganga River, a tributary of the Brahmaputra, Bangladesh, were also used by Philip et al. (1989). They applied linear contrast stretching and band ratioing to date channel changes between 1975 and 1986 and to reconstruct palaeochannel patterns. A more rudimentary approach based on the visual interpretation of photographic images was used by Ramasamy et al. (1991) to identify former courses of the Yamuna River, Western India. More remarkably, Jacobberger (1988) mapped abandoned river channels that were active in Sahelian Mali 6000 to 8000 years ago using MSS and TM images.

In most cases the relatively coarse spatial resolution of satellite sensors limits their application to the study of channel planform and stream migration in small stream systems. In such cases, airborne platforms are required which can map channel morphological features of a few metres' width. Few examples exist of the application of multispectral imagery to the study of stream morphology although channel networks have been quantified. As part of a geomorphological study of channel change on the River Tay, Scotland, following major floods in 1990 and 1993 (Gilvear and Winterbottom, 1992; Gilvear et al., 1995a), Daedalus Airborne Thematic Mapper (ATM) data are being used to map the characteristics of a 12 kilometre reach of the River Tay and its floodplain. Comparison of imagery flown in 1992 and 1994 with a nominal spatial resolution of 2 metres is allowing changes in channel morphology to be quantified. Soil moisture variations and vegetation types on the floodplain are also revealing information about channel change over longer time-scales (see below).

Change in channel bed morphology

Recently remotely sensed data have also been applied to bathymetric mapping and the identification of instream habitats in non-turbid rivers. Under such conditions the amount of water-leaving radiance depends upon the amount of energy absorbed by the water and the reflectance of the substrate. The absorption properties of clear water are well known and vary systematically through the visible and near infra-red region. The technique has been widely used in coastal environments but rarely applied to rivers.

Recent work on the Green River, Utah, by Hardy et al. (1994) shows how multispectral video imaging may be used for this task, but even panchromatic aerial photographs may be used to provide a quantitative estimate of water depth and instream features under certain conditions. This may be shown from an application of the technique in a study of the effect of gold placer mining on Faith Creek, a small headwater stream in the Circle Mining District of Alaska (Gilvear et al., 1995). In this case, the remotely sensed data consisted of scanned panchromatic aerial photographs of Faith Creek of different dates. The scanned images were contrast enhanced after applying a logarithmic transformation to the dn values to account for the exponential decline in the levels of light reaching the channel bed with depth (Lyzenga, 1981). Comparison of the 'image processed' dn values with known water depths for the reach demonstrated that the deep water areas had highest dn values and the shallowest areas the lowest dn values, thus spatial variability in the pixel values indicated high and lows in the channel bed topography. The number of pixels in each of 15 water depth categories (0–75 cm) as predicted from the dn values was analysed for all aerial photographs available for prior (1966), during (1989) and after the cessation of mining (1993) (Figure 13.5). Statistical analyses of the dn values demonstrated that prior to mining most water depth classes were well represented indicating a diverse bed morphology and a range of water depths. The images depicted a range of medial and lateral bars and pool–riffle sequences (Figure 13.5A). During mining, when the creek was confined to an artificial channel an absence of deep water areas was apparent (Figure 13.5B). Four years after mining, the pixel value distribution indicated a few deep water areas but shallow areas still dominated (Figure 13.5C).

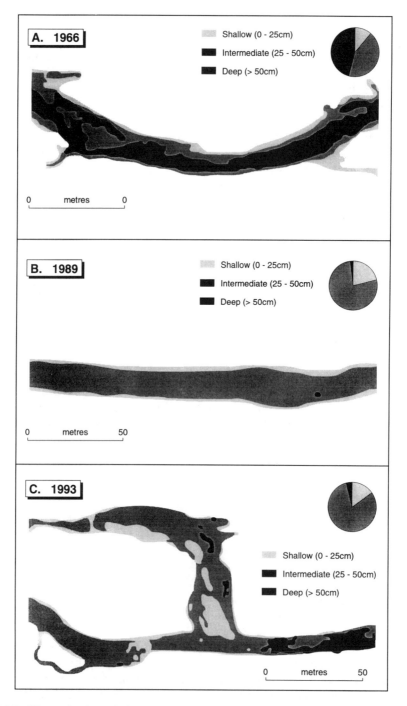

Figure 13.5 Water depth variations and channel changes along Faith Creek, Alaska, processed from aerial photography taken prior to, during, and after placer mining

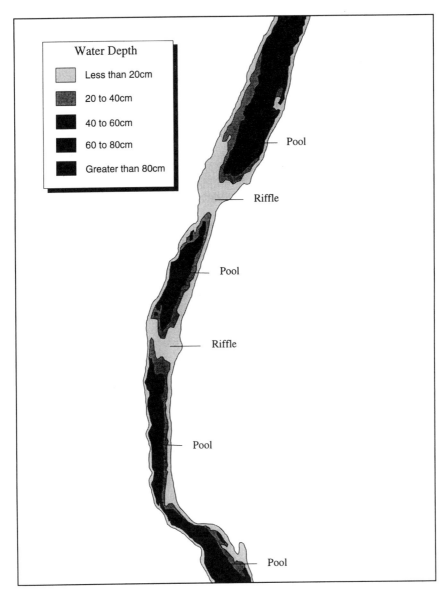

Figure 13.6 Bed morphology of the River Tummel, Scotland, mapped using ATM band 4 (0.605–0.625 μm). (Provided by S. Winterbottom)

ATM data collected in 1992 and 1994 for the catchment of the River Tay, Scotland, are also being used by S.J. Winterbottom and D.J. Gilvear (in preparation) for bathymetric mapping, based upon an observed strong negative correlation between measured water depth and radiance in the interval 0.605 to 0.625 μm. Figure 13.6 shows a bathymetric map produced from this imagery (nominal spatial resolution 2 metres) which shows a pool–riffle sequence and the location of various other bedform features over a 10 kilometre reach.

FLOOD INUNDATION MAPPING

The application of remote sensing to flood inundation has been extensively reviewed
(e.g. Salomonson et al., 1983; Barton and Bathols, 1989; Engman and Gurney, 1991)
and this will not be repeated here. One of the prime advantages of remote sensing is
its synoptic perspective, that is, it captures an image of the Earth's surface at one
instant in time. This ability to 'freeze the action' gives unique data sets, such as a
single satellite image showing the spatial consequences of the passage of an extended,
slow-moving flood, including the area inundated, the flood wave itself and regions of
localised erosion and deposition where the flood is past its peak flow (Green et al.,
1983). The extent of inundation due to flooding can easily be seen on aerial photographs taken at the time of flooding. However, rarely is photography available for
the area of concern during the height of the flood, particularly on small river systems, and it is usually the maximum extent of flooding that is of interest. Lewin and
Manton (1975) were fortunate to be on a hillside overlooking a stream during a
flood and mapped the sequence of floodplain inundation and emptying using oblique
black and white photographs. Gilvear and Davies (unpublished) conversely had to
reconstruct the maximum extent of inundation using colour aerial photography
taken 10 days after a 1:100 year flood event on the River Tay, Scotland, by the
location of strand lines (Plate 13.1). The aerial photography was also used to map
locations where flood embankment overtopping had led to failure and embankment
breaches (Gilvear et al., 1994).

Blasco et al. (1992) used SPOT XS scenes to map floods in Bangladesh and
showed that within one SPOT scene near Dhaka, the area of open water increased
from 324 km^2 to 1303 km^2 during destructive floods. Sippel et al. (1994) also used
Scanning Multichannel Microwave Radiometer data (SMMR) to map seasonal
changes in the extent of inundation on the Amazon river floodplain using a linear
mixing model to overcome the sensor's low spatial resolution. Inundation mapping
under riparian woodland and forested floodplains can be problematic. However,
Ormsby et al. (1985a) found that the L-band data from the Shuttle Imaging Radar
(SIR-A) was helpful in separating forest vegetation from shorter partially submerged grasses and shrubs and permitted a good definition of the land–water
boundary even below a forest canopy. The usefulness of L-band information in
providing a better delineation of water channels and the extent of water, even when
obscured by floating or emergent vegetation, was confirmed in Chesapeake Bay
(Ormsby et al., 1985b).

FLOODPLAIN GEOMORPHOLOGY

Floodplains are complex morphological features that relate to overbank deposition
and shifts in channel location. Many landforms, including oxbow lakes, levees, and
scroll bars can often be seen from above, identification being facilitated by variations
in vegetation and soil moisture (see also Chapter 11). Given easy identification of
morphological features, visual interpretation of floodplain morphological features
using aerial photography has been widely undertaken. Photogrammetric techniques
have also been applied. For example, Lewin and Manton (1975) used 1:5000 scale

photography of three Welsh rivers to map the floodplain morphology at a vertical resolution of 0.10 metres and horizontal resolution of 0.30 metres.

Multispectral images also show clear evidence of floodplain morphological features due to moisture differences and vegetational patterns. Plate 13.2 shows the results of a supervised minimum distance classification of ATM data from part of the River Tummel, Scotland. The correspondence between the habitat types determined by remote sensing (Plate 13.2A) and the old river channels mapped by conventional methods (Plate 13.2B) is clearly evident, and has been confirmed by field survey. A number of recent studies show that satellite imagery also allows improved mapping of floodplain geomorphology. In the Garonne river valley, France, Muller (1992) evaluated a time series of TM images for the definition of a typology of the floodplain. TM band 5 was found to be best for discriminating the floodplain from the adjacent terraces and for discriminating several alluvial deposits. Band 5 also allowed a new morphological unit to be identified, adjacent to the lowest terrace, characterised by higher clay and moisture content. Similarly Philip *et al.* (1989) used Landsat TM band 4 to identify old channels and the ratio of TM bands 2 and 3 to map meander scrolls. The newly emerging view of the patchy fluvially-induced forest structure of the Amazon floodplain has also only become evident since the use of Landsat MSS data (Salo *et al.*, 1986).

Soil moisture variations on alluvial plains are primarily a function of soil particle size, organic content and floodplain elevation. As a result soil moisture can be used as a good surrogate variable for mapping floodplain morphological and sedimentological units. Using the thermal infra-red band of ATM data for the River Tay floodplain, Scotland, a relationship between band 11 and soil moisture content determined from field samples with a correlation coefficient of 0.83 was obtained. These data were subsequently used to map soil moisture variability within three fields on the River Tay (Davidson and Watson, 1995, see Figure 13.7). Soil moisture for field 1 displayed an approximate wave-like pattern with a periodicity of 30–40 pixels (Figure 13.7a) and was related to a braided palaeochannel that cuts across the field. The depressions of former channels were distinguished by high soil moisture content. In field 2 the areas low in soil moisture identified from the imagery (Figure 13.7b) related to a former location of the main channel of the River Tay, the channel being depicted on a 1783 map of the area (Gilvear and Winterbottom, 1992). The pattern in field 3 was similar to that in field 1 (Figure 13.7c). Geostatistics were also applied to the digital data to examine spatial dependency and the size of objects in the three scenes. Fields 1 and 3 had a 'range' of approximately 108 metres whereas for field 2 the value was 177 metres. It was therefore concluded that the palaeochannel patterns in fields 1 and 3 have a different history from field 2. Canonical regression was also applied to the digital data and used to map the sand, silt and clay content of soils on the floodplain (Watson *et al.*, 1993), ground data being gained from the same samples as used for soil moisture determination.

Bands 10 and 11 of ATM data have also been used to map the depth to the water table at 2 metre resolution within the Insh Marshes (Gilvear and Watson, 1995), a floodplain mire on the River Spey, Scotland. Sixty water table depth measures in open auger holes were taken simultaneously with the airborne remote sensing campaign to produce a regression model and to allow a classified image depicting water table depth to be produced (Plate 13.3). The image allowed inferences to be made

a) FIELD 1

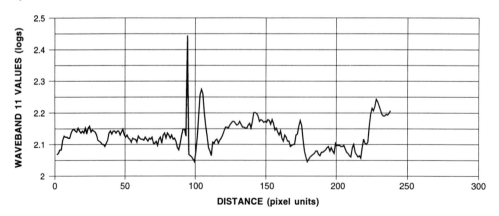

b) FIELD 2

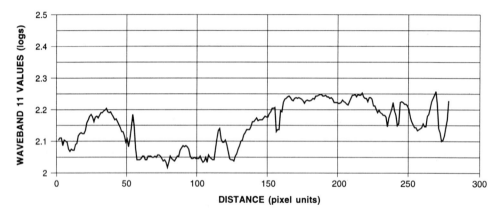

c) FIELD 3

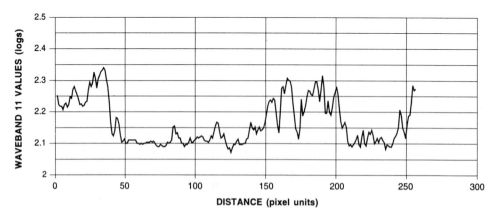

Figure 13.7 Variability in ATM band 11 dn values along transects within three fields of the floodplain of the River Tay, Scotland (from Davidson and Watson, 1995, reproduced by permission of John Wiley & Sons)

about subsurface flows across the marsh, in particular the effect of drainage ditches on adjacent water tables. Without remotely sensed data it would have been impossible to gain such detailed information on floodplain water table depths.

Not all remote sensing takes place from airborne or satellite platforms. Ground-penetrating radar (GPR) also has the ability to detect changes in water content and thus the reflection image can be related to small changes in the degree of sediment saturation and in sediment composition. Thus, Huggenberger (1993) used GPR to distinguish facies patterns with Pleistocene gravel laid down by the River Rhine.

VEGETATION IN THE FLUVIAL ENVIRONMENT

Over much of the Earth's surface river channels are embedded within, and themselves contain, patterns of vegetation which interact with the fluvial system through a complex set of feedback mechanisms and threshold effects (Chapter 11). The synoptic, large-area capability of remote sensing offers the potential to link patterns of vegetation in fluvial systems, so that vegetation change up- and down-stream may be related to changes in the fluvial system, and so that lateral variations in vegetation communities between the active channel, riparian zone and floodplain may be seen in context. Furthermore, the entire system may be placed within the broader context of the river catchment.

Remote sensing of aquatic vegetation

Several researchers have measured the spectral reflectance of aquatic macrophytes, either in broad bands (Best *et al.*, 1981; Ihse and Graneli, 1985) or in many contiguous bands (Hooper, 1992; Malthus and George, 1992; Peñuelas *et al.*, 1993). Although the green leaves and stalks of aquatic plants reflect in much the same way as their terrestrial counterparts, the presence of water as a background to the canopy causes a strong absorption in the near infra-red region. This effect is most pronounced for those canopies in which the leaves are predominantly vertical and is less evident in horizontally leaved canopies. The effect of increasing depth of submergence may be seen in Figure 13.8, which shows the results of a laboratory experiment by Hooper (1992) in which the spectral reflectance of a canopy of *Rorippa nasturtium-aquaticum* (water-cress) was measured at different depths. It is clear from these data that the main effect of submergence is a depth-dependent reduction of spectral reflectance in the near infra-red region, especially around 760 nm and 870 nm, leading to a slight reflectance peak around 800 nm. The visible region is much less altered.

Aquatic macrophytes are often classified according to whether they are submerged, floating or emergent. Spectral information from the infra-red region would clearly be useful in discriminating submerged plants from the others, but may not provide enough information by itself to discriminate between emergent and floating species. Narrow wavelength spectral differences in the visible region may be more useful for this purpose and Peñuelas *et al.* (1993) have demonstrated the utility of a Normalized Total Pigment to Chlorophyll *a* ratio index (*NPCI*) defined as

$$NPCI = (R_{680} - R_{430})/R_{680} + R_{430})$$

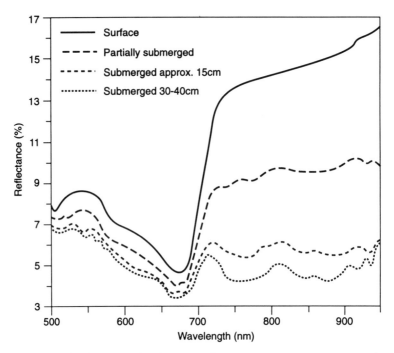

Figure 13.8 Changes in the spectral reflectance of *Rorippa nasturtium-aquaticum* (water-cress) as it becomes progressively submerged. (Modified from Hooper, 1992)

for additional discrimination between aquatic macrophyte species, where R_{680} is the reflectance at 680 nm and R_{430} is the reflectance at 430 nm.

Often, stands of aquatic macrophytes are too small to be detected reliably with existing satellite sensors and airborne sensors must be used. An airborne multi-spectral scanner was considered by Welch *et al.* (1988) for this task but rejected in favour of false colour aerial photography because of its variable geometry and high cost. In this study an interpretation key was devised which allowed emergent, sub-merged and floating macrophytes to be discriminated based on their colour, texture and height on 1:10 000 scale false colour aerial photographs. Malthus *et al.* (1990) also found height information to be useful when classifying emergent and floating species, but in contrast they used 1:3000 scale true colour aerial photographs and were able to discriminate six individual species.

In later work Malthus and George (1993) used a field spectroradiometer to mea-sure the spectral reflectance of several species of aquatic macrophytes in two lakes on the Isle of Anglesey, Wales, and found that the growth habit as well as the depth and turbidity of the surrounding water affected the reflectance. In the same study the authors also used summer data from an airborne multispectral scanner to attempt to classify lake vegetation communities but with limited success, and they suggest that multi-date imagery may be better suited to this task.

Remote sensing of marginal vegetation

The vegetation communities bordering rivers and those found on point bars and islands are important topics for study. They are sometimes referred to as occupying the 'riparian zone' (Hewitt, 1990) although a strict geomorphological definition of this term would confine it to river bank communities (Hooper, 1992). Features of marginal zone vegetation of importance in river channel change studies include the presence of trees and large shrubs and the proportion of ground covered by vegetation. Such zones are also of considerable ecological interest, for which purposes information on the plant species present and on the connectivity and shape of different habitats would be useful.

In the semi-arid zone, marginal vegetation communities may be sufficiently distinct from their surroundings to be mapped with relative ease, even with coarse resolution Landsat Thematic Mapper data, as demonstrated by Hewitt (1990) who found a supervised maximum likelihood classification of TM bands 2 (green), 4 (near infra-red) and 7 (short-wave infra-red) to be adequate to map the marginal habitats of the Yakima River in central Washington. Provided the marginal zone is large enough, and the classification simple enough, even visual analysis of single band, 1:1 million scale Landsat MSS images is adequate to provide useful information for some purposes as shown by Salo *et al.* (1986). In this case, remote sensing provided the only practical method of mapping forest on previous floodplains in an area of 5×10^5 km^2 in the headwaters of the Peruvian Amazon. The time of year when the imagery is flown can also be important. In the Savannah River swamp, South Carolina, Jensen *et al.* (1986) demonstrated that, while MSS data acquired during the spring could differentiate large vegetated wetlands, a TM image obtained late in the growing season was of limited use. Similarly in a comparison of SPOT XS, SPOT Pan and Landsat TM for riparian forest survey in the Garonne Valley, France, Muller (1990) found data collected in April to be better than that collected in July, although neither date was particularly good for this task.

At a more local scale airborne remote sensing methods again prove to be very useful. Hooper (1992) investigated the potential of panchromatic aerial photography and ATM data for assessing the vegetation in and around the River Exe north of Exeter and the River Teme north-west of Worcester. Part of this study also evaluated different methods of automated digital classification for this purpose. The best results were found using a supervised maximum likelihood classification, modified such that the *a priori* probability of a pixel being classified as shadow was reduced from 0.05 to 0.01, and the probabilities of pixels being allocated to water, tree and herb classes were marginally increased. The modified probabilities were derived from interpretation of aerial photographs, colour composites and field knowledge. Hooper also considered the visual interpretation of false-colour composites produced from the multispectral scanner data and identified a number of band combinations which highlighted selected features of marginal or aquatic vegetation (Table 13.3). Successful application of these band combinations also requires that the data be selectively contrast enhanced so as to maximise the information conveyed to a human interpreter. Examples of a reach on the River Teme processed by M. Hunt following Hooper's method are shown in Plate 13.4 and clearly reveal the wealth of vegetation and channel information available from such sources.

Table 13.3 Combinations of Daedalus ATM bands identified by Hooper (1992) as being particularly useful in studies of riverine environments using remote sensing. The spectral intervals are listed in Table 13.1 and examples of three of these colour composites are shown in Plate 13.4

	R	G	B	
Conventional composite for vegetation mapping. Similar to a colour infra-red aerial photo	7	5	3	Good for aquatic vegetation and riffles, but herbs, shrubs, trees and emergent aquatics not separated (Plate 13.4a)
Short-wave infra-red band introduced and one of the visible bands omitted	7	10	5	Good discrimination between trees, submerged aquatics and herbs/shrubs. Emergent aquatics and herbs/shrubs not separated (Plate 13.4b)
Combination of a band in the thermal infra-red with two visible bands	11	5	3	Discrimination between trees, herbs, shrubs and emergent aquatics but low visual contrast
One band from each of the visible, near infra-red and thermal regions	7	11	3	Good discrimination between all vegetation types. High visual contrast (Plate 13.4c)
One band from each of the visible, short-wave infra-red and thermal regions	3	10	11	Discrimination between herbs and trees but low visual contrast

Where high resolution panchromatic data are available for the same area as lower resolution multispectral data, the opportunity arises to use the former to 'sharpen up' the latter. Several techniques have been developed to achieve this (Chavez *et al.*, 1991; Munechika *et al.*, 1993), and, although none is without problems, they can achieve a worthwhile improvement, particularly to an image intended for visual interpretation. Plate 13.4d shows a sharpened {7,11,3} composite which was achieved by transforming the contrast-enhanced composite into intensity-hue-saturation space, applying a high pass ('edge enhancement') filter to the intensity component, and then transforming the resulting image back to RGB colour space for display. Other possibilities investigated by Hunt (1994) included the substitution of a digitised aerial photograph for the intensity component of the multispectral data set. Techniques such as these, which develop the synergy between traditional photographic products and digital multispectral scanner data have much to offer the fluvial geomorphologist, and have an even greater potential application to river corridor surveys and to riverscape assessment at the catchment scale.

CONCLUSIONS

The 'ideal' remote sensing system for monitoring change in fluvial systems does not exist, but a number of operational and forthcoming systems have much to offer the fluvial geomorphologist. For the survey of large areas satellite sensors are the logical choice and for areas where cloud cover is a problem, synthetic aperture radars (SAR) have much to offer. Satellite radar data are available from ERS-1, ERS-2 and the Japanese Earth Resources Satellite, Fuyo-1, and these will soon be joined in space by the Canadian satellite, Radarsat. Although of great interest by themselves, these data are even more useful for change detection and analysis when combined with archival Landsat or SPOT data, or when two SARs with different technical characteristics are combined (e.g. those on ERS-1 and Fuyo-1). Although the Landsat and SPOT systems have long dominated the provision of optical wavelength data from space, other sources should not be overlooked, notably the Indian remote sensing satellites (IRS-1A and 1B) which carry high-performance linear array sensors known as LISS-I and LISS II, and the very high resolution (up to 2 metre ground resolution) photographic data available from the Russian DD-5, KVR-1000 and KFA-3000 systems.

At the opposite extreme, where the focus of interest is small areas, perhaps in headwater regions or riparian vegetation communities, airborne sensors are the logical choice. Aerial photographs have very great value for interpreting the fluvial environment, and the existence of historic archives of photographs for many parts of the world guarantees their continued utility for studies of river channel change. However, techniques of digital image processing and, in particular, digital photogrammetry have had a major impact on the methods used to analyse such photographs and the widespread adoption of GIS for mapping applications has encouraged this trend. This provides the opportunity to apply techniques of geostatistics and spatial analysis to digital images of fluvial environments, for purposes such as the quantification of floodplain vegetation patch size, braid bar morphology, and habitat diversity and connectivity. As the world 'goes digital' it becomes easier to integrate such digitised aerial photographs and maps with digitised images from video sensors, and with pure digital images from scanners and imaging spectrometers. Much attention is now being directed to the problem of geometrically correcting data from airborne multispectral scanners, and technological advances in real-time GPS should contribute to making these sensors viable alternatives to aerial photography for many purposes. Clearly, the additional cost of a multispectral scanner would not be justified for many applications, but, where good discrimination within water bodies, or between plant communities is required, the extended wavelength range of the scanner may be invaluable, as is evident from Plate 13.4. Increasingly, some airborne sensors (e.g. CASI) are either offering large numbers of spectral bands (typically several hundred) or allowing the user to choose the position and bandwidth of the spectral regions sensed, whilst others offer the possibility of stereo multispectral images (e.g. the Caesar instrument, van Persie, 1994). Finally, we should not overlook the potential of the latest generation of videographic sensors (e.g. Lyon et al., 1994; Sun and Anderson, 1994), which offer good image geometry, multispectral sensing, and a convenience in deployment and data processing unmatched by more traditional systems.

Remote sensing has wide application to the study of change in fluvial systems. It is useful both as a means of detecting and mapping change and as a means of measuring some of the physical variables which promote, sustain and control change in the fluvial environment. Use of remote sensing for change detection is based on the assumptions previously described and may adopt a purely empirical approach in which images are co-registered, compared either digitally or manually, and conclusions drawn regarding changes to the fluvial system. Change in fluvial systems often consist of changes both in the spectral properties of objects and in their spatial arrangement, as channels shift or bars emerge, for example. Analysis of this type of change requires the merging of approaches based on the recognition of object boundaries (image segmentation) and the analysis of their spectral properties. In order to progress from the detection and mapping of change to the analysis of those changes it is necessary to understand more about the scene properties which make up the images being compared. This understanding may be gained from semi-empirical methods, such as the use of training sites located at known pools and riffles, or may be deterministic, such as the use of a physical model of the interactions between electromagnetic radiation and the components of the fluvial system. Whether the methodology adopted is empirical or deterministic, the latest generation of remote sensing systems offer the fluvial geomorphologist a rich set of tools with which to visualise, classify and analyse river channel change over spatial scales from the reach to the catchment, and over temporal scales from days to decades.

ACKNOWLEDGEMENTS

The UK Natural Environment Research Council (NERC) supported some of the work described here through the provision of airborne data, instrument loans from the NERC Equipment Pool for Field Spectroscopy and through studentship GT4/88/AAPS/44 awarded to Ian Hooper.

Ted Milton would like to acknowledge the provision of data from Michael Hunt, formerly of the Department of Geography, University of Southampton, and Chris Hill, GeoData Institute, University of Southampton. David Gilvear would like to acknowledge research support from Sandra Winterbottom and Tertia Waters.

REFERENCES

Babey, S.K. and Anger, C.D. (1993) Compact airborne spectrographic imager (casi): a progress review. In: *Proceedings of the Society of Photo-Optical Instrumentation Engineers (SPIE)* **1937**, 152–163.

Barton, I.J. and Bathols, J.M. (1989) Monitoring floods with AVHRR. *Remote Sensing of Environment* **30**, 89–94.

Best, R.G., Wehde, M.E., *et al.* (1981) Spectral reflectance of hydrophytes. *Remote Sensing of Environment* **11**, 27–35.

Blasco, F., Bellan, M.F., *et al.* (1992) Estimating the extent of floods in Bangladesh using SPOT data. *Remote Sensing of Environment* **39**, 167–178.

Blyth, K. and Biggin, D.S. (1992) Monitoring floodwater inundation with ERS-1 SAR. *Earth Observation Quarterly* **42**, 6–8.

Boxall, S.R., Chaddock, S.E. *et al.* (1993) *Airborne Remote Sensing of Coastal Waters.* Project Report (R&D Report 4), National Rivers Authority, Bridgewater.

Brivio, P.A., Desseana, M.A., *et al.* (1988) Detection of geomorphological units on Landsat images over the inland delta of the Niger river for the assessment of regional hydrology. In: *Proceedings of the 22nd International Conference on Remote Sensing of the Environment*, Environmental Research Institute of Michigan (ERIM), Ann Arbor, 399–407.

Chavez, P.S., Sides, S.C., *et al.* (1991) Comparison of three different methods to merge multi-resolution and multispectral data: Landsat TM and SPOT panchromatic. *Photogrammetric Engineering and Remote Sensing* **57**, 295–303.

Christensen, E.J., Jensen, J.R., *et al.* (1988) Aircraft MSS Data Registration and Vegetation Classification for Wetland Change Detection. *International Journal of Remote Sensing* **9**, 23–38.

Clark, C.D. (1990) Remote sensing scales related to the frequency of natural variation: an example from paleo-ice-flow in Canada. *IEEE Transactions on Geoscience and Remote Sensing* **28**, 503–508.

Cosandier, D., Ivanco, T.A., *et al.* (1994) The integration of a digital elevation model in casi image geocorrection. In: *Proceedings of the First International Airborne Remote Sensing Conference and Exhibition*, Strasbourg, France, Environmental Research Institute of Michigan (ERIM), Ann Arbor, Michigan, 515–529.

Davidson, D.A. and Watson, A. (1995) Soil moisture variability on alluvial soils: River Tay, Scotland. *Earth Surface Processes and Landforms* (in press).

Devereux, B.J., Fuller, R.M., *et al.* (1989) The geometric correction of airborne thematic mapper imagery. In: *Proceedings of the NERC Workshop on Airborne Remote Sensing*, Institute of Freshwater Ecology, Windermere/NERC, Swindon, 19–33.

Diakite, M., Yergeau, M., *et al.* (1986) The utility of Landsat TM imagery in the inland delta cartography of Mali. In: *Proceedings of the 20th International Symposium on Remote Sensing of the Environment*, Nairobi, Kenya, Environmental Research Institute of Michigan (ERIM), Ann Arbor, 567–574.

Duggin, M.J. and Robinove, C.J. (1990) Assumptions implicit in remote sensing data acquisition and analysis. *International Journal of Remote Sensing* **11**, 1669–1694.

Engman, E.T. and Gurney, R.J. (1991) *Remote Sensing in Hydrology*. Chapman & Hall, London.

Ferguson, R.I. and Werrity, A. (1983) Bar development and channel changes in the gravelly River Feshie, Scotland. In: J. Collinson and J. Lewin (eds) *Modern and Ancient Fluvial Systems*, Blackwell Scientific Publications, Oxford, 181–194.

Gilvear, D.J. and Watson, A. (1995) The use of remotely sensed imagery for mapping wetland water table depths: Insh Marshes, Scotland. In: J. Hughes and L. Heathwaite (eds) *British Wetlands*, Wiley, Chichester (in press).

Gilvear, D.J. and Winterbottom, S.J. (1992) Channel change and flood events since 1783 on the regulated River Tay, Scotland: Implications for flood hazard management. *Regulated Rivers* **7**, 247–260.

Gilvear, D.J., Davies, J., *et al.* (1994) Mechanisms of floodbank failure during large events on the River Tay and Earn, Scotland. *Quarterly Journal of Engineering Geology* **27**, 319–332.

Gilvear, D.J., Waters, T., *et al.* (1995) Image analysis of aerial photography to quantify changes in channel morphology and instream habitat following placer mining in interior Alaska. *Freshwater Biology* (in press).

Green, A.A., Whitehouse, G., *et al.* (1982) Causes of flood streamlines observed on landsat images and their use as indicators of floodways. *International Journal of Remote Sensing* **4**, 5–16.

Hardy, T.B., Anderson, P.C., *et al.* (1994) Application of multispectral videography for the delineation of riverine depths and mesoscale hydraulic features. In: R. Marston and V. Hasfurther (eds) *Effects of human-induced changes on hydrologic systems*, American Water Resources Association, Maryland, 445–454.

Hewitt, M.J. (1990) Synoptic inventory of riparian ecosystems: the utility of Landsat thematic mapper data. *Forest Ecology and Management* **33/34**, 605–620.

Hockey, B., Richards, T., *et al.* (1990) Prospects for operational remote sensing of surface water. *Remote Sensing Reviews* **4**, 265–283.

Hooper, I.D. (1992) Relationships between vegetation and hydrogeomorphic characteristics of British riverine environments: A remotely sensed perspective. PhD thesis, University of Southampton.

Huggenberger, P. (1993) Radar facies: recognition of facies pattern and heterogeneities within Pleistocene Rhine gravels, N.E. Switzerland. In: J.L. Best and C.S. Bristow (eds) *Braided Rivers*, The Geological Society, London, 163–177.

Hunt, M. (1994) The application of remote sensing to river corridor surveys. Undergraduate Dissertation, University of Southampton, Department of Geography.

Ihse, M. and Graneli, W. (1985) Estimation of reed (*Phragmites australis*) biomass through spectral reflectance measurements. *Biomass* **8**, 59–79.

Jacobberger, P.A. (1988) Mapping abandoned river channels in Mali through directional filtering of thematic mapper data. *Remote Sensing of Environment* **26**, 161–170.

Jensen, J.R., Hodgeson, M.E., *et al.* (1986) Remote sensing inland wetlands—a multispectral approach. *Photogrammetric Engineering and Remote Sensing* **52**, 87–100.

Lewin, J. and Manton, M.M.M. (1975) Welsh floodplain studies: the nature of floodplain geometry. *Journal of Hydrology* **25**, 37–50.

Lewin, J. and Weir, M.J.C. (1977) Monitoring river channel change. In: *Monitoring Environmental Change by Remote Sensing*, University of Reading, Remote Sensing Society, 23–27.

Li, Z., Hill, C.T., *et al.* (1993) Exploiting the potential benefits of digital photogrammetry: some practical examples. *Photogrammetric Record* **14**, 469–475.

Lyon, R.J.P., Honey, F.R., *et al.* (1994) Second generation airborne digital multispectral video: evaluation of a DMSV for environmental and vegetation assessment. In: *Proceedings of the First International Airborne Remote Sensing Conference and Exhibition*, Strasbourg, France, Environmental Research Institute of Michigan (ERIM), Ann Arbor, 105–116.

Lyzenga, D.R. (1981) Remote sensing of bottom reflectance and water attenuation in shallow water using aircraft and Landsat data. *International Journal of Remote Sensing* **2**, 71–82.

Malthus, T.J. and George, D.G. (1993) Remote sensing of aquatic macrophytes in lakes. In: *Proceedings of the NERC Symposium on Airborne Remote Sensing 1993*, University of Dundee, NERC, Swindon, 59–69.

Malthus, T.J., Best, E.P.H., *et al.* (1990) An assessment of the importance of emergent and floating-leaved macrophytes to trophic status in the Loosdrecht lakes (The Netherlands). *Hydrobiologist* **191**, 257–263.

Mertes, L.A.K., Smith, M.O., *et al.* (1993) Estimating suspended sediment concentrations in surface waters of the Amazon River wetlands from Landsat images. *Remote Sensing of Environment* **43**, 281–301.

Mouchot, M-C. (1991) Monitoring the water bodies of the Mackenzie delta by remote sensing methods. *Arctic* **44**, 21–28.

Muller, E. (1990) Discrimination des boisements dans la moyenne vallée de la Garonne a l'aide des données SPOT P, SPOT XS et Landsat TM. *Troisiemes Journees Scientifiques du risea de teledetection de l'UREf*, Toulouse, France.

Muller, E. (1992) *Evaluation de la bande TM5 pour la cartographie morpho-hydrologique de la moyenne vallée de la Garonne* (No. Project SPOT4/MIR), CNES, Paris.

Muller, E., Décamps, H., *et al.* (1993) Contribution of space remote sensing to river studies. *Freshwater Biology* **29**, 301–312.

Munechika, C.K., Warnick, J.S., *et al.* (1993) Contribution of space remote sensing to river studies. *Freshwater Biology* **29**, 301–312.

Munechika, C.K., Warnick, J.S., *et al.* (1993) Resolution enhancement of multispectral image data to improve classification accuracy. *Photogrammetric Engineering and Remote Sensing* **59**, 67–72.

Nagarajan, R., Marathe, G.T., *et al.* (1993) Identification of flood prone regions of Rapti river using temporal remotely-sensed data. *International Journal of Remote Sensing* **14**, 1297–1303.

Ormsby, J.P., Blanchard, B.J., *et al.* (1985a) Detection of lowland flooding using active microwave systems. *Photogrammetric Engineering and Remote Sensing* **51**, 317–328.

Ormsby, J.P., Gervin, J.C., *et al.* (1985b) Wetland physical and biotic studies using multispectral data. In: *Proceedings of the 19th International Symposium on Remote Sensing of the Environment*. Environmental Research Institute of Michigan (ERIM), Ann Arbor, 799–807.

Peñuelas, J., Gamon, J.A., *et al.* (1993) Assessing community type, plant biomass, pigment composition, and photosynthetic efficiency of aquatic vegetation from spectral reflectance. *Remote Sensing of Environment* **46**, 1–25.

Philip, G., Gupta, R.P., *et al.* (1989) Channel migration studies in the middle Ganga basin, India, using remote sensing data. *International Journal of Remote Sensing* **10**, 1141–1149.

Ramasamy, S.M., Bakliwal, P.C., *et al.* (1991) Remote sensing and river migration in Western India. *International Journal of Remote Sensing* **12**, 2597–2609.

Rudant, J-P. (1994) French Guiana through the clouds: first complete satellite coverage. *Earth Observation Quarterly* **44**, 1–6.

Salo, J., Kalliola, R., *et al.* (1986) River dynamics and the diversity of Amazon lowland forest. *Nature* **322**, 254–258.

Salomonson, V.V., Jackson, T.J., *et al.* (1983) Water resources assessment. In: *Manual of Remote Sensing*, American Society of Photogrammetry, Falls Church, Virginia, 1497–1570.

Schumann, R.R. (1989) Morphology of Red Creek, Wyoming, an arid-region anastomosing channel system. *Earth Surface Processes and Landforms* **14**, 277–288.

Sippel, S.J, Hamilton, S.K., *et al.* (1994) Determination of inundation area in the Amazon river floodplain using the SMMR 37GHz polarization difference. *Remote Sensing of Environment* **48**, 70–76.

Strahler, A.H., Woodcock, C.E., *et al.* (1986) On the nature of models in remote sensing. *Remote Sensing of Environment* **20**, 121–139.

Sun, X. and Anderson, J.M. (1994) An easily-deployable miniature, airborne imaging spectrometer. In: *Proceedings of the First International Airborne Remote Sensing Conference and Exhibition*, Strasbourg, France, Environmental Research Institute of Michigan (ERIM), Ann Arbor, 178–189.

Thorne, C.R., Russell, A.P.G., *et al.* (1993) Planform pattern and channel evolution of the Brahmaputra River, Bangladesh. In: J.L. Best and C.S. Bristow (eds) *Braided Rivers*, The Geological Society, London, 277–290.

Townshend, J.R.G. and Justice, C.O. (1988) Selecting the spatial resolution of satellite sensors required for global monitoring of land transformations. *International Journal of Remote Sensing* **9**, 187–236.

van Persie, M. (1994) Airborne multispectral pushbroom scanner CAESAR. In: *Proceedings of the First International Airborne Remote Sensing Conference and Exhibition*. Strasbourg, France, Environmental Research Institute of Michigan (ERIM), Ann Arbor, I 93–I 101.

Wagner, M.J. (1994) ERS-1 images of the Christmas flood over Europe. *Earth Observation Quarterly* **43**, 12–13.

Warburton, J., Davies, T.R.H., *et al.* (1993) A meso-scale investigation of channel change and floodplain characteristics in an upland braided gravel-bed river, New Zealand. In: J.L. Best and C.S. Bristow (eds) *Braided Rivers*, The Geological Society, London, 241–256.

Watson, A., Gilvear, D.J., *et al.* (1993) Examination of the use of imagery for mapping river channels and floodplain sediments. In: *Proceedings of the NERC Symposium on Airborne Remote Sensing*, University of Dundee, Scotland/NERC, Swindon, 117-136.

Welch, R., Remillard, M.M., *et al.* (1988) Remote sensing and Geographic Information System techniques for aquatic resource evaluation. *Photogrammetric Engineering and Remote Sensing* **54**, 177–185.

Wilson, A.K. (1988) The effective resolution element of Landsat Thematic Mapper. *International Journal of Remote Sensing* **9**, 1303–1314.

Wilson, A.K. (1994) The NERC integrated ATM/CASI/GPS system. In: *Proceedings of the First International Airborne Remote Sensing Conference and Exhibition*, Strasbourg, France, Environmental Research Institute of Michigan (ERIM), Ann Arbor, 249–259.

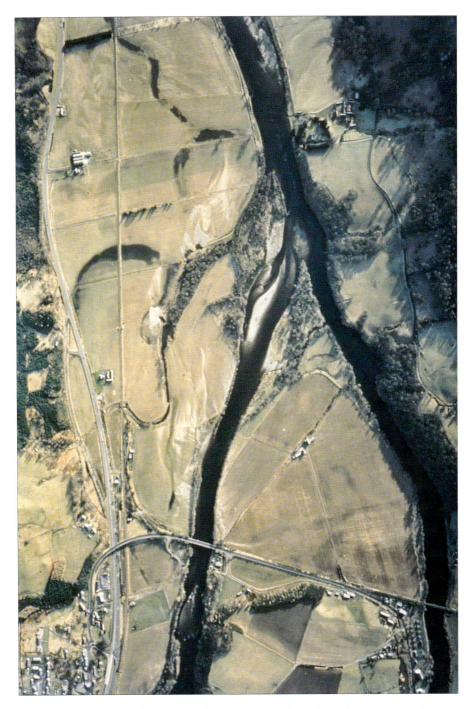

Plate 13.1 Colour aerial photograph of the River Tay floodplain 10 days after a 1:100 recurrence interval flood event. (Photo reproduced courtesy of the NERC)

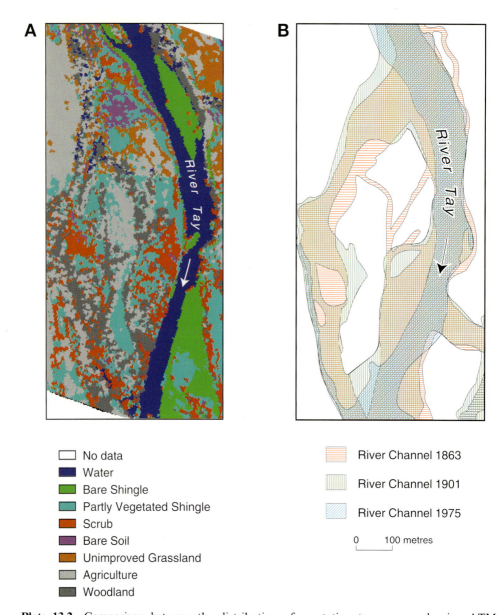

A

No data
Water
Bare Shingle
Partly Vegetated Shingle
Scrub
Bare Soil
Unimproved Grassland
Agriculture
Woodland

B

River Channel 1863
River Channel 1901
River Channel 1975

0   100 metres

Plate 13.2 Comparison between the distribution of vegetation types mapped using ATM data (A) and channel changes identified using maps of various dates (B) on the floodplain of the River Tummel. (Provided by S. Winterbottom). Airborne multispectral scanner data provided by the NERC

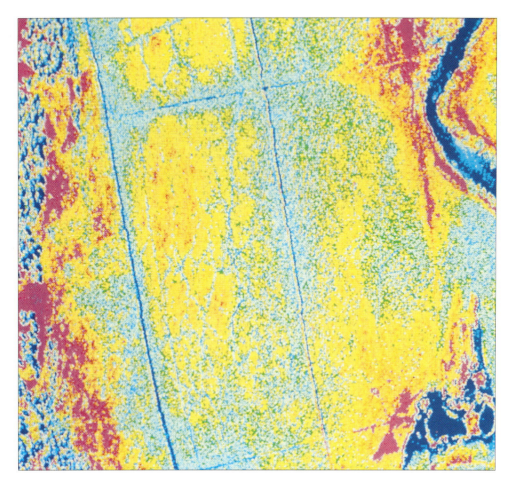

Plate 13.3 Classified image of the depth to the water table for a 1km × 1km area of the Insh Marshes, Scotland (7 cm class intervals, red = wettest, yellow = driest). Airborne multispectral scanner data provided by the NERC

(a)

(b)

(c)

(d)

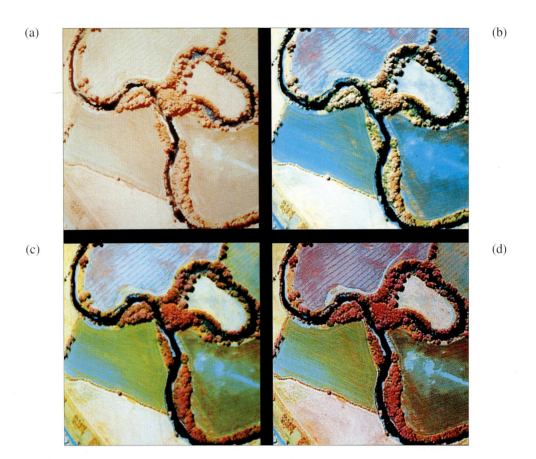

Plate 13.4 Mosaic of three of the Daedalus ATM colour composite combinations recommended by Hooper (1992) for visual interpretation of riverine environments. The image shows a reach of the River Teme north-west of Worcester. Quadrant (a) is enhanced for instream features, (b) and (c) for different aspects of riparian vegetation, whilst (d) shows a sharpened intensity-hue-saturation rendition of the image in quadrant (c). See text and Table 13.3 for explanation of the bands used and the processing involved. (Taken from Hunt, 1994, reproduced by permission of the author). Airborne multispectral scanner data provided by the NERC

14 Information from Topographic Survey

S.R. DOWNWARD

School of Geography, Kingston University, UK

INTRODUCTION

The contemporary river landscape is one that has been shaped and modified through history. Historical documentation has sought to record its form and nature at discrete intervals in time by various means. As a result the geomorphologist may draw upon a diversity of historical records to reconstruct river channel changes. Historical sources of information take many different formats which have been extensively reviewed in terms of their use for river channel change studies (e.g. Hooke and Kaine, 1982; Trimble and Cook, 1991). This chapter focuses specifically on historical topographic information which typically seeks to represent the landscape in planform (e.g. map and aerial photographic imagery) and in cross-section. It presents a critical examination of the potential use of topographic information to discriminate river channel change. First, issues of scale are considered by relating the scale of adjustment in the fluvial landscape to the scales of representation in the topographic information sources. Secondly, the process of transcription of information from field observation via topographic sources to the quantitative estimation of river channel changes is considered, by identifying the errors propagated through the analysis procedure to generate a total error estimate. Finally, Total Mapping Error estimates are presented for two river channels: error associated with the estimation of topographic planform adjustments from 1:10 000 and 1:10 560 scale maps are presented for the River Towy, Dyfed, and errors associated with the estimation of topographic cross-section adjustment from surveyed channel cross-sections are presented for the River Sence, Leicestershire.

TOPOGRAPHIC SCALE AND RIVER CHANNEL CHANGE

Physical systems operate over different scales in both space and time. Topographic information seeks to record the form of the landscape at a given moment in time within three-dimensional space at a variety of scales of representation. The ability of this information to reconstruct the changes occurring in the landscape is dependent upon the relative sizes of the features, the magnitude of the changes occurring, and the scales of topographic representation (Schumm, 1991). The ability to resolve change is determined by resolution. Resolution differs from scale in that it defines the smallest individual landscape elements that may be separated from each other.

Changing River Channels. Edited by Angela Gurnell and Geoffrey Petts.
© 1995 John Wiley & Sons Ltd.

For example, an aerial photograph of a river has a spatial resolution defined by the individual film elements of the photographic image. The scale of the image may be altered by enlarging its size from the negative, but the smallest defined elements remain the same. As resolution increases so the level of generalisation decreases and the representation of landscape complexity increases (Mandelbrot, 1982). For example, an atlas-scale representation of a river may show a single line with no distinction of width. Generalisations employed at this scale of representation mean that complex planforms will be simplified and smaller tributaries selectively omitted. As map scales decrease then smaller river channels will increasingly be represented by the position of both boundaries representing the 'channel/non-channel' interface. In the British context, for example, river channels greater than 10 m in width are mapped by plotting both banks on 1:10 000 scale Ordnance Survey maps. By increasing resolution further still the same river channel can be represented not by a single generalised line depicting a single boundary location but by a complex number of breaks of slope defining bank features in detail and thus any number of possible topographic boundaries for the channel.

In seeking to reconstruct channel change it is important to specify the nature of the changes to be observed and to select the most appropriate topographic scales to achieve these objectives. For example, Noorbergen (1993) describes a multi-temporal analysis to quantify planform changes on the Jamuna (Brahmaputra) River in Bangladesh. The methodology employs satellite imagery with a spatial resolution which defines landscape units of approximately 80 m by 80 m in size. This level of resolution is justified in this case given that the width of the Jamuna River channel varies between 2.5 km and 12 km, but clearly this resolution would be inappropriate to discriminate boundary changes on many Western European rivers. As the size of the landscape unit decreases, or the net change between any two observations of form decreases, then successively finer topographic resolution is required to observe these changes. For example, Hooke and Redmond (1989) compared maps at approximately 1:10 000 scale for a sample of rivers across England, noting that from the planform evidence alone, most of the lowland rivers were stable over the study period. However, geomorphological investigations and documented examples of river channel management problems for lowland rivers in England illustrate that significant changes are occurring on these lesser energy rivers, but at a scale which is inappropriate for detection from the 1:10 000 scale planform record alone. Therefore, when attempting to define the stability of the river channel, or fluctuations in the rate of change at specified locations over time, it is vital to consider the appropriate spatial scale at which to investigate such changes and to support statements on the degree and nature of stability.

The estimation of change from topographic survey information involves the measurement of the difference in form between like indices recorded at defined points in time. Temporal resolution is defined by the time that elapsed between individual observations (the sample interval). The rate of change is the net recorded change in a particular variable over the period separating the individual observations. Net rates of change will always be an underestimate of the total change or gross change. The deviation between the two values is dependent upon the spatial resolution and scale of the sources and temporal sampling interval. Errors in the estimation of the rate of change may be generated in two principal ways: the accuracy with which the time of survey of river channel form is recorded, and the assumptions made concerning the period separating any two given samples in the historical record. For example, aerial photographs will typically depict time to the nearest minute of observation while map evidence, in the absence of detailed surveyors' notes, will typically define time according to the year in which the observation or revision was made. Therefore, the resolution with which the period separating any two maps is defined could, in theory, be almost two years. Variation in the temporal resolution of a study may have a marked influence upon an understanding of the stability of the river channel and it is, therefore, important that the temporal resolution and sampling interval adopted is appropriate to successfully discriminate the changes which occur. Typically, topographic information used to reconstruct river channel change is dependent upon the available historical information and is thus not specified by a pre-determined sampling interval. The sampling intervals may be irregular and may not be the most suitable for the study objectives. Under such circumstances, it is important to ensure that interpretations are valid and take account of artefacts induced by the temporal resolution and sampling intervals associated with the historical sources.

THE TRANSCRIPTION OF TOPOGRAPHIC INFORMATION

No methodology for identifying the character of the environment from topographic information can ever be error free. Error is a fundamental element in all topographical information and this error is incorporated when topographic information is employed to identify river channel form. If the total sum of all the errors involved in any analysis of river channel form exceeds an acceptable level of tolerance then the identified changes have little significance for developing an understanding of the fluvial system. The process by which landscape information is recorded at the moment of observation and communicated to the 'user' (e.g. the geomorphologist) concerned with reconstructing river channel change is called data transcription. Errors are incorporated at each stage of data transcription and are propagated throughout any analysis procedure, so affecting the overall accuracy of any conclusions drawn from the analysis. The component errors thus accumulate to a total error estimate, the Total Mapping Error. Two general error classes may be identified: *inherent errors* are those present within the topographic source information prior to analysis, and *operational errors* are those produced by the operational process used to estimate river channel change from the topographical source information (Figure 14.1). Each of these categories of error are examined below.

Inherent error

Inherent errors are incorporated whenever we define space in geographical analysis. Inherent errors are incorporated at each individual stage of the mapping process: collection, representation and storage.

Gross errors of data collection are those errors resulting from mistakes by the surveyor. These may manifest themselves obviously within the data (e.g. a mid-channel bar that suddenly rises outrageously above the floodplain), or they may be subtle

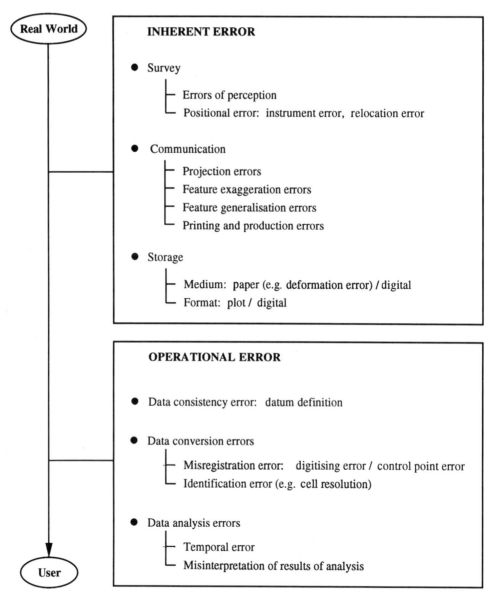

Figure 14.1 Components of the transcription of topographic information

and remain undetected. Gross errors may be minimised by adhering to routines and checks which identify obvious mistakes, but errors of this type can never be completely discounted.

Systematic errors are errors introduced by the system employed to transform information from the landscape into a topographic format. Thapa and Bossler (1992) note that the presence of systematic errors may produce measurements that are precise but not necessarily accurate. For example, successive surveys of a river channel cross-section may be undertaken by several different surveyors employing variations of the basic surveying equipment. Errors arising from these individual surveys are unlikely to be the result of differences in the surveying equipment, because typically the working 'precision' of the instrumentation used will far exceed the surveyors' ability to finely interpret the cross-section form. Systematic errors are more likely to result from the mis-interpretation of channel form by the surveyors. Different surveyors (typically without a geomorphological background) will have different interpretations on how to represent river-channel form most effectively. Remotely-sensed topographic information (including aerial photographs) is effectively free of the perceptual errors introduced by the surveyor, but it is subject to systematic errors which determine its planimetric accuracy (e.g. photogrammetric and relief scale distortions, see Chapter 13).

Errors resulting from the communication of topographic information result from shortcomings in the representation of the survey information. Information may be communicated in graphical, written or statistical formats (Hooke and Kain, 1982). Graphical communication is achieved through the mapped image as a scale-dependent representation of form in either a vector format (e.g. map or cross-section plot) or cellular format (e.g. the remote-sensed image). Maffini *et al.* (1989) note that for a given scale of presentation, vector representations assume a precision that does not occur in reality, whereas cellular formats provide a generalisation of nature. In terms of river channel form, the vector model of the landscape is one in which the boundaries dividing discrete entities are represented by points, lines and polygons. The 'line' therefore is representative of nature as an absolute division between two entities. In reality the river channel boundary locations between different entities may be described as 'fuzzy', that is to say they lie within a given zone of the line, and the line itself may be said to be representative of a location of maximum probability that a transition between two entities exists for a given scale of representation. The determination of indices of river channel form from vector representations is dependent upon the definition of common datums of measurement of the river channel boundary locations in order to provide standardisation for comparison. For example, map sources commonly identify the channel/non-channel boundary as a datum defining channel width. In the absence of corroborative evidence, there is little basis to suggest the consistency with which this boundary is represented across the map sheet. By contrast, remotely sensed information (including aerial photographs) has a specified spatial resolution which represents a pre-determined generalised image of the landscape. Remotely sensed topographic information is thus essentially free from surveyors' perceptual errors because the information content of the image conforms to a rationalised set of criteria regarding resolution and scale of presentation. The decision that determines the location of the boundary of an object is evaluated either manually (on the basis of the user's definitions of form) or by substituting a computer-based algorithm to undertake the determination automatically on the basis of a set of given 'rules' (Hooper, 1992, Chapter 13).

Computer-based (e.g. Geographical Information Systems (GIS)) approaches to the discrimination of river channel change typically require that data be entered in a statistical or digital format prior to analysis. Topographical information that may be obtained in this format is therefore ideally suited to computer analysis. Statistical and vector–digital topographic representations are not scale-dependent, so comparisons of river channel form can be made on the basis of the original surveyor's values and do not require modification to a common scale prior to analysis and transcription. In cases where vector-map representations must be transcribed to a digital format prior to analysis (e.g. digitising or scanning the defined river channel boundary locations) then a further stage of data transcription is included. Errors which result from the conversion of inherent topographic information are considered to be operational errors and so are considered later in this chapter. Statistical and digitally recorded topographic information is also less likely to suffer degradation through storage than paper copies of topographic information. Typically, paper has been the primary medium for storage of graphically-based topographic information, and thus a legacy of the 'paper-age' of data storage is the extent to which the image representation may be distorted differentially across the map sheet by the expansion and contraction of the sheet over time. This problem may be exacerbated where original copies of maps are not available, and when the information has been transferred as a photocopy or tracing.

One of the main advantages of a digitally-based representation in the analysis of river channel change is that information may be registered at a common scale and projection from control points identified across the image using GIS software. For example, river channel planform information may be spatially corrected using 'rubber-sheet' algorithms (Burrough, 1986) to transform each individual map to a commonly defined base-map scale and projection. In doing so, not only are different maps registered to a common map base for comparison, but additionally the deviation of control point values between the original topographic source and the digital base-map provides an indication of the (non-perceptual) inherent planform error. By obtaining a large sample of residual values between control points on the map source and the base-map, positional errors introduced by digitising the locations of the control points are minimised because these digitising errors are assumed to occur randomly.

Operational error

The nature of the operation undertaken to identify the character of river channel from topographic information is largely dependent upon the specific nature of the river channel change study and the availability of topographic source information to meet such requirements. Techniques fall into three categories: planform, cross-section and volumetric analysis. Planform and cross-section techniques consider the river-channel in two-dimensional space (plan and section, respectively). Volumetric techniques are more reliant upon computer software to model the channel in three-dimensional space. Regardless of the precise nature of the technique employed, errors of data transcription introduced by the operational procedure increase the estimate of total error. For example, Vitek *et al.* (1984), p. 296) note:

> Operational error increases the total error based on the premise that inherent error cannot be eliminated but only enhanced by the operational procedures.

In order to make observations of channel change, it is important to identify common indices of channel form. The selection an index or 'datum' must be stan-dardised so that the methodology employed to discriminate change is able to com-pare like with like. To achieve a level of standardisation, definitions for comparison should be based upon common scales of representation. For example, maps at a scale of 1:10 000 will commonly depict the channel/non-channel boundary location as the representation of channel width with no other supporting information. Selecting this datum level assumes that this width definition is standardised for all similar sheets at a similar scale of representation. As the available resolution increases (e.g. the representation of the same reach from an aerial photograph or in cross-section) so there are any number of datums that may be taken to represent channel width, each of which may be selected by several different means. Topographic information is commonly provided without supporting sedimentary or vegetational character-istics, so typically the definition of the datum must be made solely upon the basis of channel geometry. The minimum width–depth ratio (Wolman, 1955), or the change in the relationship between the geometry of the channel banks and the immediate floodplain surface level (Leopold *et al.*, 1964; Woodyer, 1968; Harvey, 1969; Riley, 1972) provide examples of such definitions.

In order to undertake a given analysis procedure, information may have to be converted from a cellular/digital to vector format (e.g. to undertake manual compar-isons of river channel change) or, vector to digital or cellular format (e.g. in the case of computer-based analysis techniques). The transcription procedure itself will involve the incorporation of positional systematic error. For example, digitising is commonly used as a method for converting map information to allow GIS-based analysis. Burrough (1986) notes that errors resulting from the digitising process are generated from the user's inability to perceive and digitise representations of objects. Digitising error will, therefore, affect not only the ability to represent the river chan-nel boundary locations, but also the ability to represent control points used in the calibration of the map sheet to a common reference base-map. The conversion of topographic information from vector to cellular format will also involve the further generalisation of topographic information. Cells which may be positioned to within a given level of confidence may be misidentified or be misrepresentative of the infor-mation they represent. This is particularly true for cells which lie on the river chan-nel boundary. The greater the magnitude of channel change between any two representations, the lesser the likelihood that these misrepresentation errors will lie at the same location (Walsh *et al.*, 1987; Newcomer and Szajgin, 1989).

ESTIMATING TOTAL MAPPING ERROR

Two methodologies are widely employed in the analysis of river channel change from topographic information. They are appropriate to river channel reaches repre-sentative of different scales of adjustment within the context of the scale and resolu-tion of the mapped information. The first approach estimates change from planform information. This is exemplified by applying a GIS-based procedure to an actively migrating river channel, the River Towy in Dyfed. The second approach uses cross-section information for discriminating river channel change for a regraded river channel, the River Sence in Leicestershire. In each example, component errors of

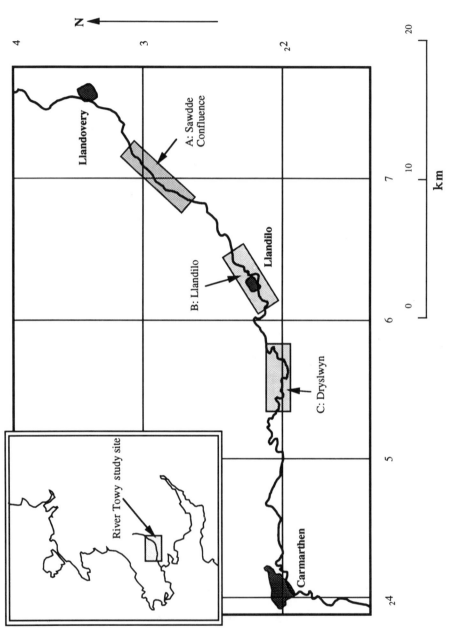

Figure 14.2 The River Towy study sites, Dyfed

both the inherent information and the operational procedures are considered, as is the way they propagate to a Total Mapping Error estimate.

Planform adjustments

A GIS-approach to estimating channel planform changes has been implemented using the TYDAC Technologies product SPANS to quantify river channel planform change for the River Towy in Dyfed (Figure 14.2). The River Towy represents an ideal study site to demonstrate the methodology as it has a well-documented history of channel migration (Lewin and Brindle, 1977; Lewin, 1984; Lewin and Chisholm, 1986; Newson and Leeks, 1986; Smith, 1987, 1989). Seven dates were selected from the historical topographic information depicting the river channel planform at scales up to 1:10 560 for the period 1842 to 1986.

The vector boundaries depicting the channel/non-channel interface were imported to the GIS by digitising and then transcribing the information to a common map base by selecting control points over the image. Control point locations were chosen at the corners of buildings and at consistently defined field boundaries, locations for which there was confidence of non-movement. Control points are digitised and regis-

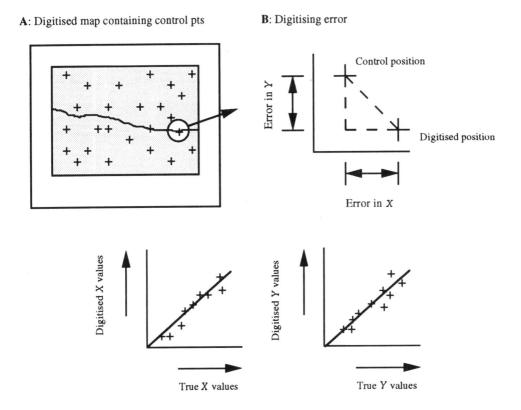

A: Digitised map containing control pts   B: Digitising error

C: Residual values estimated for planform correction

Figure 14.3   Estimation of residual values for the determination of inherent positional errors

tered against their location relative to the Ordnance Survey National Grid coordinates obtained from the most recent map sheet at the largest available scale using a micrometer to obtain the 'reference' values for the eastings and northings. These values will not be absolute as systematic error is introduced as a result of the depiction of the control points from the reference map (itself the subject of inherent error) and the systematic error involved in obtaining the easting and northing values. What is important, however, is that these reference values are used consistently at all dates in the topographic record. A linear transformation of the topographic planform information is undertaken to correct locational displacements in the inherent data relative to the defined control point network. This procedure places the residual values for each control point as estimates of easting and northing displacements between the digitised value and reference value (Figure 14.3). The values can be combined to give the deviation in metres between the digitised and reference locations. The frequency distribution of these values may then be used to estimate total inherent positional error as they represent the mean positional error for each topographic source prior to operational analysis. For example, using data obtained from a 1:10 000 scale 1979 Ordnance Survey map and a 1:10 560 scale photocopy of a 1876 Ordnance Survey map, the frequency distribution of deviations was found to be normally distributed. The mean and the standard deviation of the residual deviation values (sample sizes 39 and 38, respectively) were used to estimate the probability of specific deviations. At the 95% level of confidence, limits of 2.33 m and 6.77 m were estimated for the 1979 and 1876 maps, respectively.

The conversion of information from the historical sources to the GIS was achieved by digitising the channel/non-channel boundary location as indicated on the map sheets, or by estimating the location of the 'normal winter level' from aerial photographs at a similar scale of representation. The scale of representation for the River Towy aerial photographs (1:7500 to 1:10 000), combined with the steep angles of the banks, allows the image to be interpreted with confidence to define the channel/non-channel boundary location.

A test was conducted to establish the magnitude of likely digitising errors involved in the transcription of information to the GIS. A chosen river channel boundary location representing a reach of the River Towy study site was repeatedly digitised from a 1:10 000 scale sheet in order to observe the variability of this representation. Figure 14.4 demonstrates how the test procedure involved the re-digitising of a string of arcs between fixed nodes representing each end of the reach and a fixed arc representing a 'control' boundary line. By digitally overlaying the vector representations of the same boundary the result was to produce 'slithers' representing the displacement between the two lines. By repeating this process 50 times a frequency distribution of the mean individual slither displacement was estimated. The normal distribution of these errors is estimated with a standard deviation of 0.101 mm. Therefore at scale representations of 1:10 000 and 1:10 560, respectively, this translates to a standard deviation value of 1.01 m or 1.06 m and thus errors of approximately 2.02 m and 2.12 m ($P < 0.05$), respectively.

Analysis procedures undertaken using the GIS-based software are unlikely to further increase the positional error in the topographic information. Errors incurred in the analysis are more likely to be the result of identification errors from the generalisation of the topographic information. For example, vector representations of the

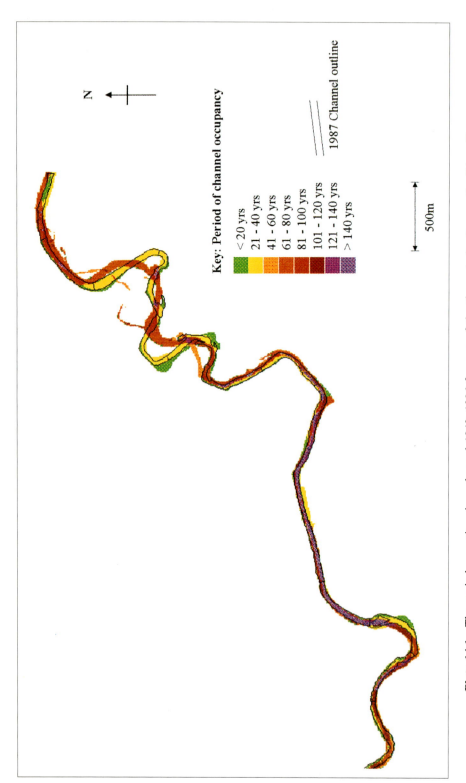

Plate 14.1 The period as active river channel 1842–1986 for one of the three study sites on the River Towy

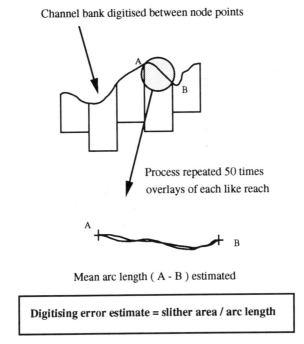

Channel bank digitised between node points

A

B

Process repeated 50 times
overlays of each like reach

A

B

Mean arc length (A - B) estimated

Digitising error estimate = slither area / arc length

Figure 14.4 Estimation of digitising errors from repeat digitising of the river channel boundary location

channel boundary locations were converted to a cellular format. As each cell has a
unique spatial location, then the same landscape element is compared for each historic source. The outcome of this comparison or overlay can be represented as a
reclassified map showing the planform extents of, for example, areas of net erosion
or deposition or, by overlaying several maps, can be used to produce a map representing the total period as active channel for the time-scale of the historical investigation (Plate 14.1). This information may then be used to identify areas of relative
stability and instability in relation to the contemporary river channel. The size
chosen for these cells must be appropriate to the size of the river channel, the magnitude of adjustment and the spatial errors involved in the data being analysed.
Figure 14.5 illustrates this: GIS-vector information for a map selected from the
topographic planform information from reaches within a study site for the River
Towy were converted to cell information representing classes that define areas of
channel occupancy and non-occupancy. This conversion was undertaken at a range
of spatial resolutions defining cell sizes from 0.75 m by 0.75 m to 80 m by 80 m. At
each spatial resolution the total channel area for each reach, defined by the sum of
cells within each reach assigned 'channel occupancy', was estimated. As the 0.75 m
by 0.75 m cells represent the finest spatial resolution identified, and thus define the
most accurate description of the reach areas available, the percentage variation in
area between these estimates for identical reaches and all other cell resolutions was
estimated. In the case of the River Towy cells with a spatial dimension of 2.5 m by
2.5 m were found to represent an appropriate resolution. By increasing the cell size

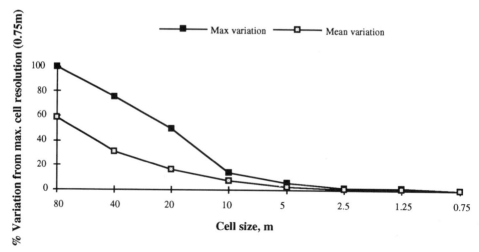

Figure 14.5 Estimating the optimum spatial resolution for digital processing of topographic information. In this example for a 1:10 000 scale map of the River Towy, a cell size of 2.5 m by 2.5 m represents the optimum balance between generalisation (decreasing spatial resolution) and data quantity (increasing spatial resolution)

the level of generalisation is increased and the total area of cells which lie at the river channel boundary location increases in proportion to the total area of the study reach. In contrast, increasing the spatial resolution provides little practical benefit to the overlay analyses and significantly increases the processing time involved in handling the topographic information. In this example, choosing a spatial resolution with cell sizes 0.75 m by 0.75 m would increase the number of cells to be processed by a factor of 16 and would offer little benefit in terms of overall analysis accuracy.

The estimation of the Total Mapping Error is based upon the propagation of spatial errors derived from cartographic theory (Muller, 1987; Lodwick, 1989). The Total Mapping Error is considered to be a function of the component inherent and operational errors, such as that for a given line representation of the river channel boundary location, the true position of this boundary may be estimated as the probability of the true value lying at a distance from this line representation (Chrisman, 1982). Because we may use the residual error estimates and the digitising error estimates as surrogates for inherent and operational errors respectively, and because these estimates both display a normal distribution about a mean then we may attach confidence limits to both estimates. Using the residual error as a surrogate for inherent error then the exceedance probabilities provide an estimate of the likelihood of inherent errors exceeding a given threshold. Using the digitising error test as a surrogate to represent operational error than exceedance probability values may be estimated for any given scale of representation. A worst-case scenario may then be envisaged whereby for any given location along the line representation of the river channel boundary location, the probability of the line being displaced in the same direction by both the inherent and operational errors occurs, then:

Total Mapping Error = Σ(Inherent error + Operational error)

For example, for the 1:10 000 scale 1979 and 1:10 560 scale 1876 maps the Total Mapping Error may be estimated at 4.35 m and 8.89 m, respectively ($P < 0.025$). Therefore, when employing maps and aerial photographs at a scale of approximately 1:10 000 the results imply that channel planform changes in excess of about 5 m can be considered genuine and not the product of data transcription.

Cross-section adjustment

In view of the physical size and the scale of the planform adjustments occurring on the River Towy, it is fair to assume that a similar methodology may be successfully applied to any number of similar river channels given the supporting topographic information at an appropriate size and resolution. However, from the estimates of Total Mapping Error, it can be seen that as the size and the scale of the representation of planform changes diminish, so these channel changes become increasingly obscured by the errors inherent in the source material and through the operational procedures undertaken to estimate river channel change. Topographic information in cross-section typically provides an increased spatial resolution over planform information. However, as with planform information, estimates of river channel change

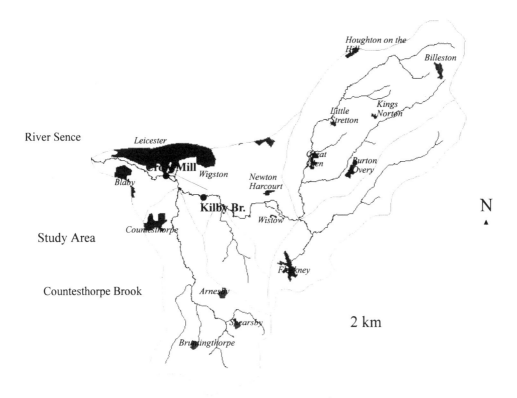

Figure 14.6 The River Sence study site, Leicestershire

derived from cross-section information is still subject to inherent and operational errors. Cross-section information obtained for a regraded reach of the River Sence in Leicestershire is used to illustrate the propagation of errors to a total error estimate. (Figure 14.6). It should be noted that while the cross-section record offers a considerably greater spatial resolution in comparison, for example, to the River Towy planform record, there is a reduction in the available time-scale of observation because of the more restricted availability of surveyed cross section information for many rivers. However, the cross-section records were analysed in this example case specifically to establish quantitative estimates of adjustments pre-regrade and post-regrade (1973) for which there is sufficient temporal coverage. Thirty-seven cross-sections were relocated for surveys undertaken in 1966 and 1971, and engineers design cross-sections for the 1973 regrade. This historical cross-section information was augmented by a re-survey in 1993. Cross-section information relocated at specific sites allows the direct comparison of river channel cross-section form at each individual location. The indices of change may then be observed in planform to observe their spatial distribution along the study reach.

Inherent errors in the cross-section information result from errors in the relocation of the cross-sections and errors in obtaining the cross-sectional information. Lawler (1993) suggests that, ideally, cross-sections should be marked out with pegs to aid relocation but, when working retrospectively, this may not be possible. In the absence of markers of the cross-section locations, the field relocation is achieved through accompanying cartographic evidence depicting the locations of the sections relative to the channel planform. Whilst not the case with the River Sense information, problems may arise if the scale of representation of the planform is too crude to confidently locate the cross-section, or in instances where the river channel planform has adjusted significantly between surveys, causing the new channel location to bear little resemblance to that at the time of survey. In the case of the River Sence study, relocation in the field was achieved using a compass and bearing from the map sheet to fixed features in the surrounding landscape. Repeat attempts by different surveyors demonstrated that the maximum variance in relocating cross-section locations was of the order of 5 m for the 1993 re-survey.

Errors in obtaining the cross-section information are most likely to result from the surveyor's perception of cross-section form in the field rather than through systematic errors in the equipment used to obtain the topographic information. For example, the equipment used in the 1993 re-survey was able to delimit vertical changes to within a tolerance of approximately 5 mm over a distance of approximately 20 m. The ability of the surveyor to detect vertical changes as subtle as 5 mm is limited, given vegetation cover or channel bedforms which are submerged. It is thus more likely that the error element introduced by the surveyors' perception of the breaks of slope is the more significant in accurately defining channel cross-section form. The nature of breaks of slope will influence the degree to which they may be identified, and there is greater confidence in the ability of the surveyor to locate the boundaries between the floodplain surface and a near-vertical bank than, say, a gently sloping bar assemblage. This type of error may be compounded by seasonal variations in vegetation cover and flow depth which may obscure topographic features. There is no clear ruling to determine the horizontal spacing between points on a channel cross-section survey. In order to preserve continuity with the earlier surveys for the

River Sence, the 1993 repeat survey chose unequal horizontal divisions instead of a fixed survey interval, thus identifying breaks of slope in the field. Lawler (1993) notes that this method is more effective in the description of channel profiles which commonly have locally well-defined breaks of slope separated by small distances which may otherwise be missed by surveys using an equal horizontal spacing approach.

When quantitative indices of the channel cross-section are not required, then the vector representation of the cross-section form plotted at a common horizontal and vertical scale will serve to manually overlay historical plots to indicate change. Quantitative estimates require a fuller understanding of operational errors inherent in obtaining the information. The cross-section records available for the River Sence were available in a statistical format, so that the individually surveyed points for each cross-section were referenced in terms of their horizontal and vertical separation. Therefore, unlike the planform record, which suffered from the transcription of information through digitising, the cross-section information could be analysed directly to obtain indices of cross-sectional form. In order to estimate cross-section change it was necessary to define a common datum for comparison. This was defined by the channel bankfull level which was represented morphometrically from the topographic information. Minimum width–depth values were found to be inappropriate in the case of the River Sence because, at the level of resolution provided by the cross-section information, the difference in width–depth ratios obtained at individual breaks of slope within the section were very small and extremely sensitive to small changes in the channel width. As a result, small scale surveying errors (as little as 5 mm) could in some cases give rise to a completely different estimate of the minimum width–depth ratio. Instead, the bankfull level was determined on the basis of the relationship between the individually surveyed points defining changes in slope angle between the channel banks and the floodplain. Field observations for the River Sence demonstrated that the level of the floodplain may itself be uncertain immediately adjacent to the channel, especially where the breaks of slope are poorly defined and the bank tops grade gently to the floodplain surface. Cross-section information derived for the River Sence also demonstrated that the floodplain levels themselves may vary on either side of the channel, compounded in this instance by the frequent recording of a 'lip' of material at the banktop level, which is believed to result from material dumped from the channel regrade. Riley (1972) provided a refined approach to define the bankfull level with the estimation of a 'bench index'. The bench index is defined as:

$$\text{Bench Index} = BI = ((w^i) - (w^{i+1}))/((d^i) - (d^{i+1}))$$

The index considers the relative slope between every surveyed point in the cross-section (Figure 14.7) where w = width, d = depth and i = rank order of depth of each bank. The channel bankfull level is defined by the first bench index minimum, the bench index value with the highest elevation. The accuracy of this approach in part depends upon the spacing of the survey coordinates, but does not suffer from such a high degree of error sensitivity as minimum width–depth value. Having established a common datum for comparison of cross-section form, statistics representing bankfull cross-section area, bankfull width, bankfull depth and other parameters of cross-section form may then be derived for each cross-section location on

Left bank data

i	Width (i)	Depth (i)	BI
1	14.68	2.56	19.8
2	12.73	2.46	5.95
3	11.6	2.27	4.44
4	8.76	1.63	5.03
5	7.2	1.32	4.03
6	5.67	0.94	3.2
7	3.78	0.35	•
8	•	•	•

Right bank data

i	Width (i)	Depth (i)	BI
16	•	•	•
15	•	•	•
14	10.05	2.02	4
13	7.73	1.44	4
12	6.33	1.09	2.92
11	4.78	0.56	4.96
10	3.64	0.33	10.05
9	1.93	0.15	•
8	•	•	•

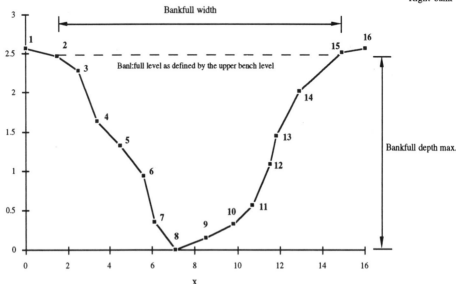

Bankfull level defined at $i=2$, from which statistics obtained :

Bankfull cross-section area = 17.09 square metres
Bankfull width = 12.73 metres
Bankfull depth maximum = 2.46 metres

Figure 14.7 Estimation of the channel bankfull level employing Riley's bench index (Riley, 1972) in order to provide a consistent datum to obtain indices of channel form. Example for cross-section number 67 on the River Sence from a re-survey in 1993

each survey date. Comparison of like indices, for the same cross-section location, selected at different dates from the historical record may then be used to discriminate cross-section change. For example Figure 14.8 illustrates the spatial distribution of bankfull cross-section area change pre-regrade (1966–1971) and post-regrade (1973–1993) for the River Sence study site.

    Cross-section records do not offer the same flexibility for determining absolute quantitative indices of error as planform information. Whereas the adoption of residual values and digitising errors offer surrogates for the estimation of inherent and operational errors, respectively, for historical planform information no such surrogates exist for cross-section error estimation. Because data are typically obtained in statistical or digital format, they do not undergo positional transcription error prior to analysis; therefore, positional errors can only be attributed to those from the ori-

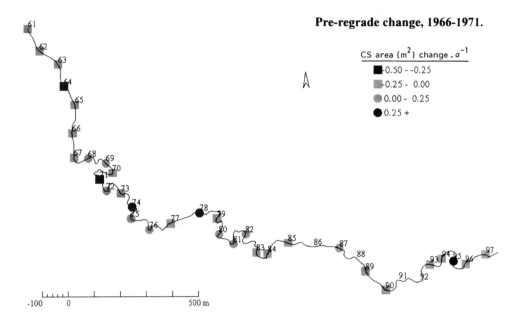

Pre-regrade change, 1966-1971.

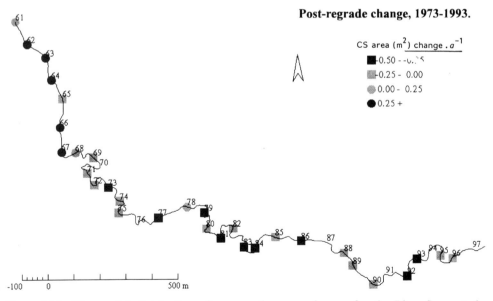

Post-regrade change, 1973-1993.

Figure 14.8 The spatial distribution of cross-section area changes for the River Sense study site. Channel change estimates are spatially referenced and analysed here employing SPANS MAP GIS

320

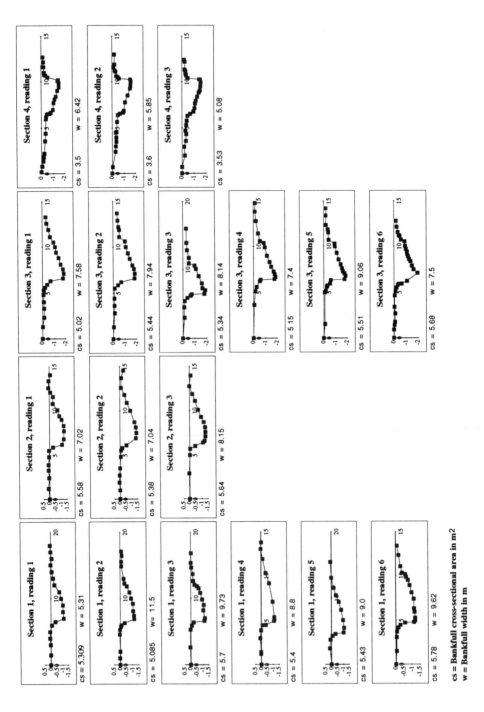

cs = Bankfull cross-sectional area in m2
w = Bankfull width in m

Figure 14.9 Estimates of channel form indices for the repeated survey of four cross-sections on the Highland Water, Hampshire

ginal survey. In the absence of corroborating evidence to compare with the historical cross-section information for the River Sence at a similar resolution, there is no immediate indication of the likely magnitude of surveyor error. The likely magnitude of surveying error can be estimated from repeat surveys of the same cross-section by several surveyors. In order to provide an indication of these errors an experiment was conducted on the Highland Water in the New Forest, Hampshire, to obtain and compare cross-section form measurements by repeat survey. As the experiment was designed to compare surveyor's relative error, four channel cross-sections were each surveyed independently from three to six times by three different pairs of surveyors. The survey methods employed the same equipment as the 1993 resurvey of the River Sence but interpretation of the cross-section form was determined solely by the individuals undertaking the survey. Bankfull width and bankfull cross-section area are estimated in a consistent manner for each cross-section and are presented in Figure 14.9. From these repeat surveys, it can be seen that estimation of the channel bankfull width is the most error-sensitive of the measured variables. For example, Section 1 shows a variation in width estimate ranging from 5.31 m to 11.5 m. The estimates of bankfull cross-section area appear to be less sensitive to surveying errors.

The estimation of Total Mapping Error from cross-section information, as with planform information, can be considered a function of the sum of the inherent and operational errors. Provided a constant definition is used to define channel form indices, operational errors cannot be quantified as there is no loss of positional information when data is obtained and analysed in a digital format. The estimation of Total Mapping Error must therefore be made on the basis of the likely positional errors incorporated at survey. The test conducted on cross-sections of the Highland Water provides an indication of the likely magnitude of these errors.

CONCLUSIONS

Topographic information provides an invaluable input to the reconstruction of historical river channel changes. In view of the spatial and temporal variability in the scales of adjustment that may occur within the fluvial landscape, it is important to closely match the resolution of observation to the scale of the phenomenon being studied. Topographic information and the operations employed to discriminate changes introduce scale-dependent errors to the information which are propagated through all stages of data handling from collection to analysis. An awareness of such errors, and where possible, error quantification is important if estimates of river channel change are to be sensibly identified and interpreted.

ACKNOWLEDGEMENTS

S.R. Downward gratefully acknowledges the receipt of a NERC research studentship CASE with the National Rivers Authority, Thames Region. Jim Milne and James Feaver are acknowledged for their advice in tricky numerical matters, Angela Gurnell for reading an earlier draft of this chapter, and Ken Gregory for the initial inspiration, guidance and direction of this research.

REFERENCES

Burrough, P.A. (1986) *Principles of Geographical Information Systems for Land Resources Assessment.* Monographs on Soil and Resources Survey Number 12, Oxford University Press, Oxford.

Chrisman, N.R. (1982) A theory of cartographic error and its measurement in digital data bases: *Proceedings AUTO-CARTO* **5**, 159–168.

Harvey, A.M. (1969) Channel capacity and the adjustment of streams to hydrologic regime. *Journal of Hydrology* **8**, 82–98.

Hooke, J.M. and Kain, J.P. (1982) *Historical Changes in the Physical Environment: A Guide to Sources and Techniques.* Butterworth, London.

Hooke, J.M. and Redmond, C.E. (1989) River-channel changes in England and Wales. *Journal of the Institution of Water and Environmental Management* **3**, 328–335.

Hooper, I.D. (1992) Relationships between vegetation and hydrogeomorphic characteristics of British riverine environments: A remotely sensed perspective. Unpublished PhD thesis, University of Southampton.

Lawler, D.M. (1993) The measurement of river bank erosion and lateral channel change: A review. *Earth Surface Processes and landforms* **18**, 777–821.

Leopold, L.B., Wolman, M.G. and Miller, J.P. (1964) *Fluvial Processes in Geomorphology.* Freeman, San Francisco.

Lewin, J. (1984) River Towy erosion study. A report to the west Wales division, Welsh Water Authority.

Lewin, J. and Brindle, B.J. (1977) Confined meanders. In Gregory, K.J. (ed.), *River Channel Changes*, Wiley, Chichester, 221–233.

Lewin, J. and Chisholm, N.W.T. (1986) Fluvial geomorphology effects of gravel extraction from the River Towy. Technical Supplement: Rates of river migration on the River Towy (Llywnjack and Llanwrda sites). Scientific Services, Welsh Water Authority.

Lodwick, W.A. (1989) Developing confidence limits on errors of suitability analysis in geographical information systems. In: Goodchild, M. and Gopal, S. (eds), *Accuracy of Spatial Databases*, Taylor & Francis, London, 69–78.

Maffini, G., Arno, M. and Bitterlich, W. (1989) Observations and comments on the generation and treatment of error in digital GIS data. In Goodchild, M. and Gopal, S. (eds), *Accuracy of Spatial Databases*, Taylor & Francis, London, 55–67.

Mandelbrot, B. (1982) *The Fractal Geometry of Nature*, Freeman, New York.

Muller, J. (1987) The concept of error in cartography. *Cartographica* **24**, 1–5.

Newcomer, J.A. and Szajgin, J. (1984) Accumulation of thematic map errors in digital overlay analysis. *The American Cartographer* **11**, 58–62.

Newson, M.D. and Leeks, G.J. (1986) Fluvial geomorphological effects of gravel extraction from the River Towy. Technical Supplement. Scientific Services, Welsh Water Authority.

Noorbergen, H.H.S. (1993) Multitemporal analysis of a braiding river. In: *Proceedings International Symposium, Operationalisation of Remote Sensing*, vol. 3, pp. 161–168.

Riley, S.J. (1972) A comparison of morphometric measures of bankfull. *Journal of Hydrology* **17**, 23–31.

Schumm, S.A. (1991) *A Scientific Approach to Earth Science: Ten Ways to be Wrong.* Cambridge University Press, Cambridge.

Smith, S.A. (1987) 'Gravel counterpoint bars: examples from the River Towy, South Wales. In Ethridge, F.G., Flores, R.M. and Harvey, M.D. (eds), *Recent Developments in Fluvial Sedimentology*, Society of Economic Paleontologists and Mineralogists Special Publication **39**, pp. 75–81.

Smith, S.A. (1989) Sedimentation in a meandering gravel bed river: the River Towy, South Wales. *Geological Journal* **24**, 193–204.

Thapa, K. and Bossler, J. (1992) Accuracy of spatial data used in geographic information systems. *Photogrammetric Engineering and Remote Sensing* **58**, 835–841.

Trimble, S.W. and Cooke, R.U. (1991) Historical sources for geomorphological research in the United States. *Professional Geographer* **43**, 212–228.

Vitek, J.D., Walsh, S.J. and Gregory, M.S. (1984) Accuracy in Geographic Information Systems: An assessment of inherent and operational errors. In: *Proceedings PECORA IX Symposium*, pp. 296–302.

Walsh, S.J., Lightfoot, D.R. and Butler, D.R. (1987) Recognition and assessment of error in Geographic Information Systems. *Photogrammetric Engineering and Remote Sensing* **53**, 1423–1430.

Wolman, M.G. (1955) *The Natural Channel of the Brandywine Creek, Pennsylvania.* United States Geological Survey Professional Paper 271.

Woodyer, K.D. (1968) Bankfull frequency in rivers. *Journal of Hydrology* **6**, 114–142.

15 Information from Channel Geometry–Discharge Relations

GERALDENE WHARTON

Department of Geography, Queen Mary and Westfield College, University of London, UK

INTRODUCTION

Rivers are open systems whose form and behaviour are a response to environmental factors (climate, geology, land use and basin physiography) with direct control primarily from water and sediment discharge (Knighton, 1984, p. 2). The human impact is also significant through river channelisation (Winkley, 1982; Brookes, 1988), river regulation (Petts, 1984) and land use changes (Hammer, 1972; Hollis, 1979). River channel adjustments are achieved through the processes of entrainment, transport and deposition of sediment, but the changes can be described in terms of nine categories of possible adjustments (Gregory, 1976) or four interdependent degrees of freedom: cross-sectional form; bed configuration; channel pattern; and channel bed slope (Knighton, 1984, p. 89). The dominant controls on cross-sectional form, the component of channel change on which this chapter focuses, are discharge (Leopold and Maddock, 1953; Knighton, 1987), hydrologic regime (Harvey, 1969; Stevens *et al.*, 1975; Osterkamp, 1980), the absolute and relative amounts of bedload transport (Pickup, 1976; Kirkby, 1977; Parker, 1979), and the composition of the channel boundary (Schumm, 1960, 1971; Ferguson, 1973), including the role of vegetation, particularly as it relates to bank stability (Charlton *et al.*, 1978).

Of great significance in river channel management is the ability of natural rivers to develop characteristic forms for relatively constant conditions of the controlling variables (Brunsden and Thornes, 1979), and to return approximately to their previous state following disturbance. This self-regulatory behaviour is dramatically illustrated by artificially straightened channels which regain their former sinuosity in the absence of bank protection measures (Brookes, 1987). Although true stability (static equilibrium) never exists in natural rivers, except over very short periods of time, a steady-state equilibrium may be achieved over the short to medium term (10–100 years) with adjustments occurring about an average channel geometry. This state of balance, described by Schumm (1977) as when there is *no progressive change* in channel form through aggradation or degradation, is the design objective of river engineers and the equilibrium of 'regime theory' may be classified as this type. In contrast, changes in the environmental controls over the longer term can lead to a state of dynamic equilibrium (Chorley and Kennedy, 1971) in which river channel

Changing River Channels. Edited by Angela Gurnell and Geoffrey Petts.
© 1995 John Wiley & Sons Ltd.

changes may be more dramatic, such as pronounced channel incision leading to the development of terrace sequences (Womack and Schumm, 1977). Equilibrium concepts imply that a variety of stable river forms exist which are adjusted to a particular set of environmental controls. For example, slope–discharge relationships (Lane, 1957; Leopold and Wolman, 1957; Henderson, 1963; Ackers and Charlton, 1970b; Osterkamp, 1978b) have been widely used to discriminate between channel pattern types (straight, meandering and braided) and a picture emerges of a continuum of channel pattern from straight to meandering to braided associated with increasing slope, stream power and bedload transport (Bagnold, 1977).

Attempts to identify stable river forms for the management of river channel change are complicated by the fact that the channel cross-section, planform, bed configuration and longitudinal profile are adjustable over a range of spatial and temporal scales which implies a different ability to absorb change and assume an equilibrium form (Knighton, 1984, p. 90). Thus, present channel form may be the product of past as well as present processes. For example, rivers directly affected by events during and immediately after the last glaciation may have retained some features from that time. The fact that slope correlates well with residual bed material size but not average bed material size along the River Hodder in Lancashire (Wilcock, 1967) suggests that the river is graded to conditions which are a legacy of the Pleistocene and that slope is largely an imposed characteristic. The length of this 'memory' for past events depends on the inherent resistance of the system, or any component of that system, to respond to change, on the sequence of past events and their magnitude and frequency (Wolman and Miller, 1960). The size of a flood's impact and its persistence also depends partly on whether or not a significant threshold is exceeded (Schumm, 1973). Thus, sensitivity to change varies from one environment to another, and from one channel form to another, in a way that is not completely understood. However, the evidence of rivers subject to altered hydrologic conditions, for example Burkham's (1972) study of historical changes along the Gila River, indicates that cross-sectional form is one of the most adjustable components of channel geometry, at least in the width dimension, and should indicate present flow conditions. This fact has underpinned the development of relations between channel cross-sectional form and discharge.

Traditional river engineering practices, in a bid to exert control over natural river behaviour, have frequently necessitated further costly construction and maintenance procedures to combat the river's desire for self-adjustment to imposed conditions. Straightened river channels frequently require bank reinforcements to restrain river channel erosion, particularly in high energy environments, whilst in over-enlarged channel sections, regular dredging operations may be necessary to remove excess sediment (Brookes, 1985). River managers are now increasingly concerned to use the knowledge of river channel adjustment and equilibrium in alleviating river erosion and flooding whilst minimising the environmental impact on the fluvial system, as advocated by the Design with Nature school of thought (McHarg, 1969), and embodied in the concept of 'geomorphic engineering' (Coates, 1976), and Leopold's (1977) 'reverence for rivers' (see also Chapter 17). Recent studies of channel form adjustments have shown, for example, how meandering channels are inherently more stable than straight ones in addition to being ecologically more acceptable and aesthetically more pleasing (Hey, 1982). Furthermore, the use of river corridors may be the sensible economic and ecological river management option in less densely populated drainage basins.

To inform river channel design and the management of river channel changes an understanding of the relationship between river channel morphology and flow characteristics is paramount. Although Schumm's (1969, 1977) qualitative process–response equations give the most likely channel adjustments to be expected as a result of changes in discharge and bedload, such descriptive relationships are far from adequate. This chapter considers two sets of empirical regression-type equations (see Appendices 1, 2 and 3) which have attempted to quantify the nature of the channel form–discharge relation. Hydraulic geometry and regime equations can be used to predict stable channel dimensions necessary for river channel design and restoration, whereas channel geometry equations can assist river management through the prediction of flood discharge characteristics from channel dimensions at ungauged sites and by indicating the sensitivity of particular channel reaches to change and the likely direction of channel form adjustment.

HYDRAULIC GEOMETRY EQUATIONS

Until the pioneering work of Anglo-Indian engineers in the late 19th century, irrigation canals were built with arbitrarily selected channel dimensions, with the consequence that they often suffered severe sedimentation during self-adjustment. Some of the canals though were observed to have remained stable, evidently having been built to regime dimensions intuitively, and they formed the basis for the development of regime equations to facilitate the design of stable channels under similar geotechnical conditions.

Derived initially for canals with a steady flow and fine sediment, 'regime theory' consists of a set of empirical equations which can be manipulated to give the width, depth and slope of an approximately stable channel whose cross-sectional form is maintained by a local balance between erosion and deposition. The relationship between the geometric and hydraulic characteristics can be expressed most simply by the form

$$y = aQb$$

where y is some channel dimension (e.g. width or depth) dependent upon a reference discharge Q (e.g. bankfull flow), whilst a and b are constants derived to achieve the best fit to measured data. The approach recognises discharge as the most important parameter responsible for the geometrical shape of the channel. Consequently, the influence of other controlling factors, such as sediment discharge and slope, is effectively incorporated in the coefficient a and the exponent b, giving a potentially greater variance to the errors incurred for wide-ranging conditions. Over a narrow range of environmental conditions, the single-variable equation is satisfactory as the range of variance of the other independent variables is minimised and their combined influence is suitably reflected by the constants a and b.

Appendix 1 illustrates the variety of regime equations that have been developed and the progression from 'classical' regime studies based on straight, alluvial, irrigation canals to equations for the design of natural rivers. Lacey (1930) was the first to introduce explicitly into the equations an external sedimentological constraint, 'the silt factor', a coefficient dependent upon the size of sediment transported, which offered the opportunity to develop equations that had a wide applicability. The relationships produced by Lacey (1930) are written in the form derived by Inglis (1949) for design purposes, with each dependent variable being expressed as a function of discharge and the silt factor.

Blench (1969) later postulated that the stable channel geometry must be dependent upon not only the sediment in transport, which would determine the composition of the bed, but also upon the materials composing the banks. In attempting to incorporate these factors he introduced a bed factor (F_b) and a side factor (F_s) to the equations defining width, depth and slope. However, if F_b and F_s cannot be obtained from measurements from similar channels there is difficulty in quantifying sedimentological conditions and F_s and F_b may be poorly defined. In 1963, Simons and Albertson partially overcame the problems associated with the quantification of sedimentological factors by classifying channels into five categories on the basis of a qualitative description of bed and bank composition. The result is a set of five regime equations that together cover a wide range of conditions.

The extension of regime theory from alluvial irrigation canals to natural river channels suffers from a number of difficulties in that conditions may be far from those originally specified. The main problems, as summarised by Blench (1969), are that fundamental differences exist between the properties of sand and gravel, the range of fluctuation of water and sediment flow may be relatively large, suspended load may be both large and permanent, and natural rivers are considered to have a further degree of freedom, namely plan configuration (Ackers, 1972), which is interrelated with cross-sectional geometry and precludes the simple geometric sections designed for canals. River regime is thus more complex than canal regime and we should not expect the same equations to apply without the development of new concepts.

Implicit in adopting the framework of classical regime formulae for the development of equations applicable to natural rivers is the need to relate geometric properties to a single characteristic discharge. Although natural rivers experience a range of flows, it has been argued that there is a steady discharge which produces the same channel dimensions as the natural sequence of flow events and as such can be regarded as the dominant or channel-forming discharge. This concept was first put forward by Inglis (1949) and has made possible the application of regime-type formulae to rivers. However, a difficulty arises in selecting the channel-forming discharge from the wide range of flows experienced by natural rivers.

Dominant discharge can be defined in various ways: as the flow which determines particular channel parameters, such as meander wavelengths (e.g. Ackers and Charlton (1970a) defined the dominant discharge as the steady flow that would yield the same meander wavelength as the observed range of varying flows within which the steady flow lies); or as the flow which performs most work, where work is defined in terms of sediment transport (Wolman and Miller, 1960). This is the flow which cumulatively transports the most sediment. Although flood flows may individually transport greater loads they are so infrequent that the smaller, more frequent flows are cumulatively responsible for transporting the most sediment.

Evaluating the most effective flow for sediment transport requires both flow and sediment discharge records. Whilst flow data are readily available from surface water archives, sediment discharge is seldom measured at gauging stations in the UK (Hey and Thorne, 1986). Although, as an alternative to measurement, sediment discharge may be estimated using theoretical expressions, such as the Meyer–Peter–Müller equations (Hey and Thorne, 1986), this is generally unsatisfactory. These difficulties have meant that other, less absolute, definitions of dominant discharge are more widely used.

Since it seems reasonable to suppose that river channels are adjusted on average to a flow which just fills the available cross-section, dominant discharge has been equated with bankfull flow. This assertion was based on an apparent consistency in the frequency with which bankfull discharge occurs along streams (Wolman and Leopold, 1957) and an approximate correspondence between the frequency of bankfull discharge and the frequency of the flow which cumulatively transports most sediment (Wolman and Miller, 1960). A link was thus established between dominant discharge, most effective discharge and bankfull discharge (see also Andrews, 1980) with an approximate recurrent interval of 1–2 years. This association is not without its problems (see Knighton, 1984, pp. 95–96), not least being the realisation that a range of flows are effective in determining channel geometry (Pickup and Warner, 1976). This limitation of the dominant discharge concept implies that regime equations should not be used to describe mean channel behaviour where large fluctuations in discharge are common (Stevens et al., 1975; Pickup and Reiger, 1979). The 1.5 year flood, annual series, closely responds to the bankful discharge and has also been suggested as the design discharge (Hey, 1982).

Despite the difficulties, the representation of the range of flows by a characteristic discharge has facilitated the generation of a large number of regime equations for the design of stable natural river channels (Appendix 1). Although the early regime studies of natural rivers continued to focus on straight alluvial channels, the equations presented by Bray (1982), Hey (1982) and Hey and Thorne (1986) extended the range of applicability of regime equations to the design of meandering mobile-bed river channels with a riffle–pool bed topography. They recognised the need to develop equations that can be used to determine the plan shape of a river for design purposes and which explicitly define the effect of sediment load on the hydraulic geometry of the channels. Regime equations for gravel-bed rivers are more complex because gravel-bed rivers possess seven degrees of freedom. They are free to adjust their velocity, hydraulic radius, slope, wetted perimeter, maximum flow depth, sinuosity and meander arc length through erosion and deposition (Task Committee, 1971; Hey, 1978) in response to the discharge, sediment load, bed sediment size, bank material type and valley slope. It follows that for stable channels the former are dependent variables and the latter independent variables (Hey, 1978).

Bray (1975, 1982) developed regime equations for gravel-bed rivers from a data set of 70 gravel-bed reaches in Alberta, Canada. Equations for width, depth, velocity and slope were developed as power functions of the two-year flood flow and a characteristic bed material size. The generalised hydraulic geometry exponents, $b = 0.527$, $f = 0.333$ and $m = 0.140$, for the downstream analysis of the Alberta gravel-bed data are in, or very close to, the modal class of the values for b, f and m, reported by Park (1977) for the downstream analyses of 72 rivers from around the world. In addition, the values of b, f, and m plot at the centre of the grouping on a triaxial graph of humid temperate climates as presented by Park. They also agree with the findings of Leopold and Maddock (1953) and suggest some consistency in the rates of width and depth adjustment. Certainly, width seems to vary approximately as the square root of discharge.

For the UK, Hey and Thorne (1986) have provided design equations to predict the bankfull dimensions of stable mobile gravel-bed rivers. The existing equations enable the plan geometry of the channel and the pool–riffle dimensions to be determined and make due allowance for the effect of vegetation on channel geometry. Furthermore, the relations indicate that bedload discharge has an important influence on channel slope, and bank vegetation has a major control on width, wetted perimeter and velocity. Conversely, width and wetted perimeter are unaffected by bed material size, and bank shear strength has no apparent influence on channel geometry.

Although the equations presented in Appendix 1 can, with certain limitations, be used to define the stable dimensions of gravel-bed rivers, they cannot be used to accurately predict the response of a channel to changes in flow regime, sediment supply or calibre of the bedload. Change in any of these factors can trigger erosional and depositional activity and eventually a new stable channel geometry will result. During the period of instability all the variables, both geometric and those relating to flow and sedimentary characteristics, are interdependent. Consequently, any change in the flow regime that produces instability will also affect sediment supply and the size of the bed material (Hey, 1982). Without any information on their new equilibrium values it is not possible to predict the new regime geometry of the channel. The equations can be used only to identify the general *direction of change*.

Perhaps the major problem of regime equations, in common with all empirical relationships, is that they are only applicable over the range of conditions from which they were derived. For instance, Charlton (1975) has shown that regime equations developed by Lacey and others can only be applied to straight channels, where sediment transport rates are low, width–depth ratios are > 5 and relative roughness lies between 3 and 80. Similarly, the design equations developed by Hey (1982) and Hey and Thorne (1986) are only applicable to gravel-bed rivers with characteristics lying within certain criteria. To ensure that equations can have general applicability it is necessary to obtain field data from a wide range of river environments in order to maximise the variance within and between variables (Hey, 1982).

As a first step towards further increasing the range of conditions covered by regime equations in the UK, Reeves (1994) developed a set of hydraulic geometry relations (Appendix 1) from a data set of channel geometry and flood discharge information for stable channels in equilibrium with their flow regime (Wharton, 1989). The criteria employed by Wharton (1989) for the selection of stream sites compares with the specifications proposed by Hey and Thorne (1986). The highest correlations were associated with the width expressions (Figure 15.1 is given as an example). This corresponds with conclusions of previous researchers (Nixon, 1959; Kellerhals, 1967; Parker, 1979; Hey, 1982; Hey and Thorne, 1986) who have found it most appropriate to represent width as independent of sedimentological factors. For this reason, width has often been expressed purely as a function of discharge. Comparison with results of previous studies shows how the value of the exponent for the overtopping equation derived for the width equation of 0.49, is close to the widely attained value of 0.5 that is now accepted as the norm (the exponent for the bank-

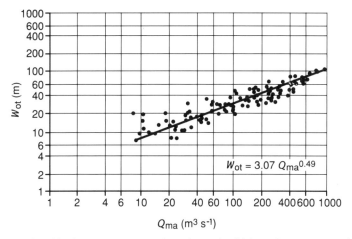

Figure 15.1 Relationship between overtopping channel width and mean annual flood. *Note:* All width values refer to those measured at the overtopping level. A distinction is drawn between the level of overtopping, defined as the level of incipient flooding or top overflow surface, and the level of the active floodplain or bankfull level (see Wharton, 1992, for a fuller explanation)

full equation is a little lower at 0.39). The values of the regression constants fall within the range of those for equations previously developed.

The low R^2 values for the depth relations developed by Reeves (1994) confirm that depth is poorly represented exclusively as a function of discharge. The difficulty of assigning a single variable to represent the depth of a reach arises because depth can be highly variable along reaches and across sections because of transient bedforms such as ripples and dunes in sand-bed channels and more permanent features such as riffles, pools and bars in gravel-bed streams. Consequently, bed configuration and depth values will reflect the magnitude and competence for bed material transport of the flows experienced (see Pickup and Warner, 1976), and channel depth is best defined by various measures of sediment (e.g. D_{50}) in conjunction with both discharge and slope.

Reeves' (1994) analysis of regression residuals (Figure 15.2) included visits to 12 sites exhibiting residuals between 0 and 50% and suggested that a baseflow regime may lead to high positive residuals. The observation that equations appear to overpredict for tree-lined channels corresponds to the findings of Hey and Thorne (1986), who report a reduction in the regression constant from 4.33 to 2.34 between 'no trees' and 'dense trees'. A similar lowering of the regression constants could overcome this problem with Reeves' (1994) equations.

The behaviour of natural channels is highly complex, and possibly indeterminate, and the problem of specifying the complete equilibrium geometry of channels with mobile beds remains. Although regime-type analysis continues to be important, other approaches include Lane's (1953, 1955) tractive force theory, the use of minimum variance theory (Langbein, 1964), the concept of minimum stream power (Chang, 1979, 1980), a principle of maximum transport efficiency (Kirkby, 1977) and a principle of lateral diffusion of turbulent momentum (Parker, 1979), formulated often in

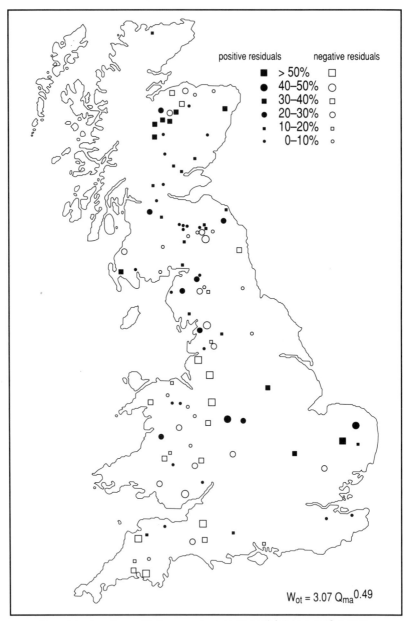

Figure 15.2 Map of % residual values for the relation between overtopping channel width and mean annual flood (after Reeves, 1994). See note to Figure 15.1

terms of hydraulic geometry relations. Indeed threshold theory has also been extended to derive exponent values in such relations (Li *et al.*, 1976).

CHANNEL GEOMETRY EQUATIONS

Relationships between channel dimensions as the independent variables and discharge as the dependent variable were first employed by Langbein (1960) to estimate flow geometry directly from channel geometry. Langbein initially estimated mean flow from average width as a check on extrapolations of streamflow data made from climatologic data and later devised relations for estimating mean flow from channel width and mean depth for ungauged streams in Nevada (Riggs, 1978). The underlying rationale of the channel geometry method is that the size and shape of the channel cross-section are the integrated resultant of all discharges conveyed by that channel. The equations take the form of power function relations (Osterkamp, 1978a)

$$Q = aW^b \quad \text{and} \quad Q = cA^d$$

and, once defined for a country or region, can be used to compute flow characteristics from channel size measured at a standard geomorphic reference level such as bankfull. Wharton (1992) has described in detail the procedures employed to develop channel geometry equations for British rivers.

Nixon's (1959) research on 22 rivers in England and Wales was also significant for the early development of the channel geometry method. Nixon's equations are listed in both Appendices 1 and 2 because, although he developed relations between bankfull discharge dimensions and bankfull discharges for the purpose of estimating discharge, the equations adopted the form of hydraulic geometry and regime equations in that channel dimensions are expressed as the dependent variables.

In subsequent decades, other investigators have extended Langbein's work, first by developing channel geometry relations for a range of environments beyond the arid regions studied initially and, secondly, by developing relations to estimate a variety of flood frequency characteristics (Appendix 2). The extension of the channel geometry method from arid to humid environments has been successful (Wharton *et al.*, 1989) and is consistent with observations that river channels in humid environments are adjusted to flood discharges of shorter return periods and are more likely to maintain average channel dimensions.

Thus, in the USA channel geometry equations have been developed for the following: four regions within the western United States (Hedman and Osterkamp, 1982); perennial streams in the Missouri river basin, with separate equations being developed for seven sediment categories (Osterkamp and Hedman, 1982); high gradient stream channels in Montana, Wyoming, South Dakota, Colorado and New Mexico (Osterkamp and Hedman, 1977); streams in the arid and subhumid parts of California (Hedman, 1970); streams in Kansas (Hedman *et al.*, 1974); streams in New Mexico (Scott and Kunkler, 1972); streams in Wyoming (Lowham, 1976); perennial streams in the mountain region of Colorado (Hedman *et al.*, 1972); and perennial, intermittent and ephemeral streams in south-eastern Montana (Omang *et al.*, 1983). Beyond the USA, channel geometry relations have been successfully

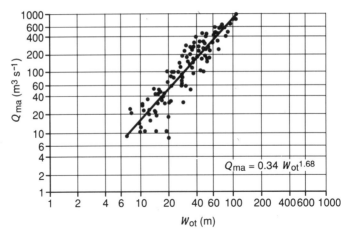

Figure 15.3 Relationship between mean annual flood and overtopping channel width. See note to Figure 15.1

defined for gravel-bed rivers in Canada (Bray, 1975); rivers in the Canadian North (Pitchen and Jolly, 1977); alluvial river channels in the South Island, New Zealand (Mosley, 1979); Piedmont, north-west Italy (Caroni, 1982); the Po river basin, northern Italy (Caroni and Maraga, 1984); British rivers (Wharton, 1992), as illustrated in Figure 15.3, and rivers in Java, Burundi and Ghana (Wharton and Tomlinson, 1995).

Whereas the first channel geometry equations were designed for the estimation of mean flows (Langbein, 1960), subsequent workers have developed equations to estimate mean annual runoff (Moore, 1968; Hedman, 1970; Omang et al., 1983); the mean annual flood (Osterkamp and Hedman, 1977, 1982; Mosley, 1979; Wharton, 1992; Wharton and Tomlinson, 1995); the two-year flood (Hedman et al., 1972, 1974; Hedman and Osterkamp, 1982; Omang et al., 1983); and the median annual flood (Caroni and Maraga, 1984). Few channel geometry studies have defined relations for ephemeral channels due to the limited number of ephemeral streams which have been gauged long enough to provide reliable streamflow records. However, initial results suggest that ephemeral channel widths are more indicative of unusual discharges than they are of mean discharge and such channels appear to be wider relative to total discharge the more highly ephemeral are the flow events. Thus, Osterkamp and Hedman (1979) have developed an equation relating the 10-year flood to the channel width for ephemeral streams (defined as channels in which discharge occurs on 10% or less of days) in strippable coal areas of arid and semi-arid regions, western USA; and Caroni (1982) has developed a relation for estimating the 50-year flood from the bankfull channel width of rivers in Piedmont, north-west Italy.

Channel geometry equations now exist for a range of river channel types (perennial, intermittent, ephemeral, gravel-bed and alluvial) in a variety of environments (arid, semi-arid, tropical and humid temperate) as illustrated in Appendix 2. To develop accurate channel geometry equations and achieve reliable discharge predictions it is essential to follow consistent procedures for the selection of measurement sites and the survey of river channel dimensions, as Wahl (1976, 1977) has demonstrated and Wharton (1992) has detailed. If inter-regional consistency is achieved it will be possible to directly compare channel geometry equations developed in different environments. This will help to improve our understanding of the relationship of river channel geometry to flow and ultimately the design and management of river channels.

A further potential contribution of channel geometry equations in river management was first recognised by Osterkamp and Hedman (1979) when they observed how changes in the discharge characteristics along a stream or over time, that otherwise might pass unnoticed, can be identified by channel geometry techniques if the departure is greater than the error associated with the model. For British rivers, local deviations from the national channel geometry relations (as shown graphically in Figure 15.3) were studied by examining the residuals from regression analyses. Maps of percentage residuals indicate not only the river locations where the worst and best flood discharge estimates are obtained but also the magnitude of local deviations from the national/average channel form–discharge relations. Thus, positive residuals, as mapped in Figure 15.4, result from an under-prediction of the mean annual flood by channel width implying either that the mean annual flood is larger, or the channel smaller, than one would expect given the national relation. Conversely, negative residuals result from an over-prediction of the mean annual flood, suggesting a situation where the observed flood discharge is smaller, or the channel wider, than one would predict from the national equation. Figure 15.4 reveals a concentration of high relative residuals in southern England and East Anglia, whereas the smallest residuals are recorded throughout the remainder of river sites in Britain. The exact locations of the river sites can be obtained by reference to the maps in Wharton (1992) giving details of gauging station numbers.

Thus, the residual sign may indicate whether the channel reach is under- or oversized and, therefore, the likely direction of channel changes as the river cross-section adjusts towards the national (average) channel form–discharge relation characteristic of the particular set of environmental conditions. In addition, the magnitude of the residual value could provide a measure of the sensitivity of the channel to change in the near future, in that a large percentage residual value reflects a large local deviation and may indicate a greater instability and an increased likelihood of channel change. The potential of this information on regression residuals to contribute to river channel management depends, however, upon the extent to which a national or regional channel geometry equation is indicative of the equilibrium or regime conditions for that environment. Further research is required.

SUMMARY

Studies of river channel changes and the controls on river channel adjustment have resulted in the development of two basic types of empirical relationships. First, a number of investigations have been devoted to establishing the discharges which have a significant influence upon river channel dimensions and have considered channel dimensions primarily as a response to the discharge regime. Hydraulic geometry relations, following the pioneering work of Leopold and Maddock (1953), have resulted in numerous power function equations showing the nature of the

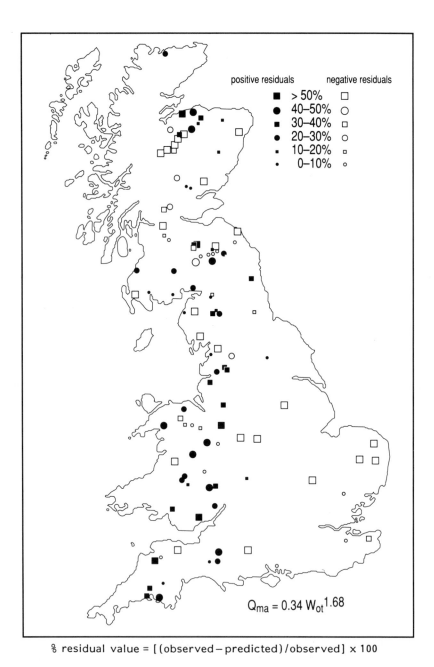

Figure 15.4 Map of % residual values for the relation between mean annual flood and over-topping channel width. See note to Figure 15.1

adjustment of channel width and depth to increasing discharge in the downstream direction for a variety of rivers in different environments. Relations between channel geometry and discharge developed for channels in equilibrium (regime equations) are employed in the design of stable river channels. Hey (1982) and Hey and Thorne (1986) have undertaken the most substantive work to date on regime equations for natural British rivers. The equations which they developed for meandering mobile gravel-bed rivers were important in reversing the previous concentration of regime studies on straight, alluvial channels arising from the early research on irrigation canals. However, to ensure that regime equations have general application it is necessary to obtain the field data from a wide range of river environments in order to maximise the variance within and between the variables (Hey, 1982). The simple equations developed by Reeves (1994) represent a first stage in an attempt to expand the coverage of UK river environments.

A second group of studies has led to the development of equations with channel dimensions employed as independent variables to estimate parameters of the discharge regime. Channel geometry equations now exist for a range of river channel types in a variety of environments and have been widely utilised. Flood discharge estimates are necessary for the design and appraisal of a variety of engineering structures used in the management of river channel changes and the channel geometry method provides a valuable reconnaissance technique or an alternative approach to traditional catchment-based methods of flood discharge estimation. However, of growing interest is the information which can be gained from the residuals of such regression relations. The channel geometry equations describe the average form of the relationship between selected flood discharge characteristics and river channel dimensions, under the particular set of environmental conditions for which the equations were developed. Thus residuals provide a basis for assessing local deviations in the national or regional channel form–discharge relations, deviations which could be employed as indicators of both the sensitivity of channel stretches of river to channel change and the direction of likely river channel adjustments.

ACKNOWLEDGEMENTS

The author is grateful to Anthony Reeves for his contribution to the section on hydraulic geometry and to Ed Olivier for drawing the diagrams.

REFERENCES

Ackers, P. (1972) River regime: research and application. *Journal of the Institution of Water Engineers and Scientists* **26**, 257–281.

Ackers, P. and Charlton, F.G. (1970a) Dimensional analysis of alluvial channels with special reference to meander length. *Journal of Hydraulics Research* **8**, 287–316.

Ackers, P. and Charlton, F.G. (1970b) The slope and resistance of small meandering channels. *Proceedings of the Institution of Civil Engineers* **47**, Supplementary Paper 7362-S, 349–370.

Andrews, E.D. (1980) Effective and bankfull discharges of streams in the Yampa river basin, Colorado and Wyoming. *Journal of Hydrology* **46**, 311–330.

Bagnold, R.A. (1977) Bed load transport by natural rivers. *Water Resources Research* **13**, 303–312.

Blench, T. (1969) *Mobile-bed Fluviology, A Regime Theory Treatment of Canals and Rivers for Engineers and Hydrologists*, 2nd edition. Alberta University Press, Edmonton.

Bray, D.I. (1975) Representative discharge for gravel-bed rivers in Alberta, Canada. *Journal of Hydrology* **27**, 143–153.

Bray, D.I. (1982) Regime equations for gravel-bed rivers. In Hey, R.D., Bathurst, J.C. and Thorne, C.R. (eds), *Gravel-bed Rivers*, Wiley, Chichester, 517–542.

Brookes, A. (1985) River channelization: traditional engineering methods, physical consequences and alternative practices. *Progress in Physical Geography* **9**, 44–73.

Brookes, A. (1987) The distribution and management of channelized streams in Denmark. *Regulated Rivers: Research and Management* **1**, 3–16.

Brookes, A. (1988) *Channelized Rivers: Perspectives for Environmental Management*. Wiley-Interscience, New York.

Brunsden, D. and Thornes, J.B. (1979) Landscape sensitivity and change. *Transactions of the Institute of British Geographers* NS**4**, 463–484.

Burkham, D.E. (1972) *Channel Changes of the Gila River in Safford Valley, Arizona, 1846–1970*. United States Geological Survey Professional Paper 655-G.

Caroni, E. (1982) I metodi empirici per la valutazione della portate. In: Marchi, E. (ed.), *Valutazione delle Piene*, Consiglio Nationale dell Ricerche, Conservazione del sulo, Dinamica Fluvial, Pubicazione 165, Rome.

Caroni, E. and Maraga, F. (1984) Flood prediction from channel width in the Po river basin. In: *Progress in Mass Movement and Sediment Transport Studies: Problems of Recognition and Prediction*, Torini, 5–7 December 1984, pp. 265–276.

Chang, H.H. (1979) Geometry of rivers in regime. *Journal of the Hydraulics Division, American Society of Civil Engineers* **106** (HY6), 691–706.

Chang, H.H. (1980) Geometry of gravel streams. *Journal of the Hydraulics Division, American Society of Civil Engineers* **106**, HY9, 1443–1456.

Charlton, F.G. (1975) *An Appraisal of Available Data on Gravel Rivers*. Report No. INT 151, Hydraulics Research, Wallingford, UK.

Charlton, F.G., Brown, P.M. and Benson, R.W. (1978) *The Hydraulic Geometry of Some Gravel Rivers in Britain*. Report No. IT 180, Hydraulics Research, Wallingford, UK.

Chorley, R.J. and Kennedy, B.A. (1971) *Physical Geography: A Systems Approach*. Prentice-Hall, London.

Coates, D.R. (ed.) (1976) *Geomorphology and Engineering*. Allen & Unwin, London.

Ferguson, R.I. (1973) Channel pattern and sediment type. *Area* **5**, 38–41.

Gregory, K.J. (1976) Changing river basins. *Geographical Journal* **142**, 237–247.

Hammer, T.R. (1972) Stream channel enlargement due to urbanization. *Water Resources Research* **8**, 1530–1540.

Harvey, A.M. (1969) Channel capacity and the adjustment of streams to hydrologic regime. *Journal of Hydrology* **8**, 82–98.

Hedman, E.R. (1970) *Mean Annual Runoff as Related to Channel Geometry of Selected Streams in California*. United States Geological Survey Water Supply Paper 199-E.

Hedman, E.R. and Osterkamp, W.R. (1982) *Streamflow Characteristics Related to Channel Geometry of Streams in Western United States*. United States Geological Survey Water Supply Paper 2193.

Hedman, E.R., Kastner, W.M. and Hejl, H.R. (1974) *Selected Streamflow Characteristics as Related to Active Geometry of Streams in Kansas*. State of Kansas Water Resources Board Technical Report No. 10.

Hedman, E.R., Moore, P.O. and Livingston, R.K. (1972) *Selected Streamflow Characteristics as Related to Channel Geometry of Perennial Streams in Colorado*. United States Geological Survey Open-File Report (200), H358s.

Henderson, F.M. (1963) Stability of alluvial channels. *Transactions of the American Society of Civil Engineers* **128**, 657–686.

Hey, R.D. (1978) Determinate hydraulic geometry of river channels. *Journal of the Hydraulics Division, American Society of Civil Engineers* **104** (HY6), 869–885.

Hey, R.D. (1982) Design equations for mobile gravel-bed rivers. In: Hey, R.D., Bathurst, J.C. and Thorne, C.R. (eds), *Gravel-bed Rivers*, Wiley, Chichester, 553–574.

Hey, R.H. and Thorne, C.R. (1986) Stable channels with mobile gravel beds. *Journal of Hydraulic Engineering* **112**, 671–689.

Hollis, G.E. (ed.) (1979) *Man's Influence on the Hydrological Cycle in the UK*. Geobooks, Norwich.

Inglis, C.C. (1949) *The Behaviour and Control of Rivers and Canals*. Research Publication 13, Central Water Power, Irrigation and Navigation Research Station, Poona, India.

Kellerhals, R. (1967) Stable channels with gravel-paved beds. *Journal of the Waterways and Harbors Division, American Society of Civil Engineers* **93** (WW1), Proc. Paper 5091, 63–84.

Kennedy, R.G. (1895) The prevention of silting in irrigation canals. *Proceedings of the Institution of Civil Engineers* **119**, 281–290.

Kirkby, M.J. (1977) Maximum sediment efficiency as a criterion for alluvial channels. In: Gregory, K.J. (ed.), *River Channel Changes*, Wiley, Chichester, 429–442.

Knighton, D. (1984) *Fluvial forms and Processes*. Edward Arnold, London.

Knighton, A.D. (1987) River channel adjustment: the downstream dimension. In: Richards, K.S. (ed), *River Channels: Environment and Process*, Blackwell, Oxford, 95–128.

Lacey, G. (1930) Stable channels in alluvium. *Proceedings of the Institution of Civil Engineers* **229**, 259–284.

Lane, E.W. (1953) Design of stable channels. *Proceedings of the American Society of Civil Engineers* **79**, 280-1–280-31.

Lane, E.W. (1955) Design of stable channels. *Transactions of the American Society of Civil Engineers* **120**, Paper 2776, 1234–1279.

Lane, E.W. (1957) *A Study of the Shape of Channels formed by Natural Streams Flowing in Erodible Material*. MRD Sediment Series 9, United States Army Engineer Division, Missouri River, Corps Engineers, Omaha, Nebraska, 106pp.

Langbein, W.B. (1960) *Hydrologic Data Networks and Methods of Extrapolating or Extending Available Hydrologic Data*. Hydrologic Networks and Methods, Flood Control Series No. 15. United Nations Economic Commission for Asia and the Far East, Bangkok.

Langbein,W.B. (1964) Geometry of river channels. *Journal of the Hydraulics Division, American Society of Civil Engineers* **90**(HY2), 301–312.

Leopold, L.B. (1977) A reverence for rivers. *Geology* **5**, 429–430.

Leopold, L.B. and Maddock, T. (1953) *The Hydraulic Geometry of Stream Channels and some Physiographic Implications*. United States Geological Survey Professional Paper 252.

Leopold, L.B. and Wolman, M.G. (1957) *River Channel Patterns—Braided, Meandering and Straight*. United States Geological Survey Professional Paper 282B, 39–85.

Li, R.M., Simons, D.B. and Stevens, M.A. (1976) Morphology of cobble streams in small watersheds. *Journal of the Hydraulics Division, American Society of Civil Engineers* **102**(HY8), 1101–1117.

Lindley, E.S. (1919) Regime channels. *Proceedings of the Punjab Engineering Congress* **7**, 63.

Lowham, H.W. (1976) *Techniques for Estimating Flow Characteristics of Wyoming Streams*. United States Geological Survey Water Resources Investigations 76–112.

McHarg, I.L. (1969) *Design with Nature*. Doubleday, Garden City, New York.

Moore, D.O. (1968) Estimating mean runoff in ungauged semi-arid areas. *International Association of Scientific Hydrology Bulletin* **13**(3), 66–76.

Mosley, M.P. (1979) Prediction of hydrologic variables from channel morphology, South Island rivers. *Journal of Hydrology (NZ)* **18**(2), 109–120.

Nixon, M. (1959) A study of bankfull discharges of the rivers in England and Wales. *Proceedings of the Institution of Civil Engineers* **12**, 157–174.

Omang, R.J., Parrett, C. and Hull, J.A. (1983) *Mean Annual Runoff and Peak Flow Estimates Based on Channel Geometry of Streams in Southeastern Montana*. United States Geological Survey Water Resources Investigations, 82–4092.

Osterkamp, W.R. (1978a) *Bed- and bank-material sampling procedures at channel-geometry sites*. Paper presented at the National Conference on Quality Assurance of Environmental Measurements, Denver, Colorado.

Osterkamp, W.R. (1978b) Gradient, discharge, and particle-size relations of alluvial channels in Kansas, with observations on braiding. *American Journal of Science* **278**, 1253–1268.

Osterkamp, W.R. (1980) Sediment-morphology relations of alluvial channels. In: *Proceedings of the Symposium on Watershed Management, American Society of Civil Engineers, Boise 1980*, 188–199.

Osterkamp, W.R. and Hedman, E.R. (1977) Variation of width and discharge for natural high-gradient stream channels. *Water Resources Research* **13**, 256–258.

Osterkamp, W.R. and Hedman, E.R. (1979) Discharge estimates in surface-mine areas using channel-geometry techniques. In: *Proceedings of the Symposium on Surface Mining Hydrology, Sedimentology and Reclamation*. University of Kentucky, Lexington, Kentucky.

Osterkamp, W.R. and Hedman, E.R. (1982) *Perennial-streamflow Characteristics Related to Channel Geometry in Missouri River Basin*. United States Geological Survey Professional Paper 1242.

Park, C.C. (1977) World-wide variations in hydraulic geometry exponents of stream channels: an analysis and some observations. *Journal of Hydrology* **33**, 133–146.

Parker, G. (1979) Hydraulic geometry of active gravel rivers. *Journal of the Hydraulics Division, American Society of Civil Engineers* **105**(HY9), 1185–1201.

Petts, G.E. (1984) *Impounded Rivers: Perspectives for Ecological Management*. Wiley, Chichester.

Pickup, G. (1976) Adjustment of stream-channel shape to hydrologic regime. *Journal of Hydrology* **30**, 365–373.

Pickup, G. and Reiger, W.A. (1979) A conceptual model of the relationship between channel characteristics and discharge. *Earth Surface Processes* **4**, 37–42.

Pickup, G. and Warner, R.F. (1976) Effects of Hydrologic regime on magnitude and frequency of dominant discharge. *Journal of Hydrology* **29**, 51–75.

Pitchen, M.J.R.G. and Jolly, J.P. (1977) *Flood Magnitudes in the Canadian North from Channel Geometry Measurements*. Department of Indian and Northern Affairs, Canada.

Reeves, A. (1994) *Hydraulic geometry equations for UK rivers*. Unpublished BSc thesis, Queen Mary and Westfield College, University of London.

Riggs, H.C. (1978) Streamflow characteristics from channel size. *Journal of the Hydraulics Division, American Society of Civil Engineers* **104**(HY1), 87–96.

Schumm, S.A. (1960) *The Shape of Alluvial Channels in Relation to Sediment Type*. United States Geological Survey Professional Paper 353B, 17–30.

Schumm, S.A. (1969) River metamorphosis. *Journal of the Hydraulics Division, American Society of Civil Engineers* **95**(HY1), 255–273.

Schumm, S.A. (1971) Fluvial geomorphology: the historical perspective. In: Shen, H.W. (ed.), *River Mechanics*, vol. 1, H.W. Shen, Fort Collins, Colorado, 4-1–4-30.

Schumm, S.A. (1973) Geomorphic thresholds and complex response of drainage systems. In: Morisawa, M. (ed.), *Fluvial Geomorphology*, New York State University Publications in Geomorphology, Binghamton, New York, 299–309.

Schumm, S.A. (1977) *The Fluvial System*. Wiley-Interscience, New York.

Scott, A.G. and Kunkler, J.L. (1972) *Flood Discharges of Streams in New Mexico as Related to Channel Geometry*. United States Geological Survey Open-file Report 76-414.

Simons, D.B. and Albertson, M.L. (1963) Uniform water conveyance channels in alluvial material. *Transactions of the American Society of Civil Engineers* **128**(1), Paper 3399, 65–167.

Stevens, M.A., Simons, D.B. and Richardson, E.V. (1975) Non-equilibrium river form. *Journal of the Hydraulics Division, American Society of Civil Engineers* **101**(HY5), 557–566.

Task Committee for Preparation of Sediment Manual (1971) Sediment transportation mechanics: H. Sediment discharge formulas. *Journal of the Hydraulics Division, American Society of Civil Engineers* **97**(HY4), 523–567.

Wahl, K.L. (1976) Accuracy of channel measurements and the implications in estimating streamflow characteristics. In: *Modern Developments in Hydrometry*, 2, World Meteorological Organisation, Padua, Italy, 311–319.

Wahl, K.L. (1977) Accuracy of channel measurements and implications in estimating streamflow characteristics. *United States Geological Survey Journal of Research* **15**, 811–814.

Wharton, G. (1989) *River discharge estimated from river channel dimensions in Britain*. Unpublished PhD thesis, University of Southampton.

Wharton, G. (1992) Flood estimation from channel size: guidelines for using the channel-geometry method. *Applied Geography* **12**, 339–359.

Wharton, G. and Tomlinson, J. (1995) Flood discharge estimation from river channel size in Java, Ghana and Burundi (in preparation).

Wharton, G., Arnell, N.W., Gregory, K.J. and Gurnell, A.M. (1989) River discharge estimated from river channel dimensions. *Journal of Hydrology* **106**, 365–376.

Wilcock, D.N. (1967) Coarse bedload as a factor determining bed slope. *Publication of the International Association of Scientific Hydrology* **75**, 143–150.

Winkley, B.R. (1982) Response of the Lower Mississippi to river training and realignment. In: Hey, R.D., Bathurst, J.C. and Thorne, C.R. (eds), *Gravel-bed Rivers*, Wiley, Chichester, 652–681.

Wolman, M.G. and Leopold, L.B. (1957) *River Flood Plains: Some Observations on their Formation*. United States Geological Survey Professional Paper 282C, 87–109.

Wolman, M.G. and Miller, J.P. (1960) Magnitude and frequency of forces in geomorphic processes. *Journal of Geology* **68**, 54–74.

Womack, W.R. and Schumm, S.A. (1977) Terraces of Douglas Creek, northwestern Colorado: an example of episodic erosion. *Geology* **5**, 72–76.

APPENDICES

Appendix 1. Hydraulic geometry equations

(a) Classical regime equations derived for design of straight irrigation canals with sandy beds and sandy or cohesive banks

Kennedy (1895) +	(a)	Manning-type resistance equation
	(b)	$V_o = 0.55\, D^{0.64}$

Lindley (1919) +	(a)	Manning-type resistance equation
	(b)	$V_o = 0.57\, D^{0.57}$
	(c)	$V_o = 0.28\, W^{0.56}$

Lacey (1930) +
$$P = 4.84\, Q^{0.5}$$
$$R = 0.47\, Q^{1/3}\, F_L^{-1/3}$$
$$V = 0.44\, Q^{-1/6}\, F_L^{-1/3}$$
$$S = 0.0003\, Q^{-1/6}\, F_L^{5/3}$$

Simons and Albertson (1963) +
$$W = 0.9\, k_1\, Q^{0.5}$$
$$D = 1.21\, k_2\, Q^{0.36} \qquad \text{for } R \leqslant 7 \text{ ft}$$
$$D = 2.0 + 0.93\, k_2\, Q^{0.36} \qquad \text{for } R > 7 \text{ ft}$$
$$S = 0.0000028 \text{ to } 0.71\, Q^{0.341} \qquad \text{depending upon nature of channel}$$

	k_1	k_2
sand-bed channels	3.5	0.52
sand-bed and cohesive banks	2.6	0.44
cohesive bed and banks	2.2	0.37
coarse non-cohesive material	1.75	0.23
sand-bed and banks with heavy sediment load	1.70	0.34

+ indicates Imperial units (discharge values expressed in cubic feet per second and channel dimensions expressed in feet).

Blench
(1969) +

$$W = \frac{F_b^{0.5}}{F_s^{0.5}} Q^{0.5}$$

$$D = \left[\frac{F_s}{F_b}\right]^{1/3} Q^{0.33}$$

$$S = \frac{F_b^{5/6} F_s^{1/12} x^{1/4}}{3.63 \ g \ Q^{1/6}}$$

F_s = 0.004 for loam of very slight cohesiveness
   = 0.018 for loam of medium cohesiveness
   = 0.027 for loam of high cohesiveness

(b) Regime equations derived for natural rivers

Leopold and Maddock (1953) +	river systems across the USA (average exponent values)	$W = * Q_{ma}^{0.50}$ $D = * Q_{ma}^{0.40}$	*coefficients vary for individual streams
Nixon (1959) +	22 gravel-bed rivers, England and Wales (29 river sites)	$W = 1.65 \ Q_b^{0.5}$ $D = 0.545 \ Q_b^{0.33}$	
Kellerhals (1967) +	Seven river reaches of the Quesnel, Cariboo, Taseko, Cilko and Thompson gravel rivers, in south-central British Columbia	$W = 1.8 \ Q_d^{0.5}$ $D = 0.166 \ Q_d^{0.40} D_{90}^{-0.12}$ $S = 0.120 \ Q_d^{-0.40} D_{90}^{0.92}$	

Charlton et al. (1978)   23 gravel-bed rivers, UK

(a) channels with negligible sediment load
$$W_b = 3.74 \ Q_b^{0.45}$$
$$D_e = 0.066 \ D_{65} \ S^{-1}$$
(b) channels with appreciable sediment load
$$W_b = 2.43 \ Q_b^{0.41} \ S^{-0.098}$$
$$D_e = 0.24 \ Q_b^{0.30} \ D_{90}^{0.24} \ S^{-0.20}$$

Bray (1982)	70 gravel-bed river reaches in Alberta, Canada	$W = 2.38 \ Q_2^{0.527}$ $D = 0.266 \ Q_2^{0.333}$ $S = 0.0354 \ Q_2^{-0.342}$
Paker (reported in Bray, 1982, p.543)	23 gravel-bed rivers, UK	$W = 3.73 \ Q_b^{0.446}$ $D = 0.308 \ Q_b^{0.398}$ $S = 0.00910 \ Q_b^{-0.244}$

	21 gravel-bed rivers, Alberta USA (single channel)	$W = 5.86\,Q_b^{0.441}$ $D = 0.308\,Q_b^{0.398}$ $S = 0.219\,Q_b^{0.331}$	
	30 gravel-bed rivers, Alberta (braided)	$W = 7.08\,Q_b^{0.417}$ $D = 0.292\,Q_b^{0.331}$ $S = 0.0081\,Q_b^{-0.0197}$	
Hey (1982)	66 sites on gravel-bed rivers, UK (for gravel-bed rivers in the UK the 1.5 year flood on the annual series can be used as the dominant discharge for design purposes)	$P = 2.20\,Q^{0.54}\,Q_s^{-0.05}$ $R = 0.161\,Q^{0.41}\,D_{50}^{-0.15}$ $D_m = 0.252\,Q^{0.38}\,D_{50}^{-0.16}$ $S = 0.679\,Q^{-0.53}\,Q_s^{0.13}\,D_{50}^{0.97}$ $p = S_v/S$ $Z = 2\,\pi W$	
Hey and Thorne (1986)	62 gravel-bed rivers, UK (hydraulic geometry equations— reach average)	$W = 3.98\,Q_b^{0.52}\,Q_s^{-0.01}$ $W = 3.08\,Q_b^{0.52}\,Q_s^{-0.01}$ $W = 2.52\,Q_b^{0.52}\,Q_s^{-0.01}$ $W = 2.17\,Q_b^{0.51}\,Q_s^{-0.01}$ $D = 0.16\,Q_b^{0.39}\,Q_s^{-0.02}\,D_{50}^{-0.15}$ $D = 0.19\,Q_b^{0.39}\,Q_s^{-0.02}\,D_{50}^{-0.15}$ $D = 0.20\,Q_b^{0.39}\,Q_s^{-0.02}\,D_{50}^{-0.15}$ $S = 0.087\,Q_b^{-0.43}\,Q_s^{0.10}\,D_{50}^{-0.09}\,D_{84}^{0.84}$	[Veg I] [Veg II] [Veg II] [Veg IV] [Veg I] [Veg II/III] [Veg IV]
	Practical design equations	$W = 4.33\,Q_b^{0.5}$ $W = 3.33\,Q_b^{0.5}$ $W = 2.73\,Q_b^{0.5}$ $W = 2.34\,Q_b^{0.5}$ $D = 0.22\,Q_b^{0.37}\,D_{50}^{-0.11}$ $S = 0.087\,Q_b^{-0.43}\,D_{50}^{-0.09}\,D_{84}^{0.84}\,Q_s^{0.10}$ $Z = 6.31\,W$ $p = S_v/S$ $RW = 1.034\,W$ $RD = 0.951\,D$ $RD_m = 0.912\,D_m$	[Veg I] [Veg II] [Veg III] [Veg IV] [Veg I–IV] [Veg I–IV] [Veg I–IV] [Veg I–IV]
Reeves (1994)	75 UK rivers (bankfull) 109 UK rivers (overtopping)	$W_b = 3.42\,Q_{ma}^{0.39}$ $D_b = 0.59\,Q_{ma}^{0.16}$ $W_{ot} = 3.07\,Q_{ma}^{0.49}$ $D_{ot} = 0.58\,Q_{ma}^{0.28}$	$R^2 = 0.77$ $R^2 = 0.26$ $R^2 = 0.83$ $R^2 = 0.53$

Appendix 2. Channel geometry equations

I   USA

i   Missouri river basin [Osterkamp and Hedman, 1982]
$$Q_{ma} = 0.027 \ W_a^{1.71} \qquad n = 252 \qquad R^2 = 0.93$$

ii   Western United States [Hedman and Osterkamp, 1982] 151 streamflow sites were divided
     into four regions for the development of channel geometry equations:
     Alpine and pine-forested + $Q_2 = 1.3 \ W_a^{1.65}$
     Northern plains and intermontane
       areas east of Rocky Mountains + $Q_2 = 4.8 \ W_a^{1.60}$
     Southern plains east of Rocky
       Mountains + $Q_2 = 7.8 \ W_a^{1.70}$
     Plains and intermontane areas
       west of Rocky Mountains + $Q_2 = 1.8 \ W_a^{1.70}$

iii Surface-mine areas of arid and semi-arid regions, western US [Osterkamp and Hedman,
     1979] +
     $$Q_{10} = 4.14 \ W_a^{1.63}$$

iv   High-gradient streams in Montana, Wyoming, South Dakota, Colorado and New
     Mexico [Osterkamp and Hedman, 1977]
     $$Q_{ma} = 0.017 \ W_a^{1.98} \qquad n = 32$$

v   Kansas [Hedman et al. 1974] +
     $$Q_2 = 22.3 \ W_a^{1.4} \qquad n = 120 \qquad R^2 = 0.88$$

vi   Perennial streams in the mountain region of Colorado [Hedman et al., 1972] +
     $$Q_2 = 0.991 \ W_d^{1.797} \qquad n = 53 \qquad R^2 = 0.89$$

vii Arid and subhumid parts of California [Hedman, 1970] +
     Perennial streams $Q_r = 186 \ W_d^{1.54} \ D_d^{0.88} \qquad n = 28$
     Ephemeral streams $Q_r = 258 \ W_d^{0.80} \ D_d^{0.60} \qquad n = 20$

viii Perennial, intermittent and ephemeral streams in south-eastern Montana [Omang et al.,
     1983]. South-eastern Montana was divided into three regions and separate relations
     developed for a range of peak flow discharges.

     Region 1: generally flat plains land and the area most affected by intense summer
     thunderstorms. Runoff is largely variable, with most smaller streams flowing only
     intermittently. Flood peaks are produced by prairie snowmelt and rainfall.
     $$Q_2 = 10.0 \ W_a^{1.16} \qquad n = 38$$

     Region 2: similar in topography to Region 1 but intense thunderstorms are not as
     prevalent. Flood peaks are not as variable or as large as in Region 1.
     $$Q_2 = 3.52 \ W_a^{1.59} \qquad n = 28$$

     Region 3: contains mountainous area and is generally forested. Annual precipitation is
     large, resulting in accumulated snowpack, and runoff occurs primarily as a result of
     snowmelt.
     $$Q_2 = 10.7 \ W_a^{1.14} \qquad n = 12$$

II   Great Britain

i   Rivers in England and Wales [Nixon, 1959] +
$$W_b = 1.65 \, Q_b^{0.05} \qquad n = 29$$

ii   Rivers in England, Scotland and Wales [Wharton, 1992]

$$
\begin{aligned}
Q_{ma} &= 0.20 \, W_b^{1.97} & n &= 75 & R^2 &= 0.78 \\
Q_{ma} &= 0.34 \, W_{ot}^{1.68} & n &= 109 & R^2 &= 0.83 \\
Q_{ma} &= 1.16 \, A_b^{1.31} & n &= 75 & R^2 &= 0.73 \\
Q_{ma} &= 1.20 \, A_{ot}^{1.07} & n &= 109 & R^2 &= 0.83
\end{aligned}
$$

III   Java [Wharton and Tomlinson, 1995]
$$Q_{ma} = 0.87 \, W_b^{1.60} \qquad n = 24 \qquad R^2 = 0.915$$

IV   Burundi [Wharton and Tomlinson, 1995]
$$Q_{ma} = 0.96 \, W_b + 1.44 \qquad n = 8 \qquad R^2 = 0.94$$

V   Ghana [Wharton and Tomlinson, 1995]
$$Q_{ma} = 1.8 \, W_b^{1.50} \qquad n = 12 \qquad R^2 = 0.71$$

VI   Canada: gravel-bed rivers in Alberta, Canada [Bray, 1975]
$$W_b = 4.75 \, Q_2^{0.527} \qquad n = 71 \qquad R^2 = 0.962$$

VII   New Zealand: alluvial river channels, South Island [Mosley, 1979]
$$Q_{ma} = 1.600 \, A_b^{0.900} \, ASPRAT^{-0.376} \, SLOPE^{-0.392} \, DMEAN^{0.287}$$
$$[R^2 = 0.903; \, n = 73]$$

VIII   Italy: Po river basin, northern Italy [Caroni and Maraga, 1984]
$$Q_{med} = 0.0450 \, W_b^{2.09} \qquad n = 30 \qquad R^2 = 0.95$$

Appendix 3. List of symbols

Q	water discharge (m^3 s^{-1} or ft^3 s^{-1})
Q_{ma}	mean annual flood (m^3 s^{-1} or ft^3 s^{-1})
Q_{med}	median annual flood (m^3 s^{-1})
Q_r	mean annual runoff (acre-feet per year)
Q_2	two-year return period flood (m^3 s^{-1} or ft^3 s^{-1})
Q_{10}	10 year return period flood (m^3 s^{-1} or ft^3 s^{-1})
Q_s	sediment (bedload) discharge (m^3 s^{-1} or ft^3 s^{-1})
Q_d	dominant discharge (m^3 s^{-1} or ft^3 s^{-1})
Q_b	bankfull discharge (m^3 s^{-1} or ft^3 s^{-1})
A	channel cross-sectional area (m^2 or ft^2)
W	mean waterline width (m or ft)
D	mean flow depth ($=A/W$) (m or ft)
D_m	maximum flow depth (m or ft)
D_e	effective depth of flow ($=D+D_{50}$ (ie D$_{50z}$) (m or ft)
W_b	channel width at bankfull level (m or ft)
A_b	channel cross-sectional area at bankfull (m^2 or ft^2)
D_b	mean channel depth at bankfull, calculated as A_b/W_b (m or ft)
W_a	channel width at the active channel reference level (m or ft)
W_d	channel width at the depositional bar reference level (m or ft)
D_d	mean channel depth at the depositional bar reference level (m or ft)

W_{ot}	channel width at the overtopping level (m or ft)
A_{ot}	channel cross-sectional area at the overtopping level (m^2 or ft^2)
P	wetted perimeter ($P \sim W$ for large channels (m or ft)
R	hydraulic radius ($R \sim D$ for large channels) (m or ft)
S	channel slope
S_v	valley axis slope
RW	mean waterline width, riffles (m or ft)
RD	mean flow depth, riffles (m or ft)
RD_m	maximum flow depth, riffles (m or ft)
Z	meander arc length (m or ft)
ρ	channel sinuosity
F_L	Lacey's silt factor
F_b	channel bed factor defined as V^2/D ($F_b = 0.58 D_{50}^{0.50}$)
F_s	channel side factor defined as V^3/W
x	kinematic viscosity of fluid ($cm^2 s^{-1}$)
g	acceleration due to gravity
V_o	Kennedy's non-silting velocity ($m s^{-1}$ or $ft s^{-}$)
V	flow velocity ($m s^{-1}$ or $ft s^{-1}$)
D_{50}	particle size of bed material, intermediate axis, such that 50% are finer
D_{65}	particle size of bed material, intermediate asix, such that 65% are finer
D_{84}	particle size of bed material, intermediate axis, such that 84% are finer
D_{90}	particle size of bed material, intermediate axis, such that 90% are finer
D_{50z}	particle size of bed material defined by the length of the minor axis such that 50% respectively of the particles by number are smaller
Veg I	grassy banks with no trees or shrubs
Veg II	1–5% tree/shrub cover
Veg III	5–50% tree/shrub cover
Veg IV	> 50% shrub cover or incised into floodplain
R^2	coefficient of determination
n	number of gauging station sites employed in the development of the channel geometry equations
$ASPRAT$	Aspect ratio ($DMAX/R$) [where $DMAX$ is maximum depth (m) and R is the hydraulic radius measured at the bankfull level]
$SLOPE$	bankfull channel slope
$DMEAN$	mean diameter of bed sediment (mm)

16 River Channel Classification for Channel Management Purposes

PETER W. DOWNS

Department of Geography, University of Nottingham, UK

INTRODUCTION

The desire of geomorphologists to classify river channels can be explained as a means of reducing an extremely complex environmental feature into a series of discrete units which facilitate further study or help organise management operations. Classifications provide a weak form of explanation because all schemes involve a set of criteria which relate to an *a priori* expectation of the way in which researchers believe their river channels to be distinguished. As the criteria for any one classification scheme are unlikely to be generally applicable for numerous uses, designs for classification tend to be specific to the intended *purpose* of that scheme, and this is one of a series of fundamental attributes of classification outlined by Grigg (1967), and summarised by Mosley (1987). Within fluvial geomorphology, the majority of river classifications have concerned natural river channel patterns, sub-dividing according to distinct morphological characteristics. These characteristics may indicate discrete physical processes for the particular channel category, thus facilitating an explanation of the resulting morphology. Since the middle of the 20th century, classification schemes, starting with Horton's (1945) ordering of river networks, and Leopold and Wolman's (1957) division of streams into 'straight', 'meandering' and 'braided' on the basis of their sinuosity, have become progressively more complex, utilising increasing numbers of criteria, involving multiple levels of study and resulting in greater numbers of class divisions. For instance, a recent scheme by Alabyan (1992) incorporates a hierarchy of structural, planform, and limiting conditions within which each of several channel types can exist. Recent reviews of geomorphologically-based natural river classifications include those by Church (1992), who discusses classification according to channel size, and Mosley (1987) who stresses the overlap between geomorphological and ecological classifications. Mosley (1987) also notes the interest in schemes of river *characterisation* rather than *classification*, stemming from the interest in the river continuum concept (Vannote *et al.*, 1980) and the transitional nature of morphological change in river channels (e.g. Ferguson, 1987). River classifications involve a sequential subdivision according to designated criteria whereas characterisations concurrently use multiple criteria which allow the formation of statistically distinct groupings. Growth in the number of hierarchical river classification schemes, especially those with an ecological basis or purpose, are reviewed in Naiman *et al.* (1992).

Changing River Channels. Edited by Angela Gurnell and Geoffrey Petts.
© 1995 John Wiley & Sons Ltd.

Now that the legislation associated with river channel management in many countries has become more environmentally aligned, the desirability of retaining or re-creating natural river features has permitted fluvial geomorphology to become integrated more fully into river channel management. With this integration has been an increase in geomorphological river classifications which are designed for, or have application to, river channel management. This may be because, as Rosgen (1994, p. 195) suggests,

> Rivers are complex natural systems. A necessary and critical task towards the understanding of these complex systems is to continue the river systems research. In the interim, water resource managers must often make decisions and timely predictions without the luxury of a complete and thorough data base. Therefore, a goal for researchers and managers is to properly integrate what has been learned about rivers into a management decision process that can effectively utilize such knowledge.

Clearly, river classifications are appropriate tools in this context, as they can be designed according to the level of available knowledge or data, allowing managers to act upon summaries of this information, and then be readily re-iterated as understanding improves.

RIVER CHANNEL MANAGEMENT CLASSIFICATIONS

The distinguishing feature of river channel classifications applied to management purposes is an emphasis on the processes which have created and act to maintain the channel, rather than on the river morphology. Thus, in opposition to Cowardin's (1982) assertion that classifications simply label objects without producing any information, it is the express purpose of applied schemes to inform river management decisions, for instance, to guide engineering designs or to assess the river's conservation value. Management applications not only explain the pre-eminence of process inference in the classification, but help to delimit a restricted range of temporal and spatial scales over which schemes normally operate. Classification schemes of natural river morphology have to reconcile interconnectivity of catchment-scale structural controls, reach-level channel pattern differences and micro-scale variations in channel bedforms, each of which varies over different time periods (see Frissell *et al.*, 1986; Naiman *et al.*, 1992) and it is these challenges which have led to interest in hierarchical classification schemes. Conversely, applied geomorphological classifications have tended to adopt both a meso-scale spatial outlook (i.e. a 'reach' level in the approximate range of 10^1-10^3 metres) with a meso-scale temporal concern (changes over the 10^0-10^2 year period) as a consequence of the link with practical management concerns and with civil engineering. Some channel management classifications have therefore developed as a particular level of a hierarchical classification scheme (see below). These spatial and temporal restrictions should not, however, be regarded as making channel management classification simple. Frissell *et al.* (1986) note that, of the five spatial hierarchies in their classification scheme, 'reach' systems can be the least physically discrete and, as reaches can display an inter-dependency of process and form, an understanding of both features is necessary for a successful classification. Furthermore, the potential influence of catchment characteristics (including human influences) and position in the river network means that reaches may appear to exhibit an homogeneity over two orders of magnitude between 10^1 and 10^3 metres in length (cf. example by Downs below). Therefore, it can be difficult for a surveyor to retain scale-independency when delimiting appreciable differences between reaches. However, because applied classification schemes tend to focus upon 'disturbed' (previously altered) rivers rather than on natural ones, the problem of delimiting boundaries between reaches is, in some cases, simplified from natural river classifications in which the geomorphologically progressive change in river channel characteristics inhibits recognition of homogeneous river reaches. This is because piecemeal channelisation works along with land use pressures which result in construction at or near to the channel edge help to segment the channel network.

Although they are designed for relatively restricted spatial and temporal interpretation, river channel management classification schemes show a wide variety of forms. This chapter illustrates a number of these differences. For instance, a major distinction may be drawn between schemes which proceed on the basis of existing features (morphological), in which information about current processes is inferred at a later stage, and methods which explicitly distinguish active processes from facets of the existing morphology and thus classify on the basis of river channel adjustment. Also, some schemes are designed to provide information about the river channel conservation value (whereby differences are noted between the morphology of the existing river channel and the morphology of the channel which would exist naturally without human disturbances), whereas others more directly serve the requirements for a channel management authority. In the section focusing on classification of adjustment processes, a further division is suggested between procedures for classifying prevailing changes which, in the absence of changes in controlling factors, may be expected to continue in the short-term (perhaps 10^1 years depending on the dynamism of the system), from those which focus on the sequence of change. Classifications of sequential alterations often use location-for-time methods in which empirical evidence of spatial segregations in adjustment processes are used to infer a relative temporal sequence of adjustment; lack of knowledge concerning time periods of geomorphological adjustment precludes a more exact sequencing. Finally, attention is turned to the way in which a management-style classification can be used to increase geomorphological understanding of river channel adjustments. This is achieved by transforming the classification into a characterisation scheme which indicates the sensitivity of the river channels to adjustment. Understanding the sensitivity of river channels potentially provides an extremely appropriate tool by which to design environmentally sympathetic river channel management options.

CLASSIFICATION OF EXISTING FEATURES

Channel management classification methods based on the existing channel morphology are a natural progression from geomorphological attempts to typify natural rivers. Two examples are given below: the first obtains management information directly from a classification based on natural features, and the second collates an inventory of natural features which are used to score the environmental value of the channel and thus indicate, for example, its susceptibility to degradation by channel management procedures.

As a consequence of natural stream channels being governed by catchment characteristics which vary over a wide variety of spatial and temporal scales (see Frissell *et al.*, 1986), natural channel classifications are increasingly utilising a hierarchical approach. The scheme developed by Rosgen (1985, 1994) identifies four inventory levels ranging from a broad morphological classification (level I) to monitoring (IV). In between these are 'morphological description' (II) and 'stream state or condition' (III). The level of morphological description exists to subdivide stream channels into homogeneous reaches based upon reference to values derived from five delineative criteria. The criteria consist of an entrenchment ratio to describe the relationship between the river and its valley, the width:depth ratio to summarise channel shape, the degree of meandering through measurement of sinuosity, in-channel particle size analysis as an indication of sediment transport processes, and water surface slope as a summary of 'sediment, hydraulic and biological function' (Rosgen, 1994). Figure 16.1 provides the bounding value ranges for each criteria, derived from empirical investigations on 418 rivers, and the resulting designation of 41 stream types. Channel management applications are based upon the assumption that rivers within each type class react similarly to particular human disturbances. Therefore, case studies can be used to determine the potential response behaviour of numerous other, similar, rivers. Rosgen uses this logic to develop guidelines for installing fish habitat improvement structures which are physically suited to their river environment. In conjunction with temporal series of aerial photographs, the classification can also provide the basis for specifying channel evolution sequences occurring in response to changes in their controlling variables. Management interpretations of sensitivity to disturbance, recovery potential, sediment supply, bank erosion potential, the influence of vegetation on bank stability and river restoration techniques can be also defined according to stream type.

Rosgen's scheme provides a management tool that could be adapted for use in many areas of the world. However, the scheme has been in development since 1973 and makes heavy demands on empirical data. In many cases of channel management application, information is required rapidly, often from one site visit. In response to such a requirement, the Thames Region of the UK's National Rivers Authority have developed a system for classifying river channel susceptibility to disturbance (NRA, 1990). The site visit utilises a reconnaissance evaluation scheme to help score the environmental condition of a stream in comparison with an ideal 'natural' channel; land use pressures within the Thames catchment mean that very few streams are likely to comply with this ideal. Table 16.1 summarises the resulting six-fold classification which ranges from almost natural channels which are highly susceptible to disturbance to culverted reaches with no geomorphological value. Channel reaches delimited by this scheme are entered into a Geographical Information System to provide background information for judging development proposals requiring NRA consent and to help the NRA plan its own channel management operations. In relation to this latter use, the classification can aid the design of channel enhancement or restoration schemes, as well as being useful for scheduling maintenance operations. Decisions will be based around the classified conservation value. For instance, the rarity of highly susceptible channels implies a high con-

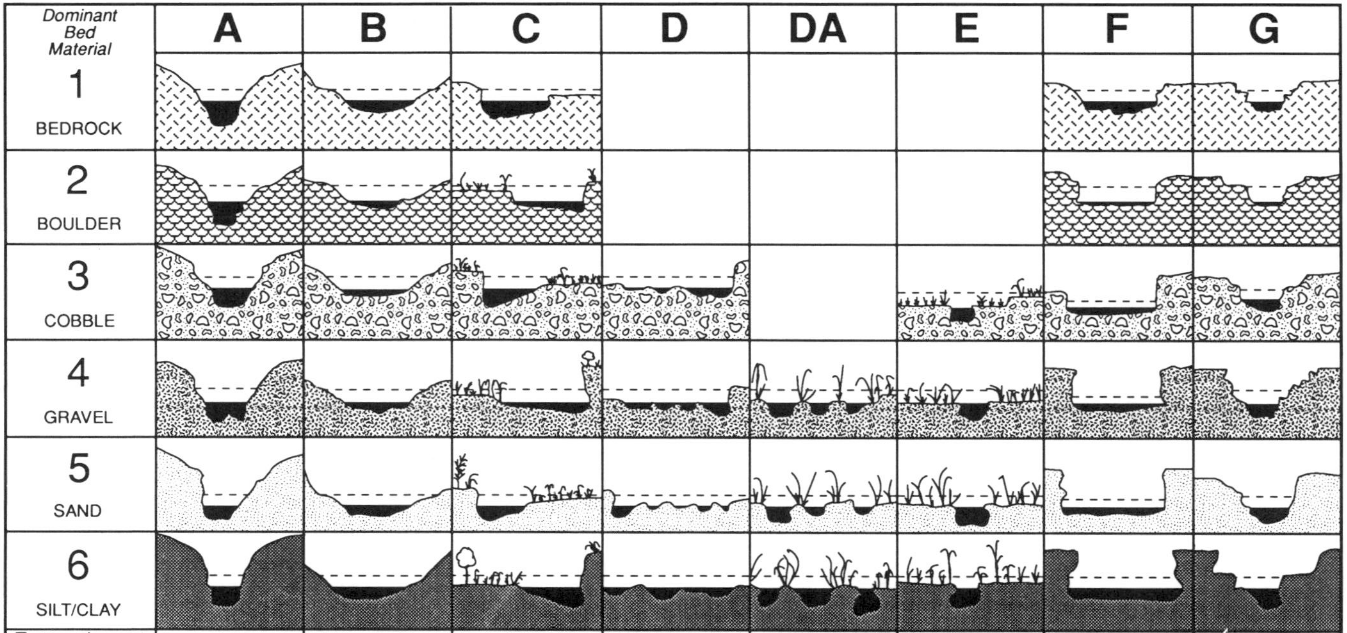

	A	B	C	D	DA	E	F	G
Entrenchment ratio	< 1.4	1.4 - 2.2	> 2.2	N/A	> 2.2	> 2.2	< 1.4	< 1.4
Sinuosity	< 1.2	> 1.2	> 1.4	< 1.1	1.1 - 1.6	> 1.5	> 1.4	> 1.2
Width : depth ratio	< 12	> 12	> 12	> 40	< 40	< 12	> 12	< 12
Water surface slope	0.04 - 0.099	0.02 - 0.039	< 0.02	< 0.02	< 0.005	< 0.02	< 0.02	0.02 - 0.039

Figure 16.1 Forty-one stream types at the scale of morphological description distinguished by Rosgen (1994). Redrawn from Rosgen (1994) by permission of Elsevier Science Publishers

Table 16.1 Summary of NRA (1990) scheme for classifying river channel susceptility to disturbance

Susceptibility to degradation	Score	Description
High	8–10	Conform most closely to a natural, unaltered, state and will often exhibit signs of free meandering and possess well-developed bedforms (point bars and pool–riffle sequences)
Moderate	5–7	Show signs of previous alteration but still retain many natural features or may be recovering towards conditions indicative of the high category
Low	2–4	Substantially modified by previous engineering works and are likely to possess an artificial cross-section (e.g. trapezoidal) and will probably be deficient in channel bedforms and bankside vegetation
Channelised	1	Awarded to reaches whose banks and/or bed have been subject to hard protection (e.g. concrete walls, sheet steel piling)
Culverted	0	Totally enclosed by hard protection
Navigable	–	Classified separately due to their high levels of flow regulation and bankside protection, and their probable strategic need for maintenance dredging

servation value and engineering works should be strongly opposed. Engineering works should also be resisted in moderately susceptible channels because, depending on stream power and sediment budget, they are likely to have the highest recovery potential (ability to return to a natural form). Where recovery is evident, then in-channel maintenance operations should also be resisted unless unavoidable (i.e. genuine need for flood defence operations). Conversely, reaches of low susceptibility have the highest enhancement potential, and engineering works should be encouraged to incorporate restoration measures.

CLASSIFICATION OF ADJUSTMENT PROCESSES

As river managers increasingly recognise that rivers are not simply 'water courses', but involve the interrelated movement of water and *sediment* within a dynamic river channel, there has developed an aspiration to comprehend river channel dynamics prior to undertaking river management operations or land use developments close to the channel edge. In response to this demand, and to simplify the specific nature of individual river channel adjustment processes, a number of methods have been produced which classify adjustments. Unlike geomorphological classifications of river channel patterns based on (hierarchies of) morphological features, classifications of river channel adjustment are explicitly value laden; adjustment processes are inferred directly from morphological features. Essentially, therefore, the resulting classifications represent a simple summary of an evaluation procedure.

Prevailing adjustments

In the absence of repeat river channel cross-section and planform surveys, absolute measures of river channel adjustment over the 'recent' period (arguably 10^0–10^2 years) are very difficult to ascertain, especially if changes in the channel depth are required, which negates the use of map comparisons. As a consequence, assessments of river channel adjustment required for management purposes have often used programmes of field observations. Some of these programmes have proceeded to classify the observed adjustments as prevailing changes which, unless their controlling parameters are altered, should be expected to continue into the near future. Examples of such schemes include those developed for relocated channels in the United States by Brice (1981), for channelised rivers in Denmark by Brookes (1987) and for catchment-wide surveys of rivers within the Thames basin, UK, by Downs (1992). Each method separates channels at the 'reach' level (of Frissell *et al.*, 1986) and classifies river channel activity according to cross-sectional adjustments. The resultant classifications by Brice and Brookes are illustrated in Figure 16.2 and details of the scheme by Downs is given below (and see Figure 16.4). Not surprisingly, many similar modes of adjustment are recognised, including processes of lateral migration (erosion of outer bank, deposition on inner; cross-sectional dimensions preserved) which is given as B by Brice, W4 and D2 by Brookes, and as **M** by Downs, and processes whereby a sinuous low flow channel is reforming within an overwide straightened reach (C in Brice, W5 in Brookes, **R** in Downs). The condition of morphological inactivity (short-term absolute stability) may be contrasted to the named categories of Brice and Brookes, and is given explicitly by Downs.

Classification of river channel adjustments in the Thames basin, UK, reconciled field survey evidence amassed from morphological indicators of change with a logical set of river channel adjustment categories. After development and testing of a reconnaissance evaluation procedure for a pilot catchment, four river networks were assessed for indicators of river channel changes (Downs, 1992). Grouping the individual indicators of change suggested over 20 types of prevailing adjustment. The total number of categories was reduced by developing a matrix centring on four basic forms of change in the river channel cross-section. These basic categories described channels which were *Stable* (no observable indicators of recent river channel change), cross-sections where indicators of erosion demonstrated that the cross-section was *Enlarging*, cross-sections which appeared to be contracting in area (*Depositional*) and cross-sections which were undergoing planform shift whilst approximately maintaining their dimensions (*Laterally Migrating*). A non-directional two-dimensional matrix of these four basic conditions results in 10 classes of river channel adjustment (Figure 16.3). In this classification, the three basic active categories are subdivided by their rate of change, and the three types of compound activity are recognised as indicative of complex channel adjustments following human disturbance of the river channel. Some of the adjustment categories represent channels which are in dynamic equilibrium, whereas others may represent non-equilibrium conditions. Figure 16.4 illustrates each class with a summary description.

The purpose of classifying modes of river channel change is to guide planners and engineers involved in the management of the river or its surrounding corridor. For instance, in the case of the Danish survey (Brookes, 1987), recommendations were

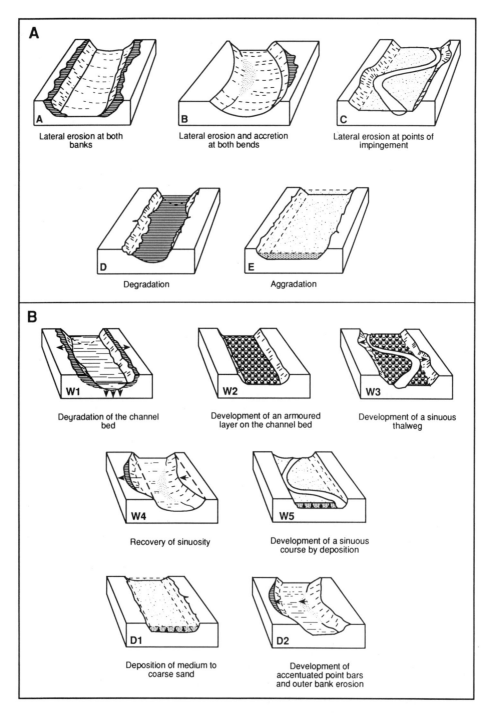

Figure 16.2 Comparison of river channel adjustment classifications of (A) Brice (1981) and (B) Brookes (1987). Redrawn from Brice (1981), and from Brookes (1987) with permission of J. Wiley & Sons

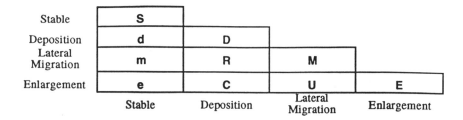

Channel Adjustment Categories:

S -	Stable	**m** -	less severe lateral migration
D -	Deposition	**e** -	less severe enlargement
M -	Lateral Migration	**R** -	'Recovering' reach
E -	Enlargement	**U** -	'Undercutting' reach
d -	less severe deposition	**C** -	'Compound' reach

Figure 16.3 Classification of river channel adjustments based on reconnaissance surveys undertaken in the basins within the Thames catchment, UK. From Downs, 1992

made for the introduction of environmentally sympathetic channel management techniques, as required by Danish law, according to channel change type. The assessment of Brice (1981) was designed to indicate the degree of instability induced by relocating stream channels for the purposes of highway construction. Thus, the classification scheme becomes the basis of a stability assessment which combines four spatial frequencies of bank erosion (*rare*; *local*; *local and severe*; *general*) and four extents of dimensional change (*> 5%*; *5–20%*; *local changes > 20%*; *general changes > 20%*). From the potential of 16 stability classes, 13 are recognised in the 103 study sites. The largest class group is the 'stable' channels (*rare* bank erosion and *<5%* dimensional change; 31 cases, 30.1%), followed by two classes indicating local changes (20 sites with *local* erosion and *5–20%* dimensional change; 15 cases *local* erosion with *<5%* dimensional change). In comparison, Downs (1992) found that the proportion of 'stable' Thames channels is 37% when defined by type **S** and the culverted reaches, or 77% if slow rate of change categories (**d**, **m**, **e**) are also incorporated. However, this overall figure masks the fact that the percentage of 'active' channels (those not of type **S** or culverted) varied between catchments from nearly 83% (92.4 km) of the Roding to only 16.6% (6.4 km) of the Lambourn. The Ravensbourne, in which 12% of the river network is culverted, has 28.5% (8.3 km) of active channels while the Sor has 73% (47.8 km). Clearly, the inter-basin differences in adjustment revealed by the classification scheme highlights the fact that Thames channels should not be viewed simply as presenting one single channel management challenge and that geomorphological appraisal is a necessary pre-requisite for environmentally-aligned management approaches. Additional information processing from this scheme is outlined later in this chapter.

A related technique for assessing the prevailing stability of river channels has been developed by Simon *et al.* (1989) and is summarised in Simon and Downs (1995).

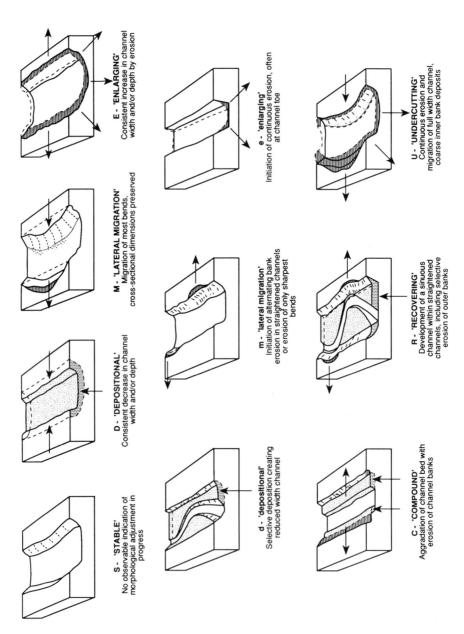

Figure 16.4 Diagrammatic representation and description of river channel adjustment classification from Downs (1992)

Table 16.2 Summary of scheme for assessing the potential instability of river channels (Reproduced from Simon and Downs (1995), with permission of Elsevier Science Publishers

1.	Bed material					
	bedrock	boulder/cobble	gravel	sand	unknown alluvium	silt/clay
	0	1	2	3	3.5	4
2.	Bed protection					
	yes	no	(with)	1 bank protected	2 banks protected	
	0	1		2	3	
3.	Stage of channel evolution (see Figure 16.5B)					
	I	II	III	IV	V	VI
	0	1	2	4	3	1.5
4.	Percent of channel constriction					
	0–5	6–25	26–50	51–75	76–100	
	0	1	2	3	4	
5.	Number of piers in channel					
	0	1–2	<2			
	0	1	2			
6–8.	Percent of blockage: horizontal (6), vertical (7), total (8)					
	0–5	6–25	26–50	51–75	76–100	(divide values
	0	1	2	3	4	by three)
9.	Bank erosion for each bank					
	none	fluvial	mass-wasting			
	0	1	2			
10.	meander impact point from bridge (in feet)					
	0–25	26–50	51–100	> 100		
	3	2	1	0		
11.	Pier skew for each pier (sum for all piers)					
	yes	no				
	1	0				
12.	Mass wasting at pier (calculated for each pier)					
	yes	no				
	3	0				
13.	High-flow angle of approach (in degrees)					
	0–10	11–25	26–40	41–60	61–90	
	0	1	2	2.5	3	
14.	Percent woody vegetative cover					
	0–15	16–30	31–60	61–99	100	
	3	2.5	2	1	0	

This scheme, which is now being utilised in 11 States of the USA to assess the magnitude and distribution of bridge scour problems, scores potential instability as the total of marks obtained from individual questions in a reconnaissance evaluation survey. Table 16.2 shows how questions score the degree of instability implied by either an observed morphological feature of the channel or its vegetation, or from structural features of the bridge. Higher overall scores indicate a more unstable channel and field experience suggests that the threshold classification for channel instability which may threaten the crossing structure occurs where scores exceed 20. When tested on 1100 West Tennessee sites, 13% scored more than this critical value, with a mean of 12.3 (Simon and Downs, 1995). Sites which are identified as critically unstable are then assessed for their socio-economic and strategic value prior to making recommendations for engineering mitigation works.

Sequences of adjustment

If channel management strategies are to be designed which manage the dynamics of river channels then, alongside the procedures outlined above for assessing *existing* changes, it is desirable to have knowledge of the temporal *sequence* of changes which is probable at individual sites. With this knowledge, it becomes possible to manage probable future adjustments of the channel. One example, using a classification scheme alongside aerial photographs to induce stereotypical sequences of changes, has already been mentioned (Rosgen, 1994). However, in the absence of detailed archival evidence, and because full sequences of changes are likely to exceed a geomorphologist's working life, many schemes utilise knowledge of river channel processes in conjunction with location-for-time substitution to translate spatially segregated modes of change into projected temporal sequences. Study is usually facilitated by distinct human actions which modify the river channel or aspects of its water and sediment discharges.

The model of bed-level changes following channel straightening forwarded by Parker and Andres (1976), where degradation and then aggradation follow the passage of a knickpoint from the base of the affected reach, is an early example of this type of study, and classification of post-channelisation changes is still the most fully developed. Two similar models of adjustment now exist; the first is a five-stage evolution model developed for the passage of a knickpoint through Oaklimiter Creek, Mississippi (Schumm *et al.*, 1984; Harvey and Watson, 1986), and the second is a six-stage model based on channel bank changes following channelisation for rivers in West Tennessee (Simon, 1989, 1995). Figure 16.5 illustrates the two classification schemes. Both schemes begin with a pre-modified category which, in the Simon classification, is followed by the channel immediately following re-sectioning. Degradation of the channel bed follows due to stream power increases in the steepened reach (caused either by knickpoint progression or the reduction in channel length). Eventually, deepening of the channel leads to oversteepening of the banks and, when critical bank heights are exceeded, bank failure occurs primarily through mass-wasting processes, and a channel widening stage ensues. An aggradational phase follows in which a new low-flow channel may begin to form in the sediment deposits. Instability of upper channel banks may continue during this period. The final stage indicated by both schemes is the restabilisation of the channel banks and the development of a channel within the deposited alluvium which is of a similar capacity to the pre-modification channel. However, the new channel may possess a higher width:depth ratio than the original (Schumm *et al.*, 1984), and may be multi-staged in cross-sectional form.

In a straightened channel, fitting an observed change into the appropriate stage in either of these schemes would help to guide a proactive channel management strategy. However, both the schemes rely on the homogeneity of the underlying alluvial sediments and the uniformity of the channelisation procedure to produce a response sequence which is suitable for classification. Classification of river channel changes resulting from, for instance, urbanisation are much more difficult to classify due to

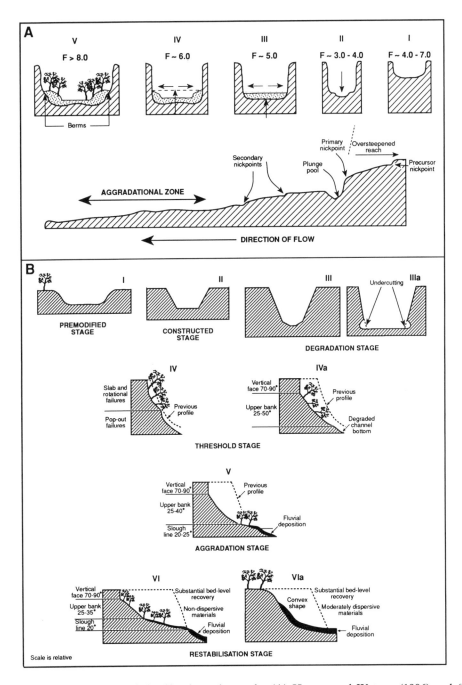

Figure 16.5 Comparison of classification schemes by (A) Harvey and Watson (1986) and (B) Simon (1989) for sequences of river channel adjustment following channelisation. Redrawn from Schumm *et al.* (1984), by permission of American Water Resources Association, and from Simon (1989) by permission of J. Wiley & Sons. F = form ratio (width:depth)

the number of interacting factors. For instance changes may depend on the degree of urbanisation (Leopold, 1968) and its position within the drainage basin (Ebisemiju, 1989), the type of drainage system installed (Roberts, 1989), local variations in the underlying rock type and channel gradient (Neller, 1988), whether or not the urban hydrograph exceeds the critical threshold of stream power required for bed erosion, whereby channel incision rather than general cross-sectional enlargement may occur (Booth, 1990), and whether concurrent channelisation works take place (Neller, 1989; Gregory et al., 1992). Indeed, for the Monks' Brook in Hampshire, Gregory et al. (1992) found that location-specific factors did not allow a spatially segregated model of river channel adjustment to be recognised from the urban expansion of Chandler's Ford. Instead six styles of river channel change were agreed (Figure 16.6) which included active and inactive channels, cross-sectional contraction or enlargement, and different styles of erosion.

FROM CLASSIFICATION TO CHARACTERISATION: RIVER CHANNEL SENSITIVITY TO ADJUSTMENT

Classification schemes which predict spatial or temporal sequences of channel adjustment are not yet used routinely in river channel management. They require further validation by empirical study and should, ideally, indicate the time periods involved. Furthermore, contrast between the examples classifying post-channelisation adjustments in homogeneous alluvial sediments with those following urbanisation demonstrates that prediction of change needs to assess confounding environmental factors as well as general models of adjustment. This notion demands an understanding of the impact of individual parameters and a method which assesses their cumulative effect on the river channel. Gregory (1987) suggests that some knowledge exists about causes of river channel change, how this change is manifested in the channel, and how much change may be expected, but the specific nature and extent of changes brought about by individual influences will vary significantly according to their design, site and situation (the 'singularity' of Schumm, 1991). This variability is illustrated by the range of channel capacity changes in response to dam construction, urbanisation, land use and other changes documented by Brookes and Gregory (1988, pp. 152–153).

Recognising specific causes of river channel change implies knowledge of the *sensitivity* of river channel change to each individual parameter. However, although geomorphologists have shown increasing interest in the concept of sensitivity (recent advances in Thomas and Allison, 1993), difficulties in obtaining suitable data has prevented geomorphological sensitivity approaching the level of sophistication common in neighbouring disciplines (Downs and Gregory, 1993). Data deficiencies include those of defining the complete set of controlling environmental mechanisms which are responsible for river channel adjustment (input data), and the paucity of long-term monitoring of adjustments with which to validate models (response data). To reduce these problems, Downs (1992) based an empirical assessment of the sensitivity of Thames river channel adjustments on a classification scheme derived from Figure 16.4 to provide the response data. For purposes of statistical validity, the number of types of adjustment was reduced from the 10 in Figure 16.4 into the four

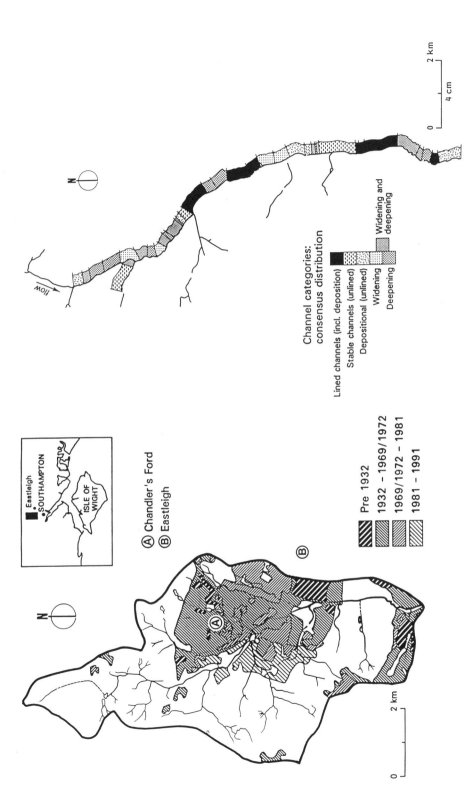

Figure 16.6 Consensus distribution of river channel adjustments in the Monks' Brook, Hampshire, UK. The Monks' Brook has been affected by the extensive urban area growth of Chandler's Ford and channelisation measures. Modified from Downs and Gregory (1993) with permission of J. Wiley & Sons

basic types in which *Enlargement* included channel adjustment classes **E**, **e** and **U**, *Lateral Migration* comprised **M**, **m** and **R** and *Deposition* consisted of channel adjustment types **D**, **d** and **C**. Input data was compiled from maps and field survey. Information was processed via multivariate logistic regressions equations (details in Downs, 1994a, 1994b and 1995) so that the parameter estimate accompanying statistically significant variables facilitated the calculation of a percentage probability of obtaining a particular adjustment style for single or combined environmental parameters within the class groups *channel gradient*, *rock type*, *land use* and *channel management* (Downs, 1994b). In utilising attainable data, this procedure converts the meaning of sensitivity from '... the ratio of the response of a device to the stimulus causing it.' (*Oxford English Dictionary*, 1989, vol. 14, p. 986) to 'the likelihood of obtaining a particular response (style of river channel adjustment) relative to the existence of an input parameter (catchment characteristic)'.

The best-fit equations produced by this procedure are multivariate linear characterisations of the classified adjustment (Downs, 1994b). Channels in the Thames basin are found to be sensitive to variables within all four categories of environmental parameter. By broadening the study to consider how river channel adjustment may be associated with drainage basin characteristics upstream and downstream of the adjustment, as well as alongside it, Downs (1995) demonstrates that channel adjustment is primarily conditioned by regional natural characteristics (*gradient* and *rock type*) while the specific nature of an adjustment category is likely to result from human actions (*land uses* and *channel management*).

CONCLUSION

Examples in this chapter have illustrated how river channel classifications are providing information for the management of dynamic river channels. Developing from geomorphological classifications of natural river channels, a focus on meso-scale spatial and temporal concerns has led to a range of management-related (applied and applicable) classifications whose strength lie in segregating river channel (adjustment) types as the basis for spatially differentiated, but systematic and integrated, management operations. Uses for the schemes include providing inventories of conditions for prioritising river conservation efforts, supplying information about adjustment processes so that management designs may account for the prevailing river channel dynamics, and contributing stereotypical sequences of change so that channel management measures may pre-empt future changes.

Classifications schemes are not ideal, they contain aspects of subjectivity, of generalised judgement and of qualitative interpretation when, for proactive river channel management purposes, a comprehensive physical model of river channel sensitivity to adjustment would provide management certainty. Unfortunately, the gap between conceptual understanding of river channel behaviour and modelling ability (Anderson and Sambles, 1988) ensures river channels will continue to be managed without perfect understanding. At a time when public opinion, enshrined now in legislation, is putting a high value on environmentally sympathetic river management, leading to a demand for geomorphological information, classification schemes provide an attainable target, the 'intermediate technology' with which to communicate geomorphological information of use to river managers. Additionally, the value of a classification may be enhanced further when the scheme is used in extended analysis, as illustrated in the characterisation of adjustment sensitivity. In this mode, classification provides a basis for increasing basic geomorphological understanding of the spatial variability of river channel adjustment. Indeed, linking cause with effect within the context of the drainage basin has broader environmental management implications, particularly within the context of schemes of integrated river basin management which may become more prominent in the 21st century.

ACKNOWLEDGEMENTS

I am indebted to the Cartographic Unit at The University of Nottingham, Department of Geography, for re-drafting Figures 16.1, 16.2, 16.4 and 16.5.

REFERENCES

Alabyan, A.M. (1992) Plain river channel patterns and factors of their forming. *Geomorphologiya* **2**, 37–42 (in Russian).

Anderson, M.G. and Sambles, K.M. (1988) A review of the bases of geomorphological modelling. In: Anderson, M.G. (ed.), *Modelling Geomorphological Systems*, Wiley, Chichester, 1–32.

Booth, D.B. (1990) Stream channel incision following drainage basin urbanization. *Water Resources Bulletin* **26**, 407–418.

Brice, J.C. (1981) *Stability of Relocated Stream Channels*. Technical Report No. FHWA/RD-80/158, Federal Highways Administration, US Dept. of Transportation, Washington, DC, 177pp.

Brookes, A. (1987) The distribution and management of channelized streams in Denmark. *Regulated Rivers: Research and Management* **1**, 3–16.

Brookes, A. and Gregory, K.J. (1988) Channelization, river engineering and geomorphology. In: Hooke, J.M. (ed.), *Geomorphology in Environmental Planning*, Wiley, Chichester, 145–168.

Church, M. (1992) Channel morphology and typology. In: Calow, P. and Petts, G.E. (eds), *The Rivers Handbook: Hydrological and Ecological Principles*, vol. 1, Blackwell, Oxford, 126–143.

Cowardin, L.M. (1982) Wetlands and deepwater habitats: a new classification. *Journal of Soil and Water Conservation* **37**, 83–85.

Downs, P.W. (1992) Spatial variations in river channel adjustments: implications for channel management in south-east England. Unpublished PhD thesis, University of Southampton, 340pp.

Downs, P.W. (1994a) Estimating the probability of river channel adjustments. Paper presented at *IBG'94*, 4–7 January 1994, University of Nottingham.

Downs, P.W. (1994b) Characterization of river channel adjustments in the Thames basin, south-east England. *Regulated Rivers: Research and Management* **9**, 151–175.

Downs, P.W. (1995) River channel adjustment sensitivity to drainage basin characteristics: implications for channel management planning in south-east England. In: McGregor, D. and Thompson, D. (eds), *Geomorphology and Land Management in a Changing Environment*, Wiley, Chichester, 247–264.

Downs, P.W. and Gregory, K.J. (1993) The sensitivity of river channels in the landscape system. In: Thomas, D.S.G. and Allison, R.J. (eds), *Landscape Sensitivity*, Wiley, Chichester, 15–30.

Ebisemiju, F.S. (1989) Patterns of stream channel response to urbanization in the humid tropics and their implications for urban land use planning: a case study from southwestern Nigeria. *Applied Geography* **9**, 273–286.

Ferguson, R.I. (1987) Hydraulic and sedimentary controls of channel pattern. In: Richards, K.S. (ed.), *River Channels: Environment and Process*, Blackwell, Oxford, 129–158.

Frissell, C.A., Liss, W.J., Warren, C.E. and Hurley, M.D. (1986) A hierarchical framework for stream habitat classification: viewing streams in a watershed context. *Environmental Management* **10**, 199–214.

Gregory, K.J. (1987) Environmental effects of river channel changes. *Regulated Rivers: Research and Management* **1**, 358–363.

Gregory, K.J., Davis, R.J. and Downs, P.W. (1992) Identification of river channel change due to urbanisation. *Applied Geography* **12**, 299–318.

Grigg, D.B. (1967) Regions, models and classes. In: Chorley, R.J. and Haggett, P. (eds), *Models in Geography*, Methuen, London, 461–509.

Harvey, M.D. and Watson, C.C. (1986) Fluvial processes and morphological thresholds in incised channel restoration. *Water Resources Bulletin* **22**, 359–368.

Horton, R.E. (1945) Erosional development of streams and their drainage basins: hydrophysical approach to quantitative morphology. *Bulletin of the Geological Society of America* **56**, 275–370.

Leopold, L.B. (1968) *Hydrology for Urban Land Planning—A Guidebook on Hydrological Effects of Urban Land Use.* United States Geological Survey, Circular 554.

Leopold, L.B. and Wolman, M.G. (1957) River channel patterns, braided, meandering and straight. *United States Geological Survey, Professional Paper* **282**, 39–84.

Mosley, M.P. (1981) Delimitation of New Zealand hydrologic regions. *Journal of Hydrology* **49**, 173–192.

Mosley, M.P. (1987) The classification and characterisation of rivers. In: Richards, K.S. (ed.), *River Channels: Environment and Process*, Blackwell, Oxford, 295–320.

Naiman, R.J., Lonzarich, D.G., Beechie, T.J. and Ralph, S.C. (1992) General principles of classification and the assessment of conservation potential in rivers. In: Boon, P.J., Calow, P. and Petts, G.E. (eds), *River Conservation and Management*, Wiley, Chichester, 93–123.

NRA (National Rivers Authority) (1990) *River Stort Morphological Survey: Appraisal and Watercourse Summaries*, compiled by Brookes, A. and Long, H., September, 1990.

Neller, R.J. (1988) Complex channel response to urbanisation in the Dumaresq Creek Drainage Basin, New South Wales. In: Warner, R.F. (ed.), *Fluvial Geomorphology of Australia*, Academic Press, Sydney, 323–341.

Neller, R.J. (1989) Induced channel enlargement in small urban catchments, Armidale, New South Wales. *Environmental Geology & Water Science* **14**, 167–171.

Oxford English Dictionary, 2nd edition (1989) Clarendon Press, oxford.

Parker, G. and Andres, D. (1976) Detrimental effects of river channelization. In: *Proceeding of Conference "Rivers '76"*. American Society of Civil Engineers, New York, 1248–1266.

Roberts, C.R. (1989) Flood frequency and urban induced channel change: some British examples. In Beven, K. and Carling, P.A. (eds), *Floods: Hydrological, Sedimentological and Geomorphological Implications*, Wiley, Chichester, 57–82.

Rosgen, D.L. (1985) A stream classification system. In: Johnson, R.R., Zeibell, C.D., Patton, D.R., Pfolliott, P.F. and Hamre, R.H. (eds), *Riparian Ecosystems and their Management: Reconciling Conflicting Uses*, United States Forest Service Technical Report M-120, Fort Collins, Colorado, Rocky Mountain Experimental Forest and Range Experimental Center.

Rosgen, D.L. (1994) A classification of natural rivers. *Catena* **22**, 169–199.

Schumm, S.A. (1991) *To Interpret the Earth: Ten Ways to be Wrong.* Cambridge University Press, Cambridge, 131pp.

Schumm, S.A., Harvey, M.D. and Watson, C.C. (1984) *Incised Channels: Morphology, Dynamics and Control.* Water Resources Publications, Littleton, Colorado, 200pp.

Simon, A. (1989) A model of channel response in disturbed alluvial channels. *Earth Surface Processes and Landforms* **14**, 11–26.

Simon, A. (1995) Geomorphology and landscape response of the Toutle River System in the aftermath of the 1980 eruption of Mount St Helens, *United States Geological Survey, Professional Paper* **1470** (in press).

Simon, A. and Downs, P.W. (1995) An inter-disciplinary approach to evaluation of potential instability in alluvial channels. *Geomorphology* (in press).

Simon, A., Outlaw, G.S. and Thomas, R. (1989) Evaluation, modeling, and mapping of potential bridge scour, West Tennessee. In: *Proceedings of the National Bridge Scour Symposium*, Federal Highways Administrative Report FHWA-RD-90-035, 112–119.

Thomas, D.S.G. and Allison, R.J. (eds) (1993) *Landscape Sensitivity*, Wiley, Chichester.

Vannote, R.L., Minshall, G.W., Cummins, K.W. and Sedell, J.R. (1980) The river continuum concept. *Canadian Journal of Fisheries and Aquatic Sciences* **37**, 130–137.

Part IV

MANAGEMENT FOR CHANGE

17 River Channel Restoration: Theory and Practice

ANDREW BROOKES

National Rivers Authority—Thames Region, Reading, UK

INTRODUCTION

To date perhaps the most impressive restoration efforts have been related to water quality improvement of some catchments (Cairns *et al.*, 1977; Jordan *et al.*, 1987). By contrast restoring the natural morphology to channels previously managed for drainage or flood control has been site-specific and geographically more restricted. Geomorphology should be applied to the appraisal and design of projects which attempt to emulate or restore natural channel characteristics. This is essential to ensure a sustainable and cost-effective approach. Geomorphology should therefore also be applied to set a site-specific project in the broader floodplain and catchment contexts. This chapter examines some of the current technical issues and future challenges surrounding river channel restoration from the perspective of a UK river manager. The discussion concentrates on restoring the natural morphological characteristics to streams and small rivers and evaluates the problems of attaining sustainable designs.

DEFINITIONS

There are a large number of terms used to describe intervention by river managers to improve the riverine environment and a widely accepted definition for these has not yet been devised. Measures are increasingly implemented to mitigate or compensate directly for damage caused by a recent or current development. Practice with such measures has often provided insight into how disturbed systems may be restored. However, there are perhaps four distinct management terms which can be used to describe the restoration of disturbed river systems namely, restoration, rehabilitation, enhancement and creation.

Table 17.1 provides some definitions of the term 'river restoration'. Perhaps one of the most widely accepted is that by Cairns (1991) which describes restoration as 'the complete structural and functional return to a pre-disturbance state'. Both natural recovery and enhanced recovery could be regarded as processes leading to restoration.

By contrast 'rehabilitation' has been defined as 'the partial structural and functional return to a pre-disturbance state' (Cairns, 1982). It involves the selection of a

Changing River Channels. Edited by Angela Gurnell and Geoffrey Petts.
© 1995 John Wiley & Sons Ltd.

Table 17.1   Some definitions of river restoration

Author	Definition
Cairns (1991)	The complete structural and functional return to a pre-disturbance state.
Gore (1985)	In essense, river restoration is the process of recovery enhancement. Recovery enhancement enables the river or stream ecosystem to stabilise (some sort of trophic balance) at a much faster rate than through the natural physical and biological processes of habitat development and colonisation. Recovery enhancement should establish a return to an ecosystem which closely resembles unstressed surrounding areas.
Osborne *et al* (1993)	Restoration programmes should aim to create a system with a stable channel, or a channel in dynamic equilibrium that supports a self-sustaining and functionally diverse community assemblage.
Herricks and Osborne (1985)	Implicit in the concept of water quality restoration is some knowledge of the undisturbed or natural state of the stream system. Restoration of water quality can be defined as returning the concentration of substances to values typical of undisturbed conditions.

limited number of attributes and is a pragmatic option for the management of many disturbed systems. 'Enhancement' is regarded as 'any improvement of a structural or functional attribute' (National Research Council, 1992) and is it not taken to refer to the pre-disturbance condition. For example, instream flow devices may artificially improve the river environment. Finally 'creation' is seen as 'the birth of a new (alternative) ecosystem that previously did not exist at the site' (National Research Council, 1992).

DETERMINING THE PRE-DISTURBANCE CONDITION

Restoring rivers to their original or pre-disturbance condition is perhaps an ideal towards which river managers could aspire but in practice is not a viable option. In reality there may be a whole range of options for a site depending on land use and other constraints. These may include re-creating a semi-natural channel, which existed perhaps 100 to 500 years ago, or even a modified channel form created only a few decades ago. Geomorphological investigation may allow an attempt to be made to re-create an historical channel form, but in practice restoration efforts are likely to be hindered by lack of knowledge of previous conditions.

Cairns (1987) suggested that because all ecosystems are dynamic and continually changing, then restoration to a pre-disturbance condition may not be as appropriate as restoration to the present condition of a comparable undisturbed system. This view should also apply to geomorphological reconstruction. Restoration of the river morphology to a pre-disturbance condition may not be appropriate due to climatic change or to subsequent catchment land use changes, leading to altered sediment and water discharges, which it may not be possible to control as part of a restoration project. Examining adjacent 'natural' sections of river channel, and interpreting how they might have changed in response to catchment influences, may lead to the most appropriate channel dimensions for a length of river to be restored. Equally, if designs are to be cost-effective and sustainable it may be necessary to anticipate how further or future catchment land use change will affect the restored river channel.

An objective of the application of geomorphology to river channel restoration should be to develop projects which are sustainable. It may not be cost effective to replace large quantities of gravel to a river channel for fish spawning purposes if, over the duration of the next flood, these are washed out and transported away from the site. Although a restoration project may seek to re-create a more natural channel, which adjusts through time, it may be necessary to address the question of life expectancy of a project.

Figure 17.1 depicts some of the key contributions which can be made by a geomorphologist at the option evaluation stage of a project to help ensure a more sustainable design for the river channel and/or adjacent floodplain. This should be preceded by a preliminary appraisal of catchment and historical changes using maps and other records, and a survey and characterisation of the morphology of the reach to be restored.

ISSUES SURROUNDING THE RE-CREATION OF SUSTAINABLE MORPHOLOGICAL FEATURES

Catchment context

Figure 17.2 depicts the channel, riparian river corridor, floodplain and valley floor in the context of the catchment. In restoring a river channel these are important elements to consider: because of floodplain development it may only be possible in many cases to restore the channel and a river corridor of limited width. With regard to the three primary zones within fluvial systems (Schumm, 1977), the dominance of supply, storage or transport within each zone produces a variety of river morphologies. The piedmont and upland zone typically has large volumes of recently eroded sediment with channel bars and braids storing sediment. The transfer zone is therefore dynamic as sediment passes through the reach, with storage of sediment in mid-channel and point bars and erosion of the outside of bends (Figure 17.2). In Zone 3 the deposition of fine sediments may occur on the adjacent floodplain and on berms within the channel. Extensive catchment changes have occurred in Europe over both the Holocene and recent time periods (see Chapters 2 and 3). Land use changes would have caused soil erosion and initiated channel sedimentation through time (see also Chapters 9 and 19).

Understanding these different processes is necessary when considering the types of morphological feature to be restored to a particular length of river: reinstating gravels in a lowland river may not be appropriate because of the dominance of fine sediments which would tend to smother the new substrate, whilst channel narrowing devices installed in the transfer zone may have a limited life-span because of the dynamic nature of the channel bed and banks. A well-known problem arising from river engineering for flood control or navigation is that for the purpose of design the river is considered to be unchanging in dimensions, shape or planform. The fact that in practice rivers do change has led to recurrent maintenance of such channels. Like-

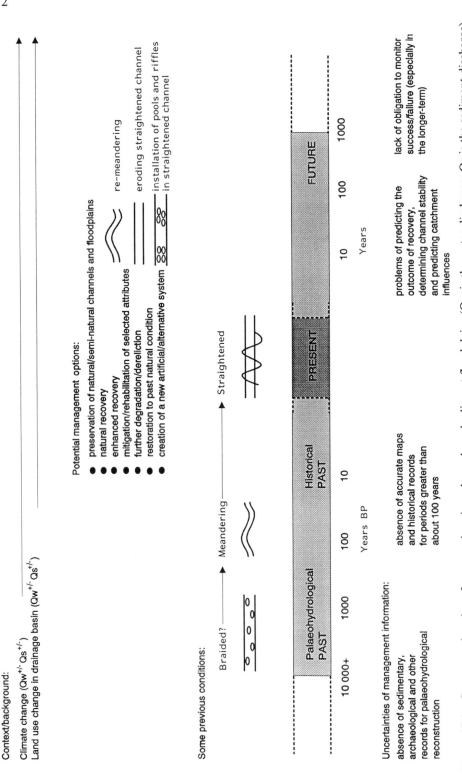

Figure 17.1 Context and options for managing river channels and adjacent floodplains (Qw is the water discharge; Qs is the sediment discharge)

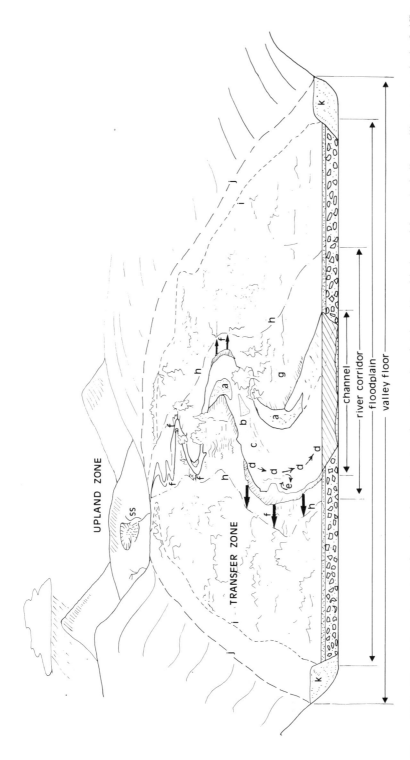

Figure 17.2 The catchment context for river channel restoration. *Key*: a) point bar; b) mid-channel bar; c) shallow still water, typical of riffle; d) deeper, convergent flow corresponding to pool; e) secondary current; f) erosion of bank as point bars form (arrow indicates direction of change); g) vegetation in river corridor increases hydraulic roughness; h) boundary of river corridor; i) edge of floodplain; j) limit of valley floor; k) terrace; ss) zone of sediment supply and storage

wise an eroding bank or river bend may not necessarily be indicative of instability. Thus, key issues for sustainable channel restoration design are (i) recognition that channels change either as a direct consequence of instability or of dynamic equilibrium, and (ii) understanding the spatial controls on the different types of channel change.

Attempts have been made to evaluate the success of restoration projects in terms of the energy or stream power of the river occupying the channel (Brookes, 1990, 1992). Other variables which affect success or failure are the bed and bed stability and the availability of sediment from the upstream catchment. High-energy gravel-bed rivers in the UK with a supply of sediment may adjust rapidly over the duration of one flood and reinstated features may not be sustainable. However, by contrast, these were also found to be the rivers which tended to recover naturally from management intervention (Brookes, 1992). In exceptional circumstances rivers which had been artificially straightened regained a meandering course in about 100 years. Many rivers may have insufficient stream power to erode non-alluvial beds and banks: natural recovery in these instances may involve reworking fine-sediment derived from upstream.

Site-specific design issues

In the UK, at least, the demand for morphological information to restore river channels is rapidly increasing. The National Rivers Authority has spent in excess of £300 000 over a five-year period (1989–1994) on geomorphological research and development. At one extreme it is possible to review the scientific literature and information available for the river environments under consideration, and to produce general guidance or reports for specific projects detailing design principles. The type of information available is summarised in Table 17.2, together with some of the limitations. This information may be particularly useful at the option evaluation and outline design stages of a project. For many countries much of the technical information specifically on river restoration is not widely available and has been relegated to unpublished technical reports and other government documents. Table 17.3 attempts to summarise some of the recent key documents originating in the UK.

To some extent, knowledge may be built up as river managers attempt to restore reaches by mimicking adjacent natural reaches. This may be a 'trial and error' approach and such information is likely to be sporadic and largely unpublished. Problems may arise if very detailed guidance developed for one set of environments is applied to other channel types. Detailed design using guidance may be particularly difficult in some circumstances because of a lack of understanding of the natural form and functioning of particular river types. For example, relatively little morphological research has been undertaken on lowland non-alluvial rivers. Clearly if sustainable river restoration is to be widely applied to many different river environments then further pure research on the processes and function of stream systems needs to be initiated and funded by research or river management organisations. Whilst morphological information is incomplete, this should not preclude an attempt at applying geomorphological knowledge to all river channel and floodplain restoration proposals. It has been shown in many instances that, despite a lack of a robust theoretical base, many projects based on general geomorphological principles have been a success (Brookes, 1988, 1995).

Table 17.2 Types of general geomorphological guidance of importance to river restoration

Type	Content	Limitations
Channel profiles (long-section and cross-section)	Knowledge of variation of channel morphology along a reach, particularly in relation to pools and riffles and planform	Sparse data on typical width–depth ratios for a wide range of channel types
Low-flow width in channel design	Best obtained from neighbouring natural section of same slope and geology	Site-specific measurement required. Natural widths for a range of channel types related to catchment area largely unavailable
Design and location of pools and riffles	Information on topographical, sedimentary and flow characteristics, size, location, spacing and slope values at which they occur	Knowledge-base particularly for gravel-bed rivers. Limited knowledge of adjustments during and after flood flows
Substrate reinstatement	Reinstatement of gravels for different channel types (either to remain static (e.g. armoured/segregated bed) or to be mobile)	Most knowledge is for mobile gravel-bed rivers
Prediction of channel changes	Use of historical records, maps or surveys to predict nature and locations of future channel change (including lateral and vertical change)	Imperfect knowledge for a wide range of channel types (e.g. sand-bed rivers). Assumes change at a site will be an ongoing process
Bank erosion/protection locations	Good knowledge of location of natural bank erosion mechanism, especially for gravel-bed rivers. Some understanding of how artificial influences (e.g. boat-wash erosion) modify or initiate patterns of erosion	Bank erosion mechanism not fully understood for all key channel types

Table 17.3 Some recent important technical publications pertinent to river restoration in the UK

Publication	Client	Authors	Guidance for river restoration
River Restoration Project (1993a) *Phase I. Feasibility Study*	River Restoration Project (RRP)	ECON (University of East Anglia)	A review, description and assessment of current restoration projects, measures and techniques, accompanied by a list of contacts and an extensive bibliography. Recommendations made on the scope of river restoration in the UK, including site selection
Stort Catchment Morphological Survey: Appraisal Report and Watercourse Summaries (1990)	National Rivers Authority (NRA)	Brookes and Long (NRA)	Method for geomorphological selection of reaches for river restoration in a catchment. Applied to six lowland catchments in UK so far
Sediment and Gravel Transport in Rivers, Including the Use of Gravel Traps (1994a)	NRA (R&D) R&D Report 384	Newson and Sear (University of Newcastle)	Analysis of problems from a wide range of environments in the UK. Includes a method for geomorphological audit of a catchment (enabling link of cause and effect) and input to management solutions
The New Rivers and Wildlife Handbook (1994)	NRA, Royal Society for the Protection of Birds, The Wildlife Trusts	Ward, Holmes and Jose	A practical conservation guide to techniques of river management, integrating the requirements of flood defence, wildlife and other interests. The success or otherwise of over 40 case studies are evaluated *but not* generally in terms of the stability of the river channels
Development of Geomorphological Guidance Notes (1994b)	NRA	Sear (University of Southampton)	Guidance on location of pools and riffles, substrate reinstatement, channel profiles, planform and low-flow widths, etc
Draft Guidelines for the Design and Restoration of Flood Alleviation Schemes (1993)	NRA (R&D) R&D Note 154	Hey and Heritage (University of East Anglia)	Guidelines for the design of UK steep gravel-bed rivers, based on the regime approach

A further potential problem is the translation of geomorphological information into the final design drawings for a scheme and ultimately construction of a project. The traditional technical engineering approach to river management has been to represent channel cross-sections as trapezoids, with a uniform long section. This has required minimal topographic data. Reconstruction of geomorphological features, such as pools and riffles and varied channel symmetries (Table 17.2) is not widely practised and represents far more complexity for the draughtsperson. Whilst not advocating that the 'nth' degree of detail is necessary for all projects, considerably more detail may be required to show the variation of micro-topography. Translation of the design drawings into a contract document which specifies the works is a further challenge which requires careful checking.

Finally, there may be a problem in the way the project is constructed. In practice the contractor will need considerable skill, perhaps with little prior experience, in re-creating features such as pools and riffles (Brookes, 1987a). Supervision by a 'Clerk of Works' with sufficient knowledge of the morphology may be essential. In any case, if the constructed project is close to the design then it is possible that subsequent adjustment will result in a more natural channel form (Brookes, 1991).

THE NEED FOR ASSESSMENT OF ENVIRONMENTAL IMPACTS

Although in many countries Environmental Assessment (EA) for restoration schemes may not be formally required under legislation, it is recommended that as a matter of best practice the environmental impacts should be anticipated as early as possible in the project management programme and documented where necessary. Whilst the intention of a restoration scheme may be to re-create natural characteristics, potential impacts may need to be scrutinised to a similar level warranted by an engineering scheme to increase flood conveyance. Table 17.4 is a checklist of some of the potential impacts which might arise from restoration, including hydraulic, morphological, groundwater, surface water and ecological impacts. In addition impacts on human issues such as flood risk, aesthetic and recreation values and land use should be considered.

Clearly, a multi-functional view should be taken of restoration projects. Restoration of morphological characteristics cannot be undertaken without reference to wider issues (Rhoads and Miller, 1990). There may be direct conflicts between restoration and development on the floodplain. For example, restoration measures may increase the hydraulic roughness and lead directly to a reduction in the standard of service for flood protection (Brookes, 1995). Similarly, extractions of gravels from the floodplain for the construction industry may have led to straightening of a watercourse. If the gravel pit has subsequently been filled with waste material, attempting to re-meander the channel can have potentially serious consequences for surface and groundwater quality, even if the material is documented as being inert. Other impacts may be potentially beneficial, such as the intentional rise of the water-table and re-creation of wetland habitats.

The value of Environmental Assessment is in scoping the key issues/constraints for a site, assessing the impacts using appropriate methods, identifying mitigation measures and preparing the appropriate documentation. Beyond implementation, allow-

Table 17.4 Restoration of river channels: some sources of impact and their potential effects

	Sources of impact				
	Bank reprofiling	Instream habitat devices	Pool and riffle reconstruction	Substrate reinstatement	Re-meandering
1. Surface water hydrology/hydraulics	Changed flow velocities	Increased hydraulic roughness Convergence/divergence of flow Increased flooding	Increased hydraulic roughness Convergence/divergence of flows Increased flooding	Increased hydraulic roughness Changed flow velocities Increased flooding	Increased hydraulic roughness Changed flow velocities Increased frequency of flooding
2. Channel morphology	Increased stability Deposition/siltation	Erosion of bed and/or banks Deposition/siltation Downstream erosion Upstream deposition/siltation Decreased bed load	Change of slope Deposition/siltation Reduced channel size	Reduced channel size Change of slope Deposition/siltation Increased bedload	Reduced channel size Increased stability Erosion of bed and/or banks Change of slope Change of planform/pattern Decreased bedload
3. Groundwater hydraulics				Rise in water-table Loss of infiltration	Rise in water table
4. Surface water quality		Improved quality Oxygenation Decreased turbidity	Oxygenation	Oxygenation	Oxygenation/eutrophication
5. Aquatic ecology	Improved habitat Increased species diversity	Improved habitat Increased fish biomass Increased species diversity Increase of invertebrates Effects on fish spawning	Improved habitat Effects on fish spawning Increased species diversity	Improved habitat Effects on fish spawning Change in the fish community	Improved habitat Increased fish biomass Increase of invertebrates Increased plant biomass Change in the fish community Increased species diversity Effects on fish spawning

ance should also be made for monitoring and audit (Wathern, 1988). From a practical river manager's point of view, it will be necessary to collect sufficient data within the programme and time-scale for the project. Each restoration project should state what the end product is likely to be.

THE SCIENTIFIC VERSUS PRAGMATIC APPROACHES

Perhaps one of the key ongoing challenges is to ensure that river managers make decisions from a sound enough base of scientific knowledge.

Monitoring and audit

One key issue which has been talked about for many years in various countries but apparently with relatively little action is the need for post-project appraisal of schemes (Gardiner, 1991). This also requires adequate baseline data collected prior to construction to compare with post-project data. In relation to restoration projects this issue has been highlighted many times in the scientific literature in the 1980s (Gore, 1985; Brookes, 1990) and has been re-stated more recently (Osborne et al., 1993; NRA, 1994b). Successful stream restoration requires monitoring the outcome of past, existing and future projects to gather information on the feasibility and success or failure of particular techniques and approaches. Finance appears to be rarely available for monitoring most projects. A major failing of EA practice generally has been the common use of EA to obtain a development permit, rather than as a tool for achieving sound environmental management either within the project objectives or on a broader regional and national basis (Bisset and Tomlinson, 1988). Currently emphasis is directed to the approval procedure with little attention given to the post-approval stage. Perhaps one way to maximise the use of any resources which may be available for monitoring would be to concentrate on one or two key representative sites.

Since relatively little published data is available on the impacts and success/failure of restoration projects it is probably true to say that the discipline can be viewed in its infancy. From an immediate practical management point of view, short-term monitoring may be the most important issue (Iversen et al., 1993) and allowance should be made to adjust the morphology if necessary after construction. However, post-project evaluation after 10–20 years may also be required to determine the success of habitat enhancements under a variety of hydrological regimes (Osborne et al., 1993). Longer-term monitoring may not be pragmatic from a river manager's point of view. Certain aspects of post-project appraisal may be more appropriate to scientific research. However, a typical research degree (e.g. MSc or PhD) may take less than three years to complete and within that period a maximum of 18 months is likely to be devoted to the collection of data. Another problem perceived by some scientists is that because proper evaluation of restoration projects is an interdisciplinary effort, individual scientists may anticipate little reward or recognition in their area of expertise (Osborne et al., 1993). These academics also feel strongly that post-project evaluation efforts need to attain the status of a valid scientific endeavour worthy of funding and publication (Osborne et al., 1993).

The River Restoration Project

Within the UK, at least, the need to increase scientific knowledge is demonstrated by the River Restoration Project (RRP), which was formed in 1990 by seven enthusiastic river managers. The RRP is a non-profit-making company officially launched in December 1992 by the British Prime Minister to promote the restoration of rivers for conservation, recreation and amenity. It uses the expertise of river ecologists, engineers, planners, fisheries biologists and geomorphologists drawn from both the scientific and river management communities. RRP is advised by a Steering Group with representatives drawn from major UK organisations with an interest in land and water management (RRP, 1993a and b).

The aims are to establish demonstration projects which show how state-of-the-art restoration techniques can be used to re-create natural ecosystems in damaged river corridors, to improve understanding of the effects of restoration work on nature conservation value, water quality, visual amenity, recreation and the views of the public, and to encourage others to restore streams and rivers. The latter aim can be achieved by disseminating information and guidelines about river restoration methods and by developing model partnerships between institutions and landholders who have a common aim to improve rivers but different powers, resources, responsibilities and interests.

Opportunities for river restoration in the UK are seen by RRP to be greater than ever before. For the first time in many decades large areas of land are coming out of intensive agriculture as part of the managed reduction in productivity. This may provide newly formed space for channel restoration as well as the potential for reinstatement of areas of fully functioning floodplain with a variety of semi-natural habitats.

RRP is independent and aims to be a catalyst for river restoration in the UK through partnerships that provide the opportunity for organisations to come together to progress initiatives that are of common interest (RRP, 1994). Many public sector bodies and voluntary organisations have a measure of responsibility or interest in the advancement of river restoration. The availability of finance further restricts the progress that can be made; the pooling of both powers and resources offers a way ahead.

One of the key objectives of RRP is the construction of two Demonstration Sites in the UK. Funded by the European Community Life Fund, programmes for appraisal, monitoring and construction of the two sites are currently (autumn 1994) being devised and implemented. The two sites selected are the River Skerne through Darlington (an urban channel) and the River Cole near Swindon (a rural location). Both can be described as lowland stream channels, previously channelized by straightening, widening and deepening, and with bankfull widths in the range 4.0 to 12.0 metres. The lengths of watercourse under consideration are of the order 2 to 5 kilometres. A further site is being monitored in Denmark as part of the Life Project. It is intended that a wide range of environmental attributes will be monitored over at least a 3-year period (1994–1997), including geomorphology, hydrology/hydraulics, water quality and sediment monitoring, vegetation, river and floodplain vegetation, invertebrate ecology, fisheries and birds. This will provide valuable management information for sustainable and cost-effective restoration at many other sites.

Opportunities created by development

In practice, restoration opportunities in many countries may arise from development applications. Typical developments might include housing, business parks and infrastructure projects, road, rail and airport construction and other types of land use change. Any substantial attempts at channel restoration, such as restoring bends and at least partially re-creating floodplains will require land-take. Restoration may be carried out as an integral part of a development, or at adjacent sites along the same river, or in a nearby catchment, as a direct mitigation of impact. Figure 17.3 demonstrates the principles whereby development of the wider floodplain for housing is to be undertaken, leaving a residual river corridor of up to 60 metres width and a more natural sinuous low-flow channel. The river corridor also accommodates the 100-year flood discharge with a multi-stage channel planted with appropriate vegetation. In this example the channel was previously degraded by straightening and deepening through an agricultural drainage scheme, allowing the adjacent land to be used for arable production. A similar approach has now been adopted for about 30 km of river in the UK. In practice this concept of partial restoration or rehabilitation of a channel and floodplain is likely to be one of the most widely used because of the many development applications which occur.

Restoration of morphological characteristics has been undertaken for both urban and rural situations. The use of geomorphology in the design of channels receiving urban runoff has been demonstrated by Brookes (1991). For example, a key issue in many urban channels may be establishing a more natural low-flow width. Low-flow widths for urbanised channels in the Thames River basin have been established by examining the affected reach as well as adjacent natural reaches, both above and below the urban area, which are considered to be in regime. If there is an inadequate depth of water to allow passage of fish at low-flow, or to maintain a gravel-bed free from silt, then the channel can be narrowed to increase the depth and velocity. Figure 17.4 is a photograph of a newly formed channel with pools, riffles and point bars in southern England. This is prior to the flow being diverted as a result of development away from a previously straightened alignment. A more natural sinuous course has been created as a direct result of the development.

The water quality issue

Bradshaw (1987) argues that the underlying logic behind restoration must be ecological. Restoration is seen to provide an opportunity to test in practice the current understanding of ecosystem development and functioning. From a scientific viewpoint, Statzner and Sperling (1993) argue that many stream management approaches are ecologically unwise. A sounder, more scientific, approach would be to scrutinise knowledge available for the catchment or system, stream habitat, organismic responses and self-purification (Statzner and Sperling, 1993). A key issue in urban areas may be that the water quality is too poor to support a diverse flora and fauna. Stream water quality cannot be dissociated from existing conditions in the watershed (Herricks and Osborne, 1985). In fact, restoration of water quality is so dependent on catchment conditions that any water quality restoration or protection activity must include remedial action on the catchment. In urban areas there are disturbances which create non-point sources of water quality degradation (pollution),

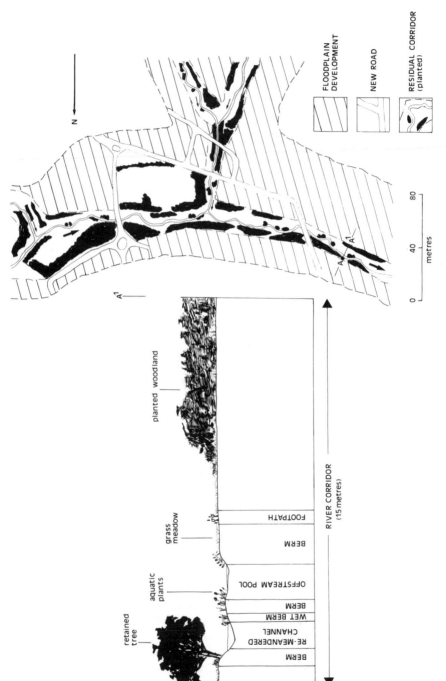

Figure 17.3 Taking advantage of development opportunities to restore natural morphological characteristics to river channels

Figure 17.4 Restoration of pools, riffles, point bars and varied channel symmetry to the Redhill Brook, Surrey, UK (1991)

while industrial development and population growth produce point source pollutants that are highly concentrated. From a pragmatic point of view it is never going to be possible to identify and treat all these sources for a substantial urban area. Also costs are often prohibitive because substance concentrations are low and associated treatment costs high (Herricks and Osborne, 1985). However, even with no water quality improvement it could be argued that increasing the depth of flow by manipulating the morphology leads to a more aesthetically desirable channel. Narrowing the channel also discourages the deposition of urban silts by locally increasing the velocities of low-flows. Aesthetics may be a key issue for urban residents.

Petersen *et al.* (1992) recommended a series of 'building blocks' for stream restoration, based principally on lowland agricultural catchments but applicable to all streams and rivers. This model was prompted by the deterioration of surface water quality by non-point source pollution and is centred on a decrease in the transport of nutrients and sediments to river channels. The restoration measures may be used individually, or in combination, depending on the local circumstances and include buffer strips (or riparian zones or river corridors). This involves set-aside of a minimum of 10 m-wide buffer strips along both banks of the entire length of a stream to trap dissolved and particulate nutrients moving from agricultural land and sediments. Ideally buffer strips should be planted with fast-growing woody shrubs (e.g. alder, willow or hybrid aspen) to further reduce sediment input to the channel.

These measures are in addition to re-creating in-channel morphological character-istics. Whilst an extremely useful concept for landowners and government agencies, the effectiveness of the measures recommended for many geographical settings remains largely unproven because of the lack of monitoring. Undoubtedly buffer strips trap sediment but they may only serve as a temporary sink (Osborne and Kovacic, 1993) unless a long-term maintenance programme is initiated.

Land availability

River restoration efforts may depend on the negotiating ability of individual river managers and the goodwill and cooperation of landowners. Clearly some parcels of land in public rather than private ownership may be easier to negotiate. Even in Denmark where a watercourse restoration law was implemented in 1982, obtaining land for re-meandering of substantial lengths of watercourse can be a difficult pro-cess (Brookes, 1987a; Iversen et al., 1993). It involves the purchase and controlled interchange of fields, based on voluntary participation by farmers.

Particularly important for restoration is the existence of schemes promoting change in land use. These offer major opportunities for restoration of floodplain habitats for which financial assistance may be available under a number of schemes including, for the UK (RRP, 1994):

- Set-aside, established by the Ministry of Agriculture, Fisheries and Food to reduce agricultural production by removing land from agricultural use.
- A Countryside Stewardship Scheme promoted by the Countryside Commission, which includes incentives for the reversion of arable land to grassland in waterside landscapes in certain areas.
- Environmentally Sensitive Areas, which offer tiered payments to farmers who undertake prescribed forms of environmentally beneficial management, aimed at protecting and enhancing the characteristic landscapes and wildlife habitats of the selected areas. Payments differ between areas.
- Woodland grants, which are available from the Forestry Authority for areas of 0.25 hectares and above.

Most of these schemes involve agreements covering periods of only 5–10 years.

NATURAL CHANNEL RECOVERY

Direct channel modification alone has led to the channelisation of the majority of lowland rivers in Europe over the last 2000 years. For example up to 98% of Danish streams may have been modified (Brookes, 1987b; Iversen et al., 1993) and up to 96% of channels in lowland river catchments in the UK (Brookes and Long, 1990). However, even despite a rolling programme of river restoration in Denmark for a decade, less than 1% (200 km) of channelised rivers have been restored. The figure for UK lowland rivers is currently even less impressive. The task is therefore enor-mous and constraints for all countries are likely to include the costs of the works in addition to political and landownership issues. Perhaps the most pragmatic approach for widespread restoration would be to allow natural recovery to occur where practicable, and where the process is negligible, or slow, to install low-cost devices to enhance recovery. In this way many thousands of kilometres of watercourse could be improved over relatively short time-scales, perhaps ranging from 1 to 150 years (Brookes, 1992). Clearly this requires consideration of the impact of recovery features on flood conveyance and arterial drainage and the drawing up of plans for the alterative management and maintenance of watercourses.

CONCLUSIONS

To conclude, it is essential that projects which attempt to restore the natural characteristics of river channels and floodplains are based on the best available geomorphological information. At the very least this will involve general guidance, but ideally should be based on a geomorphological survey of the catchment and reach affected, although the degree to which it will be necessary to provide a geomorphological input will vary from site to site. If geomorphology is not adequately applied then sustainable designs, based either on stable channels, or channels in dynamic equilibrium, will not be achievable. Considerable monies may be wasted on projects which will subsequently fail. Further pure research is ideally required to enhance knowledge of the form and process of a wide range of natural river channel types, especially lowland rivers (Osborne *et al.*, 1993). Although general principles exist, restoration of many rivers remains very much trial and error because re-creating natural characteristics can be complex. Due to the general lack of predictive capability, outcomes remain uncertain. Further advances in understanding catchment influences, and floodplain processes, will lead to more sustainable projects. Undoubtedly description of the entire system can never be complete, but the knowledge should be good enough to enable intelligent management decisions (Statzner and Sperling, 1993). Another assumption in most approaches to river management is that improvement in ecological integrity will follow re-creation of physical characteristics. Data is now being accumulated to suggest that re-creation of habitat features is fundamental, although further scientific work is clearly needed to answer biological questions (Cowx, 1994).

A number of representative sites should be monitored closely because they may have unanticipated impacts on ecological, hydrological and geomorphological systems. Monitoring the outcome of past, existing and future stream restoration projects is required. Evaluation of restoration efforts has so far been limited because agencies that fund the infrastructure of channel restoration may be unable for economic reasons to support post-project evaluation. Future expenditure on monitoring needs to be weighed against the risks of wasted financial, intellectual and natural resources on projects which fail. However, it should be remembered that restoration projects themselves provide opportunities for basic research (Cairns, 1987) and this is likely to provide an ongoing means of increasing knowledge. Publication and distribution of results is essential for the benefit of river managers worldwide. Training of river managers in restoration techniques may also be desirable. This should be based as far as possible on geomorphological/scientific principles.

Complete restoration of catchment processes to a pre-disturbance state is in reality unattainable. Restoration of a reach will usually have to be considered in the context of changed sediment and water discharges as a result of land use changes. For the majority of rivers it will not be possible to determine the pre-disturbance conditions. In reality the channel which is constructed will imitate the morphology of a reference channel, ideally within the same catchment. Whilst attempts may be made to mimic an adjacent channel, it should always be stressed to the public, and to those involved in the project, that the precise outcome is likely to be unknown in many instances. Allowance should be made for adjustments of restored channels following restoration, either natural or managed. Perhaps one of the greatest challenges during the next decade will be a move away from the channel to restoration of riparian areas and wetlands. On an extensive scale this is likely to be restricted by the difficulties of negotiating land use change.

Finally, it is argued that restoration of channels for aesthetic reasons may be just as valid in certain circumstances as ecological improvement. Land use change is likely to provide a far greater opportunity for restoring some of the characteristics of previously degraded river channels, than scientific demonstration projects. In practice only a partial solution will be achievable and, even in countries where watercourse restoration legislation exists, restoration depends very much on the negotiating ability of individual river managers and the cooperation and goodwill of landowners.

ACKNOWLEDGEMENTS

The views expressed in this paper are those of the author and not necessarily those of his employer, the National Rivers Authority. Permission of the Regional General Manager (Thames Region) to publish this paper is appreciated. Ken Gregory is thanked for the inspiration which he has provided to undertake scientific research on rivers and to subsequently become a river manager.

REFERENCES

Bisset, R. and Tomlinson, P. (1988) Monitoring and auditing of impacts. In: Wathern, P. (ed.), *Environmental Impact Assessment; Theory and Practice*, Unwin Hyman, London, 117–128.

Bradshaw, A.D. (1987) Restoration: an acid test for ecology. In: Jordan, W.R., Gilpin, M.E. and Aber, J.D. (eds), *Restoration Ecology*, Cambridge University Press, Cambridge.

Brookes, A. (1987a) Restoring the sinuosity of artificially straightened stream channels. *Environmental Geology and Water Science* **10**, 33–41.

Brookes, A. (1987b) The distribution and management of channelised streams in Denmark. *Regulated Rivers: Research and Management* **1**, 3–16.

Brookes, A. (1988) *Channelised Rivers: Perspectives for Environmental Management*. Wiley, Chichester, 336pp.

Brookes, A. (1990) Restoration and enhancement of engineered river channels: some European experiences. *Regulated Rivers: Research and Management* **5**, 45–56.

Brookes, A. (1991) Design practices for channels receiving urban runoff: examples from the River Thames Catchment, UK. In: *Proceedings of American Society of Civil Engineers Conference*, August 1991, Mt Crested Butte, Colorado.

Brookes, A. (1992) Recovery and restoration of some engineered British river channels. In: Boon, P.J., Calow, P. and Petts, G.E. (eds), *River Conservation and Management*, Wiley, Chichester, 337–352.

Brookes, A. (1995) The importance of high flows. In: Harper, D. and Ferguson, A. (ed.), *The Ecological Basis for River Management*. Wiley, Chichester, 33–49.

Brookes, A. and Long, H.J. (1990) *Stort Catchment Morphological Survey: Appraisal Report and Watercourse Summaries*. National Rivers Authority, Reading, UK.

Cairns, J. (1982) Restoration of damaged ecosystems. In: Mason, W.T. and Iker, S. (eds), *Research and Wildlife Habitat*, Office of Research and Development and US Environmental Protection Agency, Washington, DC.

Cairns, J. (1987) Disturbed ecosystems as opportunities for research in restoration ecology. In: Jordan, W.R., Gilpin, M.E. and Aber, J.D. (eds), *Restoration Ecology*, Cambridge University Press, Cambridge, 307–320.

Cairns, J. (1991) The status of the theoretical and applied science of restoration ecology. *The Environmental Professional* **13**, 186–194.

Cairns, J., Dickson, K.L. and Herricks, E.E. (eds) (1977) *Recovery and Restoration of Damaged Ecosystems*. University Press of Virginia, Charlottesville.

Cowx, I.G. (ed.) (1994) *Rehabilitation of Freshwater Fisheries*. Blackwell Scientific, Oxford, 486pp.

Gardiner, J.L. (ed.) (1991) *River Projects and Conservation: A Manual for Holistic Appraisal*. Wiley, Chichester.

Gore, J.A. (ed.) (1985) *The Restoration of Rivers and Streams: Theories and Experience*. Butterworth Publishers, Boston, 280pp.

Herricks, E.E. and Osborne, L.L. (1985) Water quality restoration and protection in streams and rivers. In Gore, J.A. (ed.), *The Restoration of Rivers and Streams: Theories and Experience*, Butterworth Publishers, Boston, 1–20.

Iversen, T.M., Kronvang, B., Madsen, B.L., Markmann, P. and Nielsen, M.B. (1993) Re-establishment of Danish streams: restoration and maintenance measures. *Aquatic Conservation: Marine and Freshwater Ecosystems* **3**, 1–20.

Jordan, W.R., Gilpin, M.E. and Aber, J.D. (eds) (1987) *Restoration Ecology*. Cambridge University Press, Cambridge.

National Research Council (1992) *Restoration of Aquatic Ecosystems—Science, Technology and Public Policy*. National Academy Press, Washington, DC.

National Rivers Authority (1993) *Draft Guidelines for the Design and Restoration of Flood Alleviation Schemes*. Report prepared by Hey, R.D. and Heritage, G. of University of East Anglia, R&D Note 154, NRA, Bristol, 98pp.

National Rivers Authority (1994a) *Sediment and Gravel Transportation in Rivers including the Use of Gravel Traps*. Report prepared by Newson, M.D. and Sear, D.A. of University of Newcastle, R&D Report C5.384/1, NRA, Bristol, 42pp.

National Rivers Authority (1994b) *Development of Geomorphological Guidance Notes*. Project undertaken by GeoData Institute, University of Southampton, for NRA, Reading.

Newson, M.D. (1992) *Land, Water and Development: River Basin Systems and their Sustainable Management*. Routledge, Chapman and Hall, London, 351pp.

Osborne, L.L. and Kovacic, D.A. (1993) Riparian vegetated buffer strips in water quality restoration and stream management. *Freshwater Biology* **29**, 243–258.

Osborne, L.L., Bailey, P.B., Higler, L.W.G., Statzner, B., Triska, F. and Moth Iversen, T. (1993) Restoration of lowland streams: an introduction. *Freshwater Biology* **29**, 187–194.

Petersen, R.C., Petersen, L.B.M. and Lacoursiere, J. (1992) A building block model for stream restoration. In: Boon, P.J., Calow, P. and Petts, G.E. (eds), *River Conservation and Management*, Wiley, Chichester, 293–309.

Rhoads, B.L. and Miller, M.V. (1990) Impact of riverine wetlands construction and operation on stream channel stability: conceptual framework for geomorphic assessment. *Environmental Management* **14**(6), 799–807.

River Restoration Project (1993a) *Phase I. Feasibility Study: Final Report*. The River Restoration Project, PO Box 127, Huntingdon, PE18 8QB, UK, 187pp.

River Restoration Project (1993b) *Business Plan*. The River Restoration Project (see above), 23pp.

River Restoration Project (1994) *Partnership for River Restoration, Part I: Institutional Aspects of River Restoration in the UK*, Summary, March 1994. The River Restoration Project (see above), 11pp.

Schumm, S.A. (1977) *The Fluvial System*. Wiley-Interscience, New York.

Statzner, B. and Sperling, F. (1993) Potential contribution of system-specific knowledge (SSK) to stream management decisions: ecological and economic aspects. *Freshwater Biology* **29**(2), 313–342.

Ward, D., Holmes, N. and Jose, P. (1994) *The New Rivers and Wildlife Handbook*. Royal Society for the Protection of Birds, Sandy, Bedfordshire, UK, 426pp.

Wathern, P. (1988) An introductory guide to EA. In: Wathern, P. (ed.), *Environmental Impact Assessment: Theory and Practice*, Unwin Hyman, London, 3–30.

18 Towards a Sustainable Water Environment

JOHN L. GARDINER

National Rivers Authority—Thames Region, Reading, UK

INTRODUCTION—THE CHALLENGE TO PROGRESS

For many years, in many countries around the world, decision-making over land use has paid limited attention to the interests of the water environment. Short-term and piecemeal commercial interests have often overlooked the argument that a river system, with its catchment, floodplains and groundwaters, should be treated with respect as a hydrological, geomorphological and ecological continuum—a complex system from the watershed to the sea (Naiman, 1992). Similar considerations therefore extend to esturial and coastal waters (Townend and Fleming, 1994).

The global debate over 'sustainable development' (SD), focused by the 1992 United Nations Conference on Environment and Development (UNCED) 'Earth Summit' at Rio de Janeiro, has served to underline this fact; the continuing welfare of mankind can be seen as dependent on the sustainable management of water and the supporting natural environment (NRA, 1994a). Issues of scale and complexity need to be understood sufficiently to guide strategic and local decision-making and improve ways to encouraging sustainable activities while discouraging unsustainable ones (DOE, 1994a, p.32).

The last decade has seen considerable changes in the United Kingdom and elsewhere in the attitude towards, and treatment of, the water environment and rivers in particular. These changes have been brought about by the promotion of new directions in legislation and institution cultures (Newson, 1992). The following discussion shows that the roots of change run deep, but the growth that they sustain is now very visible and helpful in the worlds of river and catchment management, on which any form of sustainable development must surely depend.

New ideas spring from philosophers, theorists and practising managers of the water environment. It would be comforting to believe that philosophic discourse were continuously mingled with theoretical constructs and practical trialling of new ideas in a virtuous circle of progressive thought and activity extending among all philosophers, theorists and practitioners. A small but 'critical mass' of such activity would hasten the pace of change, but it has to be faced that the sheer volume of literature, coupled with a general diffidence towards reading widely, creates a significant inertia among practitioners. Philosophers pursue traditional approaches until

Changing River Channels. Edited by Angela Gurnell and Geoffrey Petts.
© 1995 John Wiley & Sons Ltd.

forced to do otherwise, and theorists elaborate hypotheses which may relate to either but rarely to both the other groups.

So how is further progress to be made? According to Bernard Shaw (in *Man and Superman*), it depended on the 'unreasonable man' who, trying to adapt the world to himself is at once questioning, willing to analyse and synthesise old and new ideas, and pragmatic in response to need. But gone are the days when professional leadership was judged adequate to provide the needs of society, when one person could stamp their authority sufficiently to implement new methods without wide consultation, endless bureaucratic form-filling and justification to the nth degree. Widespread education has created a world where knowledge of uncertainty now matches the innate fear of the unknown, destroying a willingness to believe in a 'benevolent technocracy'. Those who can cope with the widespread knowledge of uncertainty, who are able to facilitate decision-making among diverse groups and individuals, are relatively rare and would probably acknowledge the challenge of weaning functional experts away from turf protection and on to positive, cross-functional teamwork. Certainly such teamwork could not operate successfully without some supportive institutional framework to encourage not only implementation of change, but its absorption into national, if not international culture. Under the influence of an increasing rate of change, will that culture be prepared to accept the resource implications of achieving the consensus among stakeholders which appears central to the attainment of SD?

If consensus among stakeholders is indeed necessary, then a new 'common language' is needed for them to communicate effectively with each other. SD is a concept which could promote such a language, leading in time from decisions which are less sustainable to those that are more sustainable. It is through such debate that managers may be helped to identify new targets, select new routes towards them, and sense the broader background to the changes which they are experiencing at a very practical level.

THE EARTH SUMMIT—AGENDA 21

At the 'Earth Summit' governments, environmentalists, development experts and representatives of all sectors of society from around the globe considered the progress made in achieving SD. As a result of this conference, an action plan called Agenda 21 was drawn-up which identifies the various ways in which SD can be achieved. Acceptance of Agenda 21 by most countries has given legitimacy to the concept of SD as a platform from which bridges between philosophy, theory and practice—across the sciences and the arts—can be built. There is therefore much scope and opportunity for exploring the potential of 'Sustainability in Action', particularly in identifying 'model' opportunities and their management. This is at least as true for the water environment as any other area of interest, as indicated by the following passage from Agenda 21:

> Water resources must be planned and managed in an integrated and holistic way to prevent shortage of water, or pollution of water sources, from impeding development. Satisfaction of basic human needs and preservation of ecosystems must be the priorities; after these, water users should be charged appropriately.

By the year 2000 all states should have national action programmes for water management, based on catchment basins or sub-basins, and efficient water-use programmes. These could include integration of water resource planning with land use planning and other development and conservation activities, demand management through pricing or regulation, conservation, reuse and recycling of water.

To become an implementable reality, SD based on such action plans must satisfy criteria in at least three major dimensions: Ecological–Social–Economic (ESE). It is also heavily dependent on the 'Institutional' influence over the decision-making process. In assessing whether a development is sustainable, its profile in terms of institutional acceptability must therefore be recognised as well as its ESE profile. Institutional constraints are likely to be centred on economic issues, and often lead to an imbalance in decision-making between the cost–benefit analysis and less tangible ecological/social criteria.

Local planning authorities (LPAs) were seen at Rio to be in the prime position to deliver SD, given their role in strategic and local planning of land use. LPAs co-ordinated by the Local Government Management Board (LGMB) in the United Kingdom (UK) have taken up the challenge of *Local Agenda 21* (LGMB, 1992), and pioneered an appraisal framework for plans which seeks to introduce sustainability criteria at strategic levels. The framework which has emerged (DOE, 1994b) offers an evolutionary approach for any group of stakeholders to develop their common language through agreement over sustainability criteria and their implications for policy, strategy and tactical decision-making.

This chapter will look globally at some of the aspects of recent change in strategic decision-making for the water environment, focusing then on the National Rivers Authority (NRA) in the UK and looking at prospects for progress towards a sustainable water environment.

INSTITUTIONAL RESPONSES—CATCHMENT PLANNING

Within the holistic approach to SD required for the water environment are many overlapping fields or functions (e.g. tourism, agriculture, forestry, urban planning and design, transport, water quality, flood defence, water resources, water supply and sewerage) which, nevertheless, are further fragmented in the current institutional structures of most countries.

The new Regional Councils of New Zealand offer an example of a determined attempt to reduce the institutional underpinning of fragmentation, having combined land use planning with water and soil conservation after 40 years' experience of separation. It is, however, the concept of 'sustainable management'—seeking environmental management through constraints on the use, development and protection of natural resources—which is consistent with (but not equivalent to) SD, that has been put into action (Scott, 1993). In Ontario, the requirement to co-ordinate surface water management at all scales from major river catchment to local urban catchment (Gardiner *et al.*, 1994) has been recognised through provision for 'Watershed Plans', Sub-Watershed Plans and Master Drainage Plans (MEE/MNR, 1993). In California, the Urban Streams Restoration Programme operated by the

Department of Water Resources in Sacramento has shown how the community can participate in the initiation, planning and implementation of environmental improvements, a process which has strong parallels in the Murray–Darling basin in Australia. Australia has also made determined efforts to engage in 'Total Catchment Management'; useful distinctions have been made between 'technical' (integrating functional strategies), 'bureaucratic' (co-ordination between authorities) and 'participating' (involving stakeholders in decision-making and implementation) versions (Collett, 1992).

However, many countries share river catchments with neighbouring countries who are not always in political accord. The emerging system for environmental management of the River Rhine (Schulte-Wulver-Leidig, 1993) may provide a translatable model for international co-ordination. Despite the issues of real-politik, much progress remains to be made towards integrated management of land and the water environment in terms not only of science and technology, but also of the process of decision-making in natural resource management, including its social aspects at local, sub-catchment and catchment scales.

At a time when attention is being focused on major issues as diverse as the sustainability of cities (Girardet, 1993) and biosafety protocols to control release of genetically modified organisms into the environment (Munson, 1994), it can prove difficult to attract and maintain the interest of a relatively large number of organisations in such a broad, complex and yet vital subject as sustainable management of river catchments. Yet this is a major goal of the UK's National Rivers Authority's corporate plan (NRA, 1994a), and is now recognised in many countries as the necessary way forward to underpin SD. It was, for example, central to the findings of the 1993 International Symposium 'Management of Rivers for the Future' in Kuala Lumpur (Bramley, pers. comm.) and was held to provide a sound framework for the management of the Upper Mississippi floodplains to ensure their ecological integrity (Galloway, 1994).

THE UK APPROACH TO CATCHMENT PLANNING

In the UK, the NRA was set up by the Water Act (1989) as a regulator of the Water Services Public Limited Companies, which had been privatised by the same Act. Following earlier initiatives to produce catchment plans (Gardiner, 1991a; Chandler, 1994), the NRA introduced the production of catchment management plans (CMPs) as a strategic component of the Corporate Plan (NRA, 1990). A set of guidelines was produced which have been modified in the light of subsequent experience (NRA, 1993a).

Viewed rather superficially, catchment management planning seeks to ensure that activities in the NRA functions of flood defence, water resources, water quality, fisheries, recreation, navigation and conservation do not conflict with each other but rather identify opportunities for multi-functional benefit to be gained (Davies, 1992). Through consultation with local authorities and other interested parties, including the public, this principle is extended to relate NRA activities and concerns to those of other organisations or individuals (Woolhouse, 1994).

Historically, the problem has been to achieve multifunctional expenditure when each function has its own independent budget. The solution has been partially achieved through legislation; in the Water Resources Act (1991), Section 16 requires the NRA to conserve and enhance (the water environment) when carrying out any of its functions. Quite apart from these aspects, CMPs are seen as a means of identifying statutory water quality objectives (SWQOs), and present an opportunity to assess the entire catchment in relation to 'generic' issues. These might include:

- Identification of economic incentives driving development (e.g. economic policies in development plans, full structural/regional funding; Countryside Stewardship; Long-term Set-aside; Environmentally Sensitive Areas (ESAs); Groundwork Trust initiatives, etc.)
- Identification of environment capacities (carrying/absorptive).
- Identification of sources and sinks (e.g. contaminants).
- Identification of river corridors (for inclusion in local planning authority documents).
- Requirements and priorities regarding buffer zones.
- Identification of catchment areas vulnerable to erosion.
- Appropriate areas for source control (infiltration not exfiltration areas).
- Role of master drainage plans for future development planning and control.
- Planned interface between river and sewered systems (flood defence and water quality interests).
- Opportunities for demand management activities.
- Opportunities for retrofitting balancing/storage ponds for water quality, etc.
- Prevention, minimisation, reduction, recycling and re-use of waste.
- Opportunities for public access to river and new public rights of way.
- Involving schools in management of the water environment as part of the curriculum.
- Empowering the community to develop a strong sense of place and involvement in its river environment.
- Considering issues of health and safety in relation to the water environment.

Many of these generic issues imply strong partnership with other agencies such as local (government) planning authorities. The full effectiveness of CMPs can be seen within the wider context involving not only existing use of land and water but development planning and control as well (Gardiner and Cole, 1992). In the UK, there is a well-developed system of Town and Country Planning which is under the control of local authorities.

Thus, even if existing uses cannot currently be influenced, the policy framework should be clearly in place in order to influence the inevitable change of use in the future. This relies on achieving a level of competence (Figure 18.1) which is able to identify issues related to what exists and its state (inventory and legacy). The approach to natural resource management and planning involving CMPs can be seen as a form of strategic environmental assessment, particularly when the entire Thames catchment is appraised in terms of resource sensitivities and development pressures as in the *Thames 21* strategic planning document (NRA, 1994b). To appreciate fully how this fits into the approach to decision-making for SD, the economic approach needs to be outlined.

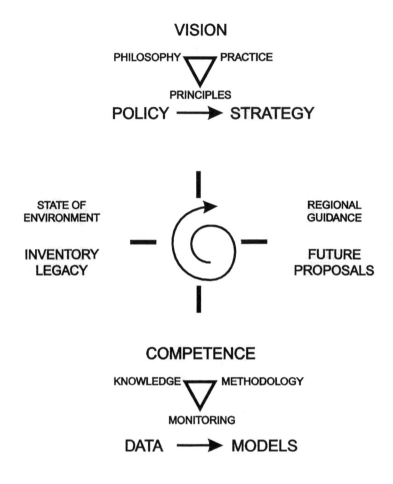

Figure 18.1 Dynamic iteration in regional strategic planning

APPROACHES TO SUSTAINABLE DEVELOPMENT

If it is accepted that SD is strongly related to what happens 'on the land' (including the water environment), then it is the influences on decisions over land use which are of particular interest, primarily: *Land Use Planning—Legislation—Economic Instruments*. The purpose of this chapter is to outline an approach to decision-making which will take in these influences, but be particularly focused on achieving environmental and social sustainability. However, while neither land use planning nor legislation are specifically addressed here, the use of economics as applied to valuation (rather than incentives such as tradeable permits or grants associated with UK designations such as ESAs and incentive schemes such as Countryside Stewardship or Set-aside) will be briefly examined.

Economic valuation

Examining the methods of economic valuation proposed to measure sustainability (e.g. in *Blueprint 3*, Pearce *et al.*, 1993), among the many arguments it is noted that both damage and mitigation measures can be given a value by their subsequent cost. For example, the monetary value of acid rain damage in the UK can be calculated from the damage done to buildings, health, crops and trees. This value could therefore be set against the cost of defensive measures against acid rain. However, it is clear that damage exposed in such terms may ignore damage to species and ecological systems which is not immediately apparent as damage to human interests, and can therefore at best be regarded as a surrogate—and probably a highly conservative—valuation. Its use, then, is limited by our ability to assess the total damage and hence 'value the environment'. Similar problems emerge from consideration of contingent valuation, or public willingness to pay (e.g. for a better environment) or accept a loss (e.g. of environment).

The four basic categories of environmental valuation techniques include:

- market price approaches (i.e. dose–response and replacement cost)
- hedonic pricing methods
- household production function approaches (avertive expenditure and travel cost)
- experimental markets (contingent valuation and contingent ranking)

The last two techniques are centred on public participation, and underline the need for greater public awareness of the issues surrounding water and the natural environment. For example, work carried out recently in the NRA on the alleviation of low flows (River Darent) and rehabilitation of a channellised reach of the River Ravensbourne included surveys of public willingness to pay involving use values and, in the former case, non-use values. The results indicated that the public are keen to regain their rivers in as natural a state as possible, and that their perception of the value of doing so is several times greater than the estimated cost.

This area has been subject to comprehensive UK overviews (CNS Scientific and Engineering Services, 1992; NRA, 1993b and c). Development of the approach, including the derivation of 'standard values' in the case of proposals to rehabilitate damaged rivers for example, while regarded by many as necessary to achieve SE in practice, may not prove sufficient.

To what extent can we therefore rely on economic arguments and legislation to secure SD? It is already acknowledged by Pearce *et al.* (1993, p. 62) that environmental objectives are currently driven by political sensitivities rather than economics. Politics is subject to many influences, not least the attitude of the electorate for which refined environmental economics may prove inaccessible. Issues which are brought to the attention of the public can therefore achieve politically what may otherwise fail when based solely on theoretical argument.

Carrying capacity

According to the definition proposed by the World Conservation Union and others in *Caring for the Earth* (IUCN/UNEP/WWF, 1991), SD is 'improving the quality of human life while living within the carrying capacity of supporting ecosystems'. The approach to sustainability which is based on identification of carrying capacity, involving critical natural capital (to be protected) and constant natural assets (which may be substituted under certain conditions) may therefore prove a more effective overall strategy, particularly when supported by economic argument. This is the direction of recent initiatives in NRA Thames Region, in which the national NRA catchment planning programme, together with a cohesive programme of influencing development planning and control, is being supported by an applied definition of critical/constant natural capital (Gardiner, 1994a).

For example, the carrying capacity of a river and its floodplain has a local value, which can be modified locally by floodplain development. But this carrying capacity can also be effectively reduced by development in the catchment which accelerates runoff into the river system, often leading to increased erosion and deposition as the morphological regime is disturbed. So the catchment itself has a 'carrying capacity' dependent on natural capital which includes not only the entire watercourse system and its storage components of floodplains, ponds and lakes, but also the catchment's surface and groundwater system and its management for agriculture, forestry, urban settlement, etc. For a given local carrying capacity related to any NRA function, there is likely to be a definable 'critical capital' required in the catchment to support the quality of human life. For purely flood defence or water quality purposes this could be a mixture of natural and 'built' capital assets. However, to maintain the ecological continuum of the river and its corridor, a similar argument will require the 'built capital' to be of a certain type (e.g. vegetative treatment of urban runoff in local wetlands rather than at a distant sewage treatment works, usually downstream).

NATURAL CAPITAL

Natural capital consists of renewable and non-renewable resources, each of which can be affected in reversible and irreversible ways. The question is whether the loss of such capital may be effectively substituted, and whether such a loss would be critical to the quality of human life (according to the IUCN definition). Where the loss is deemed critical, the capital resource can be regarded as *critical natural capital*.

Clearly, protecting critical natural capital should include *protecting the system on which it may depend*; lowering of groundwater levels through deepening a river may destroy a dependent wetland. The Norfolk Broads may be a valuable habitat formed by traditional peat cutting, but modern methods of mechanised peat stripping effectively 'mines' the peat resource with no concurrent asset creation to maintain the overall *stock of natural capital* at a constant level. Thus, recognition of the need for species (and therefore habitat) diversity to maintain a genetic pool and play a role in environmental systems causes concern over shrinking areas of, say, wetlands and peat bogs. There would perhaps be some *threshold* below which stakeholders can agree (based on available evidence with a margin of precaution applied) that the resource should no longer be substitutable as part of the area's '*constant*' *natural assets*, but identified as critical natural capital.

PRECAUTION—PREVENTION RATHER THAN CURE

The threshold or (*capacity limit*) between critical natural capital and substitutable natural assets should take into account that scientific knowledge of ecosystems is incomplete; uncertainties over whether the asset can be re-created or replaced by one of equal ecological value leads to application of the *precautionary principle*—comparable perhaps to a factor of safety. In surface water terms, the precautionary principle for example may include definition of environmentally acceptable flows (Newson, 1993) which allow some 'freeboard' or safety margin above the figure suggested by investigations. Flows which may be ecologically acceptable in a 'natural' river may be quite unacceptable if it is enlarged into a high flow capacity trapezoidal channel (typical of traditional flood defence designs). Also, since there is current uncertainty over the effects of climate change on the water environment (Beven, 1993) and the state of the environment is not always as could be wished, adoption of the precautionary principle would be to assess and manage the overall trajectory of environmental wellbeing in the direction of sustainable conservation and enhancement, if not rehabilitation.

At a strategic level, a crucial element in the calculation of threshold values or conditions concerns the space and time frames which should be taken into account. In flood defence, current efforts at river rehabilitation can be wasted if there is no appreciation of *geomorphological processes at catchment scale* (Brookes, 1992; Sear, 1994). This example of what the holistic approach can mean in the geophysical sense has analogies in the social, political and economic areas.

For example, decision-making based on benefit–cost analysis (BCA) as the dominant factor in assessment of options often leads to an imbalance between the economic and the less tangible environmental/social criteria, with the option offering short-term, financial gain being preferred to options with longer-term benefits, often of a less tangible nature. This may owe as much to the fact that the financial value of some environmental investments (such as woodlands) may reach their maximum after their discounted value becomes insignificant, as to the difficulty in valuing natural 'assets' against the known value of the built environment.

It is rather ironic that economic incentives to plant woodlands should lead to destruction of part of the flow country in Scotland, until brought to an end by environmental protest. This was just one example of the many unfortunate results of inadequately focused economic incentives being applied without environmental assessment of either carrying capacities or likely impacts. If, in any situation, the carrying capacities cannot be maintained for *all* the functions involved, the question arises whether environmental interests should not take precedence in man's longer-term interests.

A significant challenge lies in the field of water resources. Eagerness to clinch the argument as one which begins and ends with economic factors can only lead to assertions that water demand should be met out of effective runoff only (Dubourg, 1992), i.e. the stock of fresh water from precipitation left after evapotranspiration. This assertion would only hold if 'water demand' includes maintenance of the stock required to support the carrying capacity of ecological systems (on which the quality of human life depends). The assumption made over environmental ethics will determine whether the words in brackets are included, although some would argue that the ecocentric version (without the bracketed words) implies the apparently qualified anthropocentric version (with the bracketed words), in the end. Recent work points to biodiversity being a vital factor in the resilience of ecosystems (Cherfas, 1994), a strong message for mankind faced with the possibility of climatic change.

From this standpoint, recognition of the needs of the ecological systems, of their interconnectedness, is lacking from current tests on the economic issues of sustainability. Curiously, from the basic Brundtland definition of SD (in terms of inter-generational equity) it has been suggested that water quality should not decline over time, so as not to impose costs on future generations (Pearce *et al.*, 1993, p. 68). This appears to assume that present levels are in all cases adequate for ecological sustainability, and ignores the linkage between water quality and the quantitative carrying capacity of the hydrological system required to support the ecosystem and levels of human use.

Continuing efforts are being made by the NRA to identify the criteria for 'minimum environmentally acceptable flows'. There are many other such carrying capacities to be investigated, not least associated with maintaining the stock of topsoil (Brown, 1994) under modern farming regimes. However, there is a further major issue involved, related to the catchment's carrying capacity, which should be examined.

It is only recently that the 'working with nature' principle has been applied to all aspects of water use. The use of hydroponics or vegetative treatment, coupled with devices such as vortex separators (Andoh, 1994), may allow smaller sewage treatment works to escape being 'rationalised' into bigger sewerage systems which feed large treatment works, often at the expense of the river's ecological carrying capacity. Urban and road runoff are also being seen as essential elements of the river catchment's carrying capacity, control of its quantity and quality achieved through using vegetated storage systems, perhaps with vortex separation assisting in management of surface water to attenuate or retain runoff.

In many areas of 'surface water management'—which must of course be in tune with the needs of groundwater management—there are opportunities for bringing stakeholders together in the interests of improving the carrying capacity of the water environment. There may need to be legislative change to adjust from the current paradigm of 'surface water disposal', but the basic change will be in the hearts and minds of the professionals involved. Their willingness to consider change from the traditional, fragmented approaches will help towards a more general understanding of the issues. This may well be achieved through a combination of sustainable development with the communication needs implied by CMP approach. Stakeholders can be brought together by both elements, with sustainable development encouraging the 'common language' and CMP providing the subject area.

It is suggested that adopting the principle of 'working with nature' with a multifunctional remit will lead to 'value for money' for the community, when calculated from an holistic (or sustainability) viewpoint and therefore including intangible benefits. However, it is also possible that, for a single function, this option may not be the cheapest, nor even justifiable in strict economic terms for that one function; the marginal extra cost may need to be considered as a 'sustainability premium'.

TOWARDS CONSENSUS

Where *environmental assessment* has not been used as the central method for project development (and the *'best practicable environmental option'* therefore remains unknown as such), unforeseen adverse impacts have often resulted in expensive 'cure' procedures being required. To move towards *'prevention'* as the more sustainable solution, the environmental argument need to be strengthened in the decision-making process. The following procedural argument needs to be strengthened in the decision-making process. The following procedural 'hierarchy' (to involve all the *stakeholders*, i.e. those individuals or organisations with legitimate interests in the outcome of the decisions being made) would facilitate such strengthening:

1. Establish ecological/social/economic principles related to sustainability: *'what do we believe and value?'*
2. Establish methodology for iterative appraisal of policies/strategies (i.e. generic options), using sustainability criteria: *'how should we apply our beliefs and values?'*
3. Derive environmental objectives from (2); review catchment issues (Thames 21); *'what guidance can be applied to functional decision-making'*
4. Apply strategic environmental assessment to preferred strategic options in CMPs: *'what should be done where?'* (long-term effectiveness/value-for-money).
5. Apply environmental assessment to individual projects: *'how should it be done?'* (efficiency and short-term economy).
6. Monitor/do post-project appraisal, using a range of environmental indicators to provide feedback to (1) to (5): *'how do we progress?'*

The word 'environment' in steps (4), (5) and (6) includes here all three dimensions of sustainability, assuming the guidance provided by steps (1) and (2) is adhered to. With such a regime in place, regular procedural review as in (5) would allow progress to be made through internal and external feedback. It is recognised that 'iteration' as identified in the recent UK Department of the Environment (DOE) guidance on *Environmental Appraisal of Development Plans* (DOE, 1994b) is needed between the procedural steps. At the same time, it is clear that 'key issues' arising from multifunctional appraisal in (2) would have a significant influence on the assessment in (3), which in turn would influence (4) and (5); this is often referred to as 'tiering' (Therivel *et al.*, 1992). Steps 1 and 2, which provide opportunities for a new 'common language' of sustainability to be developed in support of decision-making through CMPs, will now be briefly examined. The first question is then: what are the principles on which the discussion should be based? Acknowledging that economic principles are already established, at least in traditional terms, and that to address social principles as in Clark and Gardiner (1994) would make this chapter over-long, examples of environmental principles follow (adapted from Gardiner, 1994b).

1. Environmental sustainability—some general principles

The NRA's basic functions are directed towards protection and conservation of the water environment, and under Section 16 of the Water Resources Act (1991), it also has duties to conserve and enhance the natural environment. A strong sustainability stance is implied, requiring protection of critical natural capital at least, and maintenance of the overall stock of capital with special attention to the natural environment (Turner, 1993). From the foregoing, some of the general principles to be drawn out might include:

Conservation, in the sense of protection, is better than re-creation.
(b) The precautionary principle should be applied in any decision.
(c) The stock of environmental capital should be kept constant or increased.
(d) Unavoidable use of unrecreatable resources should be as efficient as possible.
(e) Thresholds of environmental capacities should be established (implying also the identification of critical natural capital).
(f) System (as well as local) criteria must be satisfied, implying a holistic approach.
(g) Change and its effects should be localised ('source control').
(h) Change should be incremental and monitored to allow adaptive management.
(i) Waste that cannot be safely returned to the environment should be minimised in the interests of efficient recycling of resources.
(j) Working with nature is more sustainable than trying to overcome nature.

As examples of these principles, (a) may be likened to the overall principle that 'prevention is better than cure'; (b) may present some difficulty in interpretation, but can perhaps be likened to the design safety factor, reinforcing (a) by weighting the value of what exists higher than what might be created. In urban areas, the precautionary principle could dictate that a river 'corridor' should be conserved, free from buildings; the width of the corridor will depend on the size of the river, but should perhaps be no less than 20 metres on either side in any case.

Looking at the effects of development either locally or globally, the third principle (c) is clearly not being upheld, unless the idea that natural stock can be replaced by human knowledge and man-made assets ('weak' sustainability) is accepted. Decision-makers will need to decide whether a natural river bank or one made of concrete or steel meets this principle best; given the great habitat value of a natural bank, what is the marginal benefit of the concrete version against such value and against its cost of construction? To what extent can a length of natural river bank be lost before what remains could be regarded as critical capital? How would this argument apply to floodplains or minor watercourses?

When such a loss is permanent, because the environmental asset is 'unrecreatable', the lack of knowledge over what effect the loss may have should alone strengthen the likelihood of its being categorised as critical capital. In cases where the loss is 'avoidable' (e.g. gravel extraction), the degree of unsustainability incurred may be related to the rate of loss or the efficiency of use, as in principle (d).

Of all the challenges implicit in the above, perhaps the identification and definition of (e) is the greatest challenge (long recognised in water management) in meeting the inter-generational equity identified by the Brundtland definition of sustainable development. Consideration of these principles leads to the conclusion (f) that decisions require knowledge not only of the environmental assets of the site being assessed but also their role in wider environmental systems and in the possible scenarios of climate change. Estuaries and inter-tidal areas, for example, may play crucial roles in sustaining bird migration patterns, and are themselves dependent on the river catchment and inter-dependent with the processes affecting the coastal region, which may include sea-level rise. Principle (f) is important overall, in the sense that it will assist in determining threshold values and provide information over the adequacy of the present stock, while reinforcing the need for application of the precautionary principle.

Added to these constraints, principle (g) also suggests that inter-generational equity implies an areal limit which could be applied to development, ensuring there is no downstream impact from development, for example, or pulling buildings back from the natural line of the river bank in order to secure the river corridor as an environmental resource.

The 'source control' (HRDL, 1993) referred to in (g) means the practice of meeting both systemic and local needs of the water environment through local water management. This is a culture change from the traditional 'surface water disposal', supported by a range of enabling techniques (Urbonas and Stahre, 1993) and implemented largely through development planning and control. Under this principle, runoff from urban areas should be controlled at or near the point of rainfall before being discharged into the river. Such surface water management has water quality and flood defence as two of the important objectives.

Principle (h) says that investment involving ecological change should ideally be carried out incrementally to assess the accuracy of the prediction of change, and in any case monitored to allow adaptive management to be applied.

Principle (i) confirms the need to minimise products that cannot be returned safely to the environment, and is relevant to recycling 'waste' streams. There have been notable successes when this principle has been directly applied, such as the Aire and Calder Project (Johnston, 1994), in which commerce and regulators came together to carry out audits of industrial processes with the aim of preventing problems at source rather than at the end of pipe or instream, with relatively low investment or rapid payback. This can be seen as analogous to a surface water management system; how much water need be polluted (rather than be conserved) and how best may it be treated at or near source to limit problems of quantity and quality downstream? Controlling pollution through 'natural' processes such as vegetated buffer zones rather than chemical treatment may be regarded as one form of working with nature. A better example of principle (j) would be the use of natural floodplains rather than channellisation to achieve flood alleviation.

2. Environmental appraisal of policy and strategy

Within the UK, local authorities produce development plans complete with policies and strategies applied to the authority area. Environmental audits are now being carried out to prepare inventories and assess how the assets of the areas are being managed. A similar procedure could be applied in management of the water environment, and where suitable policies have not yet been identified it should be possible for stakeholders to define them by relating functional strategies to the sustainability criteria. The environmental criteria detailed in Table 18.1 might be appropriate for the assessment of policy and strategy.

These criteria used at strategic level will enable accumulative impacts to be recognised, thus dealing with one of the major problems associated with project Environmental Assessment (EA). Also, discussion of the issues at strategic level appraising

Table 18.1 Sustainability criteria for assessment of policy and strategy

Global sustainability	Local environmental quality	Natural Resources	Inter-generational equity
Resilience to climatic change	Morphological stability	Water conservation	Equitable use ('non-indigenous' resources)
Energy efficiency	Landscape	Groundwater quantity/quality	Retain strategic adaptability
Renewable energy potential	Recreation	Surface water quantity/quality	Not foreclose future options with
Biodiversity	Public access	Mineral conservation	part remedies
	Rehabilitation of river environment	Wildlife habitats	

policies or generic rather than actual (or geographically unique) strategic options can bring forward function-specific principles which can otherwise be disguised when dealing with specific projects.

Broadly, the criteria found in the otherwise global sustainability of inter-generational equity headings will not be found in project EA, but specific principles arising from their consideration, together with consideration of other criteria identified as appropriate by the stakeholder group, will influence the subsequent stages of SEA (actual strategic options, see below) and project EA.

3. Function-specific principles and environmental objectives

As with many such assessment matrices, the main value of comparing, say, strategic options against the criteria lies not so much in giving 'an answer' but in providing a focus for discussion. Such discussions can perhaps improve the criteria and suggest new principles, as well as providing key issues and function specific principles for further assessment of specific strategies ('what should be done where'), to feature in CMPs. A simplified example is given in Figure 18.2, from which came several key issues, including the following for source control:

- to conserve resources (greater control over runoff from rainfall events)
- to buffer system from effects of climate change
- to conserve energy through increased retention 'at source' (reducing flow volumes in floods and at sewage works)
- to promote biodiversity (retains water, uses 'natural systems')
- to promote self-sufficiency (fully retains other strategic development options)
- to reduce stress on groundwater (grass swales/vegetative infiltration systems) while managing the time of concentration of runoff into the drainage system

Such specific principles could then be used to inform strategic assessment at regional or catchment scale. These principles may often be rephrased into environmental objectives, the better to act as guidance for the next stage in the process.

4. Strategic Environmental Assessment (SEA)

Instead of perpetuating the traditional approach to water and land-use planning, which has been development-led, SEA could contribute to an environmentally-led plan (Gardiner, 1992). This implies guiding development in locality and type to ensure ecological, social and economic sustainability. The objective would be to support development of the appropriate type in an appropriate place, especially in urban areas. In this scenario, there is little doubt that the environmental requirements to be met by the promoters of the project would be mandatory, irrespective of who was paying. Mitigation, including elements of enhancement, would not be optional (as 'planning gain' may be), but obligatory, reflecting the 'real' costs of development implied by ecological and social sustainability.

The SEA process has been shown as an effective way of focusing attention, across disciplines and organisations, on strategic options in order to assist in decision-making. To be effective it should be a team process, the critical implication being that the views of each discipline can be heard, questioned and debated by colleagues

CRITERIA	Source Control	Leakage Control	Effluent Return	Reservoirs	IBT
GLOBAL SUSTAINABILITY					
Resilience to climate change	O	O x	O x	O x	o x
Energy Efficiency	o	O x	X	X	X
Biodiversity	O	O x	O X	O x	o X
NATURAL RESOURCES					
Groundwater quantity/quality	0/o	O x	o x	0/o X/x	0/o X
Surface water quantity/quality	O	O	0/o x	0/o X/x	o X/x
Wildlife Habitats	o	o x	O x	O X	X/x

CRITERIA	Source Control	Leakage Control	Effluent Return	Reservoirs	IBT
INTERGEN. EQUITY CRITERIA					
Equitable Use	O	O	O x	O	X
Retain Strategic adaptability	O	O	O	o x	X
LOCAL ENV. QUALITY					
Morphological Stability	O	O	o	X	X
Landscape	O	O	o	0/o X	X
Recreation & Amenity	O	O	o X/x	0/o X/x	0/o X/x
Enhancement/ Rehabilitation	O	O	o X/x	0/o	0/o X/x

KEY
O - Beneficial   X - Adverse   IBT - Inter-basin transfers
o - less Beneficial   x - less adverse   N - Neutral

Figure 18.2 Assessment of options against initial set of criteria for environmental sustainability

of other disciplines. The challenge may then be to establish the overall strategy in the light of sustainability principles and criteria for the key actors to agree. This agreement would be the first step to identifying and implementing the action plan implied, in the interests of future generations. The adoption of this approach could influence the appraisal methodology for individual projects from the earliest scoping stage, as indicated in Gardiner (1991b).

5. Environmental Assessment (EA) of projects

The proactive influence of the NRA on development planning and control can be traced back to the UK's response to the EC Directive 95/337 on EA, coupled with the duty to promote conservation and enhancement, etc. (Section 2.2 of the Water Resources Act 1991) and the need to support rather weak legislation on land drainage consents. A Thames Region initiative, begun in 1988, contributed through national R&D projects to NRA manuals for EA of internal and external proposals. It was shown that influence on development planning and control can contribute significantly towards environmental protection through sensitive siting and design, including the management of catchment runoff and physical use of land. This positive effect on the physical environment, often leading to restoration of canalised streams in recognition of their potential value to the ecology, recreation and amenity resource, is complementary to efforts to control abstraction and pollution through legislation. In 1991/92, this robust approach was calculated to yield £11.6 million of environmental enhancements in the Thames catchment.

6. Monitoring and post-plan/project appraisal

NRA Thames Region is working with a SERPLAN (South East Regional Planning) group to identify a range of environmental indicators which can be used at strategic level. Monitoring of these indicators and subsequent post-plan/project appraisal will inform subsequent iterations of the process from Stages 1 to 5.

COMPETENCE AND STRATEGIC PLANNING

Referring to Figure 18.1, the deficiencies in competence which have hitherto limited catchment-wide surface water management are now being rapidly overcome with the availability of more accurate aerial survey; data manipulation using Geographical Information Systems (GIS); distributed hydrological models; computational hydraulic models interfacing with Digital Terrain Models (DTMs) to provide synthetic flood envelopes for events of different frequency, and the development of common methods in such areas as river corridor wildlife survey, geomorphological and landscape assessment.

Clearly, competence in assessing current issues and the potential of development to render them more sustainable—or at least less unsustainable—is not limited to the production of facts and figures. The nature of decision-making over land use is more political than scientific; it relies much on qualitative assessments and the application of principles, policies and criteria, all of which can of course be informed by scien-

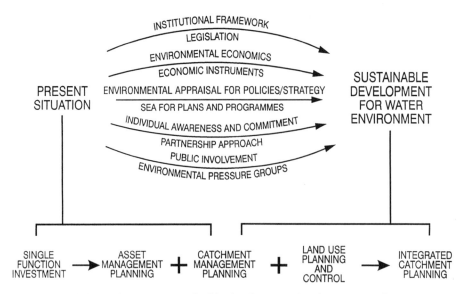

Figure 18.3 Pathways to sustainable development for the water environment

tific enquiry. However, knowledge of future development pressures is crucial for informed debate over their implications—hence the production of 'Thames 21' (NRA, 1994b).

In harmony with the principles of sustainable development and both existing and proposed European Community (EC) Directives on Environmental Assessment (EA), this approach encourages conservation and enhancement of the water environment as part of the decision-making process on land use change. Assuming a particular institutional environment and framework of law and economic instruments, the suggest equation (Figure 18.3) is:

Asset Management Plans (AMP) + Catchment Management Plans (CMP) +
Local Authority Development Plans (LADPS) =
Integrated (ICP) or Total (TCP) Catchment Planning

(where AMPS are the investment plans of the privatised water utilities).

These terms suffer from a considerable variety of usage. It is suggested that 'integrated' could imply (unrealistically) that all the many sectoral plans are welded into one (in this case the Australian terminology 'Total' may be preferable), and that 'planning' in this case subsumes (or at least, directs) 'management'. Clearly, in this interpretation, AMP would form part of the CMP where water services are not privatised.

It can be readily seen that maintaining the carrying capacity of rivers and their floodplains implies targets for surface water management in the catchment which do not exceed the threshold values of those capacities; this has already been tried successfully in Tokyo (Fujita, 1994). This management response to flood hazard is not meant to alleviate major floods, but to halt the inexorable lowering of flood thresholds caused by 'creeping urbanisation' and the cumulative affect of accelerated runoff.

As illustrated by Urbonas and Stahre (1993) in the urban context, sustainability can be served through the application of appropriate source control techniques of attenuation and infiltration at source (i.e. at point of rainfall), provided the institutional arrangement allows. The recent publication *Urban Drainage: The Natural Way* (HRDL, 1993) has illustrated the issues and some of the options, while *Tomorrow's Towns* (AME *et al.*, 1993) recognises the vital role played by the water environment in sustainable urban design. A comprehensive review of all aspects influencing the control of urban runoff is presented in CIRIA Reports 123 and 124 (CIRIA, 1992).

CONCLUSION

There are a number of general recommendations which can be made, related to Figure 18.3 and several more specific to the NRA's activities:

(A) Recommendations of a general nature related to Figure 18.3

- Actively support and promote 'sustainable development' as a paradigm of *open communication based on agreed principles, criteria and a new 'common language' for decision-making*. The interests of the water environment to thus be supported by the land use planning system and by legislation and economic instruments fully integrated with CMPs and the AMP of water utilities; implementation of *'prevention rather than cure'* begins at *policy and strategy levels*, identifying *appropriate actions in appropriate locations*.
- Communicate with other countries over their approach to sustainable development for the water environment, identifying *best management practices* and enabling progress towards sustainability on national and global scales.
- Progress partnership with other organisations such as local authorities, who have a lead role in sustainable development through the implementation of Local Agenda 21. This partnership approach (e.g. the NRA's Memorandums of Understanding with local authorities and English Nature) creates the framework for *successful influence on planning proposals through development control*.
- Obtain real value for money through co-ordination of the main sectoral plans (including CMPs) and *adoption of an SEA approach*. Issues can then be addressed within a strategic, multifunctional framework; action programmes formulated unilaterally by functions or organisations may have some difficulty in illustrating such a sound basis in relation to their sustainability.
- Progress sustainability within organisations through *environmental appraisal of policy/strategy and SEA* for present and future development needs on a *regional scale*.
- Show clear support for sustainable development in terms of the water environment at local/strategic level through establishment of *environmental carrying capacities* and techniques for achieving them, e.g. *quantitative and qualitative source control* as part of a surface/groundwater management strategy.
- Use and promote use of EA as the *procedure for project development*, rather than an 'add-on' requirement for projects or some parallel but separate activity. This allows the meeting of minds necessary to identify the *best practicable environmental*

option, rather than risking confrontation between intangible ecological needs and tangible project benefits offered by environmentally unacceptable options.

- If sustainability implies *adaptive management* and *conservation of resources*, it is easier to adapt to *incremental change* than to the imposition of large capital investment; as a corollary, *demand management and maintenance* of natural and built assets should generally take precedence over investment in capital works.
- Promote community participation and ownership over of the issues and actions (Fordham *et al.*, 1989; Lee, 1992).
- Professionals have key roles in sustainable development; *continuing professional development* is required for all who grapple with the multifunctional complexity of development proposals, total catchment planning, source control, etc.
- Professional organisations and universities have strong roles to play in this area in conjunction with environmental organisations and local authorities; *networking* among the many groups involved will lend *coherence and focus* to progress.
- Decision-makers and their advisers in land use planning and NRA functions have great opportunities to sponsor sustainability; *vision, understanding and energy* are needed to institutionalise the *holistic approach* needed.

(B) Local and specific actions related to the NRA to benefit the water environment through development planning:

- Achieve closer and more effective communication with the planning authorities (cf. Memorandum of Understanding), and other stakeholders.
- Use Catchment Management Plans as a vehicle for SEA, involving NRA activities such as water resources strategy, flood defence planning and associated conservation objectives in land use planning and control (Gardiner, 1994b).
- Promote use of NRA Guidance Notes and reference to CMPs in the production of Policy and Regional Guidance, and Local Authority Development Plans.
- Use national Manual on Planning Liaison with local planning authorities, with a standard Baseline Development Schedule and consistent responses to planning applications.
- Produce and promote a national NRA Policy for the Management of Floodplains, similar to the Groundwater Protection Policy, backed by S.105 surveys.
- Produce information showing potential opportunities for enhancement and rehabilitation as well as conservation, e.g. river corridor surveys, landscape assessments, fisheries surveys.
- use hydrological models to define runoff storage zones (Rylands and Lee, 1993) and to feed computational river models to produce Catchment/Master Drainage plans (Gardiner *et al.*, 1994) with which to assess land use proposals against river system requirements and underpin the river management process.
- Identify appropriate source control techniques for surface/ground-water management.

Meanwhile, sustainability implies a thorough review of policy, strategy and best practice. This would be an iterative process of communication, ensuring that the 'vision' can be derived from competence in assessing what resources there are (inventory) and what state they are in (legacy) and how they may be conserved and enhanced (or rehabilitated) by decisions made over development proposals. Although related to Regional Planning Policy in Figure 18.1, this process can apply at individual, team, organisation or community levels. Progress towards integrated or total catchment planning achieved in this way may prove a strong determinant in achieving sustainable development.

Achieving a vision for the sustainable development of natural resources could not be possible without some degree of consensus over principles, practice and underlying philosophy. While there is much information to be usefully gained—such as the morphological sensitivity of catchments and the ability to assess land use practices in terms of their impact on river biota—perhaps the biggest challenge lies in our ability to achieve consensus through communication of vision relying on a mix of qualitative and quantitative data; sustainability depends largely on the socio-economic and political response.

It would appear that sustainable development is the clarion call for institutional change to support the holistic approach to natural resource management. For the NRA, soon to be absorbed into the Environment Agency, there is great interest in the SD remit which may be given to the new organisation. What will evolve out of the current situation with regard to the environmental duty? Will there be equivalent duties to the social and economic dimensions, or will local authorities, water utilities and others assume the heaviest loads in those directions?

Once catchment management plans are perceived as complementary to development plans (Slater *et al.*, 1994), and the emerging environmental action plans of local authorities, there will be good reason to believe that environmental sustainability may be achieved—provided that practitioners are enabled to turn philosophy and theory into practice. Here is a 'boundary issue' between social science, environmental science and economic theory; here is the challenge to progress.

DISCLAIMER

The views expressed are those of the author, and not necessarily shared by the NRA. My thanks to Les Jones, Regional General Manager of NRA Thames Region, for his permission to publish this chapter.

REFERENCES

AME, NRA, CoL and SRC (1993) *Tomorrow's Towns: An Urban Environment Initiative.* Sponsored by the Association of Municipal Engineers, National Rivers Authority, Corporation of London and Strathclyde Regional Council. Institution of Civil Engineers, London.

Andoh, R.Y.G. (1994) Urban runoff: nature, characteristics and control. *Journal of the Institution of Water and Environmental Management* **8**(4), 371–379.

Beven, K. (1993) Riverine flooding in a warmer Britain. *The Geographical Journal* **159***(2), 157–161.*

Brookes, A. (1992) Recovery and restoration of some engineered British river channels. In: Boon, P.J., Calow, P. and Petts, G.E. (eds), *River Conservation and Management*, Wiley, Chichester, pp. 357–363.

Brown, S.A. (1994) Organic farming and water pollution. *Journal of the Institution of Water and Environmental Management* **7**(6), 586–592.

Chandler, J. (1994) Integrated catchment management planning. *Journal of the Institution of Water and Environmental Management* **8**(1), 93–97.

Cherfas, J. (1994) How many species do we need? *New Scientist*, 6 August, pp. 37–40.

CIRIA (Construction Industry Research and Information Association) (1992) *Scope for Control of Urban Runoff*. CIRIA Reports 123 and 124, London.

Clark, M.J. and Gardiner, J.L. (1994) *Strategies for handling uncertainty in river basin planning*. Paper to the international conference on Integrated River Basin Development, Wallingford, England.

CNS Scientific and Engineering Services (1992) *Economic Value of Changes to the Water Environment*, NRA R&D Project 253, NRA, Bristol.

Collett, L.C. (1992) *Total Catchment Management—A Solution to Urban Runoff?* A paper to national conference on 'Water in a Sustainable Urban Environment', Melbourne Water Corporation, Melbourne.

Davies, G.L. (1992) Catchment Management Planning in the National Rivers Authority of England and Wales. In: Saul, A.J. (ed.), *Floods and Flood Management*, Kluwer Academic Publishers, London.

DOE (Department of the Environment) (1994a) *Sustainable Development: The UK Strategy*. HMSO, London.

DOE (Department of the Environment) (1994b) *Environmental Appraisal of Development Plans*. HMSO, London.

Dubourg, W.R. (1992) *The sustainable management of the water cycle: a framework for analysis*. CSERGE Working Paper WM 92.07, Centre for Social and Economic Research on the Global Environment, University College, London, and University College, London, and University of East Anglia, England.

Fordham, M., Tunstall, S. and Penning-Rowsell, E.C. (1989) Choice and preference in the Thames floodplain: the beginnings of a participatory approach? In: *Proceedings of International Conference on Wetlands*, Leiden, The Netherlands, 791–797.

Fujita (1994) Japanese Experimental Sewer System and the Many Source Control Developments in Tokyo and other cities. In: Pratt, C.J. (1994) (ed.) *Proceedings of International Perspectives on Stormwater Management*, Coventry University.

Galloway, G.E. (1994) *Sharing the Challenge: Floodplain Management into the 21st Century*. Report of the Interagency Floodplain Management Review Committee, Washington, USA.

Gardiner, J.L. (1991a) *Influences on the development of river catchment planning in the Thames Basin*. Unpublished PhD thesis, Department of Geography, University of Southampton.

Gardiner, J.L. (1991b) (ed.) *River Project and Conservation: A Manual for Holistic Appraisal*. Wiley, Chichester.

Gardiner, J.L. (1992) Strategic Environmental Assessment and the Water Environment, *Project Appraisal*, 7(3).

Gardiner, J.L. (1994b) *Capacity planning and the water environment: experience in the Thames catchment*, paper to RSPB Annual National planners' Conference 1994, RSPB, Sandy, Beds.

Gardiner, J.L. (1994b) Sustainable development for river catchments. *Journal of the Institution of Water and Environmental Management* **8**(3), 308–320.

Gardiner, J.L. and Bolton, P. (1992) *Harmonising with the Great Forces of Nature: A New Perspective for the Professional?* International Commission for Irrigation and Drainage, Agricultural Economics Society Meeting. 'The Environment and Water Development', Institution of Civil Engineers, London.

Gardiner, J.L. and Cole, L. (1992) Catchment planning: the way forward for river protection in the UK. In: Boon, P.J., Calow, P. and Petts, G.E. (eds), *River Conservation and Management*, Wiley, Chichester, pp. 397–407.

Gardiner, J.L., Thomson, K. and Newson, M.D. (1994) Integrated watershed/river catchment planning and management: a comparison of selected Canadian and United Kingdom experiences. *Journal of Environmental Planning and Management* **37**(1), 53–69.

Girardet, H. (1993) *The Gaia Atlas of Cities: New Directions for Sustainable Urban Living*. Gaia Books, London.

HRDL (Hydro Research and Development Ltd) (1993) *Urban Drainage: The Natural Way*. HRDL, Clevedon, England.

IUCN, UNEP and WWF (The World Conservation Union, United Nations Environment

Programme and World Wildlife Fund for Nature) (1991) *Caring for the Earth: a Strategy for Sustainable Living.* Gland, Switzerland.

Johnston, N. (1994) *Waste Minimisation; A Route to Profit and Cleaner Production—An Interim Report on The Aire and Calder Project.* Centre for Exploitation of Science and Technology, London.

Lee, R.G. (1992) Ecologically effective social organisation as a requirement for sustaining watershed systems. In: Naiman, R.J. (ed.), *Watershed Management: Balancing Sustainability and Environmental Change*, Springer-Verlag, New York.

LGMB (Local Government Management Board) (1992) *Local Agenda 21: A Guide for Local Authorities in the UK.* LGMB, England.

MEE/MNR (Ministry of Environment and Energy/Ministry of Natural Resources (1993) *Integrating Management Objectives into Municipal Planning Documents,* MEE/MNR, Ontario.

Munson, A. (1994) Better safe than sorry. *New Scientist,* No. 1931, pp. 47–48.

Naiman, R.J. (ed.) *Watershed Management: Balancing Sustainability and Environmental Change.* Springer-Verlag, New York.

NRA (1990) *Corporate Plan 1990/91.* NRA, Bristol.

NRA (1993a) *Catchment Management Guidelines.* NRA, Bristol.

NRA (1993b) *Development of Environmental Economics of the NRA.* R&D Report 6, NRA, Bristol.

NRA (1993c) *Environmental Economics Manual,* NRA, Bristol.

NRA (1994a) *Corporate Plan 1994/95.* NRA, Bristol.

NRA (1994b) *Thames 21—A Planning Perspective and Sustainable Strategy for the Thames Region.* NRA Thames Region, Reading.

Newson, M.D. (1992) *Land, Water and Development.* Routledge, London.

Newson, M.D. (1993) *Environmental protection.* Paper to Institute of Hydrology, Chartered Institution of Water and Environmental Management conference: 'Sustainable Water Resources', Institution of Civil Engineers, London.

Pearce, D., Turner, K., O'Riordan, T., Adger, N., Atkinson, G., Brisson, I., Brown, K., Dubourg, R., Fankhauser, S., Jordan, A., Maddison, D., Moran, D. and Powell, J. (eds) (1993) *Blueprint 3: Measuring Sustainable Development.* Earthscan, London.

Rylands, W.D. and Lee, J.K. (1992) The possibility of an integrated approach to the provision of storage in the Thames Region of the National Rivers Authority. In: Saul, A.J. (ed.) *Floods and Flood Management,* Kluwer Academic Publishers, London.

Schulte-Wulver-Leidig, A. (1993) *'Salmon 2000': Ecological Master Plan for the Rhine.* ICPR, Koblenz, Germany.

Scott, D. (1993) New Zealand's Resource Management Act and fresh water. *Aquatic Conservation* 3(1), 53–65.

Sear, D.A. (1994) River restoration and geomorphology. *Aquatic Conservation* 4(2).

Slater, S., Marvin, S. and Newson, M.D. (1994) *Land Use Planning and the Water Environment: A Review of Development Plans and Catchment Management Plans.* Working Paper No. 24, Department of Town and Country Planning, University of Newcastle upon Tyne.

Therival, R., Wilson, E., Heaney, D., Thompson, S. and Pritchard, D. (1992) *Strategic Environmental Assessment.* Earthscan, London.

Townend, I.H. and Fleming, C.A. (1994) Planning for coastal erosion. In: *The Institution of Water and Environment Management Yearbook 1994,* FSW Group Ltd, Norwich.

Turner, K. (1993) Sustainability: principles and practice. In: turner, K. (ed.) *Sustainable Environmental Economics and Management.* Belhaven, London.

Urbonas, B. and Stahre, P. (1993) *Stormwater: Best Management Practices and Detention for Water Quality, Drainage and CSO Management.* PTR Prentice-Hall, New Jersey.

Woolhouse, C.H. (1994) *Catchment Management Plans: current successes and future opportunities.* Paper to international conference on Integrated River Basin Development, Wallingford, England.

19 Fluvial Geomorphology and Environmental Design

MALCOLM D. NEWSON

Department of Geography, University of Newcastle upon Tyne, UK

The title for this chapter is taken from Gregory (1985); in discussing the recent surge of interest in applied studies he wrote (p. 207),

> ... one of the implications of the geomorphic engineering approach is that not only is it necessary to become more familiar with the methods used by practitioners in other disciplines, but it is also desirable to assess the efficiency of alternative design strategies and it is imperative that this should proceed towards problems of environmental design.

It is my intention to look into the relationship between fluvial geomorphology, engineering science and engineering practice (principally in the UK); it is my working experience that interdisciplinary understanding is best built upon characterising the individual contributions, rather than in anticipating such fusions such as 'geomorphic engineering'. There is no doubt that river channel management now needs such an understanding between practitioners. The chapter goes on to enlarge on those developments in geomorphological theory and practice which show that potential to make a contribution to practical applications; it also identifies a number of gaps in our knowledge which need filling, either by geomorphologists or by joint action with engineers, biologists and hydrologists. It concludes with an investigation of the term 'environmental design' as applied to river channel form and process: it inevitably leads to an holistic vision of contextual design based on the whole catchment ecosystem.

FLUVIAL GEOMORPHOLOGY, ENGINEERING SCIENCE AND RIVER ENGINEERING

There is no doubt that fluvial studies have come to dominate process geomorphology in the last 30 years; explanations vary from the influence of individuals (in which case Gregory bears a considerable responsibility!) to the firm conceptual framework provided by general systems—and in particular to the contribution of catchment hydrology and its emphasis on practical field measurement. There has also been a considerable, mainly friendly, rivalry between geomorphologists and engineers over the relative roles of physical laws and ambient boundary conditions in explaining both processes and forms observed in river channels. As the pretenders, geomorphologists have had to work hard to prove their point and, in doing so, have

Changing River Channels. Edited by Angela Gurnell and Geoffrey Petts.
© 1995 John Wiley & Sons Ltd.

perhaps mistaken the *actual* approaches adopted by engineering research and applications. So, what are the true characteristics of these approaches?

Scales and styles of approach

Society has always required its river engineers to solve point problems and to solve them quickly; occasionally a fortunate engineer gets to work on a very large prestigious project but these also have tended to be at-a-point (e.g. dam schemes) or, if they are extensive, have tended to replace the natural system with an artificial one, such as a canal. Engineering *science* has supported the practical side of the subject with a theoretical basis for design, by way of empirical simplifications of river dynamics or by transferring hydraulic observations from the controlled conditions of the laboratory. Geomorphologists, however, have come from the opposite perspective: form has been studied in the large scale, historically as part of the study of the evolution of river basins as major landscape elements but, more recently, because we were interested in producing basic numerical information such as sediment yields, or downstream hydraulic geometries. Other disciplines can be fancifully pigeon-holed in this way but without such conviction (Figure 19.1).

There are further practical constraints acting to curtail the flexibility and flair of the river engineer: these can include the legal/political circumstances of the policy under which rivers are managed by the nation state, power structures acting through land ownership and other riparian interests and the existing infrastructure linked (in long-settled developed countries) to the river cross- and long-profile. The important point here is that it ill-behoves geomorphologists to criticise river engineering as a field of intellectual endeavour solely on the basis of what they see as the 'product' of river engineering. Society gets the river engineering it deserves; at the time of writing

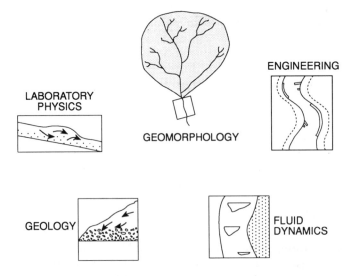

Figure 19.1 The scale-dependency inherent in modal disciplinary approaches to river studies. (From Newson, 1992, *Land, Water and Development*, Routledge, London, reproduced by permission)

Table 19.1 Characteristics of the engineering and geomorphological approaches to practical problems of river channel dynamics

Engineering	Geomorphology
Design experience	Field experience
Hydraulics	Sediment supply/transport
Project time-scales	Environmental change time-scales
Specialist role	Generalist breadth
Functional, simple channels	Complex, dynamic channels
Structural solutions	Flexible, composite solutions
Reach scale	Basin scale

Western society is demanding both the longer time-scales and spatial continuities understood by fluvial geomorphologists *but* within the financial and administrative conditions of conventional engineering projects—posing a considerable dilemma which requires an interdisciplinary solution.

Recently, a group of British fluvial geomorphologists has completed a Handbook for the US Army Corps of Engineers (Thorne *et al.*, 1995); it summarises the characteristics of the two most important groups of intellectual effort supporting river channel management (see Table 19.1).

In addition one might add that engineering solutions are traditional, 'exact' and auditable, a powerful commendation to society against the untried, more flexible and qualitative aspects of fluvial geomorphology. Hemphill and Bramley (1989) have surveyed the approaches used in practical river engineering in the UK and find that 'experience' counts as the major professional aid; information from engineering research papers is seldom used—it is not only geomorphological research which is being excluded!

An even more intriguing comparison, therefore, can be made between fluvial geomorphology and the research field of engineering science in support of practical applications (Figure 19.2). The diagram is deliberately configured to show that, in fact, both research fields tend to supply guidance from theoretical problems towards practical problems, but geomorphologists have tended to work on different topics without, until recently, gaining formal practical outlets.

BUILDING INTERDISCIPLINARY UNDERSTANDING AND IMPROVING POLICY AND PRACTICE

From the foregoing it is clear that there are two agendas facing fluvial geomorphology:

- To integrate our research endeavour with that of engineering science in a revitalised field of river dynamics.
- To work towards practical standards which can be successfully delivered within an evolving policy field, e.g. moves towards project assessment, appraisal, and holistic river basin management (e.g. Gardiner, 1991, Chapter 18).

Figure 19.2 offers a basis for the first agenda item. It is essential that geomorpholo-

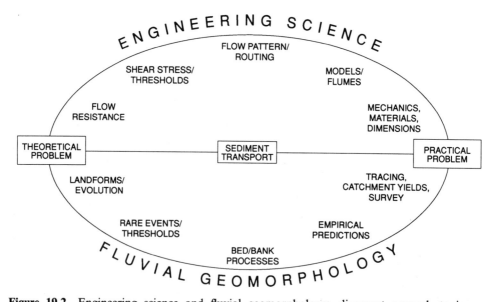

Figure 19.2 Engineering science and fluvial geomorphology—divergent research topics on routes between theory and practice in river studies. Sediment transport is a principal central thread

gists list the particular objectives of their science and the techniques they use to attain them. Too often geomorphologists, for reasons of attracting funding through academic routes (Geography Departments in Higher Education Institutions (HEIs) have special problems here), have attempted to dress up in engineering clothes; we have made significant contributions to the engineering science topics listed in Figure 19.2 (e.g. Brayshaw *et al.*, 1983), but it has clouded our singular role, i.e. our interest in landforms and their evolution as the basic and 'enchanting' theoretical problem (see Baker and Twidale, 1991). Facing even those geomorphologists who have become involved with practical channel management should be simple geomorphological questions such as, 'How do valleys form?'. This is highly distinctive compared with the neighbouring fundamental question in engineering hydraulics—that of flow resistance. However, from there on, there are opportunities for mutual cross-reference of research effort, as in the two forms of interest in threshold behaviour which are distinctive in scale. The relationship between flow pattern and bed/bank processes is highly productive as the former team at the University of East Anglia proved in the 1970s (e.g. Bathurst *et al.*, 1979). Perhaps the linkage which is most frequently ignored is that between model or flume studies and those of empirical field conditions. There is no doubt that engineering science has not spent long enough in the field but geomorphologists can hardly accuse the models used in the laboratory or on the computer of lack of *veritas* when we sit on most of the field information without seeking ways of expressing it in a transferable form. For example, we have put too little effort into process-based channel classification, an ideal interdisciplinary focus (see below).

The illustrations for this chapter are simple; the objective of technology transfer bespeaks a plain iconography. For example, Figure 19.3 shows the information

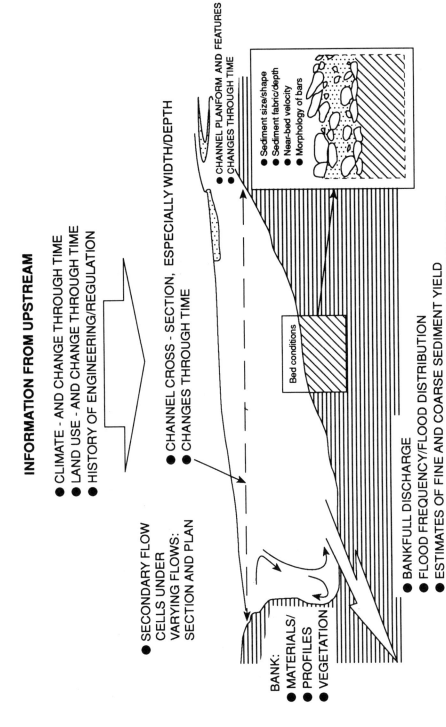

Figure 19.3 The information needs of a geomorphological assessment of river dynamics

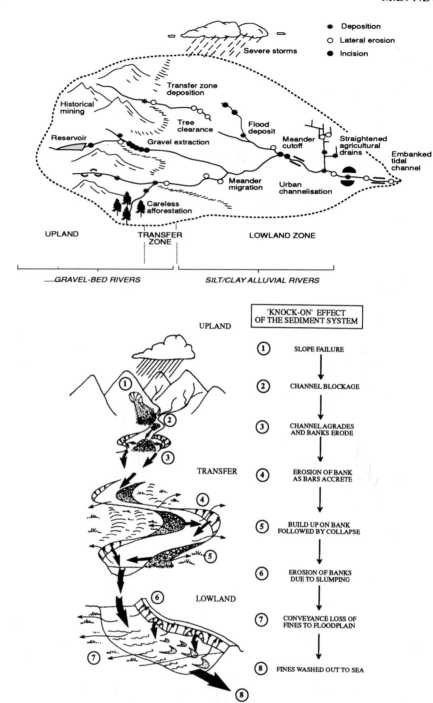

Figure 19.4 The iconography of technology transfer between fluvial geomorphology and effective catchment management. Above: 'hotspots' of geomorphological problems involving erosion or deposition; below: the continuity of process to emphasise the risks of engineering intervention (from originals by D.A. Sear)

which a geomorphologist would expect to be necessary at the reach scale when assessing any channel development or indeed any riparian or catchment development which might affect river dynamics. Newson and Sear (1994) and Sear *et al.* (1995) deliberately 'line up' the geomorphological agenda in this way, making it both open to debate within the discipline and open to scrutiny by non-specialists. Such an approach is, for example, particularly hard-hitting in stressing the historical dimension of geomorphological investigations (see Sear *et al.*, 1995).

As well as the historical dimension it is important for fluvial geomorphologists to be specific and prescriptive about the *spatial scale* of their information needs. To this end Newson and Bathurst (1990) coined the term 'fluvial audit', the catchment-scale survey via field, map and archival evidence of the past and present sources, sinks and morphological stores of sediment. The procedure is commendably clear in principle and can be carried out to different levels of detail according to basin scale and the available resources. The fluvial audit gets the river manager to a position where the 'thought experiments' (i.e. varying water discharge, sediment discharge and other variables to predict the direction of channel change) of Schumm (1977) can be embarked upon. It attempts to:

- Identify the cause of fluvial problems such as flooding, erosion, deposition or instability.
- Suggest the geomorphological constraints to a practical solution (even if the solution is a structural engineering one—though a possible outcome is 'do nothing').

Sear and Newson (1994) have prepared simple, diagrammatic guidance to auditing and have successfully demonstrated to the National Rivers Authority that it has a practical and repeatable utility. Figure 19.4 here merely rehearses the geomorphologist's view of the catchment-wide sediment system and the likely 'trouble spots' in a UK setting. The iconography of technology transfer is, once again, important; there exists considerable risk in committing simple viewpoints to illustration in this way but the two parts of this figure have acted as a catalyst to interdisciplinary discussions with practical river engineers and have opened their eyes to causal chains which they had not imagined at the reach scale!

DEVELOPMENTS IN FLUVIAL GEOMORPHOLOGY AND THEIR PRACTICAL POTENTIAL

If geomorphology is to be practically applied, is there a danger that it will turn its back on its fundamental theories and objectives? Optimism is encouraged by Clark *et al.* (1987) when they write (p. 4) that:

> Professional application of physical geography has reinforced the integrative approach of the subject and has demonstrated little support for the fear that application was somehow intellectually inferior to pure study.

In the same passage these authors also claim that applications now encourage a closer cooperation with human geographers (or at least their fields of study) and there is no doubt that, if fluvial geomorphologists wish to have their work applied, they must at least be aware of research on economic, political and policy constraints and opportunities. It is in this context that Gregory's plea for 'environmental design' must be viewed (see below).

Inevitably this section will be partial, i.e. both incomplete and representing the author's own selection! Nevertheless it can be argued that geomorphological research has, in the last decade, made substantial progress in the following areas which are now worthy of applications to practical river management:

- Long profile aspects including water and sediment storage and routing.
- Planform elements including meander development, the role of the floodplain.
- Sources of sediments, including land use and the channel bed itself.
- The role of the 'rare' event and of climate change.

Long-profile aspects

Considering the fundamental position of the question, 'How do valleys form?', it is very surprising that fluvial geomorphologists did not, until recently, begin to relate their knowledge of process to their search for the explanation of form through the storage element of the fluvial transfer system. The fact that process and morphology time-scales are out of synchroneity (Ferguson's 1981 use of the term 'jerky conveyor belt' for the fluvial sediment system is apposite) provides an explanation, but recently a number of relatively simple but thorough investigations have focused on floodplain and valley-floor storage volumes and transfer times. The work done in the north-west USA in the 1980s is very important: Swanson et al. (1982) demonstrated that gross sediment storage should form part of our process quantification in small catchments, whilst Kelsey et al. (1986) and others working in Redwood Creek prove that the results are both conceptually revealing and applicable to decisions about catchment management.

Previously, Church and Ryder (1972) introduced the shell concept of 'paraglacial sedimentation', permitting us to consider in glaciated landscapes a moving source area of sediment, together with an accompanying channel instability passing down catchments on a ten-thousand-year time-scale. Church (1983) investigated the impacts on more local patterns of instability and in the UK Macklin and Lewin (1989) have confirmed a 'beaded' downstream pattern of sedimentation, and hence channel dynamics, in a gravel-bed system. Corollaries include a very changeable perception by riparian communities (at a decadal scale) of the risk and incidence of flooding, bank erosion and depositional effects, e.g. on freeboard for land-drainage. Macklin et al. (1992) show that the detective work available to fluvial geomorphologists has predictive power to locate and specify patterns of instability.

However, perhaps the most important item for transfer to both river engineers and flood hydrologists is the vertical element of channel change. In the collection of essays on channel change gathered by Gregory (1977a), a majority of the UK contributions considered only lateral change and the emphasis was upon human impacts—from, for example, river regulation or historic mining activity. Geomorphological studies in the UK have now revealed the importance of phases of *incision* and aggradation, mainly climatically-driven (Passmore et al., 1993), and the implication is that two hallowed tenets of applications—regime theory and stationary flood-frequencies have to be proven rather than assumed (Lewin *et al.*, 1988; Newson and Macklin, 1990).

At a much smaller but no less fundamental scale, geomorphological investigation (by field process studies) of the riffle-pool sequence has recently produced applicable findings of the utmost importance. Sear (1992) records the control exercised by the riffle–pool sequence on bedload transport at both high and low flows, refining the velocity reversal hypothesis and scaling the 'jerky conveyor belt' by sediment tracing. Sear's study in a regulated river demonstrates the impact of bed material strengths and structures (see below) as well as inviting the suggestion that the unstable–stable–unstable alternation at a catchment scale (e.g. Macklin and Lewin, 1989) may be a grossed-up form of such a sediment transport control. I have stylised this suggestion in the form of a repetition of Schumm's source–transfer–sink zonation (Newson, 1992a).

There are certainly smaller analogies of the riffle–pool sequence as gradient steepens (Grant *et al.*, 1990), including fall–pool and cascade sequences. We must stress to river engineers that, far from representing sediment 'problems', such features are sediment regulators and important habitat features. When identified in channels for which capital or maintenance schemes are proposed they should be retained; in reaches which have been degraded by uniformed channel management they should be rehabilitated (see Chapter 17). In fact our dialogue as geomorphologists must also commence with freshwater ecologists who are prone to be rather free with their coinage of definitions for in-channel features; the new field of 'habitat hydraulics' has only recently been thrown open by desires for truly sustainable development of river basins.

Planform aspects

Despite the caveat about vertical channel changes raised above, the progress encouraged by Gregory's (1977a) volume has been considerable in the UK in three highly applicable areas. First, a geographical investigation into the incidence of lateral channel change (Hooke and Redmond, 1989, 1992) depicted a broadly piedmont *location* for rapidly adjusting, 'unstable' reaches and further built up our knowledge of *rates* of bank retreat. Clearly, every use of the word 'unstable' requires definition, but policy- and engineering-based decisions about intervention to curtail river bank erosion need to work from some basis of rates. Secondly, fluvial geomorphologists have continued to research the direction and style of planform change in sinuous reaches (Chapter 5), such that broad patterns can be suggested, if not predicted, for those channels in which works are proposed. In detail such an understanding of patterns of change (e.g. Hooke, 1977) can also guide the location of revetment works; a substantial number of such works in the UK have, in the past, been outflanked by inadequate upstream/downstream extent or undermined by failure to appreciate outer-bank scour processes. Geomorphologists have now integrated the simple study of planform change with process studies of meander dynamics (e.g. Hooke and Harvey, 1983; Thompson, 1986).

Thirdly, and perhaps most importantly for environmental design, geomorphologists are turning attention once more to bank erosion as a process, with direct measurements of rates (Lawler, 1991) and geotechnical analysis of the 'rival' pulls of gravity and scour (Thorne, 1982; Darby and Thorne, 1994). Too frequently the cause of bank erosion is laid, simplistically, at the door of stream power when the causes are sub-aerial weathering or geotechnical failure (both of which may be accelerated by overzealous deepening of channels for flood defence). The more we become interested in bank erosion and meander migration, the more we must become involved with the difficult and dangerous study of floodplain dynamics *in the field* (hardware models are inadequate on their own). Our increasing interest in the floodplain depositional environment is also helpful since essentially floodplains are both the source of much of the redistributed fluvially-derived material composing sediment loads (Chapter 9) and a considerable conservation and pollution control resource.

Sediment sources, including the river bed

In a nation seemingly bereft of agency interest in fluvial sediments Gregory and Walling initiated the purposeful measurement of catchment sediment yields in Exeter in the 1960s. Twenty years later Newson (1986) and Walling (1990) could pull together sufficient data for a regression and mapping approach to yields but the basic data were still coming largely from Geography Department PhD theses! The black-box yield approach continues to be useful and can point to both the disturbance effect of certain land use and land management practices (Chapter 7) and to design parameters for the design of sediment control structures such as traps (Sear and Newson, 1994).

However, linked partly to new lines of thought originating from both long-profile and planform studies, fluvial geomorphologists are now prepared to consider other causes of disturbance in the local or temporal sediment flux at sites where a stability problem is perceived. For example, engineered (channelised) reaches are a source of increased sediment—for a variety of reasons including:

- The extra stream power conveyed to the stream bed by over-deepened channels (Brookes, 1987).
- The disturbance to stream bed structures by engineering activity (Leeks *et al.*, 1988).
- The reduced storage opportunity produced by reduced diversity of channel sedimentation features and loss of floodplain deposition (Sear and Newson, 1994).

The experience of a group of UK fluvial geomorphologists in the Mimmshall Brook catchment is important (Darby and Thorne, 1992; Sear, 1994). The Mimmshall Brook is a user-stressed southern England catchment in which flood protection and conservation objectives clash, particularly because the former function has hitherto not been informed by geomorphology. Engineered reaches of the Brook are the dominant source of current management problems and a rehabilitation of the natural regimes of both morphology and flow is the most promising solution providing that the resident population of the basin identify with a more balanced political agenda (see below).

Our knowledge of the structure of river beds (aside from the riffle–pool studies described above) also now points to the importance of the bed material storages as a

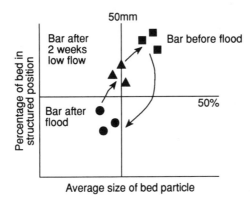

Figure 19.5 Flow and sediment size/fabric/structure on the river bed, to emphasise the potential of feedbacks in controlling river stability (especially in regulated rivers). From Sear, 1992. Reproduced by permission of John Wiley & Sons

source for the flux of sediments. Perhaps for too long, fluvial geomorphologists acknowledged the engineering science agenda of deriving increasingly powerful sediment transport prediction capabilities. Instead, they can make significant improvements to sediment transport criteria by devoted and detailed field measures of bedload flux (Carling, 1983; Bathurst *et al.*, 1986; Reid and Frostick, 1986) or by utilising geomorphological data on the heterogeneity of clast size (and relative size based on protrusion—see Ferguson, 1994; Chapter 8). Our intimate knowledge of river bed structures and the importance of flow sequences is now inviting a comprehensive feedback model of bedload transport. Initial moves have been made for gravel-bed rivers by Sear (1992 and Figure 19.5) and by Sohag (1994). Both conclude that gravel beds become less susceptible to disturbance after a lengthy duration of sub-critical flows which are important in forming bed structures. This is corroborated by the fact that ephemeral streams exhibit a high bedload transport rate (Laronne

Table 19.2. A scale approach to bed material controls on bedload

Scale	Control phenomena	Technique of study
Particle	Size/weight/shape/sorting/abrasion	Wolman surveys/tracer studies/flume studies
Fabric/strength	Vertical sorting/relative location Antecedent flow/flow duration	Bulk sampling/structure surveys/dynamic penetrometer Tracer studies/flume studies/specific flood studies
Bedform	Flow width and depth/secondary flow Riffle–pool controls/backwater controls/channel metamorphosis	Morphological mapping/Wolman surveys by location/tracer study/channel typing
Reach/megaform	Sediment flux including throughput versus storage and release/vertical controls on effective flows	Sampling fluxes/tracer study/old maps and aerial photographs/cross-section/ morphological method
System	Sediment supply/gross sediment flux/climate controls on flows	Sediment yields/channel change/ practical problems

and Reid, 1993). Also relevant to this new and exciting field of potential applications are the many records which show pulsed bedload transport (Hoey, 1992; Goff and Ashmore, 1994). Table 19.2 attempts to bring together a system-wide, modular/scaled approach to the study of gravel-bed river morphology based on these findings.

Sohag (1994) proved by using a 'blitz' of tracer studies sourced by morphological site within a wandering gravel-bed reach on the South Tyne that, whilst particle size may control both transport distance and liability to burial, bed armouring and cluster formation (both linked to characteristics of the flood and inter-flood hydrograph) controlled morphological change by switching between conditions of 'hard bed' and 'soft bed' and, thereby, between transport conditions of selective entrainment and equal mobility.

There are immediate and important applications for this body of work, for example to the prediction of conditions leading to bed scour at structures, in relation to, for example, gravel extraction upstream, and also to the reform of channel maintenance procedures so that they respect 'hard', hydraulic bedforms and protect those depositional features which stabilise reach sediment fluxes.

Finally, considerable progress has been made in geomorphological process studies of the sediment flux and hydraulic hiatus at tributary junctions. A number of applied problems result at such junctions (see Newson and Bathurst, 1990) but the work of Best (1986) on flow separation and Petts and Thoms (1987) on the phenomenon in regulated rivers has direct environmental design benefits.

It would be unwise to leave this section without a statement of the position regarding suspended sediments. The concerted effort of the last 20 years in the UK on upland and piedmont channels has left us without a good UK handle on the problem of suspended sediment delivery ratios, perhaps the most urgent applied problem of all at the world scale. The UK has its own soil erosion problems and sediment sourcing (e.g. Walling and Woodward, 1992) seems poised for a major breakthrough in applications to catchment management.

The 'rare' event

Three important aspects of flood events have been quantified and elaborated by fluvial geomorphologists in recent years. One should not underestimate the singular role of our profession in following up flood events in a number of ways (see, for example, the volume edited by Beven and Carling, 1989). Contributions have come in the form of:

- Separating flood *work* from flood *effectiveness*.
- The role of hydroclimatology—i.e. different types *and sequences* of flood event.
- Identification of threshold phenomena and feedback mechanisms.

The applied value of these research contributions lies in opening the eyes of river engineers to further complexities in conditions at-a-site (as against a tendency to consider each flood event and the channel's response as a single population—and mainly in 'work' terms) and in preparing policy for climatic change scenarios in which we feel extreme flows to be important but have little quantitative guidance from climate models.

A major question which as yet is but partially answered is the relationship between impacts of flooding as a purely natural but diverse population of events (therefore affecting different scales and locations of catchments differentially) and flooding brought about by, or exacerbated by, human-induce changes such as those of land use.

Macklin and Lewin (1993) stress the dominance of the climate signal over that of land use in the Holocene alluvial sequence, a balance which leads to relatively synchronous signals in the sediments of valley floors. Within the Tyne catchment, however, it is clear that synchroneity can be blurred as cultural effects, particularly vegetation clearance, impact on the supply of sediments (particularly the finer sizes which build up floodplains and colluvial toeslopes). There is also a problem of time-stepping resulting from the different climatic susceptibility of different sizes and locations of subcatchments in relation to subtleties of hydroclimatology (Rumsby and Macklin, 1995), subtleties substantiated by the latest work on flood seasonality. Clearly, however, fluvial climatologies are simplistic if they are only flood-driven; climate also controls flow duration (and therefore bed strength), revegetation and many other aspects controlling geomorphological response.

When it is considered that a major concentration of engineering works and maintenance occurs in the aftermath of damaging flood events, we are perhaps remiss that too few geomorphological studies of the 'rare' event have been continued through the recovery period. Re-visits for morphological studies have been occasional but sustained process investigations of sediment yield and channel dynamics are largely lacking. Pitlick (1993) demonstrates that the sorts of impression created by a flood disaster may not be a valid guide to remediation needs.

GAPS IN OUR KNOWLEDGE OR OUTPUTS IN THE QUEST FOR ENVIRONMENTAL DESIGN OF RIVER CHANNELS (AND CATCHMENTS)

At the interface between knowledge and applications one must ask two questions of information approaching the border. In the case of knowledge seeking applications one must ask whether this knowledge is adequate and realistic about its doubts, but also whether it is being presented or formatted in an appropriate fashion. Similar questions may be posed about information in the reverse direction where practical need exposes areas for research. In this direction, too, one may ask whether the perception of the problem is correct. For example, a topic posed as a problem for physical science to solve may perhaps be one instead for social science to work on (e.g. flood risk and perception in relation to flood damage).

What are the gaps?

It is the author's view that many of the questions now being posed to geomorphology by applications should initially be seen as a challenge to basic channel typology and process-based classification. Figure 19.6 shows the advantages on both 'sides', as well as a contribution to integrated management. Gregory has long backed research on broad statements of channel form and process (Gregory, 1977b, 1979). Some argue that classification represents a primitive stage of scientific evolution but one has to question the motives of a body of knowledge (i.e. fluvial geomorphology)

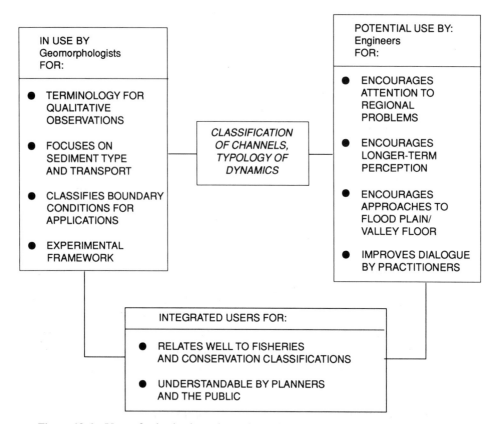

Figure 19.6   Uses of a basic river channel typology and process-based classification

which has made great research strides if it refuses to synthesise and classify that knowledge base.

There have been notable contributions such as Church and Jones' (1982) classification of bar forms, Kellerhals and Church's (1989) guidance for managers, and Nanson and Croke's (1992) classification of floodplains but, as yet, no prescriptive devices which include the heading 'What you do' for various river types and conditions. The power to incorporate a process base for classification and typology via our increasing knowledge of dynamics is beginning to show through in recent work by Whiting and Bradley (1993) and Rosgen (1994). The Geodata Institute at the University of Southampton is also working on a typology aimed purely at catchment and river managers. It is based on a fieldwork survey by the National Rivers Authority (NRA, 1994a) of the habitat characteristics (including channel morphology) at 1500 sites in England and Wales.

What techniques do we apply?

The overriding characteristic of the contribution being made to practical applications of fluvial geomorphology in the UK at present is its fieldwork basis. This does not mean that only fieldworkers and field studies are making a contribution, but rather

Table 19.3 Simple, validated field techniques in fluvial geomorphology

1.	Historical interpretations	Interpretation of relict channels, tree lines, old maps, aerial photographs, flood archives, lichenometry, boulder berms
2.	Form regularities and departures	Width/depth ratios, riffle–pool spacings and bank batter angles. Simple field tools to measure same Field classification of bar forms and hydraulic controls
3.	Signs and symptoms— stability conditions	Nick-points, bridge piers, flood channels, fresh medial bars and 'shoals'
4.	Basic sedimentary information	'Pebbleometer' technique, photographic techniques, armour ratio, bed structure surveys, bank material

that the most acceptable input in a field dominated by practical engineers is one which is distinctly different but stems from an equally practical and pragmatic approach.

Whilst, at times, this contribution has the characteristics of simple observation and recording (much akin to natural history) it is rapidly being formalised into, albeit locally applicable, 'standards'. Whilst the intellectual tradition in which we work naturally shuns unjustifiable orthodoxies, the greatest benefit of specifying techniques and standards at present is that other, neighbouring disciplines are much in need of our support, advice and cooperation. For example, in England and Wales, it is highly unlikely that a small group of academic fluvial geomorphologists can supply all the consultancy required to input geomorphology to the field of river management. We must therefore hand over, at least to derive primary field information, to those such as ecologists who have the finance and personnel power to collect it via such policy avenues as River Corridor Surveys, River Habitat Surveys or Environmental Assessment.

A justification has already been made for the derivation of a 'standard' channel typology in this respect, but there is also a need for a compendium of validated techniques which need not be sophisticated. These include the selection shown in Table 19.3.

The broader agenda

Essentially, geomorphology will provide the 'complementary medicine' for engineering in future river management but there is also a separate objective for most geomorphologists. Perhaps the most startling discovery from the application of a fundamentally catchment-scale science such as geomorphology to practical problems is that management of the *catchment* becomes a critical component, for example in the control of sediment supply or flood flows. Few geomorphological environmental designs will therefore be 'stand alone'. One finds the need for caveats such as 'this approach to channel maintenance is cost–beneficial and sustainable *IF* further afforestation is carried out to the highest protective standards' or '*IF* urbanisation follows a no-net-increase-in-runoff policy'.

It is the author's believe that the basis of Gregory's term 'environmental design' is just this: a set of clear prescriptions which, however, must be developed, presented and approved *in context*. Like natural historians, we are not located at the core of physical science, but our advantage is an acute knowledge based on the boundary conditions within which physical laws operate. The context of our practical work will often, therefore, be *local*, making the other geographical skills, available within our normal HEI working environment, a benefit not a hindrance. The local scale is, after all, where political, economic and social goals for river management are clearest, a fact recognized in Catchment Management Planning.

Inevitably, therefore, applied fluvial geomorphologists have been drawn to catchment management and planning (see Chapter 18) as a means to substantiate their view of the inseparability of catchment and channel dynamics. This is in discernible contrast to the reaction of practical engineers, who have generally been suspicious or resentful of the broadening scope. Catchment Management Plans issued by the National Rivers Authority in England and Wales have chapters detailing the 'physical attributes' of the catchment and not a few refer to problems of channel flood capacity, obstructions to flow or channel instability as issues for catchment management. For example the Plan for the Swale, Ure and Ouse (NRA, 1994b, p. 74) contains the following entries:

- Assess the extent of bankside erosion in the catchment.
- Develop an advisory strategy to combat erosion—this could include the application of techniques such as buffer zones, soft landscaping, e.g. tree and scrub cover, fencing off stretches of river from livestock.
- Expand tree planting programmes in the middle reaches of the Swale.
- Explore further landowners' support for tree planting.

As a secondary attribute of environmental design, the author would make a purely personal plea that fluvial geomorphologists working as physical geographers could take an environmental*ist* approach to their applications. The extension of our interests to catchment-wide processes is not politically neutral and requires laying out and substantiating clearly (Newson, 1992b). Clearly now, river channel management impinges on the essential future agendas of pollution control and nature conservation. Already, applied fluvial geomorphology is making contributions in support of these environmental aims. For example, the role of floodplains as pollutant sinks is well-researched (e.g. Macklin and Klimek, 1992; Marron, 1992) as is the importance of riparian vegetation (Gregory, 1992) and the desirability of conserving pollution-tolerant plant communities (Macklin and Smith, 1990), whilst an understanding of channel sediments can promote sustainable pollution control (Younger *et al.*, 1993) and channel form can be relevant to the management of contentious physiographically-justified conservation sites (Werritty and Brazier, 1991).

Perhaps the most important current role for fluvial geomorphology is in informing (and in some cases restraining!) headlong moves for river restoration. Once again a reach scale is emerging as a popular management unit, but Sear (1994) has cautioned that reach restoration which ignores the catchment sediment system is bound to be unsustainable.

Towards the end of this millennium the environmental cart may often appear to be in front of the physical science horse, as in a recent study entitled, 'Assessing the hydraulic performance of environmentally acceptable channels'; perhaps it is not such a bad thing if, under these conditions, our mutual vehicle moves very slowly backwards!

ACKNOWLEDGEMENT

As an ex-tutee of Ken Gregory I can say that, without his guidance and encouragement in my Final year, I would have never done research—preferring at that time, the lure of journalism. Apologies to readers if ambitions become, at times, confused! Having just written the words 'Wiley, Chichester' more times than I can count, I also believe it is time our publishers had a sincere note of thanks for putting British geomorphology on the world map over the last twenty years! May I also thank David Sear and Angela Gurnell for helpful editorial comments.

REFERENCES

Baker, V.R. and Twidale, C.R. (1991) The re-enchantment of geomorphology. *Geomorphology* **4**, 73–100.

Bathurst, J.C., Thorne, C.R. and Hey, R.D. (1979) Secondary flow and shear stress at river bends. *Proceedings of American Society of Civil Engineers, Journal of Hydraulics Division* **105**, 1'277–1295.

Bathurst, J.C., Leeks, G.J. and Newson, M.D. (1986) Relationship between sediment supply and sediment transport for the Roaring River, Colorado. In: Hadley, R.F. (ed.), *Drainage Basin Sediment Delivery*. IAHS Publication 159, 105–117.

Best, J.L. (1986) The morphology of river channel confluences. *Progress in Physical Geography* **10**(2), 157–174.

Beven, K. and Carling, P. (eds) (1989) *Floods: Hydrological, Sedimentological and Geomorphological Implications*. Wiley, Chichester.

Brayshaw, A.C., Frostrick, L.E. and Reid, I. (1983) The hydrodynamics of particle clusters and sediment entrainment in coarse alluvial channels. *Sedimentology* **30**, 137–143.

Brookes, A. (1987) River channel adjustments downstream from channelization works in England and Wales. *Earth Surface Processes and Landforms* **12**, 337–351.

Carling, P.A. (1983) Thresholds of coarse sediment transport in broad and narrow natural streams. *Earth Surface Processes and Landforms* **8**, 1–18.

Church, M. (1983) Patterns of instability in a wandering gravel-bed channel. In Collinson, J.D. and Lewin, J. (eds), *Modern and Ancient Fluvial Systems*, International Association of Sedimentologists, Special Publication 6, 169–180.

Church, M. and Jones, D. (1982) Channel bars in gravel-bed rivers. In: Hey, R.d., Bathurst, J.C. and Thorne, C.R. (eds), *Gravel-bed Rivers*, Wiley, Chichester, 291–338.

Church, M. and Ryder, J.M. (1972) Paraglacial sedimentation: a consideration of fluvial processes conditioned by glaciation. *Bulletin Geological Society America* **83**, 3059–3072.

Clark, M.J., Gregory, K.J. and Gurnell, A.M. (1987) *Horizons in Physical Geography*. Macmillan, Basingstoke.

Darby, S.E. and Thorne, C.R. (1992) Impact of channelization on the Mimshall Brook, Hertfordshire U.K. *Regulated Rivers: Research and Management* **7**, 193–204.

Darby, S.E. and Thorne, C.R. (1994) Prediction of tension crack location and riverbank erosion hazards along destabilized channels. *Earth Surface Processes and Landforms* **19**, 233–245.

Ferguson, R.I. (1981) Channel form and channel changes. In Lewin, J. (ed.), *British Rivers*, Allen & Unwin, London, 90–125.

Ferguson, R.I. (1994) Critical discharge for entrainment of poorly sorted gravel. *Earth Surface Processes and Landforms* **19**, 179–186.

Gardiner, J.L. (ed.) (1991) *River Projects and Conservation: A Manual for Holistic Appraisal*. Wiley, Chichester.

Goff, J.R. and Ashmore, P. (1994) Gravel transport and morphological change in braided Sunwapta River, Alberta, Canada. *Earth Surface Processes and Landforms* **19**, 195–212.

Grant, G.E., Swanson, F.J. and Wolman, M.G. (1990) Pattern and origin of stepped bed morphology in high-gradient streams, Western Cascades, Oregon. *Geological Society America Bulletin* **102**, 340–352.

Gregory, K.J. (ed.) (1977a) *River Channel Changes.* Wiley, Chichester.

Gregory, K.J. (1977b) Stream network volume: an index of channel morphometry. *Geological Society America Bulletin* **88**, 1075–1080.

Gregory, K.J. (1979) Drainage network power. *Water Resources Research* **15**(4), 775–777.

Gregory, K.J. (1985) *The Nature of Physical Geography.* Arnold, London.

Gregory, K.J. (1992) Riparian vegetation and channel processes. In: Boon, P., Petts, G.E. and Callow, P. (eds), *The Conservation and Management of Rivers*, Wiley, Chichester, 255–269.

Hemphill, R.W. and Bramley, M.E. (1989) *Protection of River and Canal Banks: A Guide to Selection and Design.* Butterworth, London.

Hoey, T. (1992) Temporal variations in bedload transport rates and sediment storage in gravel-bed rivers. *Progress in Physical Geography* **16**(3), 319–338.

Hooke, J.M. (1977) The distribution and nature of changes in river channel patterns: the example of Devon. In: Gregory, K.J. (ed.), *River Channel Changes*, Wiley, Chichester, 265–280.

Hooke, J.M. and Harvey, A.M. (1983) Meander changes in relation to bend morphology and secondary flows. In Collinson, J.D. and Lewin, J. (eds), *Modern and Ancient Fluvial Systems*, Special Publication of International Association of Sedimentologists 6, 121–132.

Hooke, J.M. and Redmond, C.E. (1989) River channel changes in England and Wales. *Journal of the Institution of Water and Environmental Management* **3**, 328–335.

Hooke, J.M. and Redmond, C.E. (1992) Causes and nature of river planform change. In: Billi, P., Hey, R.D., Thorne, C.R. and Tacconi, P. (eds), *Dynamics of Gravel-bed Rivers*, Wiley, Chichester, 557–571.

Kellerhals, R. and Church, M. (1989) The morphology of large rivers: characterization and management. In: Dodge, D.P. (ed.), *Proceedings of the International Large River Symposium*, Canadian Special Publication Fisheries and Aquatic Science **106**, 31–48.

Kelsey, H.M., Lambertson, R. and Madej, M.A. (1986) Modelling the transport of stored sediment in a gravel-bed river in Northwestern California. In: Hadley, R.F. (ed.), *Drainage Basin Sediment Delivery*, IAHS Publication 159, 367–391.

Laronne, J.B. and Reid, I. (1993) Very high rates of bedload sediment transport by ephemeral desert rivers. *Nature* **366**, 148–150.

Lawler, D.M. (1991) A new technique for the automatic monitoring of erosion and deposition rates. *Water Resources Research* **27**(8), 2125–2128.

Leeks, G.J., Lewin, J. and Newson, M.D. (1988) Channel change, fluvial geomorphology and river engineering: the case of the Afon Trannon, mid Wales. *Earth Surface Processes and Landforms* **13**, 207–223.

Lewin, J., Macklin, M.G. and Newson, M.D. (1988) Regime theory and environmental change—irreconcilable concepts? In: White, W.R. (ed.), *International Conference on River Regime*, Wiley, Chichester, 431–445.

Macklin, M.G. and Klimek, K. (1992) Dispersal, storage and transformation of metal-contaminated alluvium in the Upper Vistula basin, South-west Poland. *Applied Geography* **12**, 7–30.

Macklin, M.G. and Lewin, J. (1993) Holocene river alluviation in Britain. *Zeitschrift fur Geomorphologie* **88**, 109–122.

Macklin, M.G. and Smith, R.S. (1990) Historic riparian vegetation development and alluvial metallophyte plant communities in the Tyne Basin, Northeast England. In: Thornes, J.B. (ed.), *Vegetation and Erosion*, Wiley, Chichester, 239–256.

Macklin, M.G., Passmore, D.G. and Rumsby, B.T. (1992a) Climatic and cultural signals in Holocene alluvial sequences: the Tyne basin, northern England. In: Needham, S. and Macklin, M.G. (eds), *Alluvial Archaeology in Britain, Oxbow Monograph*, 27, Oxford, 123–139.

Macklin, M.G., Rumsby, B.T. and Newson, M.D. (1992b) Historic overbank floods and floodplain sedimentation in the lower Tyne valley, northeast England. In: Billi, P., Hey, R.D., Thorne, C.R. and Tacconi, P. (eds), *Dynamics of Gravel-bed Rivers*, Wiley, Chichester, 573–589.

Marron, D.C. (1992) Floodplain storage of mine tailings in the Belle Fourche river system: a sediment budget approach. *Earth Surface Processes and Landforms* **17**, 675–685.

Nanson, G.C. and Croke, J.C. (1992) A genetic classification of floodplains. *Geomorphology* **4**, 459–486.

National Rivers Authority (1994a) *River Habitat Survey—1994 Verification Phase*. NRA, Bristol.

National Rivers Authority (Northumberland and Yorkshire Region (1994b) Rivers Swale, Ure and Ouse Catchment Management Plan. Consultation Report, York.

Newson, M.D. (1986) River basin engineering—fluvial geomorphology. *Journal Institution of Water Engineers and Scientists* **40**(4), 307–324.

Newson, M.D. (1992a) Geomorphic thresholds in gravel-bed rivers—refinement for an era of environmental change. In: Billi, P., Hey, R.D., Thorne, C.R. and Tacconi, P. (eds), *Dynamics of Gravel-bed Rivers*, Wiley, Chichester, 3–20.

Newson, M.D. (1992b) River conservation and catchment management—UK perspectives. In: Boon, P., Petts, G.E. and Calow, P. (eds), *River Conservation and Management*, Wiley, Chichester, 385–396.

Newson, M.D. and Bathurst, J.C. (1990) *Sediment Movement in Gravel-bed Rivers*. Seminar Paper 59, Department of Geography, University of Newcastle upon Tyne.

Newson, M.D. and Macklin, M.G. (1990) The geomorphologically-effective flood and vertical instability in river channels—a feedback mechanism in the flood series for gravel-bed rivers. In: White, W.R. (ed.), *River Flood Hydraulics*, Wiley, Chichester, 123–140.

Newson, M.D. and Sear, D.A. (1994) *Sediment and Gravel Transport in Rivers: A Geomorphological Approach to River Maintenance*. National Rivers Authority, Research and Development Note 315, University of Newcastle upon Tyne.

Passmore, D.G., Macklin, M.G., Brewer, P.A., Lewin, J., Rumsby, B.T. and Newson, M.D. (1993) Variability of late Holocene braiding in Britain. In: Best, J.L. and Bristow, C.S. (eds), *Braided Rivers*, Geological Society Special Publication 75, 205–229.

Petts, G.E. and Thoms, M.C. (1987) Morphology and sedimentology of a tributary confluence basin a regulated river. *Earth Surface Processes and Landforms* **12**(4), 433–440.

Pitlick, J. (1993) Response and recovery of a subalpine stream following a catastrophic flood. *Geological Society of America, Bulletin* **105**, 657–670.

Reid, I. and Frostick. L.E. (1986) Dynamics of bedload transport in Turkey Brook, a coarse-grained alluvial channel. *Earth Surface Processes and Landforms* **11**, 143–155.

Rosgen, D.L. (1994) A classification of natural rivers. *Catena* **22**, 169–199.

Rumsby, B.T. and Macklin, M.G. (1995) Channel and floodplain response to recent abrupt climate change: the Tyne basin, Northern England. *Earth Surface Processes and Landforms* (in press).

Schumm, S.A. (1977) *The Fluvial System*. Wiley-Interscience, New York.

Sear, D.A. (1992) Impact of hydroelectric power releases on sediment transport processes in pool–riffle sequences. In: Billi, P., Hey, R.D., Thorne, C.R. and Tacconi, P. (eds), *Dynamics of Gravel-bed Rivers*, Wiley, Chichester, 629–650.

Sear, D.A. (1994) River restoration and geomorphology. *Aquatic Conservation: Marine and Freshwater Ecosystems* **4**, 169–177.

Sear, D.A. (1994) *Sediment and Gravel Transport in Rivers: A Geomorphological Approach to River Maintenance*. National Rivers Authority Research and Development Project Record 384/3/T.

Sear, D.A. and Newson, M.D. (1994) *Sediment and Gravel Transport in Rivers: A Geomorphological Approach to River Maintenance*. National Rivers Authority Research and Development Report, University of Newcastle Upon Tyne.

Sear, D.A., Newson, M.D. and Brookes, A. (1995) Sediment related river maintenance: the role of fluvial geomorphology. In: Thorne, C.R. (ed.), *Geomorphology at Work*, Wiley, Chichester (in press)

Sohag, M.A. (1994) Sediment tracing, bed structure, and morphological approaches to sediment transport estimates in a gravel bed river—the River South Tyne, Northumberland, UK. Unpublished PhD thesis, Department of Geography, University of Newcastle upon Tyne.

Swanson, F.J., Janda, R.J., Dunne,T. and Swanston, D.N. (1982) *Sediment Budgets and Routing in Forested Drainage Basins*. United States Department of Agriculture, Pacific Northwest Forest and Range Experiment Station, General Technical Report PNW-141, Portland, Oregon.

Thompson, A. (1986) Secondary flows and the pool–riffle unit: a case study of the processes of meander development. *Earth Surface Processes and Landforms* **11**, 631–641.

Thorne, C.R. (1982) Processes and mechanisms of river bank erosion. In: Hey, R.D., Bathurst, J.C. and Thorne, C.R. (eds), *Gravel-bed Rivers*, Wiley, Chichester, 227–259.

Thorne, C.R., Hey, R.D. and Newson, M.D. (eds) (1995) *Guidebook of Applied Fluvial Geomorphology for River Engineering and Management*. European Research Office of the United States Army, Project DAJA 4591MO172 (in press).

Thorne, C.R. and Tovey, N.K. (1981) Stability of composite river banks. *Earth Surface Processes and Landforms* **6**, 469–484.

Walling, D.E. (1990) Linking the field to the river: sediment delivery from agricultural land. In: Boardman, J., Foster, I.D.L. and Dearing, J.A. (eds), *Soil Erosion from Agricultural Land*, Wiley, Chichester, 129–152.

Walling, D.E. and Woodward, J.C. (1992) Use of radiometric fingerprints to derive information on suspended sediment sources. *International Association of Hydrological Sciences, Publication* 210, 153–164.

Werritty, A. and Brazier, V. (1991) *The Geomorphology, Conservation and Management of the River Feshie SSSI*. Report to Nature Conservancy Council, Department of Geography and Geology, University of St Andrews.

Whiting, P.J. and Bradley, J.B. (1993) A process-based classification system for headwater streams. *Earth Surface Processes and Landforms* **18**, 603–612.

Younger, P.L., Mackay, R. and Connorton, B.J. (1993) Streambed sediment as a barrier to groundwater pollution: insights from fieldwork and modelling in the River Thames basin. *Journal Institution of Water and Environmental Management* **7**(6), 577–585.

Index

Note: Page numbers in *italics* refer to Figures; those in **bold** refer to Tables

Acer negundo 252, 254
Acer saccharinum 254
acid rain 395
aerial photography 16, 277, 286, 297
afforestation 134
agency confidence 268
airborne multispectral scanners 297
Airborne Thematic Mapper (ATM) 280, 287
Aire and Calder Project 401
Alderney 70
alluvium 201
Alnus glutinosa 49
Alnus rhombifolia 253
Alnus serrulata 247
alternate bars 12
Amazon, River, Peru 16, 286, 295
    floodplain 290, 291
American River 205
angle of repose 179
anthropogenic channel adjustments 117
Ardour, River *249*
Armidale, New South Wales 118
armour development 182–4, 192–3, 194
    mobile armour 183
    stable armour 182, 183
arroyo cutting 207–10, *207*
Aswan Dam 150, 160
avulsions 97–8
Axe, River, Devon *92*

Baccharis bimnea 252
bank erosion 12, 88–91
    effect of vegetation on 91
    equation 98
    mechanisms **88**, 89
    on River Dane 91, 97
    on River Severn 103, 104
bar-bend theory 12
bars, types of 93–7, *93*
    alternate 95
    counterpoint 95
    evolution 97
    mid-channel 95
    point 93–4
    side 95

tributary confluence 95
beam (toppling) failure 89, *90*, 91
Bear Creek 205
Bear River, California 133
bed configuration 325
bed–grain interactions 179–80
bedload transport 177–95
    absolute size 177
    fractional bedload transport rates 184–92,
        193–4
        influence of sediment supply 190–2
        partial transport 187–90
    initiation of motion 178–84
        armour development 182–4, 192–3, 194
            mobile armour 183
            stable armour 182, 183
        bed–grain interactions 179–80
        friction angle 179
        pivoting angle 179
        reference shear stress 180–2, *181*
    normal transport 171
    relative size 177
    size effects 177, 194–5
bench index 317, *318*
bend theory 95
benefit–cost analysis (BCA) 397
best practicable environmental option 397, 399
Betula sp. 49
Betula nigra 254
Biebrza valley 35
bimodal particle size 180–2, *181*
Black Sea, sediment yields 154
Blueprint 3 395
Bollin, River
    bend development 106, 110
    rate of vertical accretion *96*, 97
Brahmaputra River, Bangladesh 286, 304
braided rivers, dynamics of 12
British Geomorphological Research Group
    (BGRG) 2
Brooks, W.A. 8
Brown Creek, woody debris in 220, *222*
Brundtland Report 400
Buchanan drainage 75
Burhi Dihing River, Khowang, India *124*,
    126–7

Caesar instrument 297
caesium-137 15
Calver, Captain E.K. 8
capacity limit 397
Carex aquatilis 254
Caring for the Earth 395
Carpathians 52
Carpinus caroliniana 254
carrying capacity 395–6
Carya spp. 246
Carya aquatica 246
Carya ovata 246
catastrophic floods 1, 13, 119, 424
catchment management plans (CMPs) 392–3,
    398, 428
catchment planning
  institutional responses 391–2
  UK approach 392–3
catchment sediment budgets and change 201–
    13
  high-energy instability, mountain and arid
    streams 212, *212*
  humid region, quasi steady state 202–3,
    203
  perturbation of humid area quasi steady
    state 203–6, *203*
  urban streams 210–12, *211*
  valley trenching or arroyo cutting 207–10,
    207
Celtis laevigata 246
Celtis occidentalis 252
Chalk dry valley formation 80
Chandler's Ford 360
channel abandonment, time of 38
channel bed slope 325
channel confluence bars 12
channel cross-section
  equilibrium channel morphology 117–20
  geometry 121–8
    channel morphometry 125–6
    discontinuous response 126–8
    hydraulic 121–5
    environmental factors and 123
channel geometry–discharge relations 135,
    325–37
channel geometry equations 333–5, 344–5
Channel Islands 78
  valleys 67, *68–9*
channel occupancy 313
channel pattern 325
channel planform change 87–111
  bank erosion 88–91
    effect of vegetation on 91
    mechanisms **88**, 89
  chute flow, cutoffs and avulsions 97–8
  classic pattern 91

deposition 92–7
events and flow regime 101–4
flow patterns and bend development 98–
    100
meander bend development 105–10, *105*,
    107–9
morphological changes 104–10
sediment supply and budgets 101
chaos theory 110
Chernobyl accident 15
Chesapeake Bay 290
Chew Valley Lake 132
chute flow 97–8
Circle Mining District, Alaska 287
Coalbrookdale 52
coarse surface layer 192–3
Cole, River, Swindon 370
colluvium 201
Colorado River 160
Colorado River Basin 119
competence and strategic planning 405–7
complex channel responses 75, 117 119, *120*
Coon Creek 57, 166
Coon Valley, Wisconsin *208*, 209
Coralville Reservoir 160
corporate productivity 273
Countryside Stewardship Scheme 384, 394
creation, definition 370
critical boundary shear stress 180, 193
critical natural capital 396
cross-sectional form 325
Cuckmere 55–7
Culm, River, floodplain *59*
cutoffs 106
  bend flattening 97
  chute 97
  mobile bar 97
  multiloop chute 97
  multiloop neck 97
  simple 97

Daedalus Airborne Thematic Mapper (ATM)
    281, 287
Dane, River 110
  bank erosion 91, 97
  bend development 98, *100*, 101, *102–3*,
    104, 106
  in late Holocene 43–4
  middle 89
Danube, River 31
Danube delta 34
Danube valley 33
Darcy–Weisbach flow resistance equation 123
Darent, River 395
data mining 266

data transcription 305
data warehouse 267
datawars 269, 274
DD-6 297
debacles *see* catastrophic floods
Dee, River (UK) 16
Deep Fork, Oklahoma 246
deltas 1
Derwent, River 127, 132
Des Plaines River 119
Deschampsia cespitosa 254
Design with Nature school 326
Digital Terrain Models (DTMs) 405
discharge relations, information from channel
    geometry 325–46
    channel geometry equations 333–5, 344–5
    hydraulic geometry equations 327–33, 341–
        3
Ditchford, River Nene 52, 53, *53*
Dnestr River, Ukraine 158, *158*
dominant discharge 328
Douglas Creek, Colorado 5
drainage density 65–81
    applications to underfit valleys 78–80
    and discharge as static index 78, **79**
    retrospect and prospect 80–1
    spatial dimension
        global scale 71–2
        local scale 66
        national scale 71
        regional scale 66–7
    temporal dimension 72–8
        longer-term network evolution 74–8
        short-term network dynamics 72–4
Dry Creek, California 252
dry valleys 43, 65
Duck River, Tennessee 44
Dunajec 35
Dvina, Western 37

Eaglehawk Creek 134
earthquakes 52
East Fork River 128
engineering
    geomorphic 326
    river 133
        engineering science and 413–15
        and stable channel design 134–5
        tradition in fluvial geomorphology 6–8
enhancement 370
Environmental Assessment (EA) 377, 399,
    401, 427
    of projects 405
environmental design and fluvial
    geomorphology 413–28

developments, and their practical potential
    419–25
    long-profile aspects 420–1
    planform aspects 421–2
    rare event 424–5
    sediment sources 422–4
engineering science and river engineering
    413–15
    scales and styles of approach 414–15,
        415
gaps in our knowledge 425–8
    broader agenda 427–8
    techniques required 426–7, **427**
    types 425–6
iconography of technology transfer *418*
information needs *417*
interdisciplinary understanding and
    improving policy and practice 415–19,
        416
spatial scale 419
environmental issues
    appraisal of policy and strategy 401–3, 402
    function-specific principles and
        environmental objectives 403
    monitoring and post-plan/project appraisal
        405
    objectives 403
    source control 401
environmental sustainability 399–401
Environmentally Sensitive Areas (ESAs) 384,
    394
equal entrainment mobility 177
equal mobility 177
equal transport mobility 177
equilibrium channel morphology 117–20
    adjustment 128–34
        to human impact 130–2
        induced channel changes 132–4
        land-use change *131*
        natural adjustments 128–30
    applications 134–6
        estimating discharge from channel
            geometry 135
        fluvial reconstruction and
            palaeohydrology 135–6
        river engineering and stable channel
            design 134–5
    fluvial equilibrium 118
        complex response 119, *120*
        instability and change 118–19
    geometry 121–8
        channel morphometry 125–6
        discontinuous response 126–8
        hydraulic 121–5
            environmental factors and 123
ERS-1 283, 296, 297

ERS-2 296
Eucalyptus camuldulensis 245, 252
Eucalyptus largiflorens 246
Eucalyptus microcarpa 246
European Community (EC) Directives on
    Environmental Assessment
    (EA) 406
European river channels
    diversity of 27–30
    evolution from late glacial to 19th century
        35–7
    factors influencing past transformation 30–
        5
        anthropogenic factor 34–5
        climatic changes 31
        deglaciation 33–4
        lithological factor 33
        marine transgression 34
        tectonic factors 31–3
    inherited landscapes 30
    longitudinal channel profile 29–30, *29*
    palaeochannels *28*
    palaeogeographic zones *28*
Exe, River 295
Exe Valley 71
Exmoor flood (1952) 74, 119

Faith Creek, Alaska 287, *288*
finite element models for flood flows 58, *59*
Flandrian transgression 34
flood geomorphology 13
floodplain 238
    classification 240, **241–3**
    definition 239
fluvial equilibrium 118
    complex response 119, *120*
    instability and change 118–19
fluvial geomorphology
    development of 1–2
    future challenges 15–18
        applications in river management 17–
            18
        linking geomorphology and ecology 16–
            17
        return to large rivers 15–16
    geographical approach 3–9
        conceptual framework 3–5
        conflict or integration 5–6
        roots of 6–9
            engineering tradition 6–8
            geological tradition 8–9
    geological tradition 4
    primary structure *11*
    recent advances 10–15
        channel dynamics 12

opportunities 13–15
palaeohydrology 12–13
process mechanics 10–12
forest clearance 34
forest drainage 134
Fort Nelson, River, British Columbia 250
Frains Lake, Michigan *155*, 156
Fraxinus lanceolata 247
Fraxinus ornus 246
Fraxinus pennsylvanica 252, 254
freshwater ecology 16
Fuyo-1 296, 297

Galileo 6
Ganga River, Bangladesh 286
Gardiner Dam 132
Garonne, River *249*
Garonne River valley, France 291, 295
Geographic Information Systems (GIS) 350,
    405
    agency confidence in 268
    data presentation 307
    estimating total mapping error using 309–
        13
    information, role in 266, 268–9
    input 268
    integrative analytical 272
    methodology in analysing channel
        planform change 16
    support level 269
geomorphic engineering 326
geomorphogenists 6
geomorphotechnicians 6
Gila River 326
Gipping 57
global denudation system 149
global warming 104
Goodwin Creek, Mississippi 182, 187
Great Eggleshope Beck 189
Great Kei Basin, Lower 121
Green River, Utah 287
ground-penetrating radar (GPR) 16, 293
Ground Resolution Element (GRE) 282
Groundwater Outcrop Erosional Model
    (GOEM) of stream network
    evolution 76

Havgårdssjön, Lake, Sweden *155*, 156
Hamemalis virginiana 247
Harris Creek, British Columbia 190
Herault River, France 246
Herm 70
Highland Water, Hampshire *320*, 321
Hocking River floodplain 246

Hodder River, Lancashire 326
Holocene channel and floodplain change 43–60
  European river channel changes 27–38
  extrinsic controls 51–5
    climatic change 54
    land use change 55
    landslides, colluviation and floodplain changes 52
    lateglacial metamorphosis of floodplains and its inheritance 52–4
  forms of change 44–6
    changing channel forms and floodplain change 44–5
    changing channel numbers and floodplain change 45–6
  history of ideas 43–4
  intrinsic 46–51
    downstream variations and flood regime 46–9
    vegetation and floodplain formation 49–51
  modelling channel and floodplain change 55–60
    channel and floodplain models 58–60
    floodplain sedimentation/budget models 57–6
    inferential stratigraphic models 55–7
  uncoupled floodplains and channels 60
Hunter River, Australia 129, 133
Hurst phenomena 60
hydraulic geometry 117, 121–5
  environmental factors and 123
hydraulic geometry equations 327–33, 341–3
  limitations of 330
hydraulic sorting 178
hydraulic stream ecology approach 17
hydro-electricity 133
hydroponics 398

ice lens 80
ice rafting 89
image analysis 16
Indian Remote Sensing satellites 297
information flow for river channel management 263–75
  communicating information 267
  data from information 268–70
    empowering the organisation 268
    empowering the system 268–9
    roots of power 269–70
  data quality 270
  information dilemma 264–5
  limits of information 270–2
    deficiencies in data acquisition 270–1

    problems in data handling 271–2
    uncertainties in information application 272
  linking information and action 263–4
  managing the unmanageable 273–4
    recasting the organisation 273–4
    recasting the scientific underpinning 274
  meeting the criterion of interest 265–7
    data and information 265–6
    exploration and analysis 266–7
  process of informing 265
  role for information 274–5
Insh Marshes 291
Instantaneous Working Area (IWA) 282
Iowa River 160
Ironbridge Gorge 52
Itres Instruments Compact Airborne Spectrographic Imager (CASI) 283, 297

Jamuna River, Bangladesh 304
jerky conveyor belt 420, 421
Jersey 70
Jessop 8
Jordan, River 212
Justicia americana 247

Kamajohka, River, Finland 254
Kennet catchment 66
Kennet Valley 70
KFA-3000 297
known-rate additions methods 57
Kolyma River, Siberia 158, *159*, 160
Kowai River 119
KVR-1000 297

Lagan, River 128
Lambourn 355
land use change 133, 170
Landsat TM 16, 286, 295, 297
Langbein-Schumm rule 169
large woody debris 217–33
  channel form and process 218–20
    dynamics of woody debris 218–20
    Prairie Creek 220, **221**
  debris removal experiments 220–9
    Larry Damm Creek 222–9, *223*, *225–9*
  large woody debris steps 229–30, *230*
    Larry Damm Creek 230–2
Larry Damm Creek, California *218*
  debris removal experiments 222–9, *223*, *225–9*
  large woody debris steps 229–32

lateral diffusion of turbulent momentum 331
levée formation 53
Lilloet River, British Columbia 246
Liquidambar styraciflua 246, 254
LISS-I 297
LISS-II 297
Llyn Peris *155*, 156
Local Agenda 21 391, 407
Local Government Management Board
    (LGMB), role of 391
local planning authorities, role of 391
log-linear model 123
log-quadratic model 123
Lyell, Charles 8
Lynmouth flood 57

Macdonald River 119
Mackenzie delta 282
Maine 46
Manawatu River, New Zealand 133
Mark valley 35
Master Drainage Plans (MEE/MNR) 391
maximum transport efficiency 331
meander bends 12
    development 105–10, *105*, *107–9*
    double-headed 106
    growth 106
    migrating 106
    new 105
meander development simulation model 98
Mediterranean, chronology of erosion and
    deposition 210, *210*
Meyer–Peter–Muller equations 329
Middle Yellow River 162, **163**, 170
Middle Yellow River Basin 153
Mienia river valley *32–3*
Mimmshall Brook 422
minimum environmentally acceptable flows
    398
minimum stream power 331
minimum variance theory 331
mining 133
    hydraulic, debris 204, 205–6
Mississippi, River 117
    Atchafalaya raft 9
    avulsions 97
    bank erosion 91
    downstream variation in channel form 9
    environmental management of floodplains
        392
    navigation 8
    suspended sediment yield 160, *161*
    woody snags 217
Mississippi Valley, Lower 244
Missouri River

channel geometry and discharge 135
    suspended sediment loads 160, *161*
Missouri River Basin 118, 135
    channel geometry equations 333
modifiable areal unit problem 271
Monks' Brook 360, *361*
Morus rubra 252
multispectral imagery 16
Murray, River, Australia 245
Murray–Darling basin, Australia 392
Murrumbidgee River, Australia 252

Narrator Brook *189*, 190
National Rivers Authority (NRA) 268, 272,
    350, 374, 391
natural capital 396
near-bank velocity 98
Neman basin 37
Nene, River 52, *53*
Nene valley 52
network typology 74–5
Neuse Basin, Lower 165
Newport Bay 211
Nile, River 150, 160
Normalized Total Pigment to Chlorophyll *a*
    ratio index (*NPCI*) 293
North Fork Toutle River, Washington 191,
    192
Nyssa aquatica 255
Nyssa sylvatica 246

Oak Creek 182, 183, 186–9, *186*, 191
Oaklimiter Creek, Mississippi 358
Obion River 220
Ohio River 8
Orange River 167, *167*, 168
Orinoco floodplain 250
Orontium aquatica 247
Otter, River 65
Otter Valley 70, 71
Oulanka river *32–3*
Oulanka valley 33, 34
Ouse, River 55
overgrazing 5, 34
overland flow generation 80

palaeohydrology 4, 12–13, 78
parcel-type stratigraphy 45
particle protrusion 179
Passage Creek, Virginia 247
Patzcuaro, Lake 157
peat accumulation 49
permafrost, impermeable 79–80, 245

Perry 46
Perry valley *47*, *48*
Photo-Electronic Erosion Pin (PEEP) 10
Physocarpus opulifolius 247
Picea glauca 245
Picea mariana 245
pivoting angle 179
plan configuration 328
Po 9
point bars 12
Poo delta 34
Populus spp. 252
Populus balsamifera 245
Populus deltoides 247, 252, 254
Populus fremontii 246
Post Project Appraisal (PPA) 272
Potomac River 247
Prairie creek
  large woody debris *218*, 220, **221**
press disturbances 13
process–response models of channel change
  128
Prosna valley 38
Pseudoraphis spinescens 245
pulse disturbances 13

Quercus spp. 246
Quercus falcata 246
Quercus lyrata 246
Quercus nuttallii 246
Quercus phellos 246

Radarsat 296
Raunds 52
Ravensbourne, River 355, 395
recirculating flume 187
Red Creek, Wyoming 286
Red River, great raft 217
Redhill Brook, Surrey *383*
Redwood Creek, California 212, 420
regime theory 117, 325, 327–8, 421
rehabilitation 369
remote sensing 16, 277–98
  analytical methods 283–5
    mathematical transformation of the
      image 283
    resampling to a new rectangular grid
      284–5
    selection of Ground Control Points
      (GCPs) 283
    temporal properties of the fluvial
      environment 283
  channel geomorphology 285–9
    change in channel bed morphology 287–9
    variability and changes in channel

    planform 285–7
  Daedalus ATM bands **281**, 287
  electromagnetic radiation (emr) 278
  flood inundation mapping 290
  floodplain geomorphology 290–3
  physical basis 9, 278–83, *279*
    spatial properties 281–2
    spectral properties 280–1
    temporal properties 283
  radiometric precision 278
  vegetation in the fluvial environment 293–6
    aquatic vegetation 293–4
    marginal vegetation 295–6
Rennie, Sir John 6, 8
Republican River floodplain 252
reservoirs
  catchments in Southeast Asia 150
  construction 132, 160
  river adjustment below 119, 125
reverence for rivers 326
Rheidol, River 286
Rhine, River 35, 293
  environmental management 392
  marine transgression 34
Rhone, River, France 240
  Upper 248
Rhone valley 33
Ringarooma River, Tasmania 133
Rio Grande 133, 160
riparian ecosystem 245
riparian zone 238, 295
river channel classification 347–63
  adjustment processes 352–60, *354–5*, *356*
    prevailing adjustments 353–8
    prevailing stability of river channels
      355–7, **357**
    sequences of adjustment 358–60, *359*,
      361
  existing features 349–52
  location–for–time methods 349
  model of bed-level changes 358
  purpose 353–5
  sensitivity to adjustment 360–2
  space-for-location methods 349
  susceptibility to disturbance 350, **352**
  in Thames basin 353, *355*
river channel restoration 369–86
  assessment of environmental impacts 377–
    9, **378**
  definitions **369–70**, *370*
  determining the pre-disturbance condition
    370–1
  natural channel recovery 384–5
  potential impacts **378**
  re-creation of sustainable morphological
    features 371–7

river channel restoration (*cont.*)
    catchment context 371–4, *372–3*
    site-specific design issues 374–7, **375**, 376
    scientific versus pragmatic approaches
      379–84
      land availability 384
      monitoring and audit 379
      opportunities created by development
        381, *382*
      River Restoration Project 380
      water quality issue 381–4
river channel transformation, past factors
    influencing 30–5
    anthropogenic factor 34–5
    climatic changes 31
    deglaciation 33–4
    evolution from late glacial to 19th century
      35–7
    lithological factor 33
    marine transgression 34
    tectonic factor 31–3
river characterisation 347
river corridor 238
River Corridor Surveys 427
river planform
    association between vegetation and
      floodplain landforms 250, *251*
    environmental design 421–2
    river bank form and process 254–5
    vegetative succession 251–4
River Restoration Project 380
river variables, changing status of **4**
Roaring River 119
Roding 355
Rorippa nasturtium-aquaticum 293
rubber-sheet algorithms 308

Sacramento River 205
Sagehen creek, California 184, *185*
Salix sp. 49, 252
Salix alaxensis 245
Salix amygdaloides 252
Salix interior 252
Salix nigra 247, 254
Salix phylicifolia 254
salt tracing experiments 228
San Diego Creek 211
San Juan River 167, *167*, 168
San valley *36*, 38
    lower 35
Sanchuanhe River, China 170, 171, **171**, 172
Sanmenxia reservoir 132
satellite sensors 296–7
Savannah River swamp, South Carolina 295
scale-linkage problems 15

Scanning Multichannel Microwave
    Radiometer data (SMMR) 290
Scheldt valley 35
Schumm's source–transfer–sink zonation 421
seed pixels 282
Segehen creek 179
Sence, River, Leicester 303, 309, *315*, 316–21,
    318–19
SERPLAN 405
set-aside 384, 394
Severn, River 46, 52, 57, 58, 127
    bank erosion 89, 91, *94*, 103, 104
    flood regime 46
    side bars 95
sewerage systems 398
shear cantilever failure 89
shear failures 89, *90*
shear stress 180–2, *181*
Shields curve 178
Shields stress 179
    critical 183
Shuttle Imaging Radar (SIR-A) 290
silt factor 328
Skerne, River, Darlington 380
slackwater bars 12
sliding failures 89
slope–discharge relationships 326
Snake River, Idaho 190, *191*
Soar valley 52, 53
soil erosion 149, **150**
    see also bank erosion
Sola 35
South Fork Quantico Creek 247
South Yuba River, California 204
SPANS 311
Spey, River, Scotland 291
SPOT 297
SPOT Pan 295
SPOT XS 284, 290, 295
St Helens, Mount, eruption of (1980) 191
St Peter's Valley, Jersey 70, *71*
stable-bed aggrading-banks (SBAB)
    conceptual model 44, 55, *56*
stationary flood-frequencies 421
statutory water quality objectives (SWQOs)
    393
stepped bed profile 229
Stour, River 49
Strategic Environmental Assessment 403–5
Sub-Watershed Plans 391
Suquehanna River 118
surface water disposal 398
surface water management 398
suspended sediment yields 149–72
    assessment of global pattern 149
    changing yields 151–2

maximum yields 149
other contemporary measurements 162–6
  catchment experiments 162–3
  space–time substitution 163–6
perspective 171–2
problems of interpretation 166–71
  linking upstream catchment behaviour
      to downstream sediment yield 166–8
  response to climatic change 168–71
reconstructing past 152–7
  evidence from lake sediments 154–7, *155*
  evidence from long-term records 157–62
  long-term geological perspective 152–4
  variations over geological time 152, *153*
  in Southeast Asia *151*
sustainable development 389, 407
  approaches to 394–6
    carrying capacity 395–6
    economic valuation 395
  pathways to *406*, 407–9
sustainable management 391
synthetic aperture radars (SAR) 296

Tamarix chinensis 246
Tana River, Kenya 170, **170**
Taxodium distichum 255
Tay, River, Scotland 14, 287, 290, 289, 291,
    292
Teign Valley 71
Teme, River 110, *280*, 295
Ter, River, Essex 132
terrace sequences 326
Thames 21 393
Thames, River 46, 57, 360
Thames River basin 381
Thiba River, Kenya 170, **170**
timber harvesting 134
Tisa river channel *32–3*
topographic survey 303–21
  advantages 308
  components *306*
  cross-section methods 308
  estimating total mapping error 305, 309–21
    cross-section adjustment 315–21
    planform adjustments 311–15
  planform methods 308
  slithers 312
  topographic scale and river channel change
      303–5
  transcription of information 305–9
    inherent errors 305–8
    operational errors 305, 308–9
    systematic errors 306–7
  volumetric analysis 308
Total Catchment Management 392

Total Mapping Error 303, 305
  estimating 305, 309–21
Towy, River, Dyfed 303, 309, *310*, 311–14,
    315–16
tractive force theory 331
tree-throw topography 53
Trent, River 8, 46
tributary junctions, changes in flow dynamics
    127, *127*
Tummel, River, Scotland *289*, 291
Tyne, River 54

Ucayali floodplain, Peru 250
Ucayali, Peru 286
Ulmus americana 252, 254
underfit valleys 43, 65
  drainage density applied to 78–80
UNESCO International Hydrological
    Programme 264
unimodal particle size 180–2, *181*
unit hydraulic geometry 123
United Nations Conference on Environment
    and Development (UNCED)
    (Earth Summit) 389
  Agenda 21 390–1
  Ecological-Social-Economic (ESE) 391
Urban Streams Restoration Programme 391
urbanisation 132

valley asymmetry, drainage and 67
Valley of the Rocks, Devon 74
valley storage fluxes, conceptual models of
    202
valley trenching 207–10, *207*
vegetation
  along river corridors 237–56
    definitions 238–9
    hillslope–floodplain types 239–44
    hydrological regime 244–9
      flooding 245–9
      topography and soil moisture and water
          table regime 244–5
    fluvial geomorphological disturbances
        250–5
    integration and application 255–6
    river planform change
      association between vegetation and
          floodplain landforms 250, *251*
      river bank form and process 254–5
    vegetative succession 251–4
  remote sensing 293–6
    aquatic vegetation 293–4
    marginal vegetation 295–6
vegetative treatment 398
velocity–discharge relationships 123

Viburnum dentatum 247
Victoria, Lake **170**
Vistula floodplain 34
Vistula River, Poland 46, 133
Vistula valley 35, *36*

Warta River, Poland 135
Warta valley 35
Washita River, Oklahoma 132
Water Act (1989) (UK) 392
water resources 397–8
Water Resources Act (1991) (UK) 393, 399
Watershed Plans 391
Weihe River, China 119, 132
Weser, River 46
White River Valley, Lower, Arkansas 246
width/depth ratio (*W/D*) 44
Wilden Marsh

floodplain succession *51*
  stratigraphic cross-section *50*
Willamette River, Oregon 217, 238
Williams River, Canada *129*, 130
Wisloka river *32–3*
Wisloka valley *36*
Witham, Lincolnshire
'trained' channel *7, 8*
woodland grants 384
working with nature principle 398
World Conservation Union 395

Yakima River, Washington 295
Yamuna River, Indua 286
Yashishti River, India 127
Yellow River, China 118, 153
Yuba River, California 205
  sediment budget *206*

Index compiled by A.J. Musker